A Textbook of ENGINEERING MATHEMATICS-II

(Uttarakhand Technical University, Dehradun)

H.S. Gangwar
M.Sc., Ph.D.
Deptt. of Mathematics
SRMS College of Engineering
and Technology, Bareilly (U.P.)

Prabhakar Gupta
M.Sc. (Math.), M.Tech., Ph.D.
Professor and Dean Academics
SRMS College of Engineering
and Technology, Bareilly (U.P.)

NEW AGE
UTU

An Imprint of

NEW AGE INTERNATIONAL (P) LIMITED, PUBLISHERS

New Delhi • Bangalore • Chennai • Cochin • Guwahati • Hyderabad
Jalandhar • Kolkata • Lucknow • Mumbai • Ranchi
Visit us at **www.newagepublishers.com**

Published by New Age International (P) Ltd., Publishers
First Edition: 2010

Branches:

- No. 37/10, 8th Cross (Near Hanuman Temple), Azad Nagar, Chamrajpet, **Bangalore**-560 018. Tel.: (080) 26756823, Telefax: 26756820, E-mail: bangalore@newagepublishers.com
- 26, Damodaran Street, T. Nagar, **Chennai**-600 017. Tel.: (044) 24353401, Telefax: 24351463 E-mail: chennai@newagepublishers.com
- CC-39/1016, Carrier Station Road, Ernakulam South, **Cochin**-682 016. Tel.: (0484) 2377004, Telefax: 4051303. E-mail: cochin@newagepublishers.com
- Hemsen Complex, Mohd. Shah Road, Paltan Bazar, Near Starline Hotel, **Guwahati**-781 008. Tel.: (0361) 2513881. Telefax: 2543669, E-mail: guwahati@newagepublishers.com
- No. 105, 1st Floor, Madhiray Kaveri Tower, 3-2-19, Azam Jahi Road, Nimboliadda, **Hyderabad**-500 027. Tel.: (040) 24652456, Telefax: 24652457, E-mail:hyderabad@newagepublishers.com
- RDB Chambers (Formerly Lotus Cinema)106A, 1st Floor, S.N. Banerjee Road, **Kolkata**-700 014. Tel.: (033) 22273773, Telefax: 22275247, E-mail:kolkata@newagepublishers.com
- 16-A, Jopling Road, **Lucknow**-226 001. Tel.: (0522) 2209578, 4045297, Telefax: 2204098 E-mail: lucknow@newagepublishers.com
- 142C, Victor House, Ground Floor, N.M. Joshi Marg, Lower Parel, **Mumbai**-400 013. Tel.: (022) 24927869. Telefax: 24915415, E-mail: mumbai@newagepublishers.com
- 22, Golden House, Daryaganj, **New Delhi**-110 002. Tel.: (011) 23262370, 23262368, Telefax: 43551305. E-mail: sales@newagepublishers.com

ISBN : 978-81-224- 3122-3

Rs. 250.00

C-10- 04-4457

Printed in India at Mohanlal Printers, Delhi.
Typeset at Aakriti Graphics, New Delhi.

PUBLISHING FOR ONE WORLD
NEW AGE INTERNATIONAL (P) LIMITED, PUBLISHERS
4835/24, Ansari Road, Daryaganj, New Delhi-110002
Visit us at **www.newagepublishers.com**

A Textbook of

ENGINEERING MATHEMATICS-II

Preface to the First Edition

This book is primarily written according to the latest unified syllabus of Mathematics of second semester of all Engineering Colleges affiliated to Uttarakhand Technical University, Dehradun.

The subject matter is presented in a systematic, illustrative and lucid manner. In each unit, all concepts, definitions and the large number of examples in the best possible way have been discussed in detail so that the students should feel no difficulty to understand the subject.

Thorough care has been taken to eradicate errors but the perfection cannot be claimed. The readers are requested to give constructive suggestions to improve the book in further editions.

We in general hope that the present book will be warmly received by the students as well as teachers. We shall indeed be very thankful to our colleagues for recommending this book to their students.

Our special thanks to Shri Dev Murti Ji, Chairman, SRMS Trust, Bareilly.

We are also thankful to Mr. Saurabha Saxena, Ms. Neha Singhal, Mr. Bharat Bushan Sharma, Mr. Kamlendra Kumar, Lecturers of Mathematics Department, SRMS College of Engineering and Technology for constructive suggestions.

The authors also wish to express their thanks to Mr. Saumya Gupta (Managing Director), New Age International (P) Limited, Publishers, as well as the editorial department for their untiring effort to complete this project within a very short period.

AUTHORS

Syllabus

Uttarakhand Technical University, Dehradun

MATHEMATICS-II (TMA-201)

L	T	P
3	1	0

Unit-1: Differential Equations **8**

Ordinary differential equations of first order, Exact differential equations, Linear differential equations of first order, Linear differential equations of *n*th order with constant coefficients, Complementary functions and particular integrals, Simultaneous linear differential equations. Solutions of second order differential equations by changing the dependent and independent variables, Method or variation of parameters.

Unit-2: Laplace Transform **10**

Laplace transform, Existence theorem, Laplace transform of derivatives and integrals, Inverse Laplace transform, Laplace transform of periodic function. Unit step function, Convolution theorem, Applications to solve simple linear and simultaneous linear differential equations.

Unit-3: Infinite Series **8**

Introduction, Sequences, Series: Convergence, Series of positive terms, Comparison tests, Integral tests, Comparison of ratio's, D'Alembert ratio test, Raabe's test, Cauchy's root test, Alternating series: Leibnitz's rule, Power series, Uniform convergence, Weierstrass's M-test, Properties of uniformly convergent series.

Unit-4: Fourier Series and Partial Differential Equations **8**

Periodic functions, Trigonometric series, Fourier series of periodic function, Euler's formula, Functions having arbitrary period, Change of intervals, Even and odd functions. Half range sine and cosine series.

Introduction to partial differential equations, Linear partial differential equations with constant coefficients of second order and their classifications: parabolic, hyperbolic and elliptic with illustrative examples.

Unit-5: Applications of Partial Differential Equations **8**

Method of separation of variables for solving partial differential equations, One dimensional wave equation, Laplace equation in two dimensions, Heat conduction equations of one dimension and two dimension.

Some Useful Formulae

Differentiation:

1. $\frac{d}{dx}(f_1(x)\cdot f_2(x))=\left\{\frac{d}{dx}f_1(x)\right\}\cdot f_2(x)+\left\{\frac{d}{dx}f_2(x)\right\}f_1(x)$

2. $\frac{d}{dx}\left(\frac{f_1(x)}{f_2(x)}\right)=\frac{\left\{\frac{d}{dx}f_1(x)\right\}f_2(x)-\left\{\frac{d}{dx}f_2(x)\right\}\cdot f_1(x)}{\{f_2(x)\}^2}$

3. $\frac{d}{dx}(a^x)=a^x\log_e a$

4. $\frac{d}{dx}(\sinh x)=\cosh x$

5. $\frac{d}{dx}(\cosh x)=\sinh x$

6. $\frac{d}{dx}(\cot x)=-\text{cosec}^2 x$

7. $\frac{d}{dx}(\log_a x)=\frac{1}{x}\log_a e$

8. $\frac{d}{dx}(\cos^{-1}x)=-\frac{1}{\sqrt{1-x^2}}$

9. $\frac{d}{dx}(\cot^{-1}x)=-\frac{1}{1+x^2}$

10. $\frac{d}{dx}(\sec^{-1}x)=\frac{1}{x\sqrt{x^2-1}}$

11. $\frac{d}{dx}(\text{cosec}^{-1}x)=-\frac{1}{x\sqrt{x^2-1}}$

Integration:

12. $\int u\cdot v\cdot dx=u\int v\,dx-\int\left\{\left(\frac{d}{dx}u\right)\cdot\int v\,dx\right\}dx+C$

13. $\int \sin x dx = -\cos x + C$

14. $\int \cos x dx = \sin x + C$

15. $\int \tan x dx = -\log|\cos x| + C$

16. $\int \cot x dx = \log|\sin x| + C$

17. $\int \sec x dx = \log|\sec x + \tan x| + C$

18. $\int \operatorname{cosec} x dx = \log|\operatorname{cosec} x - \cot x| + C$

19. $\int \frac{1}{\sqrt{x^2 - a^2}} dx = \log|x + \sqrt{x^2 - a^2}| + C$

20. $\int \sqrt{a^2 - x^2} dx = \frac{x}{2}\sqrt{a^2 - x^2} + \frac{a^2}{2}\sin^{-1}\frac{x}{|a|} + C$

21. $\int e^{ax} \cdot \sin bx\, dx = \frac{e^{ax}}{a^2 + b^2}(a \sin bx - b \cos bx) + C$

22. $\int e^{ax} \cos bx\, dx = \frac{e^{ax}}{a^2 + b^2}(a \cos bx + b \sin bx) + C$

Trigonometric formulae:

23. $2 \sin A \cos B = \sin(A + B) + \sin(A - B)$
24. $2 \cos A \sin B = \sin(A + B) - \sin(A - B)$
25. $2 \cos A \cos B = \cos(A + B) + \cos(A - B)$
26. $2 \sin A \sin B = \cos(A - B) - \cos(A + B)$
27. $\sin ix = i \sinh x$
28. $\cos ix = \cosh x$
29. $\cos x = \frac{e^{ix} + e^{-ix}}{2}$
30. $\sin x = \frac{e^{ix} - e^{-ix}}{2i}$
31. $\cosh x = \frac{e^x + e^{-x}}{2}$
32. $\sinh x = \frac{e^x - e^{-x}}{2}$

Contents

UNIT II

LAPLACE TRANSFORM

[161–269]

UNIT III

INFINITE SERIES

[270–331]

UNIT IV

FOURIER SERIES AND PARTIAL DIFFERENTIAL EQUATIONS

[332–431]

UNIT V

APPLICATIONS OF PARTIAL DIFFERENTIAL EQUATIONS

[432–516]

UNIT 1

Differential Equations

1.0 INTRODUCTION

We live in a world of interrelated changing entities. The velocity of a falling body changes with distance, the position of the earth changes with time, the area of a circle changes with size of the radius, the bending of a beam changes with the weight of the load placed on it, the path of a projectile changes with the velocity and angle at which it is fired.

In the language of mathematics, changing entities are called variables and the rate of change of one variable, with respect to another a derivative. *Equations which express a relationship among these variables and their derivatives are called differential equations*. But here we are interested to know that how the variables themselves are related. For example, from certain facts about the variable position of a particle and its rate of change with respect to time, we wish to determine how the position of the particle is related to the time so that we can know where the particle was, is, or will be at any time *t*., *Differential equations thus originate whenever a universal law is expressed by means of variables and their derivatives.*

1.1 DIFFERENTIAL EQUATION

An equation which contains the differential coefficient is called differential equation, where y is dependent and x is independent variable.

For example,

(*i*) $\dfrac{dy}{dx} + 4y = x$ (*ii*) $\dfrac{d^2y}{dx^2} - 4\dfrac{dy}{dx} - 6y = 0$

1.1.1 Order and Degree of Differential Equation

The order of the highest derivative of any differential equation is called order of the differential equation.

The degree of the highest order derivative after removing the radical sign and fraction is called degree of the differential equation.

In example (*i*), the order is one and degree is also one, and in example (*ii*) the order is two but degree is one.

There are some definitions of constants which use in the solution of differenial equation.

Constant: A symbol, which retains the same value during a set of mathematical operations, is called a constant for these operations.

Absolute Constant: A constant, which has the same value in every mathematical operation, is called an absolute constant. For example, 5, 6, π, $\sqrt{3}$, ... are absolute constants.

Arbitrary Constant: A constant, which has a fixed value throughout a particular reference and does not remain fixed in general, is called an arbitrary constant. For example, $y = mx + c$ is the equation of a straight line in which m and c have some fixed values for a particular straight line but have different fixed values for another straight line. Here m and c are called arbitrary constants.

1.1.2 Solution of Differential Equation

An equation involve dependent variable (y) as well as independent variable (x) and free from derivative, which satisfies differential equation, is called the solution of the differential equation

e.g., $y = C_1e^x + C_2e^{-x}$ is the solution of the differential equation

$$\frac{d^2y}{dx^2} - y = 0. \qquad \text{...(A)}$$

1.1.3 General Solution of Differential Equation

The solution in which the number of arbitrary constants are equal to the order of differential equation is called general solution of differential equation.

1.1.4 Particular Solution of Differential Equation

The solution for any particular value of constants is called particular solution of differential equation.

If $C_1 = 1$, $C_2 = 2$ then $y = e^x + 2e^{-x}$ is the particular solution of differential equation (*A*).

Linear Independence and Dependence

Two functions $y_1(x)$ and $y_2(x)$ are said to be linearly independent if

$$C_1y_1 + C_2y_2 = 0 \qquad \text{...}(i)$$

implies $C_1 = C_2 = 0$ *i.e.*, y_1 or y_2 cannot be expressed as proportional to other.

If $C_1 \neq 0, C_2 \neq 0$ then $y_1(x)$ and $y_2(x)$ are linearly dependent.

1.2 ORDINARY DIFFERENTIAL EQUATIONS OF FIRST ORDER

A differential equation of first order classified in the following differential equations:

(*i*) General form (Variable separation method)
(*ii*) Homogeneous differential equations
(*iii*) Linear differential equation of the first order
(*iv*) Exact differential equation.

1.2.1 Variable Separation Method

In this method we separate the variables *i.e.*, y on left hand side and x on right hand side then integrate on both sides.

Example 1: Solve the differential equation

$$(2x^2+3y^2-7)xdx-(3x^2+2y^2-8)ydy=0. \qquad (U.P.T.U.\ 2005)$$

Solution: The given differential equation can be written as

$$\frac{x\,dx}{y\,dy}=\frac{3x^2+2y^2-8}{2x^2+3y^2-7}$$

Applying componendo and dividendo rule, we get

$$\frac{xdx+ydy}{xdx-ydy}=\frac{5x^2+5y^2-15}{x^2-y^2-1}=\frac{5(x^2+y^2-3)}{(x^2-y^2-1)}$$

$$\Rightarrow \quad \left(\frac{xdx+ydy}{x^2+y^2-3}\right)=5\left(\frac{xdx-ydy}{x^2-y^2-1}\right)$$

Multiplying on both sides by 2, we get

$$\left(\frac{2xdx+2ydy}{x^2+y^2-3}\right)=5\left(\frac{2xdx-2ydy}{x^2-y^2-1}\right)$$

Integrating on both sides, we get

$$\log(x^2+y^2-3)=5\log(x^2-y^2-1)+\log C$$

$$\Rightarrow \quad x^2+y^2-3=C(x^2-y^2-1)^5. \quad \textbf{Ans.}$$

Example 2: Solve the differential equation

$$(e^y+1)\cos x dx+e^y\sin x dy=0.$$

Solution: The given differential equation can be written as

$$(e^y+1)\cos x dx=-e^y\sin x dy$$

$$\Rightarrow \quad \frac{e^y}{(e^y+1)}dy=-\frac{\cos x}{\sin x}dx$$

$$\Rightarrow \quad \int\frac{e^y}{(e^y+1)}dy=-\int\frac{\cos x}{\sin x}dx+\log C$$

$$\Rightarrow \quad \log(e^y+1)=-\log\sin x+\log C$$

$$\Rightarrow \quad \log(e^y+1)+\log\sin x=\log C$$

$$\Rightarrow \quad \log(e^y+1)\sin x=\log C \Rightarrow (e^y+1)\sin x=C. \quad \textbf{Ans.}$$

1.2.2 Homogeneous Differential Equations

A differential equation of the form $\frac{dy}{dx}=\frac{f(x,y)}{g(x,y)}$, where $f(x, y)$ and $g(x, y)$ are homogeneous functions in x and y of the same degree is called homogeneous differential equation. Such type differential equations may be solved by putting

$$y = vx, \text{ so that } \frac{dy}{dx} = v + x\frac{dv}{dx}$$

And the equation reduces to the variable separable form.

Example 3: Solve $x^3 \dfrac{dy}{dx} = y^3 + y^2\sqrt{y^2 - x^2}$.

Solution: The given equation can be written as

$$\frac{dy}{dx} = \frac{y^3 + y^2\sqrt{y^2 - x^2}}{x^3}$$

It is a homogeneous, hence putting $y = vx \Rightarrow \left(\dfrac{dy}{dx} = v + x\dfrac{dv}{dx}\right)$, we get

$$v + x\frac{dv}{dx} = v^3 + v^2\sqrt{v^2 - 1}$$

$$\Rightarrow \quad x\frac{dv}{dx} = v^3 - v + v^2\sqrt{v^2 - 1} \quad \Rightarrow \quad \frac{dv}{v(v^2-1) + v^2\sqrt{v^2-1}} = \frac{dx}{x}$$

$$\Rightarrow \quad \int \frac{dv}{v(v^2-1) + v^2\sqrt{v^2-1}} = \int \frac{dx}{x} + \log C$$

$$\Rightarrow \quad \int \frac{dv}{v\sqrt{v^2-1} \times \left(\sqrt{v^2-1} + v\right)} = \log x + \log C$$

$$\Rightarrow \quad \int \frac{\left(\sqrt{v^2-1} - v\right)}{v\sqrt{v^2-1}(v^2-1-v^2)}\,dv = \log xC$$

$$\Rightarrow \quad -\int \frac{\left(\sqrt{v^2-1} - v\right)}{v\sqrt{v^2-1}}\,dv = \log xC \Rightarrow \int \left(\frac{1}{\sqrt{v^2-1}} - \frac{1}{v}\right) dv = \log Cx$$

$$\Rightarrow \quad \log\left(v + \sqrt{v^2-1}\right) - \log v = \log Cx$$

$$\Rightarrow \quad \log \frac{\left(v + \sqrt{v^2-1}\right)}{v} = \log Cx \quad \Rightarrow \quad \frac{v + \sqrt{v^2-1}}{v} = Cx$$

$$\Rightarrow \quad 1 + \sqrt{1 - \frac{x^2}{y^2}} = Cx \quad \Rightarrow \quad Cxy = y + \sqrt{y^2 - x^2}. \quad \textbf{Ans.}$$

Example 4: Solve $\frac{dy}{dx} = \frac{y}{x} + x \tan \frac{y}{x}$.

Solution: $$\frac{dy}{dx} = \frac{y}{x} + x \tan \frac{y}{x}$$

Putting $y = vx \Rightarrow \frac{dy}{dx} = x\frac{dv}{dx} + v,$

we get, $$x\frac{dv}{dx} + v = v + x \tan v \quad \Rightarrow \quad x\frac{dv}{dx} = x \tan v \quad \Rightarrow \quad \frac{dv}{\tan v} = dx$$

$$\Rightarrow \quad \int \frac{dv}{\tan v} = \int dx + C \quad \Rightarrow \quad \int \cot v\, dv = x + C \quad \Rightarrow \quad \log \sin v = x + C$$

replace v by $\frac{y}{x}$,

So, $$\log \sin \frac{y}{x} = x + C.$$ **Ans.**

1.2.2.1 Equations Reducible to Homogeneous Form

The differential equations under this category are of the form

$$\frac{dy}{dx} = \frac{a_1 x + b_1 y + c_1}{a_2 x + b_2 y + c_2} \qquad ...(i)$$

while reducing this equation in homogeneous form there arise two cases.

Case I: When $\frac{a_1}{a_2} \neq \frac{b_1}{b_2}$,

we substitute $x = X + h,\ y = Y + k$ hence $\frac{dy}{dx} = \frac{dY}{dX}$, (i) becomes

$$\frac{dY}{dX} = \frac{a_1(X+h) + b_1(Y+k) + c_1}{a_2(X+h) + b_2(Y+k) + c_2} = \frac{(a_1 X + b_1 Y) + (a_1 h + b_1 k + c_1)}{(a_2 X + b_2 Y) + (a_2 h + b_2 k + c_2)} \qquad ...(ii)$$

where h and k are some constants choosen such that

$$a_1 h + b_1 k + c_1 = 0, \ a_2 h + b_2 k + c_2 = 0$$

Then equation (ii) reduces in the form

$$\frac{dY}{dX} = \frac{a_1 X + b_1 Y}{a_2 X + b_2 Y}$$

which is a homogeneous. The values of h, k obtained from above two equations.

Case II: If $\frac{a_1}{a_2} = \frac{b_1}{b_2}$, in this case we cannot calculate the value of the constants h and k. So, we have

$$\frac{a_1}{a_2} = \frac{b_1}{b_2} = k \Rightarrow a_1 = a_2 k \text{ and } b_1 = b_2 k$$

Hence the equation (*i*) becomes $\dfrac{dy}{dx} = \dfrac{(a_1x+b_1y)+c_1}{k(a_1x+b_1y)+c_2}$

Here we substitute $a_1x+b_1y=z$ and it can be solve by variable separation method.

Example 5: Solve $\dfrac{dy}{dx} = \dfrac{3x-2y+1}{6x-4y+1}$.

Solution: $\dfrac{a_1}{a_2} = \dfrac{3}{6} = \dfrac{1}{2}$ and $\dfrac{b_1}{b_2} = \dfrac{2}{4} = \dfrac{1}{2} \Rightarrow \dfrac{a_1}{a_2} = \dfrac{b_1}{b_2}$

The given equation may be written as

$$\frac{dy}{dx} = \frac{3x-2y+1}{2(3x-2y)+1} \qquad ...(1)$$

Let $3x-2y = z \Rightarrow 3-2\dfrac{dy}{dx} = \dfrac{dz}{dx} \Rightarrow \dfrac{dy}{dx} = \dfrac{1}{2}\left(3-\dfrac{dz}{dx}\right)$

Using this value in equation (1), we get

$$\frac{1}{2}\left(3-\frac{dz}{dx}\right) = \frac{z+1}{2z+1} \quad \Rightarrow \quad \frac{dz}{dx} = 3-\frac{2z+2}{2z+1} = \frac{4z+1}{2z+1}$$

or $$\frac{2z+1}{4z+1}dz = dx \Rightarrow \int\frac{2z+1}{4z+1}dz = \int dx + C$$

$$\Rightarrow \int\left[\frac{1}{2}+\frac{1}{2(4z+1)}\right]dz = x+C \Rightarrow \frac{z}{2}+\frac{1}{8}\log(4z+1) = x+C$$

$$\Rightarrow z+\frac{1}{4}\log 4+\frac{1}{4}\log\left(z+\frac{1}{4}\right) = 2x+2C$$

Now, replace z by $3x-2y$, we get

$$\frac{1}{4}\log\left(3x-2y+\frac{1}{4}\right) = 2x-(3x-2y)+C_1$$

or $$4x-8y+\log\left(3x-2y+\frac{1}{4}\right) = C_1'. \quad \textbf{Ans.}$$

Example 6: Solve

$$\frac{dy}{dx} = \frac{y-x+1}{y+x+5}.$$

Solution:

$$\frac{dy}{dx} = \frac{y-x+1}{y+x+5} \qquad ...(1)$$

$$\frac{a_1}{a_2} = -1, \ \frac{b_1}{b_2} = 1, \text{ so let } x = X + h \text{ and } y = Y + k$$

Then the equation (1), reduces in the form

$$\frac{dY}{dX} = \frac{(Y+k)-(X+h)+1}{(Y+k)+(X+h)+5} = \frac{(Y-X)+(k-h+1)}{(Y+X)+(k+h+5)} \quad \text{...(2)}$$

Let us choose h and k such that

$$k - h + 1 = 0, \ k + h + 5 = 0$$

which give $h = -2$, $k = -3$, then (2), becomes

$$\frac{dY}{dX} = \frac{Y-X}{Y+X} \text{ which is homogeneous.}$$

Putting $Y = vX$, we get

$$v + X\frac{dv}{dX} = \frac{X(v-1)}{X(v+1)} \Rightarrow X\frac{dv}{dX} = \frac{v-1}{v+1} - v$$

$$\Rightarrow \quad X\frac{dv}{dX} = \frac{v-1-v^2-v}{(v+1)} = -\frac{v^2+1}{v+1}$$

$$\Rightarrow \quad \left(\frac{v+1}{v^2+1}\right)dv = -\frac{dX}{X}, \text{ on integrating}$$

$$\int\left(\frac{v}{v^2+1} + \frac{1}{v^2+1}\right)dv = -\int\frac{dX}{X} + \log C$$

$$\Rightarrow \quad \frac{1}{2}\log(v^2+1) + \tan^{-1} v = -\log X + \log C$$

$$\Rightarrow \quad \tan^{-1}\frac{Y}{X} + \log\sqrt{v^2+1} + \log X - \log C = 0$$

$$\Rightarrow \quad \tan^{-1}\frac{Y}{X} + \log\sqrt{Y^2+X^2} - \log C = 0$$

$$\Rightarrow \quad \tan^{-1}\frac{y+3}{x+2} + \log\sqrt{(x+2)^2+(y+3)^2} - \log C = 0 \quad \Big| \text{As } X = x+2, Y = y+3$$

$$\Rightarrow \quad \tan^{-1}\frac{y+3}{x+2} + \log\left\{\frac{1}{C}\sqrt{(x+2)^2+(y+3)^2}\right\} = 0. \quad \textbf{Ans.}$$

1.2.3 Linear Differential Equation of the First Order

A differential equation of the form

$$\frac{dy}{dx} + Py = Q \quad \text{...}(i)$$

is said to be a linear differential equation of the first order, where P and Q are functions of x only or constants. It was introduced by Leibnitz and is therefore better known as Leibnitz's linear differential equation.

Method of Solution: To solve it, we multiply (*i*) on both sides by a factor $e^{\int P dx}$, we get

$$\frac{dy}{dx} \cdot e^{\int P dx} + Py \cdot e^{\int P dx} = Q \cdot e^{\int P dx}$$

$$\Rightarrow \qquad \frac{d}{dx}\left[y \cdot e^{\int P dx}\right] = Q \cdot e^{\int P dx}$$

Integrating both sides, we get

$$\boxed{y \cdot e^{\int P dx} = \int Q \cdot e^{\int P dx} dx + C}$$

which is the required solution where the factor $e^{\int P dx}$ is called integrating factor.

Remark: If the equation is of the form $\frac{dx}{dy} + Px = Q$, then solution of this equation, will be as follows:

$$\boxed{x \cdot e^{\int P dy} = \int Q \cdot e^{\int P dy} dy + C}$$

where P and Q are the functions of y only.

Example 7: Solve $(y \log y)\, dx + (x - \log y)\, dy = 0$. (*U.P.T.U. 2004*)

Solution: The given equation may be written as

$$\frac{dx}{dy} + \frac{1}{y \log y} . x = \frac{1}{y} \text{ which is linear in } x.$$

Here $\qquad P = \frac{1}{y \log y}, Q = \frac{1}{y}$

$$\text{I.F.} = e^{\int P dy} = e^{\int \frac{1}{y \log y} dy} = \log y$$

The solution is $\qquad x \log y = \int \frac{1}{y} \cdot \log y \, dy + C$

$$\Rightarrow \qquad x \log y = \frac{1}{2} (\log y)^2 + C. \quad \textbf{Ans.}$$

Example 8: Solve $(x+1)\frac{dy}{dx} - y = e^x (x+1)^2$. (*A.M.I.E.T.E. 2002*)

Solution: The given equation may be written as

$$\frac{dy}{dx}-\frac{1}{x+1}y=e^{x}(x+1),$$ which is the linear in y, Here, $P=-\frac{1}{x+1}$ and $Q = e^{x}(x+1)$

$$\text{I.F.} = e^{\int P dx} = e^{-\int\frac{1}{x+1}dx} = e^{-\log(x+1)} = e^{\log(1+x)^{-1}} = \frac{1}{1+x}$$

Hence, the solution is

$$y\cdot\frac{1}{1+x} = \int e^{x}(x+1)\cdot\frac{1}{1+x}dx+C$$

$$\Rightarrow \qquad \frac{y}{1+x} = e^{x}+C. \quad \textbf{Ans.}$$

1.2.3.1 Equations Reducible to Linear Form (Bernoulli's Equation)

The differential equation $\frac{dy}{dx}+Py=Qy^{n}$ ...(*i*)

where P and Q are functions of x only is known as Bernoulli's equation.

Method reduced to linear form: Dividing equation (*i*) by y^n on both sides, we get

$$\frac{1}{y^{n}}\frac{dy}{dx}+\frac{1}{y^{n-1}}P = Q \qquad ...(ii)$$

Putting $\frac{1}{y^{n-1}}=z,$ so that

$$\frac{(1-n)}{y^{n}}\frac{dy}{dx} = \frac{dz}{dx} \quad\Rightarrow\quad \frac{1}{y^{n}}\frac{dy}{dx}=\frac{1}{(1-n)}\frac{dz}{dx}$$

Substituting these in (*ii*), we get

$$\frac{1}{(1-n)}\frac{dz}{dx}+Pz = Q \quad\Rightarrow\quad \boxed{\frac{dz}{dx}+(1-n)Pz=(1-n)Q}$$

which is linear in z and can be solved easily.

> **Note:** The equations of the form
>
> $$\frac{dy}{dx}+P.f_1(y) = Qf_2(y)$$
>
> may be also solve by reducible to linear form, where P and Q are functions of x only or constant.

Example 9: Solve $\frac{dy}{dx}=\frac{y^{3}}{e^{2x}+y^{2}}.$

Solution: The given equation may be written in the form

$$\frac{dx}{dy} = \frac{e^{2x}+y^{2}}{y^{3}} \quad\Rightarrow\quad \frac{dx}{dy}=\frac{e^{2x}}{y^{3}}+\frac{1}{y}$$

$$\Rightarrow \quad \frac{dx}{dy} - \frac{1}{y} = \frac{e^{2x}}{y^3} \qquad \text{...(1)}$$

Dividing on both sides by e^{2x}, we get

$$e^{-2x}\frac{dx}{dy} - e^{-2x}\frac{1}{y} = \frac{1}{y^3} \qquad \text{...(2)}$$

Let $\quad e^{-2x} = z \quad \Rightarrow \quad -2e^{-2x}\frac{dx}{dy} = \frac{dz}{dy} \quad \Rightarrow \quad e^{-2x}\frac{dx}{dy} = -\frac{1}{2}\frac{dz}{dy}$

Substituting this value in equation (2), we get

$$-\frac{1}{2}\frac{dz}{dy} - \frac{1}{y}z = \frac{1}{y^3}$$

or $\quad \dfrac{dz}{dy} + \dfrac{2}{y}z = \dfrac{-2}{y^3}$, which is linear in z. ...(3)

Hence $\quad P = \dfrac{2}{y}, \; Q = -\dfrac{2}{y^3}$

$$I.F. = e^{\int \frac{2}{y}dy} = e^{2\log y} = e^{\log y^2} = y^2$$

The solution is $\quad y^2 \cdot z = -\displaystyle\int \frac{2}{y^3} \cdot y^2 dy + C$

$$\Rightarrow \quad y^2 \cdot z = -2\log y + C$$

$$\Rightarrow \quad y^2 z + 2\log y - C = 0$$

$$\Rightarrow \quad y^2 e^{-2x} + 2\log y - C = 0 \qquad \Big| \text{ As } z = e^{-2x}$$

$$\Rightarrow \quad y^2 e^{-2x} + 2\log y + C_1 = 0; \text{ where } C_1 = -C.$$

Example 10: Solve $x^2 dy + y(x+y)dx = 0$. (*U.P.T.U. 2006*)

Solution: The given equation may be written in the form

$$\frac{dy}{dx} + \frac{y}{x} = -\frac{y^2}{x^2}$$

Dividing on both sides by y^2, we get

$$\frac{1}{y^2}\frac{dy}{dx} + \frac{1}{y}.\frac{1}{x} = -\frac{1}{x^2} \qquad \text{...(1)}$$

Let $\dfrac{1}{y} = z$, so that $\dfrac{1}{y^2}\cdot\dfrac{dy}{dx} = -\dfrac{dz}{dx}$

Substituting in equation (1), we get

$$-\frac{dz}{dx}+\frac{1}{x}z=-\frac{1}{x^2} \quad \Rightarrow \quad \frac{dz}{dx}-\frac{1}{x}z=\frac{1}{x^2}$$

which is linear in z.

Hence, $P=-\frac{1}{x}$, $Q=\frac{1}{x^2}$

$$\text{I.F.}=e^{-\int\frac{1}{x}dx}=e^{-\log x}=e^{\log 1/x} \quad \Rightarrow \quad \text{I.F.}=\frac{1}{x}$$

The solution is $\quad z\cdot\frac{1}{x}=\int\frac{1}{x^2}\cdot\frac{1}{x}dx+C \quad \Rightarrow \quad \frac{z}{x}=\int\frac{1}{x^3}dx+C$

$$\Rightarrow \quad \frac{z}{x}=-\frac{1}{2x^2}+C$$

$$\Rightarrow \quad \frac{1}{xy}=-\frac{1}{2x^2}+C \qquad \left|\ \text{As } z=\frac{1}{y}\right.$$

$$\frac{1}{xy}=-\frac{1}{2x^2}+C. \quad \textbf{Ans.}$$

1.2.4 Exact Differential Equation

A differential equation of the form $Mdx + Ndy = 0$, where M and N are functions of x, y is said to be exact differential equation if it satisfies the condition

$$\frac{\partial M}{\partial y}=\frac{\partial N}{\partial x}$$

Method of Solution:

(*i*) First of all show that $\frac{\partial M}{\partial y}=\frac{\partial N}{\partial x}$

(*ii*) The solution is $\int\limits_{y=const.} Mdx+\int N(\text{The terms which do not contain } x)\, dy=C$

Example 11: Solve

$$(y\sin xy+xy^2\cos xy)\,dx+(x\sin xy+x^2y\cos xy)\,dy=0.$$

Solution: Comparing with $Mdx+Ndy=0$, we get

$$M=y\sin xy+xy^2\cos xy$$
$$N=x\sin xy+x^2y\cos xy$$

$$\therefore \quad \frac{\partial M}{\partial y} = \sin xy + xy\cos xy + 2xy\cos xy - x^2y^2\sin xy$$

and
$$\frac{\partial N}{\partial x} = \sin xy + xy\cos xy + 2xy\cos xy - x^2y^2\sin xy$$

This shows that $\frac{\partial M}{\partial y} = \frac{\partial N}{\partial x}$, so the given equation is exact

The general solution is

$$\int M dx + \int N(\text{terms do not contain } x)\, dy = C$$

$$\Rightarrow \quad \int (y\sin xy + xy^2\cos xy)\, dx + \int 0.dy = C$$

$$\Rightarrow \quad \frac{-y\cos xy}{y} + y^2 \int x\cos xy\, dx = C$$

$$\Rightarrow \quad -\cos xy + y^2\left(\frac{x\sin xy}{y} - \int 1\cdot\frac{\sin xy}{y}dx\right) = C$$

$$\Rightarrow \quad -\cos xy + xy\sin xy + \frac{y^2}{y^2}\cos xy = C$$

$$\Rightarrow \quad xy\sin xy = C. \ \textbf{Ans.}$$

Example 12: Solve $(1+3e^{x/y})dx + 3e^{x/y}\left(1-\frac{x}{y}\right)dy = 0.$

Solution: Comparing the given equation with $Mdx + Ndy = 0$

$$M = (1+3e^{x/y}),\ N = 3e^{x/y}\left(1-\frac{x}{y}\right)$$

$$\therefore \quad \frac{\partial M}{\partial y} = 3.e^{x/y}\left(-\frac{x}{y^2}\right) = -\frac{3x}{y^2}e^{x/y} \quad \Rightarrow \quad \frac{\partial M}{\partial y} = -\frac{3x}{y^2}e^{x/y}$$

and

$$\frac{\partial N}{\partial x} = \frac{3e^{x/y}}{y} - \frac{3x}{y^2}e^{x/y} - \frac{3e^{x/y}}{y} \quad \Rightarrow \quad \frac{\partial N}{\partial x} = -\frac{3x}{y^2}e^{x/y}$$

Since $\frac{\partial M}{\partial y} = \frac{\partial N}{\partial x}$

The solution is

$$\int (1+3e^{x/y})dx + \int 0\,dy = 0$$

$$x + 3ye^{x/y} = C. \quad \textbf{Ans.}$$

1.2.4.1 Reduction of Non-exact Differential Equations into Exact Differential Equations

If the differential equation $Mdx + Ndy = 0$ is not exact. It may become exist on multiplying by a suitable function known as integrating factor. There are some rules for finding out integrating factor.

Rule 1

For homogeneous equation

If $M\,dx + N\,dy = 0$ is a homogeneous equation in x and y then

$$\text{I.F.} = \frac{1}{Mx + Ny}$$

Example 13: Solve $y(y^2 - 2x^2)dx + x(2y^2 - x^2)dy = 0$

Solution: The given equation is homogeneous.

Here, $\quad M = y^3 - 2x^2y \quad$ and $\quad N = 2xy^2 - x^3$

$$\therefore \quad \frac{\partial M}{\partial y} = 3y^2 - 2x^2 \quad \text{and} \quad \frac{\partial N}{\partial x} = 2y^2 - 3x^2$$

Since $\dfrac{\partial M}{\partial y} \neq \dfrac{\partial N}{\partial x}$, the equation is not exact.

$$\therefore \quad \text{I.F.} = \frac{1}{Mx + Ny} = \frac{1}{3xy(y^2 - x^2)}$$

Multiplying the given D.E. by I.F., we get

$$\frac{y(y^2 - 2x^2)}{3xy(y^2 - x^2)}dx + \frac{x(2y^2 - x^2)}{3xy(y^2 - x^2)}dy = 0$$

$$\Rightarrow \quad \left(\frac{1}{x} - \frac{x}{y^2 - x^2}\right)dx + \left(\frac{y}{y^2 - x^2} + \frac{1}{y}\right)dy = 0 \qquad \text{...(1)}$$

Here, $\quad M = \dfrac{1}{x} - \dfrac{x}{(y^2 - x^2)} \quad$ and $\quad N = \dfrac{y}{(y^2 - x^2)} + \dfrac{1}{y}$

$$\therefore \quad \frac{\partial M}{\partial y} = \frac{2xy}{(y^2 - x^2)^2} \quad \text{and} \quad \frac{\partial N}{\partial x} = \frac{2xy}{(y^2 - x^2)^2} \quad \Rightarrow \quad \frac{\partial M}{\partial y} = \frac{\partial N}{\partial x}$$

$\therefore$ Equation (1) is exact

Hence, the solution is

$$\int \left(\frac{1}{x} - \frac{x}{y^2 - x^2} \right) dx + \int \frac{1}{y} dy = C$$

$$\Rightarrow \quad \log x + \frac{1}{2} \log(y^2 - x^2) + \log y = C$$

$$\Rightarrow \quad \log xy + \log(y^2 - x^2)^{1/2} = C \Rightarrow \log xy(y^2 - x^2)^{1/2} = C$$

Hence,

$$xy(y^2 - x^2)^{1/2} = e^C \quad \Rightarrow \quad xy\sqrt{y^2 - x^2} = C_1. \quad \textbf{Ans.}$$

Rule 2

For the form $yf_1(xy)\,dx + xf_2(xy)\,dy = 0$

If the equation $Mdx + Ndy = 0$ is of the form $yf_1(xy)\,dx + xf_2(xy)\,dy = 0$

then $$\text{I.F.} = \frac{1}{Mx - Ny}$$

Example 14: Solve $y(xy\sin xy + \cos xy)\,dx + x(xy\sin xy - \cos xy)\,dy = 0$

Solution: Here, $M = xy^2 \sin xy + y\cos xy$ and $N = x^2 y \sin xy - x\cos xy$

$$\therefore \quad \frac{\partial M}{\partial y} = 2xy\sin xy + x^2y^2 \cos xy + \cos xy - xy\sin xy$$

or $$\frac{\partial M}{\partial y} = xy\sin xy + x^2y^2 \cos xy + \cos xy$$

and $$\frac{\partial N}{\partial x} = 2xy\sin xy + x^2y^2 \cos xy - \cos xy + xy\sin xy$$

or $$\frac{\partial N}{\partial x} = 3xy\sin xy + x^2y^2 \cos xy - \cos xy$$

This shows that $\frac{\partial M}{\partial y} \neq \frac{\partial N}{\partial x}$

$$\therefore \quad \text{I.F.} = \frac{1}{x^2y^2 \sin xy + xy\cos xy - x^2y^2 \sin xy + xy\cos xy} = \frac{1}{2xy\cos xy}$$

Multiplying the given equation by I.F., we get

$$\left[\frac{y}{2} \tan xy + \frac{1}{2x} \right] dx + \left[\frac{x}{2} \tan xy - \frac{1}{2y} \right] dy = 0 \qquad \text{...(1)}$$

Here, $M = \frac{y}{2}\tan xy + \frac{1}{2x}$ and $N = \frac{x}{2}\tan xy - \frac{1}{2y}$

$$\therefore \quad \frac{\partial M}{\partial y} = \frac{1}{2}\tan xy + \frac{xy}{2}\sec^2 xy \quad \text{and} \quad \frac{\partial N}{\partial x} = \frac{1}{2}\tan xy + \frac{xy}{2}\sec^2 xy$$

$$\Rightarrow \quad \frac{\partial M}{\partial y} = \frac{\partial N}{\partial x}, \text{ the equation (1) is exact.}$$

Hence, the general solution is

$$\int\left(\frac{y}{2}\tan xy + \frac{1}{2x}\right)dx + \int\left(-\frac{1}{2y}\right)dy = C$$

$$\Rightarrow \quad \frac{1}{2}\log\sec xy + \frac{1}{2}\log x - \frac{1}{2}\log y = C$$

$$\Rightarrow \quad \frac{1}{2}\left[\log\frac{x\sec xy}{y}\right] = C \quad \Rightarrow \quad \log\frac{x\sec xy}{y} = 2C$$

$$\Rightarrow \quad \log\frac{x\sec xy}{y} = \log C_1 \quad \text{where, } 2C = \log C_1$$

$$\Rightarrow \quad x\sec xy = C_1 y.$$ **Ans.**

Rule 3

In the equation $Mdx + Ndy = 0$

If $\frac{1}{N}\left(\frac{\partial M}{\partial y} - \frac{\partial N}{\partial x}\right) = f(x)$ [function of x only]

then $\text{I.F.} = e^{\int f(x).dx}$

Example 15: Solve $(xy^2 - e^{1/x^3})dx - x^2 y\,dy = 0$

Solution: Here $M = xy^2 - e^{1/x^3}$ and $N = -x^2 y$

$$\therefore \quad \frac{\partial M}{\partial y} = 2xy \quad \text{and} \quad \frac{\partial N}{\partial x} = -2xy \quad \Rightarrow \quad \frac{\partial M}{\partial y} \neq \frac{\partial N}{\partial x}$$

Now $$\frac{1}{N}\left(\frac{\partial M}{\partial y} - \frac{\partial N}{\partial x}\right) = \frac{1}{-x^2 y}(2xy + 2xy) = \frac{-4xy}{x^2 y} = -\frac{4}{x}$$

$$\therefore \quad \text{I.F.} = e^{\int f(x)dx} = e^{-4\int 1/x\,dx} = e^{-4\log x} = e^{\log_e \frac{1}{x^4}} \quad \Rightarrow \quad \text{I.F.} = \frac{1}{x^4}$$

Multiplying the given equation by $\frac{1}{x^4}$, we get

$$\left(\frac{y^2}{x^3}-\frac{1}{x^4}e^{1/x^3}\right)dx-\frac{y}{x^2}dy = 0$$

Here, $M = \frac{y^2}{x^3}-\frac{1}{x^4}e^{\frac{1}{x^3}}$ and $N = -\frac{y}{x^2}$

$\therefore$ $\frac{\partial M}{\partial y} = \frac{2y}{x^3}$ and $\frac{\partial N}{\partial x} = \frac{2y}{x^3}$ $\Rightarrow$ $\frac{\partial M}{\partial y} = \frac{\partial N}{\partial x}$

The solution is $\int\left(\frac{y^2}{x^3}-\frac{1}{x^4}e^{1/x^3}\right)dx+\int 0\,dy = C$

$\Rightarrow$ $-\frac{y^2}{2x^2}+\frac{1}{3}e^{\frac{1}{x^3}} = C.$ **Ans.**

Rule 4

In the equation $M\,dx + N\,dy = 0$

If $\frac{1}{M}\left(\frac{\partial N}{\partial x}-\frac{\partial M}{\partial y}\right) = f(y)$ [the function of y only]

then I.F. $= e^{\int f(y)dy}$

Example 16: Solve $\left(\frac{y}{x}\sec y-\tan y\right)dx+(\sec y\log x-x)\,dy = 0.$

Solution: Here, $M = \frac{y}{x}\sec y-\tan y$ and $N = \sec y\log x - x$

$\therefore$ $\frac{\partial M}{\partial y} = \frac{1}{x}\sec y+\frac{y\sec y\tan y}{x}-\sec^2 y$ and $\frac{\partial N}{\partial x} = \frac{1}{x}\sec y-1$

From this $\frac{\partial M}{\partial y} \neq \frac{\partial N}{\partial x}$

Now $\frac{1}{M}\left(\frac{\partial N}{\partial x}-\frac{\partial M}{\partial y}\right) = \frac{1}{\left(\frac{y}{x}\sec y-\tan y\right)}\left(\frac{1}{x}\sec y-1-\frac{1}{x}\sec y-\frac{y}{x}\sec y\tan y+\sec^2 y\right)$

$$= \frac{-x\tan y}{x(y\sec y-x\tan y)}(y\sec y-x\tan y) = -\tan y$$

$$\therefore \qquad \text{I.F.} = e^{-\int \tan y} = e^{\log_e \cos y} = \cos y$$

Multiplying the given equation by cos y, we get

$$\left(\frac{y}{x} - \sin y\right)dx + (\log x - x\cos y)\,dy = 0$$

Here again,

$$M = \frac{y}{x} - \sin y \quad \text{and} \quad N = \log x - x\cos y$$

$$\therefore \qquad \frac{\partial M}{\partial y} = \frac{1}{x} - \cos y \quad \text{and} \quad \frac{\partial N}{\partial x} = \frac{1}{x} - \cos y \quad \Rightarrow \quad \frac{\partial M}{\partial y} = \frac{\partial N}{\partial x}$$

Hence, the solution is

$$\int\left(\frac{y}{x} - \sin y\right)dx + \int 0\,dy = C$$

$$\Rightarrow \qquad y\log x - x\sin y = C. \quad \textbf{Ans.}$$

Rule 5

If $M\,dx + N\,dy = 0$, is of another form (except in Rule 1 and Rule 2) and Rule 3, Rule 4 both failed then I.F. $= x^h y^k$ where h and k are some constants.

After multiplying the given equation by I.F. put, $\frac{\partial M}{\partial y} = \frac{\partial N}{\partial x}$ and calculate the value of h, k by equating the coefficients of like powers of xy on both sides.

Example 17: Solve the differential equation

$$y(xy + 2x^2y^2)\,dx + x(xy - x^2y^2)\,dy = 0.$$

Solution: We have

$$(xy^2 + 2x^2y^3)\,dx + (x^2y - x^3y^2)\,dy = 0 \qquad \text{...(1)}$$

Multiplying equation (1) by $x^h y^k$, we get

$$(x^{h+1}y^{k+2} + 2x^{h+2}\cdot y^{k+3})\,dx + (x^{h+2}\cdot y^{k+1} - x^{h+3}\cdot y^{k+2})\,dy = 0$$

Here, we have

$$M = x^{h+1}\cdot y^{k+2} + 2x^{h+2}y^{k+3}$$

$$N = x^{h+2}y^{k+1} - x^{h+3}\cdot y^{k+2}$$

$$\therefore \qquad \frac{\partial M}{\partial y} = (k+2)\,x^{h+1}\cdot y^{k+1} + 2(k+3)\,x^{h+2}\cdot y^{k+2}$$

and

$$\frac{\partial N}{\partial x} = (h+2)\,x^{h+1}\cdot y^{k+1} - (h+3)\,x^{h+2}\cdot y^{k+2}$$

For exact

$$\frac{\partial M}{\partial y} = \frac{\partial N}{\partial x}$$

$$(k+2)x^{h+1}y^{k+1} + 2(k+3)x^{h+2}y^{k+2} = (h+2)x^{h+1}.y^{k+1} - (h+3)x^{h+2}.y^{k+2}$$

Equating the coefficient of like powers of xy on both sides, we get

$$h + 2 = k + 2 \quad \Rightarrow \quad h = k \qquad \text{...(A)}$$

and

$$-(h + 3) = 2k + 6 \quad \Rightarrow \quad h + 2k = -9 \qquad \text{...(B)}$$

From (A) and (B), we get

$$h = -3,\ k = -3$$

Here

$$M = x^{-2}y^{-1} + 2x^{-1} = \frac{1}{x^2 y} + \frac{2}{x} \quad \Rightarrow \quad \frac{\partial M}{\partial y} = -\frac{1}{x^2 y^2}$$

and

$$N = x^{-1}y^{-2} - y^{-1} = \frac{1}{xy^2} - \frac{1}{y} \quad \Rightarrow \quad \frac{\partial N}{\partial x} = \frac{-1}{x^2 y^2}$$

$$\Rightarrow \frac{\partial M}{\partial y} = \frac{\partial N}{\partial x}$$

Hence, the solution of given equation is

$$\int\left(\frac{1}{x^2 y} + \frac{2}{x}\right)dx + \int\left(-\frac{1}{y}\right)dy = C \quad \Rightarrow \quad -\frac{1}{xy} + 2\log x - \log y = C$$

or

$$2\log x - \log y - \frac{1}{xy} = C. \quad \textbf{Ans.}$$

EXERCISE 1.1

Solve the following differential equations:

A : VARIABLE SEPARATION METHOD

1. $\dfrac{dy}{dx} - x\tan(y-x) = 1$ $\left[\textbf{Ans. } \log\sin(y-x) = \dfrac{x^2}{2} + C\right]$

2. $3e^x \tan y\, dx + (1 - e^x)\sec^2 y\, dy$ [**Ans.** $(1-e^x)^3 = A\tan y$]

3. $\dfrac{dy}{dx} = e^{x-y} + x^2 e^{-y}$ $\left[\textbf{Ans. } e^y = e^x + \dfrac{x^3}{3} + C\right]$

4. $\dfrac{dy}{dx} = (4x + y + 1)^2$ [**Hint :** Put $4x + y + 1 = t$] $\left[\textbf{Ans. } \dfrac{4x+y+1}{2} = \tan(2x + C)\right]$

5. $xy\,(1+y)\,dy-(1-x^2)\,(1-y)\,dx=0$ $\left[\textbf{Ans. } \log x+2\log(y-1)=\dfrac{x^2}{2}-\dfrac{1}{2}(y-1)^2-3(y-1)-C\right]$

B : HOMOGENEOUS METHOD

6. $x^2y\,dx-(x^3+y^3)dy=0$ $\left[\textbf{Ans. } cy=e^{x^3/3y^3}\right]$

7. $x\dfrac{dy}{dx}=y(\log y-\log x+1)$ $\left[\textbf{Ans. } y=xe^{Cx}\right]$

8. $x(x-y)\,dy+y^2dx=0$ [**Ans.** $y=x\log Cy$]

9. $x^2dy+y(x+y)\,dx=0$ $\left[\textbf{Ans. } x^2y=C(2x+y)\right]$

10. $x\sin\dfrac{y}{x}\,dy=(y\sin\dfrac{y}{x}-x)\,dx=0$ $\left[\textbf{Ans. } \cos\dfrac{y}{x}=\log x+C\right]$

C : REDUCIBLE TO HOMOGENEOUS

11. $\dfrac{dy}{dx}=\dfrac{2x+3y-4}{4x+y-3}$, given that $y\,(1)=1$ [**Ans.** $(y+2x-2)^2=(2x-2y+1)^5$]

12. $\dfrac{dy}{dx}=\dfrac{2x+3y+4}{4x+6y+5}$ $\left[\textbf{Ans. } 42y-21x-9\log(14x+21y+22)=C\right]$

13. $(x-y-2)dx-(2x-2y-3)dy=0$ $\left[\textbf{Ans. } \log(x-y-1)=x-2y+C\right]$

(U.P. B. Pharma (C.O.) 2005)

14. $(x+2y)(dx-dy)=dx+dy$ $\left[\textbf{Ans. } 3x-3y+a=2\log(3x+6y-1)\right]$

15. $(2x+4y+3)\dfrac{dy}{dx}=2y+x+1$ $\left[\textbf{Ans. } 4x+8y+5=C\,e^{4(x-2y)}\right]$

D : LINEAR DIFFERENTIAL EQUATION

16. $x(1-x^2)\dfrac{dy}{dx}+(2x^2-1)y=x^3$ $\left[\textbf{Ans. } y=x+Cx\sqrt{1-x^2}\right]$

17. $ye^y=(y^3+2xe^y)\dfrac{dy}{dx}$ $\left[\textbf{Ans. } x+y^2e^{-y}=Cy^2\right]$

18. $d+(2r\cot\theta+\sin 2\theta)d\theta=0$ $\left[\textbf{Ans. } r=-\dfrac{1}{2}\sin^2\theta+C\operatorname{cosec}^2\theta\right]$

19. $x\cos x\dfrac{dy}{dx}+y(x\sin x+\cos x)=1$ [**Ans.** $xy=\sin x+C\cos x$]

20. $(1+y+x^2y)dx+(x+x^3)dy=0$ [**Ans.** $xy = C - \tan^{-1}x$]

E : REDUCIBLE TO LINEAR EQUATION

21. $\dfrac{dy}{dx}-x^3y^3+xy=0$ $\left[\textbf{Ans. } \dfrac{1}{y^2}=1+x^2+Ce^{x^2}\right]$

22. $\sin y\dfrac{dy}{dx}=\cos x\,(2\cos y-\sin^2 x)$ $\left[\textbf{Ans. } \cos y=\dfrac{1}{4}(2\sin^2 x+2\sin x-1+4Ce^{-2\sin x}\right]$

23. $x\dfrac{dy}{dx}+y=x^3y^6$ $\left[\textbf{Ans. } \dfrac{5}{2}x^3y^5+cx^5y^5=1\right]$

24. $\dfrac{dy}{dx}-\dfrac{\tan y}{1+x}=(1+x)e^x\sec y$ [**Ans.** $\sin y=(1+x)(e^x+C)$]

25. $\dfrac{1}{1+y^2}\cdot\dfrac{dy}{dx}+2x\tan^{-1}y=x^3$ $\left[\textbf{Ans. } \tan^{-1}y=\dfrac{1}{2}(x^2-1)+Ce^{-x^2}\right]$

F : EXACT DIFFERENTIAL EQUATION

26. $(\sin x.\cos hy)\,dx-(\cos x\sin hy)\,dy=0,\ y(0)=3$ [**Ans.** $\cos x.\cos hy = 10.07$]

27. $(\sin x.\sin y-xe^y)dy=(e^y+\cos x.\cos y)dy$ [**Ans.** $xe^y+\sin x.\cos y=C$]

28. $xdx+ydy+\dfrac{xdy-ydx}{x^2+y^2}=0$ [**Ans.** $(x^2+y^2)/2+\tan^{-1}(y/x)=C$]

29. $\left[(\cos x)\log_e(2y-8)+\dfrac{3}{x}\right]+\left(\dfrac{\sin x}{y-4}\right)\dfrac{dy}{dx}=0;\ y(1)=\dfrac{9}{2}$ [**Ans.** $\sin x\log(2y-8)+3\log x=0$]

30. $(\cos x-x\cos y)\,dy-(\sin y+y\sin x)\,dx=0$ [**Ans.** $y\cos x-x\sin y=C$]

G : REDUCIBLE TO EXACT FORM

31. $(y^3-3xy^2)dx+(2x^2y-xy^2)dy=0$ $\left[\textbf{Ans. } \dfrac{x^3}{y^2}=e^Ce^{-y/x}\right]$

32. $(x^4+y^4)dx-xy^3dx=0$ [**Ans.** $y^4=4x^4\log x+Cx^4$]

33. $(x^2y^2+xy+1)ydx+(x^2y^2-xy+1)x\,dy=0$ $\left[\textbf{Ans. } xy-\dfrac{1}{xy}+\log\dfrac{x}{y}=C\right]$

34. $(2xy^2 + y)\,dx = (x + 2x^2y - x^4y^3)\,dy = 0$ $\left[\textbf{Ans. } y = Ce^{(1-3xy)/3x^3y^3}\right]$

35. $\left(y + \frac{1}{3}y^3 + \frac{1}{2}x^2\right)dx + \frac{1}{4}(1+y^2)\,xdy = 0$ $\left[\textbf{Ans. } \frac{x^4y}{4} + \frac{x^4y^3}{12} + \frac{x^6}{12} = C\right]$

36. $(2x^3y^2 + 4x^2y + 2xy^2 + xy^4 + 2y)\,dx + 2(y^3 + x^2y + x)\,dy = 0$

$\left[\textbf{Ans. } (2x^2y^2 + 4xy + y^4)e^{x^2} = C\right]$

37. $2xy\,dx + (y^2 - x^2)\,dy = 0;\ y(2) = 1$ $\left[\textbf{Ans. } x^2 + y^2 = 5y\right]$

38. $\frac{dy}{dx} = \frac{x^3 + y^3}{xy^2}$ $\left[\textbf{Ans. } \log x - \frac{y^3}{3x^3} = C\right]$

39. $(y^3 - 2x^2y)\,dx + (2xy^2 - x^3)\,dy = 0$ $\left[\textbf{Ans. } x^2y^2(y^2 - x^2) = 2C\right]$

40. $x^3y^3(2ydx + xdy) - (5ydx + 7xdy) = 0$ $\left[\textbf{Ans. } x^3y^3 + 2 = Cx^{5/3}y^{7/3}\right]$

1.3 LINEAR DIFFERENTIAL EQUATIONS OF HIGHER ORDER WITH CONSTANT COEFFICIENTS

The differential equation of the form

$$\frac{d^ny}{dx^n} + a_1\frac{d^{n-1}y}{dx^{n-1}} + a_2\frac{d^{n-2}y}{dx^{n-2}} + \cdots + a_ny = R \qquad ...(i)$$

where $a_1, a_2, ..., a_n$ are all constants and R is a function of x is called *L.D.E.* of nth order with constant coefficients.

1.3.1 Linear Differential Equations of Second Order with Constant Coefficients

Standard form of linear differential equation of second order is

$$\frac{d^2y}{dx^2} + P\frac{dy}{dx} + Qy = R \qquad ...(ii)$$

where P and Q are constants and R is a function of x. For simplicity using the differential operator

$$D = \frac{d}{dx} \text{ so that } Dy = \frac{dy}{dx} \text{ and } D^2 = \frac{d^2}{dx^2} \quad \Rightarrow \quad D^2y = \frac{d^2y}{dx^2}$$

∴ From equation (*ii*), we get

$$D^2y + P\,Dy + Qy = R$$

$$\Rightarrow \quad (D^2 + PD + Q)\,y = R \qquad ...(iii)$$

Complementary Homogeneous Equation: If $R = 0$ then equation (*ii*) is, called Complementary Homogeneous equation or Homogeneouse equeation *i.e.*,

$$\frac{d^2y}{dx^2} + P\frac{dy}{dx} + Qy = 0$$

is a complementary homogeneous equation.

Non-Homogeneous Equation: If $R \neq 0$ then the equation is called Non-homogeneous equation.

1.3.2 Method of Solution

The complete solution of such type differential equations is as follows:

Complete Solution = Complementary Function + Particular Integral

$$\Rightarrow \qquad \boxed{\text{C.S.} = \text{C.F.} + \text{P.I.}}$$

1.3.3 Complementary Function (Solution of Complementary Homogeneous Equation)

The differential equation under consideration is

$$\frac{d^2y}{dx^2} + \frac{Pdy}{dx} + Qy = 0 \qquad ...(iv)$$

i.e., R.H.S. = 0 ($R = 0$), then the solution of such equation is said to be complementary function.

1.3.3.1 Methods for Finding the Complementary Function

Let *$y = Ce^{mx}$ is the solution of equation (*iv*), then

$$\frac{dy}{dx} = Cm\,e^{mx} \text{ and } \frac{d^2y}{dx^2} = Cm^2e^{mx}$$

Substituting these values in equation (*iv*), we get

$$(Cm^2e^{mx} + PCme^{mx} + QCe^{mx})\,y = 0$$

$$Ce^{mx}(m^2 + Pm + Q)\,y = 0$$

$Ce^{mx} \neq 0$, because it is the solution of differential equation.

So $\qquad m^2 + Pm + Q = 0 \qquad ...(v)$

It is called an auxiliary equation.

Remark: On comparing equation (*iii*) and (*v*), we can say that an auxiliary equation obtained replacing D by m in given differential equation.

---

* For $y = ce^{mx}$, we solve the equation $(D - m)\,y = 0$, *i.e.*, $\frac{dy}{dx} - my = 0$

$\Rightarrow \frac{dy}{y} = m\,dx \Rightarrow \log y = mx + \log C$ or $y = Ce^{mx}$, where C is any constant.

Therefore, $(D - m_1)y = 0$, $(D - m_2)y = 0$ gives the solutions, $y = C_1e^{m_1x}$, $y = C_2e^{m_2x}$

Then solve an auxiliary equation which has two roots and arises the following cases on the bases of these two roots.

Case I: Roots of A.E. (Auxiliary Equation) are real and distinct

If m_1 and m_2 are two roots then $\boxed{\text{C.F.} = y = C_1 e^{m_1 x} + C_2 e^{m_2 x}.}$

In general if $m_1 \neq m_2 \neq m_3 \ldots \neq m_n$ then

$$\text{C.F.} = C_1 e^{m_1 x} + C_2 e^{m_2 x} + C_3 e^{m_3 x} + \ldots + C_n e^{m_n x}$$

Case II: Roots of A.E. are real and equal

If $m_1 = m_2 = m(\text{say})$, then $\boxed{\text{C.F.} = y = (C_1 + xC_2)e^{mx}.}$

In general if $m_1 = m_2 = m_3 = \ldots = m_n = m$ then

$$\text{C.F.} = (C_1 + xC_2 + x^2 C_3 + \ldots + x^{n-1} C_n)e^{mx}$$

Case III: Having Complex Roots

If $m_1 = \alpha + i\beta,\ m_2 = \alpha - i\beta$

i.e., $m = \alpha \pm i\beta$, then $\text{C.F.} = e^{\alpha x}(A\cos\beta x + B\sin\beta x)$

Proof: $\text{C.F.} = y = C_1 e^{(\alpha+i\beta)x} + C_2 e^{(\alpha-i\beta)x}$

$$\Rightarrow \quad y = e^{\alpha x}[C_1(\cos\beta x + i\sin\beta x) + C_2(\cos\beta x - i\sin\beta x)]$$

$$\Rightarrow \quad y = e^{\alpha x}[(C_1 + C_2)\cos\beta x + i(C_1 - C_2)\sin\beta x]$$

$$\Rightarrow \quad \boxed{\text{C.F.} = y = e^{\alpha x}(A\cos\beta x + B\sin\beta x)}, \text{ where } \begin{bmatrix} A = C_1 + C_2 \\ B = i(C_1 - C_2) \end{bmatrix}$$

Note: When the roots $m_1 = m_2 = \alpha + i\beta,\ m_3 = m_4 = \alpha - i\beta$ then

$$\boxed{\text{C.F.} = e^{\alpha x}[(C_1 + C_2 x)\cos\beta x + (C_3 + C_4 x)\sin\beta x]}$$

Example 18: Solve $\dfrac{d^2y}{dx^2} - \dfrac{dy}{dx} - 2y = 0,\ y(0) = 4, y'(0) = -5.$

Solution: The given equation can be written as

$$(D^2 - D - 2)y = 0$$

The A.E. is $m^2 + m - 2 = 0$

$$\Rightarrow \quad (m+1)(m-2) = 0 \quad \Rightarrow \quad m = -1, +2; \quad i.e., \quad m_1 = -1, m_2 = 2$$

Hence the required solution is

$$y = C_1 e^{-x} + C_2 e^{2x}$$

Using given conditions, at $x=0, y=4$ and $y'=-5$

$$\therefore \quad 4 = C_1 + C_2 \quad \Rightarrow \quad C_1 + C_2 = 4 \quad \ldots(1)$$

and

$$-5 = -C_1 + 2C_2 \quad \Rightarrow \quad -C_1 + 2C_2 = -5 \quad \ldots(2)$$

Solving equation (1) and (2), we get

$$C_1 = \frac{13}{3} \text{ and } C_2 = -\frac{1}{3}$$

Thus the particular solution is

$$y = \frac{13}{3}e^{-x} - \frac{1}{3}e^{2x}. \quad \textbf{Ans.}$$

Example 19: Solve $\frac{d^4y}{dx^4} - 4\frac{d^3y}{dx^3} + 8\frac{d^2y}{dx^2} - 8\frac{dy}{dx} + 4y = 0.$

Solution: The given equation can be written as

$$(D^4 - 4D^3 + 8D^2 - 8D + 4)y = 0$$

$\therefore$ The A.E. is

$$m^4 - 4m^3 + 8m^2 - 8m + 4 = 0$$

$$\Rightarrow \quad (m^2 - 2m + 2)^2 = 0$$

$$\Rightarrow \quad (m^2 - 2m + 2)(m^2 - 2m + 2) = 0$$

$$m^2 - 2m + 2 = 0 \text{ gives}$$

$$m = \frac{2 \pm \sqrt{4 - 4 \times 2}}{2} = 1 \pm i$$

Therefore, the roots of A.E. are $m = (1 \pm i)$ (twice)

Hence, $\quad y = e^x\{(C_1 + C_2x)\cos x + (C_3 + C_4x)\sin x\}$ **Ans.**

Example 20: Solve $y'' + 6y' + 9y = 0$

Solution: The given equation can be written as

$$(D^2 + 6D + 9)y = 0$$

The A.E. is $\quad m^2 + 6m + 9 = 0$

$\Rightarrow \quad (m+3)^2 = 0 \quad \Rightarrow \quad m = -3, -3$

Hence the required solution is

$$y = (C_1 + C_2x)e^{-3x}. \quad \textbf{Ans.}$$

EXERCISE 1.2

Solve the following differential equations:

1. $\frac{d^2y}{dx^2}+5\frac{dy}{dx}+12.5y=0$ $\left[\textbf{Ans. } y=e^{\frac{-5x}{2}}\left[C_1\cos\frac{5}{2}x+C_2\sin\frac{5x}{2}\right]\right]$

2. $\frac{d^2y}{dx^2}+4\frac{dy}{dx}+5y=0,\ y(0)=2$ and $y'(0)=y''(0)$ $\left[\textbf{Ans.}\ y=2e^{-2x}(\sin x+\cos x)\right]$

3. $16\frac{d^2y}{dx^2}-8\frac{dy}{dx}+y=0$ $\left[\textbf{Ans.}\ y=(C_1+C_2x)e^{x/4}\right]$

4. $\frac{d^4y}{dx^4}+\frac{d^3y}{dx^3}-6\frac{d^2y}{dx^2}-14\frac{dy}{dx}-12y=0$ $\left[\textbf{Ans.}\ y=C_1e^{3x}+C_2e^{-2x}+e^{-x}(C_3\cos x+C_4\sin x)\right]$

5. $2\frac{d^2y}{dx^2}-2\frac{dy}{dx}+y=0$ $\left[\textbf{Ans.}\ y=C_1e^{(2+\sqrt{3})x}+C_2e^{(2-\sqrt{3})x}\right]$

6. $\frac{d^3y}{dx^3}-6\frac{d^2y}{dx^2}+11\frac{dy}{dx}-6y=0, y(0)=y'(0)=0$ and $y''(0)=2$ $\left[\textbf{Ans.}\ y=e^x-2e^{2x}+e^{3x}\right]$

7. $(D^2+1)^3y=0$ $\left[\textbf{Ans.}\ y=(C_1+C_2x+C_3x^2)\cos x+(C_4+C_5x+C_6x^2)\sin x\right]$

8. $\frac{d^4y}{dx^4}-3\frac{d^3y}{dx^3}+\frac{2d^2y}{dx^2}=0, y(0)=2, y'(0)=0, y''(0)=2,\ y'''(0)=2.$ $\left[\textbf{Ans.}\ y=2(e^x-x)\right]$

9. $\frac{d^3y}{dx^3}-\frac{d^2y}{dx^2}+\frac{dy}{dx}-y=0$ $\left[\textbf{Ans.}\ y=C_1e^x+C_2\sin x+C_3\cos x\right]$

10. $\frac{d^2y}{dx^2}+2\lambda\frac{dy}{dx}+\lambda^2y=0$ $\left[\textbf{Ans.}\ y=(C_1+C_2x)e^{-\lambda x}\right]$

1.3.4 Particular Integral (P.I.)

The particular integral is also known as particular solution.

Here we have to introduce **inverse operator,** the operator $\frac{1}{D}=D^{-1}$ is known as inverse operator, thererfore,

$$\frac{1}{D}F(x) = \int F(x)\,dx+C = \int F(x)\,dx \qquad \left|\begin{array}{l}\text{choosen}\\ C=0\end{array}\right.$$

Similarly, $\frac{1}{D^2},\frac{1}{D^3}$,...... etc. denote the integration, twice, thrice etc. w.r.t. 'x'.

Hence the equation $D^2y + PDy + Qy = R$ can be written as

$$f(D)y = R$$

where $f(D) = D^2 + PD + Q$, is a function of D. Then the particular integral (P.I.) of above equation equation is

$$\boxed{P.I. = \frac{1}{f(D)} \cdot R}$$

Remark: This is particular solution for the value $C = 0$.
The value of P.I. depends upon the particular form of the function R, as follows:

1.3.4.1 When $R = e^{ax}$, $P.I. = \frac{1}{f(D)} e^{ax} = \frac{1}{f(a)} e^{ax}$, provided $f(a) \neq 0$

Consider the nth order L.D.E. with constant coefficients

$$(D^n + a_1 D^{n-1} + \dots + a_n)y = R = e^{ax} \qquad \dots(i)$$

Let $$f(D) = D^n + a_1 D^{n-1} + \dots + a_n \qquad \dots(ii)$$

$$\Rightarrow \quad f(D)\, y = e^{ax} \qquad \dots(iii)$$

$$\Rightarrow \quad y = \frac{1}{f(D)} e^{ax}$$

Now, we find $\frac{1}{f(D)} e^{ax}$ as follows:

From equation (*ii*)

$$f(D)e^{ax} = D^n e^{ax} + a_1 D^{n-1} e^{ax} + \dots + a_n e^{ax}$$

$$\Rightarrow \quad f(D)e^{ax} = a^n e^{ax} + a_1 a^{n-1} e^{ax} + \dots + a_n e^{ax}$$

$$\Rightarrow \quad f(D)e^{ax} = (a^n + a_1 a^{n-1} + \dots + a_n)e^{ax}$$

$$\Rightarrow \quad \frac{1}{f(D)} \cdot f(D)e^{ax} = \frac{1}{f(D)} \cdot f(a)e^{ax}$$

As $f(D) = D^n + a_1 D^{n-1} + \dots + a_n$
$\Rightarrow f(a) = a^n + a_1 a^{n-1} + \dots + a_n$

$$\Rightarrow \quad \boxed{\frac{1}{f(D)} e^{ax} = \frac{1}{f(a)} e^{ax}, f(a) \neq 0}$$

***Case I: If f(a) = 0* (Failure case)**

If $f(a) = 0$ it mean $D = a$ is the root of $f(D)$.
This shows that $(D - a)$ will be a factor of $f(D)$.

$\therefore \quad f(D) = (D-a)\,\phi(D)$.....(1) | $\phi(D)$ is remaining part of $f(D)$

such that $\quad \phi(a) \neq 0$

Then

$$\frac{1}{f(D)}e^{ax} = \frac{1}{(D-a)\phi(D)}e^{ax}$$

$$= \frac{1}{(D-a)}\left\{\frac{1}{\phi(D)}e^{ax}\right\}$$

$$= \frac{1}{\phi(a)(D-a)}e^{ax} \qquad \Big| \text{As } \frac{1}{\phi(D)}e^{ax} = \frac{1}{\phi(a)}e^{ax}$$

$$= \frac{1}{\phi(a)}e^{ax}\int e^{-ax}.e^{ax}dx \qquad \Big| \text{As } \frac{1}{(D-a)}e^{ax} = e^{ax}\int e^{-ax}.dx$$

| Proof of this formula is given at the last

$$\therefore \quad \frac{1}{f(D)}e^{ax} = \frac{1}{\phi(a)}x\,e^{ax} \qquad ...(iv)$$

From (1) $\quad f'(D) = \phi(D) + (D-a)\phi'(D)$

$\Rightarrow \quad f'(a) = \phi(a)$

Using this result in equation (*iv*), we get

$$\frac{1}{f(D)}e^{ax} = x\cdot\frac{1}{f'(a)}e^{ax}$$

Therefore,

$$\boxed{P.I. = \frac{1}{f(D)}e^{ax} = x.\frac{1}{f'(D)}e^{ax} = \frac{x\,e^{ax}}{f'(a)}} \qquad \Big| \text{As } \quad f'(a) \neq 0$$

Similarly for the cases $f''(a) = f'''(a) = ... = f^n(a) = 0$

We can find P.I. as follows:

$$\text{P.I.} = \frac{x^2}{f''(D)}e^{ax} = \frac{x^2 e^{ax}}{f''(a)} \text{ and so on.}$$

Example 21: Solve the equation

$$\frac{d^2y}{dx^2} - \frac{dy}{dx} - 6y = e^x \cosh 2x$$

Solution: The given equation can be written as

$$(D^2 - D - 6)y = e^x \cosh 2x$$

The A.E. is $m^2 - m - 6 = 0$ or $m = -2, 3$

$$\text{C.F.} = C_1 e^{-2x} + C_2 e^{3x}$$

$$\text{P.I.} = \frac{1}{(D^2 - D - 6)} e^x \left(\frac{e^{2x} + e^{-2x}}{2} \right) = \frac{1}{2} \cdot \frac{1}{(D^2 - D - 6)} \left(e^{3x} + e^{-x} \right) \quad \Bigg| \cosh\theta = \frac{e^{\theta} + e^{-\theta}}{2}$$

$$= \frac{1}{2(D^2 - D - 6)} e^{3x} + \frac{1}{2(D^2 - D - 6)} e^{-x} = x \cdot \frac{1}{2(2D - 1)} e^{3x} + \frac{1}{(-8)} e^{-x} \quad \Bigg| \text{Case of failure}$$

$$= x \cdot \frac{1}{2(5)} e^{3x} + \frac{1}{(-8)} e^{-x} = \frac{xe^{3x}}{10} - \frac{e^{-x}}{8}$$

Hence, the complete solution is

$$y = \text{C.F.} + \text{P.I.} = C_1 e^{-2x} + C_2 e^{3x} + \frac{1}{10} xe^{3x} - \frac{1}{4} e^{-x}. \quad \textbf{Ans.}$$

Example 22: Solve the equation $(D^2 + D + 1)\, y = 3^x$.

Solution: The A.E. is $m^2 + m + 1 = 0$

$$\Rightarrow \qquad m = \frac{-1 \pm \sqrt{1-4}}{2} = -\frac{1}{2} \pm i\frac{\sqrt{3}}{2}$$

$$\text{C.F.} = e^{-\frac{1}{2}x} \left(A \cos \frac{\sqrt{3}}{2} x + B \sin \frac{\sqrt{3}}{2} x \right)$$

$$\text{P.I.} = \frac{1}{(D^2 + D + 1)} 3^x = \frac{1}{(D^2 + D + 1)} e^{\log_e 3^x} = \frac{1}{(D^2 + D + 1)} e^{x \log_e 3}$$

$$= \frac{1}{(\log_e 3)^2 + (\log_e 3) + 1} e^{x \log_e 3} = \frac{1}{(\log_e 3)^2 + (\log_e 3) + 1} 3^x$$

Hence, the complete solution is

$$y = e^{-\frac{x}{2}} \left(A \cos \frac{\sqrt{3}}{2} x + B \sin \frac{\sqrt{3}}{2} x \right) + \frac{1}{(\log_e 3)^2 + (\log_e 3) + 1} 3^x. \quad \textbf{Ans.}$$

Example 23: Solve the equation

$$(D^4 + 5D^3 + 6D^2 - 4D - 8)\, y = e^{-2x} + 5.$$

Solution: The A.E. is $m^4 + 5m^3 + 6m^2 - 4m - 8 = 0$

$$\Rightarrow \qquad (m+2)^3 (m-1) = 0 \Rightarrow m = -2, -2, -2, 1$$

$\therefore \qquad \text{C.F.} = (C_1 + C_2x + C_3x^2)e^{-2x} + C_4e^x$

$$\text{P.I.} = \frac{1}{(D^4+5D^3+6D^2-4D-8)} \cdot (e^{-2x}+5)$$

$$= x \cdot \frac{1}{(4D^3+15D^2+12D-4)} e^{-2x} + 5\frac{1}{(D^4+5D^3+6D^2-4D-8)} e^{0x} \quad \Big| \text{ Case fail in first term}$$

$$= x^2 \cdot \frac{1}{(12D^2+30D+12)} e^{-2x} + 5\frac{1}{(0+0+0-0-8)} \quad \Big| \text{ Case fail in first term}$$

$$= x^3 \frac{1}{(24D+30)} e^{-2x} - \frac{5}{8} \quad \Big| \text{ Case fail in first term}$$

$$= x^3 \frac{1}{-18} e^{-2x} - \frac{5}{8} = -\frac{x^3}{18} e^{-2x} - \frac{5}{8}$$

Hence, the complete solution is

$$y = \text{C.F.} + \text{P.I.} = (C_1 + C_2x + C_3x^2)e^{-2x} + C_4e^x - \frac{x^3e^{-2x}}{18} - \frac{5}{8}. \qquad \textbf{Ans.}$$

Example 24: Solve $(D^2 - 12D + 32)y = (e^x + e^{-2x})^2$.

Solution: The A.E. is $m^2 - 12m + 32 = 0$

$\Rightarrow \qquad (m-4)(m-8) = 0 \qquad \Rightarrow \qquad m = 4, 8$

$\therefore \qquad \text{C.F.} = C_1e^{4x} + C_2e^{8x}$

$$\text{P.I.} = \frac{1}{(D^2-12D+32)}(e^{2x}+2e^{-x}+e^{-4x})$$

$$= \frac{1}{(D^2-12D+32)} e^{2x} + 2 \cdot \frac{1}{(D^2-12D+32)} e^{-x} + \frac{1}{(D^2-12D+32)} e^{-4x}$$

$$= \frac{1}{12} e^{2x} + \frac{2}{45} e^{-x} + \frac{1}{96} e^{-4x}$$

Hence the complete solution is

$$y = \text{C.F.} + \text{P.I.} = C_1e^{4x} + C_2e^{8x} + \frac{1}{12}e^{2x} + \frac{2}{45}e^{-x} + \frac{1}{96}e^{-4x}. \qquad \textbf{Ans.}$$

EXERCISE 1.3

Solve the following differential equations:

1. $(D^2 - D + 1)\,y = \sin hx$ $\left[\textbf{Ans. } y = e^{\frac{x}{2}}\left[C_1 \cos\frac{\sqrt{3}}{2}x + C_2 \sin\frac{\sqrt{3}}{2}x\right] + \frac{1}{6}(3e^x - e^{-x})\right]$

2. $\dfrac{d^2y}{dx^2} - 6\dfrac{dy}{dx} + 9y = 6e^{3x} + 7e^{-2x} - \log 2$

$\left[\textbf{Ans. } y = (C_1 + C_2x)e^{3x} + 3x^2e^{3x} + \frac{7}{25}e^{-2x} - \frac{1}{9}\log 2\right]$

3. $(D-1)^3\,y = 16e^{3x}$ $\left[\textbf{Ans.} (C_1 + C_2x + C_3x^2)e^x + 2e^{3x}\right]$

4. $\dfrac{d^2y}{dx^2} + 4\dfrac{dy}{dx} + 13y = 18e^{-2x},\ y(0) = 0,\ y'(0) = +4$

$\left[\textbf{Ans. } y = e^{-2x}\left(\frac{4}{3}\sin 3x - 2\cos 3x\right) + 2e^{-2x}\right]$

5. $(D^2 - 3D^2 + 3D - 1)\,y = e^x - 5$ $\left[\textbf{Ans.} (C_1 + C_2x + C_3x^2)e^x + \frac{x^3e^x}{6} + 5\right]$

6. $(D^2 - k^2)\,y = \sin hkx,$ $\left[\textbf{Ans. } y = C_1e^{kx} + C_2e^{-kx} + \frac{x}{2k}\cos hkx\right]$

7. $\dfrac{d^3y}{dx^3} + 3\dfrac{d^2y}{dx^2} + 3\dfrac{dy}{dx} + y = e^{-x}$ $\left[\textbf{Ans. } y = (C_1 + C_2x + C_3x^2)e^{-x} + \frac{x^3}{6}e^{-x}\right]$

8. $(D^2 - a^2)y = e^{2x}$ $\left[\textbf{Ans. } y = C_1e^{ax} + C_2e^{-ax} + \frac{e^{2x}}{3}\right]$

9. $(D^2 - 6D + 9)\,y = e^{3x}$ $\left[\textbf{Ans.} (C_1 + C_2x)e^{3x} + \frac{x^2}{2}e^{3x}\right]$

10. $(D^2 + 2D - 8)\,y = e^{-3x} + e^{-4x}$ $\left[\textbf{Ans. } y = C_1e^{2x} + C_2e^{-4x} - \frac{1}{5}e^{-3x} - \frac{x}{6}e^{-4x}\right]$

1.3.4.2 When $R = \sin ax$ or $\cos ax$ or $\sin(ax + b)$ and $\cos(ax + b)$

$$\text{P.I.} = \frac{1}{f(D^2)}\sin ax = \frac{1}{f(-a^2)}\sin ax \quad\text{and}\quad \frac{1}{f(D^2)}\cos ax = \frac{1}{f(-a^2)}\cos ax$$

$$D \sin ax = a\cos ax, \quad D^2 \sin ax = -a^2 \sin ax,$$

$$D^3 \sin ax = -a^3 \cos ax,\ D^4 \sin ax = a^4 \sin ax \Rightarrow D^4 \sin ax = (-a^2)^2 \sin ax$$

Here, we choose only even power of D, we get

$$(D^2)^n \sin ax = (-a^2)^n \sin ax$$

It follows that

$$f(D^2)\sin ax = f(-a^2)\sin ax$$

or

$$\frac{1}{f(D^2)} f(D^2)\sin ax = \frac{1}{f(D^2)}.f(-a^2)\sin ax$$

$$\sin ax = \frac{1}{f(D^2)} f(-a^2)\sin ax$$

$$\Rightarrow \boxed{\frac{1}{f(D^2)}\sin ax = \frac{1}{f(-a^2)}\sin x,} \qquad \text{As } f(-a^2) \neq 0$$

Similarly,

$$\boxed{\frac{1}{f(D^2)}\cos ax = \frac{1}{f(-a^2)}\cos ax,} \qquad \text{As } f(-a^2) \neq 0$$

***Case I: If $f(-a^2)$ = 0* (Failure case)**

Then

$$\frac{1}{f(D^2)}\sin ax = x \cdot \frac{1}{f'(D^2)}\sin ax = \frac{x \sin ax}{f'(-a^2)}$$

If $f'(-a^2) = 0$, then $\dfrac{1}{f(D^2)}\sin ax = \dfrac{x^2}{f''(-a^2)} \cdot \sin ax$

and so on. Similarly for cos θ.

Example 25: Solve $\dfrac{d^3y}{dx^3} - 3\dfrac{d^2y}{dx^2} + 4\dfrac{dy}{dx} - 2y = e^x + \cos x.$ *(U.P.T.U. 2001, 2006)*

Solution: The given equation can be written as

$$(D^3 - 3D^2 + 4D - 2)y = e^x + \cos x$$

The A.E. is $m^3 - 3m^2 + 4m - 2 = 0$

$$\Rightarrow \quad (m-1)(m^2 - 2m + 2) = 0$$

$$\Rightarrow \quad m = 1,\ m = 1 \pm i$$

$$\therefore \quad \text{C.F.} = C_1 e^x + e^x (A\cos x + B\sin x)$$

$$\text{P.I.} = \frac{1}{(D^3 - 3D^2 + 4D - 2)} e^x + \frac{1}{(D^3 - 3D^2 + 4D - 2)} \cos x$$

$$= \frac{x}{(3D^2-6D+4)}e^x + \frac{1}{(-1)D-3(-1)+4D-2}\cos x \quad \Bigg| \begin{array}{l}\text{Case fail}\\ \text{in first term}\end{array}$$

$$= \frac{xe^x}{3-6+4} + \frac{1}{(3D+1)}\cos x = \frac{xe^x}{1} + \frac{(3D-1)}{(9D^2-1)}\cos x$$

$$= xe^x + \frac{(3D-1)}{-9-1}\cos x = xe^x - \frac{1}{10}(-3\sin x - \cos x)$$

$$\text{P.I.} = xe^x + \frac{1}{10}(3\sin x + \cos x)$$

Hence, complete solution is

$$y = \text{C.F.} + \text{P.I.} = C_1e^x + e^x(A\cos x + B\sin x) + xe^x + \frac{1}{10}(3\sin x + \cos x). \quad \textbf{Ans.}$$

Example 26: Solve $\left(\frac{d^2y}{dx^2} - 4\frac{dy}{dx} + y\right) = \cos x.\cos 2x + \sin^2 x.$

Solution: The given equation can be written as

$$(D^2-4D+1)y = \cos x\cos 2x + \sin^2 x$$

The A.E. is $m^2-4m+1=0 \Rightarrow m = 2\pm\sqrt{3} \Rightarrow m_1 = 2+\sqrt{3},\ m_2 = 2-\sqrt{3}$

$$\therefore \qquad \text{C.F.} = C_1e^{(2+\sqrt{3})x} + C_2e^{(2-\sqrt{3})x}$$

$$\text{P.I.} = \frac{1}{(D^2-4D+1)}(\cos x\cos 2x + \sin^2 x) = \frac{1}{2(D^2-4D+1)}(2\cos x\cos 2x + 2\sin^2 x)$$

$$= \frac{1}{2(D^2-4D+1)}(\cos 3x + \cos x + 1 - \cos 2x) \quad \boxed{\begin{array}{l}2\cos A\cos B = \cos(A+B)+\cos(A-B)\\ 2\sin^2 A = 1-\cos 2A\end{array}}$$

$$= \frac{1}{2}\left[\frac{1}{(D^2-4D+1)}\cos 3x + \frac{1}{(D^2-4D+1)}\cos x + \frac{1}{(D^2-4D+1)}e^{0.x} - \frac{1}{(D^2-4D+1)}\cos 2x\right]$$

$$= \frac{1}{2}\left[\frac{1}{-3^2-4D+1}\cos 3x + \frac{1}{(-1-4D+1)}\cos x + \frac{1}{(0-0+1)}1 - \frac{1}{(-4-4D+1)}\cos 2x\right]$$

$$= \frac{1}{2}\left[\frac{1}{-4(D+2)}\cos 3x - \frac{1}{4D}\cos x + 1 + \frac{1}{4D+3}\cos 2x\right]$$

$$= \frac{1}{2}\left[-\frac{D-2}{4(D^2-4)}\cos 3x - \frac{1}{4}\sin x + 1 + \frac{(4D-3)}{(16D^2-9)}\cos 2x\right]$$

$$= \frac{1}{2}\left[-\frac{(D-2)}{4(-9-4)}\cos 3x - \frac{1}{4}\sin x + \frac{(4D-3)}{\{16(-4)-9\}}\cos 2x + 1\right]$$

$$= \frac{1}{2}\left[\frac{1}{4\times 13}(-3\sin 3x - 2\cos 3x) - \frac{1}{4}\sin x - \frac{1}{73}(-8\sin 2x - 3\cos 2x) + 1\right]$$

$$= \frac{1}{2}\left[-\frac{1}{52}(3\sin 3x + 2\cos 3x) - \frac{1}{4}\sin x + \frac{1}{73}(8\sin 2x + 3\cos 2x) + 1\right]$$

Hence the complete solution is $y =$ C.F. + P.I.

$$\therefore \quad y = C_1 e^{(2+\sqrt{3})x} + C_2 e^{(2-\sqrt{3})x} + \frac{1}{2}\left\{-\frac{1}{52}(3\sin 3x + 2\cos 3x) - \frac{1}{4}\sin x + \frac{1}{73}(8\sin 2x + 3\cos 2x) + 1\right\}$$

Ans.

Example 27: Solve $(D^4 + 10D^2 + 9)y = \cos(2x+3)$.

Solution: The A.E. is

$$m^4 + 10m^2 + 9 = 0$$

$$\Rightarrow \quad (m^2+1)(m^2+9) = 0$$

$$\Rightarrow \quad m = \pm i, \; m = \pm 3i$$

$$\therefore \quad \text{C.F.} = (C_1 \cos x + C_2 \sin x) + (C_3 \cos 3x + C_4 \sin 3x)$$

$$\text{P.I.} = \frac{1}{(D^4 + 10D^2 + 9)}\cos(2x+3)$$

$$= \frac{1}{\{(-2^2)(-2^2) + 10(-2^2) + 9\}}\cos(2x+3) \qquad \Big|\; D^2 = -2^2$$

$$= \frac{1}{(16-40+9)}\cos(2x+3) = -\frac{1}{15}\cos(2x+3)$$

Hence, complete solution is

$$y = \text{C.F.} + \text{P.I.} = C_1 \cos x + C_2 \sin x + C_3 \cos 3x + C_4 \sin 4x - \frac{1}{15}\cos(2x+3).$$ **Ans.**

Example 28: Solve $(D^2 - 4D - 5)y = e^{2x} + 3\cos(4x+3)$ where $D = \dfrac{d}{dx}$ *(U.P.T.U. 2008)*

Solution: We have $(D^2 - 4D - 5)y = e^{2x} + 3\cos(4x+3)$

The A.E. is

$$m^2 - 4m - 5 = 0$$

$$m^2 - 5m + m - 5 = 0$$

$$(m-5)(m+1) = 0$$

$\Rightarrow$ $$m = -1, 5$$

$\therefore$ $$\text{C.F.} = C_1 e^{-x} + C_2 e^{5x}$$

Now $$\text{P.I.} = \frac{1}{(D^2 - 4D - 5)}(e^{2x} + 3\cos(4x+3)$$

$$= \frac{1}{D^2 - 4D - 5} e^{2x} + 3\frac{1}{(D^2 - 4d - 5)} \cos(4x+3)$$

$$= \frac{e^{2x}}{4-8-5} + 3\frac{1}{(-16-4D-5)} \cos(4x+3)$$

$$= -\frac{e^{2x}}{9} - \frac{3}{(4D+21)} \cos(4x+3) = -\frac{e^{2x}}{9} - \frac{3(4D-21)}{(16D^2 - 441)} \cos(4x+3)$$

$$= -\frac{e^{2x}}{9} - \frac{3(4D-21)}{(-256-441)} \cos(4x+3) = -\frac{e^{2x}}{9} + \frac{3}{697}(4D-21)\cos(4x+3)$$

$$= -\frac{e^{2x}}{9} + \frac{3}{697}\{-16\sin(4x+3) - 21\cos(4x+3)\}$$

$\Rightarrow$ $$\text{P.I.} = -\frac{e^{2x}}{9} - \frac{3}{697}\{16\sin(4x+3) + 21\cos(4x+3)\}$$

Hence $$y = \text{C.F.} + \text{P.I.}$$

$$y = C_1 e^{-x} + C_2 e^{5x} - \frac{e^{2x}}{9} - \frac{3}{697}\{16\sin(4x+3) + 21\cos(4x+3)\}$$ **Ans.**

EXERCISE 1.4

Solve the following differential equations:

1. $(D^3 + D^2 - D - 1)y = \sin(2x+3)$

$$\left[\textbf{Ans. } y = C_1 e^{x} + (C_2 + C_3 x)e^{-x} + \frac{2}{25}\cos(2x+3) - \frac{1}{25}\sin(2x+3)\right]$$

2. $(D^2 + 1)y = \cos x \cos 2x$ $$\left[\textbf{Ans. } y = C_1 \sin x + C_2 \cos x + \frac{x}{4}\sin x - \frac{1}{16}\cos 3x\right]$$

3. $\frac{d^2y}{dx^2} + 3\frac{dy}{dx} + 2y = \sin 2x$ $$\left[\textbf{Ans. } C_1 e^{-x} + C_2 e^{-2x} - \frac{1}{20}(3\cos 2x + \sin 2x)\right]$$

(A.M.I.E. 2002)

4. $\dfrac{d^4y}{dx^4} - p^4 y = \cos px$ $\left[\textbf{Ans. } y = C_1 e^{px} + C_2 e^{-px} + C_3 \cos px + C_4 \sin px - \dfrac{x}{4p^3} - \sin px\right]$

5. $\dfrac{d^2y}{dx^2} + \dfrac{dy}{dx} + y = \cos 2x$ $\left[\textbf{Ans. } y = e^{-\frac{x}{2}}\left(C_1 \cos\dfrac{\sqrt{3}}{2}x + C_2 \sin\dfrac{\sqrt{3}}{2}x\right) + \dfrac{1}{13}(2\sin 2x - 3\cos 2x)\right]$

6. $(D^2 + 16)y = e^{-3x} + \cos 4x$ $\left[\textbf{Ans. } y = C_1 \cos 4x + C_2 \sin 4x + \dfrac{1}{25}e^{-3x} + \dfrac{x}{8}\sin 4x\right]$

7. $\dfrac{d^2y}{dx^2} - 2\dfrac{dy}{dx} - 3y = 2e^{2x} + 10\sin 3x, y(0) = 2$ and $y'(0) = 4$

$\left[\textbf{Ans. } y = \dfrac{29}{12}e^{3x} - \dfrac{1}{12}e^{-x} - \dfrac{2}{3}e^{2x} + \dfrac{1}{3}[\cos 3x - 2\sin 3x]\right]$

8. $(D^2 + D - 2)y = -6\sin 2x - 18\cos 2x, y(0) = 2, y'(0) = 2$ $\left[\textbf{Ans. } y = -e^{-2x} + 3\cos 2x\right]$

9. $(D^2 - 2D + 2)y = e^x + \cos x$ $\left[\textbf{Ans. } y = e^x(C_1 \cos x + C_2 \sin x) + \dfrac{1}{5}(\cos x - 2\sin x)\right]$

10. $\left(\dfrac{d^2y}{dx^2} - 4\dfrac{dy}{dx} + 13\right) = 8\sin 3x, \ y(0) = 1, y'(0) = 2$

$\left[\textbf{Ans. } y = \dfrac{1}{5}\left[e^{2x}(\sin 3x + 2\cos 3x) + \sin 3x + 3\cos 3x\right]\right]$

11. $(2D^2 - 2D + 1)y = \sin 3x.\cos 2x$

$\left[\textbf{Ans. } y = e^{\frac{x}{2}}\left[C_1 \cos\dfrac{x}{2} + C_2 \sin\dfrac{x}{2}\right] + \dfrac{10\cos 5x - 9\sin 5x}{362} + \dfrac{1}{10}(2\cos x - \sin x)\right]$

1.3.4.3 When $R = x^n$, P.I. $= \dfrac{I}{f(D)}x^n = [f(D)]^{-1}x^n$

Here expand $[f(D)]^{-1}$ in the ascending powers of D then evaluate the above expression through differentiation of x^n.

Remark: Expand $[f(D)]^{-1}$ in the powers of D till the power of x because the value of the terms which have greater power of D than of x, vanishes after differentiation.

Example 29: Solve $(2D^2 + 2D + 3)y = x^2 + 2x - 1$.

Solution: The A.E. is

$$(2m^2 + 2m + 3) = 0$$

$$\Rightarrow \qquad m = \frac{-2 \pm \sqrt{4-24}}{4} = -\frac{1}{2} \pm i\frac{\sqrt{5}}{2}$$

$$\therefore \qquad \text{C.F.} = e^{-\frac{x}{2}}\left(C_1 \cos\frac{\sqrt{5}}{2}x + C_2 \sin\frac{\sqrt{5}}{2}x\right)$$

$$\text{P.I.} = \frac{1}{(2D^2+2D+3)}x^2 + \frac{1}{(2D^2+2D+3)}2x - \frac{1}{(2D^2+2D+3)}e^{0\cdot x} \qquad \left| e^{0\cdot x} = 1\right.$$

$$= \frac{1}{3}\left\{1+\left(\frac{2D^2+2D}{3}\right)\right\}^{-1} x^2 + \frac{2}{3}\left\{1+\frac{(2D^2+2D)}{3}\right\}^{-1} x - \frac{1}{3}$$

$$= \frac{1}{3}\left(1-\frac{2D}{3}-\frac{2D^2}{9}\right)x^2 + \frac{2}{3}\left\{1-\frac{2D}{3}\right\}x - \frac{1}{3} \qquad \left| \text{As } (1+x)^{-1} = 1-x+x^2-x^3+\ldots\right.$$

$$= \frac{1}{3}\left(x^2 - \frac{4x}{3} - \frac{4}{9}\right) + \frac{2}{3}\left(x - \frac{2}{3}\right) - \frac{1}{3} = \frac{1}{3}\left(x^2 + \frac{2x}{3} - \frac{25}{9}\right)$$

Hence, the complete solution is y = C.F. + P.I.

$$\Rightarrow \qquad y = e^{-\frac{x}{2}}\left(C_1 \cos\frac{\sqrt{5}}{2}x + C_2 \sin\frac{\sqrt{5}}{2}x\right) + \frac{1}{3}\left(x^2 + \frac{2x}{3} - \frac{25}{9}\right). \quad \textbf{Ans.}$$

Example 30: Solve $(D^3 - D)y = 2x + 1 + 4\cos x + 2e^x$

Solution: The A.E. is

$$m^3 - m = 0 \quad \Rightarrow \quad m(m^2-1) = 0$$

$$\Rightarrow \qquad m = 0, +1, -1$$

$$\therefore \qquad \text{C.F.} = C_1 + C_2 e^x + C_3 e^{-x}$$

$$\text{P.I.} = \frac{1}{(D^3-D)}2x + \frac{1}{(D^3-D)}e^{0x} + \frac{4}{(D^3-D)}\cos x + \frac{2}{(D^3-D)}e^x$$

$$= -\frac{2}{D}(1-D^2)^{-1}x + \frac{x}{(3D^2-1)}e^{0x} + \frac{4}{-2D}\cos x + \frac{2x}{(3D^2-1)}e^x$$

$$\Rightarrow \qquad \text{P.I.} = -\frac{2}{D}x - x - 2\sin x + xe^x = -x^2 - x - 2\sin x + xe^x$$

Hence, the complete solution is y = C.F. + P.I.

$$\Rightarrow \qquad y = C_1 + C_2 e^x + C_3 e^{-x} - x^2 - x - 2\sin x + xe^x. \quad \textbf{Ans.}$$

EXERCISE 1.5

Solve the following differential equations:

1. $\dfrac{d^2y}{dx^2}+2\dfrac{dy}{dx}-5=2x^2-3x+1$ $\left[\textbf{Ans. } y=C_1e^{(-1+\sqrt{6})x}C_2e^{(-1-\sqrt{6})x}+\dfrac{7}{25}x-\dfrac{2x^2}{5}-\dfrac{31}{125}\right]$

2. $(D^3-1)y=x^5+3x^4-2x^3$

$$\left[\textbf{Ans. } y=C_1e^x+e^{-x/2}\left[C_2\cos\frac{\sqrt{3}}{2}x+C_3\sin\frac{\sqrt{3}}{2}x\right]-x^5-3x^4+2x^3+60x^2+72x+12\right]$$

3. $(D^2+3D+2)y=x^3+x^2$ $\left[\textbf{Ans. } y=C_1e^{-x}+C_2e^{-2x}+\dfrac{x^3}{2}-\dfrac{7x^2}{4}+\dfrac{15}{4}x-\dfrac{31}{8}\right]$

4. $(2D^2+3D+4)y=x^2-2x$

$$\left[\textbf{Ans. } y=e^{-\frac{3x}{4}}\left[C_1\cos\frac{\sqrt{23}}{4}x+C_2\sin\frac{\sqrt{23}}{4}x\right]+\frac{1}{32}\left[8x^2-28x+13\right]\right]$$

5. $(D-2)^2y=8(e^{2x}+\sin 2x+x^2)$ $\left[\textbf{Ans. } y=(C_1+C_2x)e^{2x}+4x^2e^{2x}+\cos 2x+2x^2+4x+3\right]$

6. $\dfrac{d^2y}{dx^2}-2\dfrac{dy}{dx}+3y=\cos x+x^2$

$$\left[\textbf{Ans. } y=e^x(C_1\cos\sqrt{2}x+C_2\sin\sqrt{2x})+\frac{1}{4}(\cos x-\sin\sqrt{2}x)+\frac{1}{27}(9x^2+12x+2)\right]$$

7. $(D^4+4)y=x^4$ $\left[\textbf{Ans. } y=e^x(C_1\cos x+C_2\sin x)+e^{-x}(C_3\cos x+C_4\sin x)+\dfrac{1}{4}(x^4-6)\right]$

8. $(D+1)^2y=e^{-x}+x^2$ $\left[\textbf{Ans. } y=(C_1+C_2x)e^{-x}+\dfrac{1}{2}\dfrac{x^2}{2}e^{-x}+x^2-4x+6\right]$

9. $(D^2+4D+4)y=x^2+2x, y(0)=y'(0)=0$ $\left[\textbf{Ans. } y=-\dfrac{3}{8}(1+2x)e^{-2x}+\dfrac{1}{8}(2x^2+3)\right]$

10. $\dfrac{d^2y}{dx^2}+a^2y=\dfrac{a^2R}{p}(l-x)$ where a, R, p and l are constants subject to the conditions $y=0$, $y'=0$ at $x=0$

$$\left[\textbf{Ans. } y=\frac{R}{p}\left(\frac{\sin ax}{a}-l\cos ax+l-x\right)\right]$$

1.3.4.4 When $R=e^{ax}\cdot V$, P.I. $=\dfrac{1}{f(D)}e^{ax}\cdot V=e^{ax}\cdot\dfrac{1}{f(D+a)}V$, where V is the function of x

$$f(D)y = e^{ax}\cdot V \quad ...(i)$$

where $$f(D) = D^n + a_1 D^{n-1} + \dots + a_n$$

$$\Rightarrow \quad f(D)e^{ax} \cdot V = (D^n + a_1 D^{n-1} + \dots + a_n)e^{ax} \cdot V \quad \dots(ii)$$

Now, $$D[e^{ax}V] = e^{ax}DV + ae^{ax} \cdot V = e^{ax}(D+a)V$$

$$D^2[e^{ax}V] = e^{ax}(D^2 + aD)V + ae^{ax}(D+a)V$$

$$= e^{ax}(D^2 + 2aD + a^2)V = e^{ax}(D+a)^2 V$$

Similarly, $$D^{n-1}(e^{ax}V) = e^{ax}(D+a)^{n-1}V$$

$$D^n(e^{ax}V) = e^{ax}(D+a)^n V$$

From equation (*ii*)

$$f(D)e^{ax} \cdot V = e^{ax}[(D+a)^n + a_1 (D+a)^{n-1} + \dots + a_n]V$$

$$\Rightarrow \quad f(D)e^{ax} \cdot V = e^{ax} f(D+a)V \quad \dots(iii)$$

$$\left| \, f(D+a) = (D+a)^n + a_1(D+a)^{n-1} + \dots \right.$$

Now let $$f(D+a)V = \phi(x)$$

then $$V = \frac{1}{f(D+a)} \cdot \phi(x) \quad \dots(iv)$$

Substituting this value in (*iii*), we get

$$f(D)e^{ax} \cdot \frac{1}{f(D+a)} \cdot \phi(x) = e^{ax} \cdot \phi(x)$$

$$\Rightarrow \quad \frac{1}{f(D)} \cdot e^{ax}\phi(x) = e^{ax} \frac{1}{f(D+a)} \phi(x)$$

In general $$\boxed{\frac{1}{f(D)} e^{ax} V = e^{ax} \frac{1}{f(D+a)} V}$$

Note: If $R = e^{-ax}V$, then $\frac{1}{f(D)} e^{-ax} V = e^{-ax} \frac{1}{f(D-a)} V$

Example 31: Find the complete solution for

$$\frac{d^2y}{dx^2} - 3\frac{dy}{dx} + 2y = xe^{3x} + \sin 2x \qquad \textit{(U.P.T.U. 2003)}$$

Solution: The given equation can be written as

$$(D^2 - 3D + 2)y = xe^{3x} + \sin 2x$$

The A.E. is $m^2 - 3m + 2 = 0 \quad \Rightarrow \quad (m-1)(m-2) = 0$

$$m = 1, 2$$

$$\therefore \qquad \text{C.F.} = C_1e^{x} + C_2e^{2x}$$

$$\text{P.I.} = \frac{1}{D^2 - 3D + 2}\, xe^{3x} + \frac{1}{(D^2 - 3D + 2)} \sin 2x$$

$$= e^{3x} \frac{1}{(D+3)^2 - 3(D+3) + 2}\, x + \frac{1}{(-4 - 3D + 2)} \sin 2x$$

$$= e^{3x} \frac{1}{(D^2 + 3D + 2)}\, x - \frac{1}{(3D+2)} \sin 2x$$

$$= \frac{e^{3x}}{2}\left\{1 + \frac{1}{2}(D^2 + 3D)\right\}^{-1} x - \frac{(3D-2)}{(3D+2)(3D-2)} \sin 2x$$

$$= \frac{e^{3x}}{2}\left\{1 - \frac{3D}{2}\right\} x - \frac{(3D-2)}{(9D^2 - 4)} \sin 2x \qquad \left| (1-x)^{-1} = 1 + x + x^2 + \cdots \right.$$

$$= \frac{e^{3x}}{2}\left(x - \frac{3}{2}\right) - \frac{(3D-2)}{-36-4} \sin 2x$$

$$= \frac{e^{3x}}{2}\left(x - \frac{3}{2}\right) + \frac{1}{40}(6\cos 2x - 2\sin 2x)$$

$$\Rightarrow \qquad \text{P.I.} = \frac{1}{2}e^{3x}\left(x - \frac{3}{2}\right) + \frac{1}{20}(3\cos 2x - \sin 2x)$$

Hence, complete solution is

$$y = \text{C.F.} + \text{P.I.} = C_1e^{x} + C_2e^{2x} + \frac{e^{3x}}{2}\left(x - \frac{3}{2}\right) + \frac{1}{20}(3\cos 2x - \sin 2x) \quad \textbf{Ans.}$$

Example 32: Solve $\dfrac{d^2y}{dx^2} + 4\dfrac{dy}{dx} - 12y = (x-1)e^{2x}$. *[U.P.T.U. (C.O.) 2004]*

Solution: The given equation can be written as

$$(D^2 + 4D - 12)y = (x-1)e^{2x}$$

The A.E. is

$$m^2 + 4m - 12 = 0 \quad \Rightarrow \quad m^2 + 6m - 2m - 12 = 0$$

$$\Rightarrow \quad m(m+6) - 2(m+6) = 0 \quad \Rightarrow \quad (m-2)(m+6) = 0$$

$$\Rightarrow \quad m = 2, -6$$

$$\therefore \qquad \text{C.F.} = C_1e^{2x} + C_2e^{-6x}$$

$$\text{P.I.} = \frac{1}{D^2 + 4D - 12}(x-1)e^{2x} = \frac{1}{(D-2)(D+6)}(x-1)e^{2x}$$

$$= e^{2x}\frac{1}{D(D+8)}(x-1) = \frac{e^{2x}}{8D}\left(1+\frac{D}{8}\right)^{-1}(x-1)$$

$$= \frac{e^{2x}}{8D}\left(1-\frac{D}{8}+\frac{D^2}{64}\right)(x-1) = \frac{e^{2x}}{8}\left(\frac{1}{D}-\frac{1}{8}+\frac{D}{64}\right)(x-1)$$

$$\Rightarrow \quad \text{P.I.} = \frac{e^{2x}}{8}\left(\frac{x^2}{2}-x-\frac{x}{8}+\frac{1}{8}+\frac{1}{64}\right) = \frac{e^{2x}}{8}\left(\frac{x^2}{2}-\frac{9x}{8}+\frac{9}{64}\right)$$

Hence, complete solution is

$$y = \text{C.F.} + \text{P.I.} = C_1e^{2x} + C_2e^{-6x} + \frac{e^{2x}}{8}\left(\frac{x^2}{2}-\frac{9x}{8}-\frac{9}{64}\right) \quad \textbf{Ans.}$$

Example 33: Solve $(D^2-4D+4)y = 8x^2e^{2x}\sin 2x$. *(U.P.T.U. 2004, 2005)*

Solution: The A.E. is

$$m^2-4m+4=0 \quad \Rightarrow \quad (m-2)^2=0 \quad \Rightarrow \quad m=2, 2$$

$$\therefore \quad \text{C.F.} = (C_1+C_2x)e^{2x}$$

$$\text{P.I.} = \frac{1}{(D^2-4D+4)}8x^2e^{2x}\sin 2x = \frac{1}{(D-2)^2}8x^2e^{2x}\sin 2x$$

$$= 8e^{2x}\cdot\frac{1}{(D+2-2)^2}x^2\sin 2x = 8e^{2x}.\frac{1}{D^2}x^2\sin 2x$$

$$= 8e^{2x}\cdot\frac{1}{D}\left[\int x^2\sin 2x\,dx\right] = 8e^{2x}\frac{1}{D}\left[x^2\left(-\frac{\cos 2x}{2}\right)+\frac{2}{2}\int x\cos 2x\,dx\right]$$

$$= 8e^{2x}\frac{1}{D}\left[-\frac{x^2}{2}\cos 2x+\frac{x}{2}\sin 2x+\frac{\cos 2x}{4}\right]$$

$$= 8e^{2x}\left[-\frac{1}{2}\int x^2\cos 2x dx+\frac{1}{2}\int x\sin 2x\,dx+\frac{1}{4}\int\cos 2x\,dx\right]$$

$$= 8e^{2x}\left[-\frac{x^2\sin 2x}{4}-\frac{x\cos 2x}{4}+\frac{\sin 2x}{8}-\frac{x\cos 2x}{4}+\frac{\sin 2x}{8}+\frac{\sin 2x}{8}\right]$$

$$= e^{2x}\left[-2x^2\sin 2x-2x\cos 2x+\sin 2x-2x\cos 2x+\sin 2x+\sin 2x\right]$$

$$= e^{2x}[-2x^2\sin 2x-4x\cos 2x+3\sin 2x]$$

$$= -e^{2x}[4x\cos 2x+(2x^2-3)\sin 2x]$$

Hence, complete solution is y = C.F. + P.I.

$$\Rightarrow \quad y = (C_1+C_2x)e^{2x}-e^{2x}[4x\cos 2x+(2x^2-3)\sin 2x]. \quad \textbf{Ans.}$$

Example 34: A body executes damped forced vibrations given by the equation $\frac{d^2x}{dt^2}+2k\frac{dx}{dt}+b^2x=e^{-kt}\cdot\sin\omega t.$

Solve the equation for both the cases, when $\omega^2\neq b^2-k^2$ and $\omega^2=b^2-k^2$.

(U.P.T.U. 2001, 2002, 2003, 2004)

Solution: The given equation can be written in the form

$$(D^2+2kD+b^2)x = e^{-kt}\sin\omega t$$

The A.E. is

$$m^2+2km+b^2 = 0 \qquad \Rightarrow \qquad m=\frac{-2k\pm\sqrt{4k^2-4b^2}}{2}$$

$$\Rightarrow \qquad m = -k+\sqrt{k^2-b^2} \qquad \Rightarrow \qquad m=-k\pm i\sqrt{b^2-k^2}$$

Case I: When $\omega^2\neq b^2-k^2$

$$\text{C.F.}= e^{-kt}\left\{A\cos\left(\sqrt{b^2-k^2}\right)t+B\sin\left(\sqrt{b^2-k^2}\right)t\right\}$$

$$\text{P.I.} = \frac{1}{(D^2+2kD+b^2)}\cdot e^{-kt}\cdot\sin\omega t=e^{-kt}\frac{1}{\{(D-k)^2+2k(D-k)+b^2\}}\sin\omega t$$

$$= e^{-kt}\cdot\frac{1}{(D^2-k^2+b^2)}\sin\omega t=\frac{e^{-kt}\cdot\sin\omega t}{-\omega^2-k^2+b^2}=\frac{e^{-kt}\sin\omega t}{b^2-k^2-\omega^2}$$

Hence, the complete solution is y = C.F. + P.I.

$$\Rightarrow \qquad y = e^{-kt}\left\{A\cos\left(\sqrt{b^2-k^2}\right)t+B\sin\left(\sqrt{b^2-k^2}\right)t\right\}+\frac{e^{-kt}\sin\omega t}{b^2-k^2-\omega^2}. \quad \textbf{Ans.}$$

Case II: When $\omega^2 = b^2-k^2$

$$\text{C.F.} = e^{-kt}(A\cos\sqrt{\omega^2}\,t+B\sin\sqrt{\omega^2}\,t)=e^{-kt}(A\cos\omega t+B\sin\omega t)$$

$$\text{P.I.} = e^{-kt}\frac{1}{(D^2-k^2+b^2)}\sin\omega t=\frac{e^{-kt}}{(D^2+\omega^2)}\sin\omega t \qquad \Big|\ \text{As}\ \omega^2=b^2-k^2$$

$$= \frac{te^{-kt}}{2D}\sin\omega t=-\frac{te^{-kt}}{2\omega}\cos\omega t$$

$$y = \text{C.F.} + \text{P.I.} \Rightarrow \ y=e^{-kt}(A\cos\omega t+B\sin\omega t)-\frac{te^{-kt}}{2\omega}\cos\omega t. \quad \textbf{Ans.}$$

Example 35: Obtain the general solution of the differential equation $y''-2y'+2y=x+e^x\cos x.$

(U.PT.U. 2002)

Solution: The given equation can be written as

$$(D^2-2D+2)y = x+e^x\cos x$$

The A.E. is, $m^2-2m+2=0 \Rightarrow m=\dfrac{2\pm\sqrt{4-8}}{2}=1\pm i$

$\therefore$ C.F. $= e^x(A\cos x+B\sin x)$

Now,

$$\text{P.I.} = \frac{1}{(D^2-2D+2)}x+\frac{1}{(D^2-2D+2)}e^x\cos x$$

$$= \frac{1}{2}\left(1+\frac{D^2-2D}{2}\right)^{-1}x+e^x\frac{1}{\{(D+1)^2-2(D+1)+2\}}\cos x$$

$$\Rightarrow \quad \text{P.I.} = \frac{1}{2}(1+D)x+e^x\frac{1}{(D^2+1)}\cos x=\frac{1}{2}(x+1)+\frac{xe^x}{2D}\cos x$$

$$\Rightarrow \quad \text{P.I.} = \frac{1}{2}(x+1)+\frac{xe^x}{2}\sin x$$

Hence, the complete solution is $y=e^x(A\cos x+B\sin x)+\dfrac{1}{2}(x+1)+\dfrac{xe^x}{2}\sin x.$

Example 36: Solve $\dfrac{d^2y}{dx^2}+2y=x^2e^{3x}+e^x\cos 2x.$

Solution: The given equation written as

$$(D^2+2)y = x^2e^{3x}+e^x\cos 2x$$

The A.E. is, $m^2+2=0 \Rightarrow m=\pm i\sqrt{2} \Rightarrow m=0\pm i\sqrt{2}$

$\therefore$ C.F. $= A\cos\sqrt{2}x+B\sin\sqrt{2}x$

$$\text{P.I.} = \frac{1}{(D^2+2)}x^2e^{3x}+\frac{1}{(D^2+2)}e^x\cos 2x$$

$$= e^{3x}\frac{1}{\{(D+3)^2+2\}}x^2+e^x\cdot\frac{1}{\{(D+1)^2+2\}}\cos 2x$$

$$= e^{3x}\frac{1}{(D^2+6D+11)}x^2+e^x\frac{1}{(D^2+2D+3)}\cos 2x$$

$$= \frac{e^{3x}}{11}\left(1+\frac{D^2+6D}{11}\right)^{-1}x^2+e^x\frac{1}{(-4+2D+3)}\cos 2x$$

$$= \frac{e^{3x}}{11}\left\{1+\frac{25}{121}D^2-\frac{6}{11}D\right\}x^2+e^x\cdot\frac{1}{(2D-1)}\cos 2x$$

$$= \frac{e^{3x}}{11}\left\{x^2 + \frac{50}{121} - \frac{12x}{11}\right\} + e^x \frac{(2D+1)}{(4D^2-1)}\cos 2x$$

$$= \frac{e^{3x}}{11}\left\{x^2 - \frac{12}{11}x + \frac{50}{121}\right\} + \frac{e^x(2D+1)}{-16-1}\cos 2x$$

$$= \frac{e^{3x}}{11}\left(x^2 - \frac{12x}{11} + \frac{50}{121}\right) - \frac{e^x}{17}(-4\sin 2x + \cos 2x)$$

$$\Rightarrow \quad \text{P.I.} = \frac{e^{3x}}{11}\left(x^2 - \frac{12x}{11} + \frac{50}{121}\right) + \frac{e^x}{17}(4\sin 2x - \cos 2x)$$

$\therefore$ The complete solution is y = C.F. + P.I.

$$\Rightarrow \quad y = A\cos\sqrt{2}x + B\sin\sqrt{2}x + \frac{e^{3x}}{11}\left(x^2 - \frac{12x}{11} + \frac{150}{121}\right) + \frac{e^x}{17}(4\sin 2x - \cos 2x). \quad \textbf{Ans.}$$

Example 37: Solve $(D^2+2D+1)y = \dfrac{e^{-x}}{x+7}$.

Solution: The A.E. is $m^2+2m+1=0 \quad \Rightarrow \quad m=-1,-1$

$$\therefore \quad \text{C.F.} = (C_1 + C_2x)e^{-x}$$

$$\text{P.I.} = \frac{1}{(D+1)^2}e^{-x}\cdot\frac{1}{(x+7)} = e^{-x}\frac{1}{D^2}\left(\frac{1}{x+7}\right) = e^{-x}\cdot\frac{1}{D}\left[\frac{1}{D}\left(\frac{1}{x+7}\right)\right]$$

$$\Rightarrow \quad \text{P.I} = e^{-x}\frac{1}{D}\log(x+7) = e^{-x}\int 1\cdot\log(x+7)dx$$

$$\Rightarrow \quad \text{P.I.} = e^{-x}x\log(x+7) - e^{-x}\int\frac{x}{x+7}dx$$

$$= xe^{-x}\log(x+7) - e^{-x}\int\frac{x+7-7}{(x+7)}dx$$

$$= xe^{-x}\log(x+7) - e^{-x}\int dx + 7e^{-x}\int\frac{1}{(x+7)}dx$$

$$\Rightarrow \quad \text{P.I.} = xe^{-x}\log(x+7) - e^{-x}\cdot x + 7e^{-x}\log(x+7)$$

$$\text{P.I.} = (x+7)e^{-x}\log(x+7) - e^{-x}\cdot x$$

$\therefore$ The complete solution is y = C.F. + P.I.

$$\Rightarrow \quad y = (C_1 + C_2x)e^{-x} + (x+7)e^{-x}\log(x+7) - e^{-x}\cdot x. \quad \textbf{Ans.}$$

EXERCISE 1.6

Solve the following differential equations:

1. $\dfrac{d^2y}{dx^2}-2\dfrac{dy}{dx}+4y=e^x\cos x$ $\left[\textbf{Ans. } y=e^x\left(A\cos\sqrt{3}x+B\sin\sqrt{3}x\right)+\dfrac{1}{2}e^x\cos x\right]$

2. $(D^2-3D+2)y=2x^2e^{4x}+5e^{3x}$ $\left[\textbf{Ans. } y=C_1e^x+C_2e^{2x}+\dfrac{e^{4x}}{54}(18x^2-30x+19)+\dfrac{5}{2}e^{3x}\right]$

3. $(D^2-4D+13)y=(x^2+x+9)+e^{2x}\cos 3x$

$\left[\textbf{Ans. } y=e^{2x}(A\cos 3x+B\sin 3x)+\dfrac{1}{6}xe^{2x}\sin 3x+\dfrac{1}{13}\left(x^2+\dfrac{21x}{13}+\dfrac{1579}{169}\right)\right]$

4. $(D^2-4D+3)y=2xe^{3x}+3e^x\cos 2x$

$\left[\textbf{Ans. } y=C_1e^x+C_2e^{3x}+\dfrac{e^{3x}}{2}(x^2-x)-\dfrac{3e^x}{8}(\cos 2x+\sin 2x)\right]$

5. $(D^2-4D+4)y=e^{2x}\cos^2 x$ $\left[\textbf{Ans. } y=(C_1+C_2x)e^{2x}+\dfrac{e^{2x}}{4}\left(x^2-\dfrac{1}{2}\cos 2x\right)\right]$

6. $(D^2-4)y=x\sin hx$ $\left[\textbf{Ans. } y=C_1e^{2x}+C_2e^{-2x}-\dfrac{x}{3}\sin hx-\dfrac{2}{9}\cos hx\right]$

7. $\dfrac{d^2y}{dx^2}-6\dfrac{dy}{dx}+13y=8e^{3x}\sin 4x+2^x$

$\left[\textbf{Ans. } y=e^{3x}(A\cos 2x+B\sin 2x)-\dfrac{2}{3}e^{3x}\sin 4x+\dfrac{2^x}{(\log 2)^2-6\log 2+13}\right]$

8. $\dfrac{d^3y}{dx^3}-7\dfrac{dy}{dx}-6y=(x+1)e^{2x}$ $\left[\textbf{Ans. } y=C_1e^{-x}+C_2e^{-2x}+C_3e^{3x}-\dfrac{e^{2x}}{144}(12x+17)\right]$

9. $(D^2+2D+1)y=\dfrac{e^{-x}}{x^2}$ $\left[\textbf{Ans. } y=(C_1+C_2x)e^{-x}-e^{-x}\log x\right]$

10. $\dfrac{d^2y}{dx^2}+5\dfrac{dy}{dx}+6y=e^{-2x}(\sec^2 x)(1+2\tan x)$ $\left[\textbf{Ans. } y=C_1e^{-2x}+C_2e^{-3x}+e^{-2x}\tan x\right]$

11. $D^2+2h\,D+(h^2+p^2)y=ke^{-ht}\cos pt$

$\left[\textbf{Ans. } y=e^{-ht}(A\cos pt+B\sin pt)+\dfrac{k}{2p}te^{-ht}\sin pt\right]$

12. $\frac{d^2y}{dx^2}-4\frac{dy}{dx}+4y=e^{2x}\cos^2 x$ $\left[\textbf{Ans. } y=(C_1+C_2x)e^{2x}+e^{2x}\left(\frac{x^2}{4}-\frac{1}{8}\cos 2x\right)\right]$

1.3.4.5 (*i*) When R = *xV*, *V* is a function of *x*

$$\text{P.I.}=\frac{1}{f(D)}(xV)=x\cdot\frac{1}{f(D)}V-\frac{f'(D)}{\{f(D)\}^2}V$$

(*ii*) If $R = x^n \sin ax$ or $x^n \cos ax$ then

$$\text{P.I. = I.P. (Imaginary Part) of } \frac{1}{f(D)}e^{iax}.x^n$$

or

$$\text{P.I. = R.P. (Real Part) of } \frac{1}{f(D)}e^{iax}.x^n$$

$e^{iax}=\cos ax+i\sin ax$

R.P. of $e^{iax}=\cos ax$

I.P. of $e^{iax}=\sin ax$

After this, apply the rule $R = e^{ax}.\ V$

Example 38: Solve $\frac{d^2y}{dx^2}-y=x\sin x+e^x+x^2e^x.$

Solution: The given equation can be written as

$$(D^2-1)y = x\sin x+e^x+x^2e^x$$

The A.E. is, $m^2-1=0 \Rightarrow m=\pm1$

$\therefore$ C.F. $= C_1e^x+C_2e^{-x}$

$$\text{P.I.} = \frac{1}{(D^2-1)}x\sin x+\frac{1}{(D^2-1)}e^x+\frac{1}{(D^2-1)}x^2e^x$$

or

$$\text{P.I.} = \text{I.P. of } \frac{1}{(D^2-1)}xe^{ix}+\frac{x}{2D}e^x+e^x\cdot\frac{1}{\{(D+1)^2-1\}}x^2$$

$$= \text{I.P. of } e^{ix}\cdot\frac{1}{\{(D+i)^2-1\}}x+\frac{xe^x}{2}+e^x\cdot\frac{1}{(D^2+1+2D-1)}x^2$$

$$= \text{I.P. of } e^{ix}\cdot\frac{1}{(D^2-1+2Di-1)}x+\frac{xe^x}{2}+e^x\cdot\frac{1}{(D^2+2D)}x^2$$

$$= -\text{ I.P. of } e^{ix}\cdot\frac{1}{2\left(1-\frac{D^2}{2}-Di\right)}x+\frac{xe^x}{2}+e^x\cdot\frac{1}{2D\left(1+\frac{D}{2}\right)}x^2$$

$$= -\text{ I.P. of } e^{ix}\cdot\frac{1}{2}\left\{1-\left(\frac{D^2}{2}+Di\right)\right\}^{-1} x+\frac{xe^x}{2}+e^x\cdot\frac{1}{2D}\left(1+\frac{D}{2}\right)^{-1} x^2$$

$$= -\text{ I.P. of } e^{ix}\cdot\frac{1}{2}\{1+Di\}x+\frac{xe^x}{2}+e^x\cdot\frac{1}{2D}\left\{1-\frac{D}{2}+\frac{D^2}{4}-\frac{D^3}{8}\right\}x^2$$

$$= -\text{ I.P. of } e^{ix}\cdot\frac{1}{2}\{x+i\}+\frac{xe^x}{2}+\frac{e^x}{2}\left\{\frac{1}{D}-\frac{1}{2}+\frac{D}{4}-\frac{D^2}{8}\right\}x^2$$

$$= -\text{ I.P. of } \frac{1}{2}(\cos x+i\sin x)(x+i)+\frac{xe^x}{2}+\frac{e^x}{2}\left\{\frac{x^3}{3}-\frac{x^2}{2}+\frac{x}{2}-\frac{1}{4}\right\}$$

$$= -\frac{1}{2}(\cos x+x\sin x)+\frac{e^x}{2}\left(x+\frac{x^3}{3}-\frac{x^2}{2}+\frac{x}{2}-\frac{1}{4}\right)$$

$$\text{P.I.} = -\frac{1}{2}(\cos x+x\sin x)+\frac{e^x}{2}\left(\frac{x^3}{3}-\frac{x^2}{2}+\frac{3x}{2}-\frac{1}{4}\right)$$

Hence, the complete solution is y = C.F. + P.I.

$$\Rightarrow \quad y = C_1e^x+C_2e^{-x}-\frac{1}{2}(\cos x+x\sin x)+\frac{e^x}{2}\left(\frac{x^3}{3}-\frac{x^2}{2}+\frac{3x}{2}-\frac{1}{4}\right). \textbf{ Ans.}$$

Example 39: Solve $\dfrac{d^2y}{dx^2}-2\dfrac{dy}{dx}+y=xe^x\sin x.$ *(U.P.T.U. 2005)*

Solution: The given equation can be written as

$$(D^2-2D+1)y = xe^x\sin x$$

The A.E. is $m^2-2m+1=0 \Rightarrow (m-1)^2=0 \Rightarrow m=1,1$

$\therefore$ C.F. $= (C_1+C_2x)e^x$

$$\text{P.I.} = \frac{1}{(D-1)^2}xe^x\sin x = e^x\frac{1}{D^2}(x\sin x)=e^x.\frac{1}{D}(\sin x-x\cos x)$$

$$= e^x(-\cos x-\cos x-x\sin x)=-e^x(2\cos x+x\sin x)$$

Complete solution is y = C.F. + P.I.

$\Rightarrow$ $y = (C_1+C_2x)e^x-e^x(2\cos x+x\sin x)$. **Ans.**

Example 40: Solve the differential equation

$$(D^2+4)y = x\sin^2 x$$

Solution: The A.E. is $m^2+4=0 \Rightarrow m=0\pm 2i$

$\therefore$ C.F. = ($A\cos 2x + B\sin 2x$)

$$\text{P.I.} = \frac{1}{(D^2+4)} x\sin^2 x = \frac{1}{(D^2+4)} x\left(\frac{1-\cos 2x}{2}\right)$$

$$= \frac{1}{2(D^2+4)} \cdot x - \frac{1}{2(D^2+4)} x\cos 2x$$

$$= \frac{1}{8\left(1+\frac{D^2}{4}\right)} x - \text{R.P.} \frac{1}{2(D^2+4)} xe^{2ix} \qquad \Big| \text{ As R.P.} e^{2ix} = \cos 2x$$

$$= \frac{1}{8}\left(1+\frac{D^2}{4}\right)^{-1} x - \text{R.P.} \frac{e^{2ix}}{2\{(D+2i)^2+4\}} \cdot x$$

$$= \frac{1}{8}\left(1-\frac{D^2}{4}\right) x - \text{R.P.} \frac{e^{2ix}}{2(D^2+4Di)} x$$

$$= \frac{x}{8} - \text{R.P.} \frac{e^{2ix}}{8Di}\left(1+\frac{D}{4i}\right)^{-1} x$$

$$= \frac{x}{8} - \text{R.P.} \frac{e^{2ix}}{8Di}\left(1-\frac{D}{4i}\right) x$$

$$= \frac{x}{8} - \text{R.P.} \frac{e^{2ix}}{8Di}\left(x-\frac{1}{4i}\right)$$

$$= \frac{x}{8} - \text{R.P.}\, e^{2ix}\left(\frac{x^2}{16i}+\frac{x}{32}\right)$$

$$= \frac{x}{8} - \text{R.P.}\,(\cos 2x + i\sin 2x)\left(-\frac{x^2}{16}i+\frac{x}{32}\right)$$

$$= \frac{x}{8} - \frac{x\cos 2x}{32} - \frac{x^2\sin 2x}{16}$$

$$\text{P.I.} = \frac{x}{8} - \frac{x\cos 2x + 2x^2\sin 2x}{32}$$

Hence, complete solution is y = C.F. + P.I.

$$\Rightarrow \qquad y = (A\cos 2x + B\sin 2x) + \frac{x}{8} - \frac{x\cos 2x + 2x^2\sin 2x}{32}. \qquad \textbf{Ans.}$$

Example 41: Solve $(D^2-1)y = xe^x \sin x$

Solution: The A.E. is, $m^2-1=0 \quad \Rightarrow \quad m=\pm 1$

$$\therefore \quad \text{C.F.} = C_1e^x + C_2e^{-x}$$

$$\text{P.I.} = \frac{1}{(D^2-1)} xe^x \sin x = x\cdot\frac{1}{(D^2-1)}e^x \sin x - \frac{2D}{(D^2-1)^2}e^x \sin x$$

$$= x.e^x \frac{1}{(D^2+2D-1+1)}\sin x - 2D\left[e^x \cdot \frac{1}{(D^2+2D+1-1)^2}\sin x\right]$$

$$= xe^x \frac{1}{(-1+2D)}\sin x - 2D\left[e^x \cdot \frac{1}{(-1+2D)^2}\sin x\right]$$

$$= xe^x \frac{(2D+1)}{(4D^2-1)}\sin x - 2D\left[e^x \frac{(2D+1)^2}{(4D^2-1)^2}\sin x\right]$$

$$= \frac{xe^x}{-5}(2\cos x + \sin x) - 2D\left[\frac{e^x}{25}(4D^2+4D+1)\sin x\right]$$

$$= \frac{-xe^x}{5}(2\cos x + \sin x) - \frac{2}{25}D\left[e^x(-4\sin x + 4\cos x + \sin x)\right]$$

$$= -\frac{xe^x}{5}(2\cos x + \sin x) - \frac{2}{25}\left[-4e^x \sin x - 4e^x \cos x + 4e^x \cos x\right.$$
$$\left. -4e^x \sin x + e^x \sin x + e^x \cos x)\right]$$

$$= \frac{-2xe^x}{5}\cos x - \frac{xe^x}{5}\sin x - \frac{2}{25}(e^x \cos x - 7e^x \sin x)$$

$$\Rightarrow \quad \text{P.I.} = \frac{-e^x}{25}\left[2(1+5x)\cos x + (5x-14)\sin x\right]$$

Hence, the complete solution is

$$y = \text{C.F.} + \text{P.I.}$$

$$\Rightarrow \quad y = C_1e^x + C_2e^{-x} - \frac{e^x}{25}\{2(1+5x)\cos x + (5x-14)\sin x\}. \quad \textbf{Ans.}$$

Example 42: Solve $(D^2+4)y = 3x\sin x$

Solution: The A.E. is

$$m^2+4 = 0 \quad \Rightarrow \quad m^2 = -4 \quad \Rightarrow \quad m = \pm 2i$$

$$\therefore \quad \text{C.F.} = C_1\cos 2x + C_2 \sin 2x$$

$$\text{P.I.} = \frac{1}{(D^2+4)} 3x\sin x = 3x\cdot\frac{1}{(D^2+4)}\sin x - \frac{2D}{(D^2+4)^2} 3\sin x$$

$$= 3x\cdot\frac{1}{(-1+4)}\sin x - 6D\left(\frac{1}{(-1+4)^2}\sin x\right) = x\sin x - \frac{2}{3}\cos x$$

$\therefore$ Complete solution is

$$y = \text{C.F.} + \text{P.I.}$$

$$\Rightarrow \quad y = C_1\cos 2x + C_2\sin 2x + x\sin x - \frac{2}{3}\cos x. \quad \textbf{Ans.}$$

EXERCISE 1.7

Solve the following differential equations:

1. $\dfrac{d^2y}{dx^2} - 4\dfrac{dy}{dx} + 4y = x^2 + e^x + \cos 2x$

$$\left[\textbf{Ans. } y = (C_1 + C_2x)e^{2x} + \frac{1}{8}(2x^2 + 4x + 3) - \frac{1}{8}\sin 2x + e^x\right]$$

2. $\dfrac{d^2y}{dx^2} - 2\dfrac{dy}{dx} + y = xe^x\cos x$ [*U.P.T.U. 2009*] $\left[\textbf{Ans. } y = (C_1 + C_2x)e^x + e^x(2\sin x - x\cos x)\right]$

3. $(D^3 - 3D^2 - 6D + 8)\,y = xe^{-3x}$

$$\left[\textbf{Ans. } y = C_1e^x + C_2e^{4x} + C_3e^{-2x} - \frac{e^{-3x}}{784}(28x + 39)\right]$$

4. $(D^2 - 4D - 5)\,y = xe^{-x}$; $y(0) = y'(0) = 0$, $\left[\textbf{Ans. } y = \dfrac{1}{216}(e^{5x} - e^{-x}) - \dfrac{1}{12}e^{-x}\left(x^2 + \dfrac{x}{3}\right)\right]$

5. $(D^2 - 1)\,y = x\sin x + e^x(1 + x^2)$

$$\left[\textbf{Ans. } y = C_1e^x + C_2e^{-x} - \frac{1}{2}(x\sin x + \cos x) + \frac{e^x}{2}\left(\frac{x^3}{3} - \frac{x^2}{2} + \frac{3x}{2} - \frac{3}{4}\right)\right]$$

6. $(D^2 - 1)y = (x\sin 3x + \cos x)$ $\left[\textbf{Ans. } y = C_1e^x + C_2e^{-x} - \dfrac{1}{10}\left[\dfrac{3}{5}\cos 3x + x\sin 3x + 5\cos x\right]\right]$

1.3.4.6 General Method

If $R = \phi(x)$ be any function of x, then

$$\textbf{P.I.} = \frac{1}{f(D)}\phi(x) = \frac{1}{(D+a)}\phi(x) = e^{-ax}\int e^{ax}\phi(x)\,dx$$

Consider the equation $(D+a)y=\phi(x)$ *i. e.*, $\frac{dy}{dx}+ay=\phi(x)$, which is the linear differential equation of first order

$$\therefore \qquad \text{I.F.} = e^{\int a\,dx} = e^{ax}$$

The solution of equation is

$$ye^{ax} = \int \phi(x)e^{ax}.dx + C$$

Thus a particular solution (with $C = 0$) is

$$y = e^{-ax}\int \phi(x)\cdot e^{ax}dx$$

Therefore,

$$\boxed{\text{P.I.} = \frac{1}{f(D)}\phi(x) = \frac{1}{(D+a)}\phi(x) = e^{-ax}\int e^{ax}\phi(x)\,dx}$$

Similarly, in case $f(D) = D - a$

$$\boxed{\text{P.I.} = \frac{1}{f(D)}\phi(x) = \frac{1}{(D-a)}\phi(x) = e^{ax}\int e^{-ax}\phi(x)dx}$$

Example 43: Solve $(D^2+5D+6)y = e^{-2x}\sec^2 x(1+2\tan x)$.

Solution: The A.E. is $m^2+5m+6=0$

$$\Rightarrow \qquad (m+2)(m+3) = 0 \qquad \Rightarrow \qquad m = -2, -3$$

$$\therefore \qquad \text{C.F.} = C_1e^{-2x} + C_2e^{-3x}$$

$$\text{P.I.} = \frac{1}{(D+2)(D+3)}e^{-2x}\sec^2 x(1+2\tan x)$$

$$= \frac{1}{(D+3)}\left[\frac{1}{(D+2)}\left\{e^{-2x}\sec^2 x(1+2\tan x)\right\}\right]$$

$$= \frac{1}{(D+3)}\left[e^{-2x}\int(1+2\tan x)\sec^2 x\,dx\right] = \frac{1}{(D+3)}[e^{-2x}(\tan x+\tan^2 x)]$$

$$= e^{-3x}\int e^{3x}.e^{-2x}(\tan x+\tan^2 x)\,dx = e^{-3x}\left[\int\left\{e^x\tan x + e^x(\sec^2 x-1)\right\}dx\right]$$

$$= e^{-3x}\left[\int e^x(\tan x+\sec^2 x)\,dx - \int e^x dx\right]$$

$$= e^{-3x}\left[e^x\tan x - e^x\right] = e^{-2x}(\tan x - 1)$$

Hence, the complete solution is y = C.F. + P.I.

$$\Rightarrow \qquad y = C_1e^{-2x} + C_2e^{-3x} + e^{-2x}(\tan x-1) = (C_1-1)e^{-2x} + e^{-2x}\tan x + C_2e^{-3x}$$

$$\Rightarrow \qquad y = e^{-2x}(\tan x + C) + C_2 e^{-3x},\ C = C_1 - 1. \quad \textbf{Ans.}$$

Example 44: Solve $\dfrac{d^2y}{dx^2} + 3\dfrac{dy}{dx} + 2y = e^{e^x}$.

Solution: The given equation can be written as

$$(D^2 + 3D + 2)y = e^{e^x}$$

The A.E. is, $m^2 + 3m + 2 = 0 \quad \Rightarrow \quad (m+1)(m+2) \quad \Rightarrow \quad m = -1, -2$

$$\therefore \qquad \text{C.F.} = C_1 e^{-x} + C_2 e^{-2x}$$

$$\text{P.I.} = \frac{1}{(D+1)(D+2)} e^{e^x} = \frac{1}{(D+1)}\left[\frac{1}{(D+2)} e^{e^x}\right]$$

$$= \frac{1}{(D+1)}\left[e^{-2x}\int e^{2x} \cdot e^{e^x} dx\right] = \frac{1}{(D+1)}\left[e^{-2x}\int e^x \cdot e^x e^{e^x} dx\right]$$

$$= \frac{1}{(D+1)}\left[e^{-2x}\int te^t \cdot dt\right] = \frac{1}{(D+1)}\left[e^{-2x}(t-1)e^t\right] = \frac{1}{(D+1)}\left[e^{-2x}(e^x - 1)e^{e^x}\right]$$

[Put $e^x = t, \quad e^x dx = dt$]

$$= \frac{1}{(D+1)}\left[e^{-x}e^{e^x} - e^{-2x}e^{e^x}\right] = e^{-x}\int [e^x e^{-x} e^{e^x} dx - e^{-x}\int e^x e^{-2x} e^{e^x} dx$$

$$= e^{-x}\int e^{e^x} dx - e^{-x}\int \frac{e^t}{t^2} dt \qquad \Big| \quad e^x = t$$

$$= e^{-x}\int e^{e^x} dx + \frac{e^{-x}}{t} e^t - e^{-x}\int \frac{e^t}{t} dt$$

$$= e^{-x}\int e^{e^x} dx + \frac{e^{-x}}{e^x} e^{e^x} - e^{-x}\int e^{e^x} dx$$

$$\text{P.I.} = e^{-2x} e^{e^x}$$

$\therefore$ The complete solution is

$$y = \text{C.F.} + \text{P.I.}$$

$$\Rightarrow \qquad y = C_1 e^{-x} + C_2 e^{-2x} + e^{-2x} e^{e^x}. \quad \textbf{Ans.}$$

Example 45: Solve the differential equation

$$\frac{d^2y}{dx^2} - y = \frac{1}{(1 - e^{-x})}.$$

Solution: We have $(D^2 - 1)y = \dfrac{1}{(1 - e^{-x})}$

The A.E. is $m^2 - 1 = 0 \Rightarrow m = \pm 1$

$$\therefore \quad \text{C.F.} = C_1 e^x + C_2 e^{-x}$$

$$\text{P.I.} = \frac{1}{(D-1)(D+1)} \cdot \frac{1}{(1-e^{-x})} = \frac{1}{(D-1)}\left[\frac{1}{(D+1)} \cdot \frac{1}{(1-e^{-x})}\right]$$

$$\Rightarrow \quad \text{P.I.} = \frac{1}{(D-1)}\left[e^{-x}\int e^x \cdot \frac{1}{(1-e^{-x})} \cdot dx\right] = \frac{1}{(D-1)}\left[e^{-x}\int \frac{e^x \cdot e^x}{(e^x-1)} dx\right]$$

Now

$$\int \frac{e^x e^x}{(e^x-1)} dx = \int \frac{t+1}{t} dt = \int dt + \int \frac{1}{t} dt = t + \log t = (e^x - 1) + \log(e^x - 1) \quad \left| \begin{array}{l} e^x - 1 = t \\ e^x dx = dt \end{array} \right.$$

$$\Rightarrow \quad \text{P.I.} = \frac{1}{(D-1)}\left[e^{-x}\left\{(e^x-1) + \log(e^x-1)\right\}\right]$$

$$= \frac{1}{(D-1)}\left[1 - e^{-x} + e^{-x}\log(e^x - 1)\right]$$

$$= e^x \int e^{-x} dx - e^x \int e^{-2x} dx + e^x \int e^{-2x} \log(e^x - 1) dx$$

$$= -1 + \frac{e^{-x}}{2} + e^x\left[\frac{e^{-2x}}{-2}\log(e^x-1) + \frac{1}{2}\int \frac{e^{-2x}}{e^x-1} e^x dx\right]$$

$$= -1 + \frac{e^{-x}}{2} + e^x\left[-\frac{1}{2}e^{-2x}\log(e^x-1) + \frac{1}{2}\int \frac{e^{-x}}{e^x-1} dx\right]$$

Again

$$\int \frac{e^{-x}}{e^x-1} dx = \int \frac{e^{-2x}}{1-e^{-x}} dx = \int \left(\frac{1-t}{t}\right) dt \quad \left| \begin{array}{l} 1 - e^{-x} = t \\ e^{-x} dx = dt \end{array} \right.$$

$$= \int \frac{1}{t} dt - \int dt = \log t - t$$

$$= \log(1 - e^{-x}) - (1 - e^{-x})$$

$$\Rightarrow \quad \text{P.I.} = -1 + \frac{e^{-x}}{2} - \frac{e^{-x}}{2}\log(e^x - 1) + \frac{e^x}{2}\log(1 - e^{-x}) - \frac{e^x}{2}(1 - e^{-x})$$

$$\Rightarrow \quad \text{P.I.} = -\frac{1}{2} + \frac{e^{-x}}{2}\left\{1 - \log(e^x - 1)\right\} + \frac{e^x}{2}\left\{1 + \log(1 - e^{-x})\right\}$$

Hence, the complete solution is y = C.F. + P.I.

$$\Rightarrow \qquad y = C_1e^x + C_2e^{-x} + \frac{e^{-x}}{2}\left\{1 - \log(e^x - 1)\right\} + \frac{e^x}{2}\left\{1 + \log(1 - ee^{-x})\right\} - \frac{1}{2}. \quad \textbf{Ans.}$$

Example 46: Solve $\dfrac{d^2y}{dx^2} + 3\dfrac{dy}{dx} + 2y = \sin(e^x)$.

Solution: We have $(D^2 + 3D + 2)y = \sin(e^x)$

The A.E. is

$$m^2 + 3m + 2 = 0 \quad \Rightarrow \quad (m+1)(m+2) = 0$$

$$\Rightarrow \qquad m = -1, -2$$

$$\text{C.F.} = C_1e^{-x} + C_2e^{-2x}$$

$$\text{P.I.} = \frac{1}{(D+1)(D+2)}\sin(e^x) = \frac{1}{(D+2)}\left[\frac{1}{(D+1)}\sin e^x\right]$$

$$= \frac{1}{(D+2)}\left[e^{-x}\int e^x \sin(e^x)\,dx\right] = \frac{1}{(D+2)}\left[-e^{-x}\cos(e^x)\right]$$

$$= -e^{-2x}\int e^{2x}e^{-x}\cos(e^x)\,dx = -e^{-2x}\int e^x\cos(e^x)\,dx = -e^{-2x}\sin(e^x)$$

Therefore, the complete solution is y = C.F. + P.I.

$$\Rightarrow \qquad y = C_1e^{-x} + C_2e^{-2x} - e^{-2x}\sin(e^x). \quad \textbf{Ans.}$$

Example 47: Solve the equation

$$\frac{d^3y}{dx^3} + 4\frac{dy}{dx} = 4\cot 2x. \qquad \textit{(U.P.T.U. 2007)}$$

Solution: The given equation can be written as

$$(D^3 + 4D)y = 4\cot 2x$$

The A.E. is $\quad m^3 + 4m = 0 \quad \Rightarrow \quad m(m^2 + 4) = 0$

$$\Rightarrow \qquad m(m+2i)(m-2i) = 0$$

$$\Rightarrow \qquad m = 0,\ m = 0 \pm 2i$$

$$\therefore \qquad \text{C.F.} = C_1 + C_2\cos 2x + C_3\sin 2x$$

$$\text{P.I.} = \frac{1}{D(D+2i)(D-2i)}4\cot 2x = \left[\frac{1}{4D} - \frac{1}{8(D+2i)} - \frac{1}{8(D-2i)}\right]4\cot 2x$$

$$= \frac{4}{4}\int \cot 2x\,dx - \frac{4}{8}e^{-2ix}\int e^{2ix}\cot 2x\,dx - \frac{4e^{2ix}}{8}\int e^{-2ix}\cot 2x\,dx$$

$$\text{P.I.} = \frac{1}{2}\log\sin 2x - \frac{1}{2}e^{-2ix}\int e^{2ix}\cot 2x\,dx - \frac{1}{2}e^{2ix}\int e^{-2ix}\cot 2x\,dx$$

$$\Rightarrow \quad \text{P.I.} = \frac{1}{2}\log\sin 2x - \frac{1}{2}e^{-2ix}\cdot I_1 - \frac{1}{2}e^{2ix}\cdot I_2 \quad \text{...(1)}$$

Now

$$I_1 = \int(\cos 2x + i\sin 2x)\cot 2x\,dx = \int\frac{\cos^2 2x}{\sin 2x}dx + i\int\cos 2x\,dx$$

$$= \int\frac{1-\sin^2 2x}{\sin 2x}dx + \frac{i}{2}\sin 2x$$

$$= \int\text{cosec}\,2x\,dx - \int\sin 2x\,dx + \frac{i}{2}\sin 2x$$

$$= \frac{1}{2}\log(\text{cosec}\,2x - \cot 2x) + \frac{\cos 2x}{2} + \frac{i}{2}\sin 2x$$

$$\Rightarrow \quad I_1 = \frac{1}{2}\log(\text{cosec}\,2x - \cot 2x) + \frac{1}{2}e^{2ix}$$

Similarly,

$$I_2 = \frac{1}{2}\log(\text{cosec}\,2x - \cot 2x) + \frac{1}{2}e^{-2ix}$$

From (1)

$$\text{P.I.} = \frac{1}{2}\log\sin 2x - \frac{1}{2}\left[\frac{(e^{-2ix}+e^{2ix})}{2}\log(\text{cosec}\,2x - \cot 2x) + \frac{1}{2} + \frac{1}{2}\right]$$

$$\text{P.I.} = \frac{1}{2}\log\sin 2x - \frac{1}{2}\cos 2x\log(\text{cosec}\,2x - \cot 2x) - \frac{1}{2}$$

∴ Complete solution is y = C.F. + P.I.

$$y = C_1 + C_2\cos 2x + C_3\sin 2x + \frac{1}{2}\log\sin 2x - \frac{1}{2}\cos 2x\log(\text{cosec}\,2x - \cot 2x) - \frac{1}{2}.$$ **Ans.**

EXERCISE 1.8

Solve the following differential equations:

1. $\dfrac{d^2y}{dx^2} + a^2y = \sec ax$ $\quad\left[\textbf{Ans. } y = C_1\cos ax + C_2\sin ax + \dfrac{1}{a^2}\cos ax\log\cos ax + \dfrac{x}{a}\sin ax\right]$

2. $\dfrac{d^2y}{dx^2} + y = (x - \cot x)$ $\quad\left[\textbf{Ans. } y = C_1\cos x + C_2\sin x - x\cos^2 x - \sin x\log(\text{cosec}\,x - \cot x)\right]$

3. $\dfrac{d^2y}{dx^2}+2\dfrac{dy}{dx}+2y=e^{-x}\sec^3 x$ $\left[\textbf{Ans. } y=e^{-x}\left[A\cos x+B\sin x+\dfrac{\sin x\tan x}{2}\right]\right]$

4. $(D^2+9)y=\sec 3x$ $\left[\textbf{Ans. } y=C_1\cos 3x+C_2\sin 3x+\dfrac{x}{3}\sin 3x+\dfrac{1}{9}\cos 3x.\log\cos 3x\right]$

1.4 HOMOGENEOUS LINEAR DIFFERENTIAL EQUATION (DIFFERENTIAL EQUATION WITH VARIABLE COEFFICIENTS)

There are two types of homogeneous linear differential equations which can be reduced to the linear differential equation with constant coefficient as follows:

1.4.1 Cauchy's Homogeneous Linear Equation

A differential equation of the form

$$x^n\frac{d^ny}{dx^n}+a_1x^{n-1}\frac{d^{n-1}y}{dx^{n-1}}+\dots+a_ny = F(x) \quad \dots(i)$$

where $a_1, a_2,...a_n$ are constants is called a Cauchy's homogeneous equation.

Method of Solution: Put $x = e^z, z=\log e^x$

$$\frac{dy}{dx} = \frac{dy}{dz}.\frac{dz}{dx}=\frac{1}{x}\frac{dy}{dz} \Rightarrow x\frac{dy}{dx}=\frac{dy}{dz} \Rightarrow \boldsymbol{x\frac{dy}{dx}=Dy, D=\frac{d}{dz}}$$

and

$$\frac{d^2y}{dx^2} = \frac{d}{dx}\left(\frac{dy}{dx}\right)=\frac{d}{dx}\left(\frac{1}{x}\frac{dy}{dz}\right)=\frac{d}{dx}\left(\frac{1}{x}\right).\frac{dy}{dz}+\frac{1}{x}\frac{d}{dz}\left(\frac{dy}{dz}\right)\frac{dz}{dx}$$

$$\Rightarrow \frac{d^2y}{dx^2} = -\frac{1}{x^2}\cdot\frac{dy}{dz}+\frac{1}{x}\cdot\frac{d^2y}{dz^2}\cdot\frac{1}{x}=\frac{1}{x^2}\left(\frac{d^2y}{dz^2}-\frac{dy}{dz}\right) \Rightarrow x^2\frac{d^2y}{dx^2}=\frac{d^2y}{dz^2}-\frac{dy}{dz}$$

$$\Rightarrow \boldsymbol{x^2\frac{d^2y}{dx^2} = D(D-1)y;\ D=\frac{d}{dz}}$$

Similarly, we can get $\boldsymbol{x^3\dfrac{d^3y}{dx^3}=D(D-1)(D-2)\,y}$, and so on.

The substitution of these values in (*i*) reduces the given homogeneous equation to a linear differential equation with constant coefficient.

Example 48: Solve $x^2\dfrac{d^2y}{dx^2}+x\dfrac{dy}{dx}-\lambda^2y=0.$ (*U.P.T.U. 2007*)

Solution: We have $x^2\dfrac{d^2y}{dx^2}+x\dfrac{dy}{dx}-\lambda^2y=0$ ...(1)

Let $\quad x = e^z,\ z = \log_e x$

$\Rightarrow \quad x\dfrac{dy}{dx} = Dy;\ D = \dfrac{d}{dz}$

and $\quad x^2\dfrac{d^2y}{dx^2} = D(D-1)y,\ D = \dfrac{d}{dz}$

Substituting these values in equation (1), we get

$$D(D-1)y + Dy - \lambda^2 y = 0 \quad \Rightarrow \quad (D^2 - D + D - \lambda^2)y = 0$$

$$\Rightarrow \quad (D^2 - \lambda^2)y = 0$$

The A.E. is $\quad m^2 - \lambda^2 = 0 \quad \Rightarrow \quad m = \pm\lambda$

$\therefore \quad$ C.F. $= y = C_1e^{\lambda z} + C_2e^{-\lambda z} \quad \Rightarrow \quad y = C_1x^{\lambda} + C_2x^{-\lambda}$

$\Rightarrow \quad y = C_1x^{\lambda} + \dfrac{C_2}{x^{\lambda}}.$ **Ans.**

Example 49: Solve $x^2\dfrac{d^2y}{dx^2} + x\dfrac{dy}{dx} + y = \log x\ \sin(\log x).$ *(U.P.T.U. 2002)*

Solution: We have $x^2\dfrac{d^2y}{dx^2} + x\dfrac{dy}{dx} + y = \log x \cdot \sin(\log x)$ ...(1)

Put $\quad x = e^z \quad \Rightarrow \quad z = \log x$

Also $\quad x\dfrac{dy}{dx} = Dy,\ x^2\dfrac{d^2y}{dx^2} = D(D-1)y;\ D = \dfrac{d}{dz}$

From (1), we get

$$\{(D(D-1) + D + 1)\}y = z\sin z$$

$$\Rightarrow \quad (D^2+1)y = z\sin z$$

$\therefore$ The A.E. is $\quad m^2 + 1 = 0 \quad \Rightarrow \quad m = \pm i$

$$\text{C.F.} = C_1\cos z + C_2\sin z = C_1\cos(\log x) + C_2\sin(\log x)$$

$$\text{P.I.} = \frac{1}{(D^2+1)}z\sin z = z\frac{1}{(D^2+1)}\sin z - \frac{2D}{(D^2+1)^2}\sin z$$

$$= z.\frac{z}{2D}\sin z - 2D\left[\frac{z}{4D(D^2+1)}\cdot\sin z\right]$$

$$= -\frac{z^2}{2}\cos z - \frac{D}{2}\left[\frac{z^2}{(D^2+1)+2D^2}\sin z\right]$$

$$\left|\ \frac{1}{f(D)}xV = x\frac{1}{f(D)}V - \frac{f'(D)}{\{f(D)\}^2}V\right.$$

$$= -\frac{z^2}{2}\cos z + D\left[\frac{z^2}{4}\sin z\right] = -\frac{z^2}{2}\cos z + \frac{1}{4}\left[2z\sin z + z^2\cos z\right]$$

$\Rightarrow$ $$\text{P.I.} = \frac{z}{2}\sin z - \frac{z^2}{4}\cos z = \frac{1}{2}\log x\sin(\log x) - \frac{1}{4}(\log x)^2\cdot\cos(\log x)$$

$\therefore$ $$y = \text{C.F.} + \text{P.I.}$$

$\Rightarrow$ $$y = C_1\cos(\log x) + C_2\sin(\log x) + \frac{1}{2}\log x\sin(\log x) - \frac{1}{4}(\log x)^2\cdot\cos(\log x).$$ **Ans.**

Example 50: Solve $x^2\dfrac{d^2y}{dx^2} + 4x\dfrac{dy}{dx} + 2y = e^x$ *[U.P.T.U. (C.O.) 2005]*

Solution: We have $x^2\dfrac{d^2y}{dx^2} + 4x\dfrac{dy}{dx} + 2y = e^x$ ...(1)

Let $x = e^z \Rightarrow z = \log_e x$

and $x\dfrac{dy}{dx} = Dy,\ x^2\dfrac{d^2y}{dx^2} = D(D-1)y,\ D = \dfrac{d}{dz}$

Substituting these values in equation (1), we get

$$\{D(D-1) + 4D + 2\}y = e^{e^z}$$

$\Rightarrow$ $$(D^2 + 3D + 2)y = e^{e^z}$$

The A.E. is $m^2 + 3m + 2 = 0 \Rightarrow (m+1)(m+2) = 0 \Rightarrow m = -1, -2$

$\therefore$ $$\text{C.F.} = C_1e^{-z} + C_2e^{-2z} = C_1x^{-1} + C_2x^{-2} = \frac{C_1}{x} + \frac{C_2}{x^2}$$

$$\text{P.I.} = \frac{1}{(D+1)(D+2)}e^{e^z} = \left(\frac{1}{D+1} - \frac{1}{D+2}\right)e^{e^z} = \frac{1}{(D+1)}e^{e^z} - \frac{1}{(D+2)}e^{e^z}$$

$$= e^{-z}\int e^z e^{e^z}\,dz - e^{-2z}\int e^{2z}\cdot e^{e^z}\,dz = e^{-z}e^{e^z} - e^{-2z}(e^z - 1)e^{e^z} = e^{-2z}e^{e^z}$$

$\Rightarrow$ $$\text{P.I.} = x^{-2}e^x = \frac{1}{x^2}e^x$$

Hence, the complete solution is $y = \text{C.F.} + \text{P.I.}$

$\Rightarrow$ $$y = \frac{C_1}{x} + \frac{C_2}{x^2} + \frac{1}{x^2}e^x.$$ **Ans.**

Example 51: Solve $x^2\dfrac{d^2y}{dx^2}+x\dfrac{dy}{dx}-y=\dfrac{x^3}{1+x^2}$.

Solution: Putting $x=e^z \quad\Rightarrow\quad z=\log_e x$, we have

$$x\frac{dy}{dx} = Dy,\ x^2\frac{d^2y}{dx^2}=D(D-1)y$$

The given equation reduces into the form

$$\{D(D-1)+D-1\}y = \frac{e^{3z}}{1+e^{2z}}$$

$$\Rightarrow \quad (D^2-1)y = \frac{e^{3z}}{1+e^{2z}}$$

The A.E. is $\quad m^2-1=0 \quad\Rightarrow\quad m=\pm 1$

$$\text{C.F.} = C_1e^z+C_2e^{-z}=C_1x+\frac{C_2}{x}$$

$$\text{P.I.} = \frac{1}{(D+1)}\left[\frac{1}{(D-1)}\cdot\frac{e^{3z}}{1+e^{2z}}\right]=\frac{1}{(D+1)}\left[e^z\int e^{-z}.\frac{e^{3z}}{1+e^{2z}}dz\right]$$

$$= \frac{1}{(D+1)}\left[e^z\int\frac{e^{2z}}{1+e^{2z}}dz\right]=\frac{1}{(D+1)}\left[\frac{1}{2}e^z\log(1+e^{2z})\right]$$

$$= \frac{1}{2}e^{-z}\int e^{2z}\log(1+e^{2z})dz \qquad \Big|\ \text{on putting } u=1+e^{2z}$$

$$= \frac{1}{2}e^{-z}\int\frac{1}{2}\log u\,du$$

$$= \frac{1}{4}e^{-z}(u\log u-u)=\frac{1}{4}e^{-z}\left[(1+e^{2z})\log(1+e^{2z})-(1+e^{2z})\right]$$

$$= \frac{1}{4x}\left[(1+x^2)\{\log(1+x^2)-1\}\right]=\frac{1}{4}\left(x+\frac{1}{x}\right)\{\log(1+x^2)-1\}$$

$$\therefore \quad y = \frac{C_1}{x}+C_2x+\frac{1}{4}\log\left(x+\frac{1}{x}\right)\{\log(1+x^2)-1\}. \quad \textbf{Ans.}$$

Example 52: Solve $x^3\dfrac{d^3y}{dx^3}+2x^2\dfrac{d^2y}{dx^2}=x+\sin(\log x)$.

Solution: On putting $x=e^z$ and $x\dfrac{dy}{dx}=Dy,\ x^2\dfrac{d^2y}{dx^2}=D(D-1)y\ \ x^3\dfrac{d^3y}{dx^3}=D(D-1)(D-2)y$, the given equation reduces in the form $\{D(D-1)(D-2)+2D(D-1)\}y=e^z+\sin z$

$$\Rightarrow \qquad D^2(D-1)y = e^z + \sin z$$

The A.E. is $m^2(m-1) = 0 \quad \Rightarrow \quad m = 0, 0, 1$

$$\therefore \qquad \text{C.F.} = (C_1 + C_2 z) + C_3 e^z = (C_1 + C_2 \log x) + C_3 . x$$

$$\text{P.I.} = \frac{1}{D^2(D-1)} e^z + \frac{1}{D^2(D-1)} \sin z = \frac{z}{2D(D-1) + D^2} e^z - \frac{1}{(D-1)} \sin z$$

$$= \frac{z e^z}{1} - \frac{(D+1)}{(D^2-1)} \sin z = z e^z + \frac{(D+1)}{2} \sin z$$

$$= z e^z + \frac{1}{2}(\cos z + \sin z) = x \log x + \frac{1}{2}[\cos \log x + \sin \log x]$$

Hence

$$y = (C_1 + C_2 \log x) + C_3 x + x \log x + \frac{1}{2}(\cos \log x + \sin x \log x). \qquad \textbf{Ans.}$$

Example 53: Solve $x^2 y'' - 2xy' + 2y = \log^2 x - \log x^2$.

Solution: On putting $x = e^z \quad \Rightarrow \quad z = \log x$

and $$x \frac{dy}{dx} = Dy, \; x^2 \frac{d^2 y}{dx^2} = D(D-1)y; \; D = \frac{d}{dz}$$

The given equation reduces in the form

$$\{D(D-1) - 2D + 2\} y = z^2 - \log_e e^{2z}$$

$$\Rightarrow \qquad (D^2 - 3D + 2) y = z^2 - 2z \qquad |\log_e e = 1$$

The A.E. is $m^2 - 3m + 2 = 0 \quad \Rightarrow \quad (m-1)(m-2) = 0$

$$\Rightarrow \qquad m = 1, 2$$

$$\therefore \qquad \text{C.F.} = C_1 e^z + C_2 e^{2z} = C_1 x + C_2 x^2$$

Now, $$\text{P.I.} = \frac{1}{(D^2 - 3D + 2)} z^2 - 2 \frac{1}{(D^2 - 3D + 2)} z$$

$$= \frac{1}{2} \left\{ 1 + \left(\frac{D^2 - 3D}{2} \right) \right\}^{-1} z^2 - \left\{ 1 + \frac{(D^2 - 3D)}{2} \right\}^{-1} z$$

$$= \frac{1}{2} \left\{ 1 - \frac{D^2}{2} + \frac{3D}{2} + \frac{9D^2}{4} \right\} z^2 - \left\{ 1 + \frac{3D}{2} \right\} z \qquad \text{| Neglect the higher power of } z \text{ terms}$$

$$= \frac{1}{2} \left\{ z^2 + 3z + \frac{7}{2} \right\} - \left(\frac{z}{2} + \frac{3}{2} \right) = \frac{z^2}{2} + \frac{3z}{2} + \frac{7}{4} - \frac{z}{2} - \frac{3}{2}$$

$$= \frac{z^2}{2} + z + \frac{1}{4} = \frac{\log^2 x}{2} + \log x + \frac{1}{4}$$

Hence, the complete solution is y = C.F. + P.I.

$$\Rightarrow \qquad y = C_1 x + C_2 x^2 + \frac{1}{2}\log^2 x + \log x + \frac{1}{4}. \qquad \textbf{Ans.}$$

Example 54: Solve $x^3 \frac{d^3y}{dx^3} + 3x^2 \frac{d^2y}{dx^2} + x\frac{dy}{dx} + y = x + \log x.$ (*U.P.T.U. 2001*)

Solution: Let $x = e^z \quad \Rightarrow \quad \log x = z$

and $$x\frac{dy}{dx} = Dy,\ x^2 \frac{d^2y}{dx^2} = D(D-1)y, x^3 \frac{d^3y}{dx^3} = D(D-1)(D-2)y$$

On putting these values the given equation reduces in the form

$$\{D(D-1)(D-2) + 3D(D-1) + D + 1\}y = e^z + z$$

$$\Rightarrow \qquad (D^3+1)y = e^z + z\ ;\ D = \frac{d}{dz}$$

The A.E. is $m^3 + 1 = 0 \quad \Rightarrow \quad (m+1)(m^2 - m + 1) = 0$

$$\Rightarrow \qquad m = -1,\ m = \frac{1}{2} \pm i\frac{\sqrt{3}}{2}$$

$$\therefore \qquad \text{C.F.} = C_1 e^{-z} + e^{\frac{z}{2}}\left[C_2 \cos\frac{\sqrt{3}}{2}z + C_3 \sin\frac{\sqrt{3}}{2}z\right]$$

$$\Rightarrow \qquad \text{C.F.} = \frac{C_1}{x} + \sqrt{x}\left[C_2 \cos\frac{\sqrt{3}}{2}\log x + C_3 \sin\frac{\sqrt{3}}{2}\log x\right]$$

and $$\text{P.I.} = \frac{1}{(D^3+1)}(e^z + z) = \frac{1}{(D^3+1)}e^z + \frac{1}{(D^3+1)}z$$

$$= \frac{1}{1+1}e^z + (1+D^3)^{-1}z = \frac{1}{2}e^z + (1-D^3)z = \frac{1}{2}e^z + z$$

$$\Rightarrow \qquad \text{P.I.} = \frac{x}{2} + \log x$$

$\therefore$ Complete solution is y = C.F. + P.I.

$$\Rightarrow \qquad y = \frac{C_1}{x} + \sqrt{x}\left[C_2 \cos\frac{\sqrt{3}}{2}\log x + C_3 \sin\frac{\sqrt{3}}{2}\log x\right] + \frac{x}{2} + \log x. \qquad \textbf{Ans.}$$

Example 55: Solve $\frac{d^2u}{dr^2}+\frac{1}{r}\frac{du}{dr}-\frac{u}{r^2}+kr=0.$

Solution: We have

$$r^2\frac{d^2u}{dr^2}+r\frac{du}{dr}-u+kr^3 \;=0$$

$$\Rightarrow \qquad r^2\frac{d^2u}{dr^2}+r\frac{du}{dr}-u = -kr^3 \qquad ...(1)$$

Let $r=e^z \quad \Rightarrow \quad z=\log r \quad$ and $\quad r\frac{du}{dr}=Du; D=\frac{d}{dz}$

$$r^2\frac{d^2u}{dr^2} = D(D-1)u;\; D=\frac{d}{dz}$$

Substituting these values in equation (1), we get

$\{D(D-1)+D-1\}u = -ke^{3z} \quad \Rightarrow \quad (D^2-1)u=-ke^{3z}$

The A.E. is $\quad m^2-1=0 \quad \Rightarrow \quad (m-1)(m+1)=0 \quad \Rightarrow \quad m=1,-1$

$\therefore \quad$ C.F. $= C_1e^z+C_2e^{-z} \quad \Rightarrow \quad$ C.F. $=C_1r+\frac{C_2}{r}$

and $\quad$ P.I. $=\frac{1}{(D^2-1)}(-ke^{3z})=\frac{-k\,e^{3z}}{(9-1)}=-\frac{k}{8}r^3 \qquad \left|\, e^{3z}=r^3\right.$

$\therefore \quad u=$ C.F. + P.I.

$\Rightarrow \quad u=C_1r+\frac{C_2}{r}-\frac{k}{8}r^3.$ **Ans.**

EXERCISE 1.9

Solve the following differential equations:

1. $x^2y''-xy'-3y=(x\log x)^2 \qquad \left[\textbf{Ans. } y=\frac{C_1}{x}+C_2x^3-\frac{14}{27}x^2-\frac{4}{9}x^2\log x-\frac{(x\log x)^2}{3}\right]$

2. $x^2\frac{d^3y}{dx^3}+3x\frac{d^2y}{dx^2}+\frac{y}{x}+\frac{dy}{dx}=\log x$

$$\left[\textbf{Ans. } y=\frac{C_1}{x}+\sqrt{x}\left[C_2\cos\left(\frac{\sqrt{3}}{2}\log x\right)+C_3\sin\left(\frac{\sqrt{3}}{2}\log x\right)+\frac{x}{2}\left(\log x-\frac{3}{2}\right)\right]\right]$$

3. $x^2 y'' - 4xy' + 6y = \frac{42}{x^4}$ $\left[\textbf{Ans. } y = C_1 x^2 + C_2 x^3 + \frac{1}{x^4}\right]$

4. $\frac{d^2y}{dx^2} + \frac{1}{x}\frac{dy}{dx} = \frac{12 \log x}{x^2}$ $\left[\textbf{Ans. } y = C_1 + C_2 \log x + 2(\log x)^3\right]$

5. $x^2 \frac{d^2y}{dx^2} - x\frac{dy}{dx} - 3y = x^2 \log x$ $\left[\textbf{Ans. } y = \frac{C_1}{x} + C_2 x^3 - \frac{x^2}{3}\left(\log x + \frac{2}{3}\right)\right]$

6. $x^2 y'' + 5xy' + 4y = x^2 + 16(\log x)^2$ $\left[\textbf{Ans. } y = \frac{C_1}{x^2} + \frac{C_2}{x^2}\log x + \frac{x^2}{16} + 4(\log x)^2 - 8\log x + 6\right]$

7. $x^2 \frac{d^2y}{dx^2} + 3x\frac{dy}{dx} + y = \frac{1}{(1-x)^2}$ $\left[\textbf{Ans. } y = \left(\frac{1}{x}\right)(C_1 + C_2 \log x) + \frac{1}{x}\log\frac{x}{1-x}\right]$

8. $9x^2 \frac{d^2y}{dx^2} + 3x\frac{dy}{dx} + y = 0$ $\left[\textbf{Ans. } y = (C_1 + C_2 \log x)x^{1/3}\right]$

9. $x^2 \frac{d^2y}{dx^2} - 4x\frac{dy}{dx} + 4y = 4x^2 - 6x^3, y(2) = 4, y'(2) = -1$ $\left[\textbf{Ans. } y = \frac{5x}{3} - 2x^2 + 3x^3 - \frac{23}{24}x^4\right]$

10. $x^2 \frac{d^2y}{dx^2} - 2x\frac{dy}{dx} - 10y = 0, y(1) = 5, y'(1) = 4$ $\left[\textbf{Ans. } y = \frac{3}{x^2} + 2x^5\right]$

11. $x^2 \frac{d^2y}{dx^2} - 2x\frac{dy}{dx} + 2y = x^2 + \sin(5\log x)$

$\left[\textbf{Ans. } y = C_1 x + C_2 x^2 + x^2 \log x + \frac{1}{754}[15\cos(5\log x) - 23\sin(5\log x)]\right]$

12. $x^3 y''' - 8x^2 y'' + 28xy' - 40y = -\frac{9}{x}$ $\left[\textbf{Ans. } y = C_1 x^2 + C_2 x^4 + C_3 x^5 + \frac{1}{10x}\right]$

1.4.2 Legendre's Homogeneous Linear Differential Equation

A differential equation of the form

$$(ax+b)^n \frac{d^n y}{dx^n} + a_1 (ax+b)^{n-1} \frac{d^{n-1} y}{dx^{n-1}} \ldots\ldots\ldots + a_n y = F(x) \quad \ldots(i)$$

where $a_1, a_2, \ldots, a_n$ are constants is called Legendre's homogeneous linear differential equation.

Method of Solution:

Put $ax + b = e^z \Rightarrow z = \log(ax+b) \Rightarrow \frac{dz}{dx} = \frac{a}{ax+b}$

Now
$$\frac{dy}{dx} = \frac{dy}{dz}\cdot\frac{dz}{dx}$$

$$\Rightarrow \quad \frac{dy}{dx} = \frac{dy}{dz}\cdot\frac{a}{ax+b}$$

$$\Rightarrow \quad (ax+b)\frac{dy}{dx} = a\frac{dy}{dz} \quad \Rightarrow \quad \mathbf{(ax+b)\frac{dy}{dx} = aDy\,;\, D = \frac{d}{dz}}$$

and
$$\frac{d^2y}{dx^2} = \frac{d}{dx}\left(\frac{a}{ax+b}\cdot\frac{dy}{dz}\right) = -\frac{a^2}{(ax+b)^2}\cdot\frac{dy}{dz} + \frac{a^2}{(ax+b)^2}\cdot\frac{d^2y}{dz^2}$$

$$\Rightarrow \quad (ax+b)^2\frac{d^2y}{dx^2} = a^2\left(\frac{d^2y}{dz^2} - \frac{dy}{dz}\right) = a^2D(D-1)y$$

$$\Rightarrow \quad \mathbf{(ax+b)^2\frac{d^2y}{dx^2} = a^2D(D-1)y\,;\, D = \frac{d}{dz}}$$

Similarly, $\mathbf{(ax+b)^3\frac{d^3y}{dx^3} = a^3D(D-1)(D-2)y\,;\, D = \frac{d}{dz}}$ and so on.

By substituting these values in equation (*i*), it can be reduces in the form of linear differential of constant coefficient which can be easily solve.

Example 56: Solve $(x+1)^2\frac{d^2y}{dx^2} + (x+1)\frac{dy}{dx} + y = 2\sin\{\log(x+1)\}$.

Solution: We have $(x+1)^2\frac{d^2y}{dx^2} + (x+1)\frac{dy}{dx} + y = 2\sin\{\log(x+1)\}$ ...(1)

Here, $a = 1$, $b = 1$

Let
$$x+1 = e^z \quad \Rightarrow \quad z = \log_e(x+1)$$

and
$$(x+1)\frac{dy}{dx} = Dy;\, D = \frac{d}{dz}$$

$$(x+1)^2\frac{d^2y}{dx^2} = 1^2D(D-1)y = D(D-1)y;\, D = \frac{d}{dz}$$

Using these values in equation (1), we get

$$\{D(D-1) + D + 1\}y = 2\sin\log_e(e^z)$$

$$\Rightarrow \quad (D^2+1)y = 2\sin z \qquad \Big|\log_e e^z = z$$

The A.E. is $m^2+1=0 \quad \Rightarrow \quad m=0\pm i$

$$\text{C.F.} = C_1\cos z + C_2\sin z = C_1\cos\{\log_e(x+1)\} + C_2\sin\{\log_e(x+1)\}$$

and
$$\text{P.I.} = \frac{1}{(D^2+1)}2\sin z = \frac{2z}{2D}\sin z$$

$\Rightarrow$
$$\text{P.I.} = -\ z\sin z = -\log_e(x+1)\sin\{\log_e(x+1)\}$$

Therefore, the complete solution is

$$y = C_1\cos\{\log_e(x+1) + C_2\sin\{\log_e(x+1)\} - \log_e(x+1).\sin\{\log_e(x+1)\}.$$ **Ans.**

Example 57: Solve : $(3x+2)^2y'' + 3(3x+2)y' - 36y = 3x^2+4x+1.$

Solution: $a = 3$, $b = 2$, Let $3x + 2 = e^z \quad \Rightarrow \quad z = \log_e(3x + 2)$

From this, we get $(3x+2)\dfrac{dy}{dx} = 3Dy\,;\, D = \dfrac{d}{dz}$

and
$$(3x+2)^2\frac{d^2y}{dx^2} = 3^2D(D-1)y = 9D(D-1)y$$

On putting these values in given equation, we get

$$9D(D-1)+9Dy-36y = \frac{3}{9}(e^z-2)^2 + \frac{4}{3}(e^z-2)+1$$

$$\Rightarrow \quad 9(D^2-4)y = \frac{1}{3}(e^z-2)(e^z-2+4)+1 \quad \Rightarrow \quad 9(D^2-4)y = \frac{1}{3}(e^z-2)(e^z+2)+1$$

$$\Rightarrow \quad 9(D^2-4)y = \frac{1}{3}(e^{2z}-4)+1 = \frac{e^{2z}}{3} - \frac{1}{3}$$

The A.E. is $9(m^2-4) = 0 \quad \Rightarrow \quad m^2-4=0 \quad \Rightarrow \quad m=\pm 2$

$\therefore$
$$\text{C.F.} = C_1e^{2z} + C_2e^{-2z} = C_1(3x+2)^2 + \frac{C_2}{(3x+2)^2}$$

Now
$$\text{P.I.} = \frac{1}{9(D^2-4)}\left\{\frac{1}{3}e^{2z}\right\} - \frac{1}{9(D^2-4)}\cdot\frac{1}{3}e^{0.z}$$

$$= \frac{z}{27(2D)}e^{2z} + \frac{1}{108} = \frac{z}{54}\cdot\frac{e^{2z}}{2} + \frac{1}{108} = \frac{1}{108}(ze^{2z}+1)$$

$\Rightarrow$
$$\text{P.I.} = \frac{1}{108}\left\{\log_e(3x+2)\cdot(3x+2)^2+1\right\}$$

Therefore, the complete solution is

$$y = C_1(3x+2)^2 + \frac{C_2}{(3x+2)^2} + \frac{1}{108}\left\{\log_e(3x+2)\cdot(3x+2)^2 + 1\right\}. \textbf{ Ans.}$$

EXERCISE 1.10

Solve the following differential equations:

1. $(2x+5)^2 y'' - 6(2x+5)y' + 8y = 6x$

$$\left[\textbf{Ans. } y = C_1(2x+5)^{2+\sqrt{2}} + C_2(2x+5)^{2-\sqrt{2}} - \frac{3}{4}(2x+5) - \frac{15}{8}\right]$$

2. $(2x+3)^2 \dfrac{d^2y}{dx^2} - (2x+3)\dfrac{dy}{dx} - 12y = 6x$

$$\left[\textbf{Ans. } y = C_1(2x+3)^{\left(\frac{3}{4}+\frac{\sqrt{57}}{4}\right)} + C_2(2x+3)^{\left(\frac{3}{4}-\frac{\sqrt{57}}{4}\right)} - \frac{3}{14}(2x+3) + \frac{3}{4}\right]$$

3. $(x+1)^2 \dfrac{d^2y}{dx^2} + (x+1)\dfrac{dy}{dx} + y = 4\cos\log(1+x)$

$$[\textbf{Ans. } y = C_1 \cos\log(1+x) + C_2 \sin\log(1+x) + 2\log(x+1)\sin\log(x+1)]$$

4. $(1+4x)^2 y'' + (1+4x)y' + 4y = 8(1+4x)^2$

$$\left[\textbf{Ans. } y = C_1(1+4x)^{k_1} + C_2(1+4x)^{k_2} + \frac{8}{41}(1+4x)^2 \text{ where } k_1 = 6+2\sqrt{5},\ k_2 = 6-2\sqrt{5}\right]$$

5. $(x+2)^2 \dfrac{d^2y}{dx^2} + 3(x+2)\dfrac{dy}{dx} - 3y = 0$ $\quad[\textbf{Ans. } y = C_1(x+2) + C_2(x+2)^{-3}]$

6. $(2x+3)^2 y'' + (2x+3)y' - 2y = 24x^2$

$$\left[\textbf{Ans. } y = C_1(2x+3)^{-1/2} + C_2(2x+3) + \frac{3}{5}(2x+3)^2 - 6(2x+3)\log(2x+3) - 27\right]$$

7. $(x+1)^2 \dfrac{d^2y}{dx^2} + (x+1)\dfrac{dy}{dx} - y = \log(x+1)^2 + (x-1)$

$$\left[\textbf{Ans. } y = C_1(x+1) + C_2(x+1)^{-1} - \log(x+1)^2 + \frac{1}{2}(x+1)\log(x+1) + 2\right]$$

1.5 SIMULTANEOUS DIFFERENTIAL EQUATIONS

If there are two or more dependent variables each of which is a function of one independent variable, then the system of differential equations is called a system of simultaneous linear differential equations,

e.g., $\dfrac{dx}{dt} + y = \sin t, \dfrac{dy}{dt} - 2x = t.$

Method of Solution: Such type equations can be solved by eliminating any one dependent variable i.e., by solving similar as algebraic simultaneous equations.

Example 58: Solve: $\dfrac{dx}{dt} - w^2 y = \cos at$

$$\frac{dy}{dt} + w^2 x = \sin at$$

Solution: The given equations can be written as

$$Dx - w^2 y = \cos at \qquad ...(1)$$

$$Dy + w^2 x = \sin at \qquad ...(2)$$

Operating equation (1) by D and equation (2) multiplied by w^2, we get

$$D^2 x - w^2 Dy = -\ a \sin at$$

$$w^4 x + w^2 Dy = w^2 \sin at$$

$$(D^2 + w^4)x = (w^2 - a)\sin at \quad \text{adding}$$

$$\Rightarrow \quad (D^2 + w^4)x = (w^2 - a)\sin at$$

$\therefore$ The A.E. is $\quad m^2 + w^4 = 0 \quad \Rightarrow \quad m = 0 \pm i w^2$

$$\text{C.F.} = C_1 \cos w^2 t + C_2 \sin w^2 t$$

$$\text{P.I.} = \frac{1}{(D^2 + w^4)}(w^2 - a)\sin at = \frac{(w^2 - a)}{(-a^2 + w^4)}\sin at = \frac{\sin at}{(w^2 + a)}$$

$$\therefore \quad x = C_1 \cos w^2 t + C_2 \sin w^2 t + \frac{1}{w^2 + a}\sin at. \quad \textbf{Ans.}$$

Substitute the value of x in equation (1), we get

$$D\left(C_1 \cos w^2 t + C_2 \sin w^2 t + \frac{1}{w^2 + a}\sin at\right) - w^2 y = \cos at$$

$$\Rightarrow -C_1 w^2 \sin w^2 t + C_2 w^2 \cos w^2 t + \frac{a}{w^2 + a}\cos at - w^2 y = \cos at$$

$$\Rightarrow \quad -w^2 y = \cos at + C_1 w^2 \sin w^2 t - C_2 w^2 \cos w^2 t - \frac{a}{w^2 + a}\cos at$$

$$\Rightarrow \qquad -w^2y = C_1w^2 \sin w^2t - C_2w^2 \cos w^2t + \frac{w^2}{(w^2+a)} \cos at$$

$$y = -C_1 \sin w^2t + C_2 \cos w^2t - \frac{1}{(w^2+a)} \cos at. \quad \textbf{Ans.}$$

Example 59: The equations of motion of a particle are given by $\frac{dx}{dt} + wy = 0,\ \frac{dy}{dt} - wx = 0$. Find the path of the particle and show that it is a circle. *(U.P.T.U. 2009)*

Solution: We have

$$\frac{dx}{dt} + wy = 0 \qquad ...(i)$$

$$\frac{dy}{dt} - wx = 0 \qquad ...(ii)$$

Differentiating equation (*ii*), we get

$$\frac{d^2y}{dt^2} - w\frac{dx}{dt} = 0 \Rightarrow \frac{dx}{dt} = \frac{1}{w}\frac{d^2y}{dt^2} \qquad ...(iii)$$

From equation (*i*) and (*iii*), we get

$$\frac{d^2y}{dt^2} + w^2y = 0 \Rightarrow (D^2 + w^2)y = 0$$

∴ The auxiliary equation is

$$m^2 + w^2 = 0$$

Hence, $\Rightarrow m = \pm wi$

$$y = C_1 \cos wt + C_2 \sin wt \qquad ...(iv)$$

Using (*iv*) in equation (*ii*), we get

$$\frac{d}{dt}(C_1 \cos wt + C_2 \sin wt) - wx = 0$$

$$\Rightarrow \qquad -C_1 w \sin wt + C_2 w \cos wt - wx = 0$$

$$\Rightarrow \qquad x = C_2 \cos wt - C_1 \sin wt \qquad ...(v)$$

Adding equations (*iv*) and (*v*) with square

$$x^2 + y^2 = C_2^2 \sin^2 wt + \cos^2 wt) + C_1^2(\sin^2 wt + \cos^2 wt)$$

$$\Rightarrow \qquad x^2 + y^2 = C_2^2 + C_1^2$$

$$\Rightarrow \qquad x^2 + y^2 = a^2 \qquad \text{where } a^2 = C_2^2 + C_1^2$$

which is the equation of a circle.

Example 60: Solve $\dfrac{d^2y}{dt^2} = x-2, \dfrac{d^2x}{dt^2} = y+2.$

Solution: We have $\dfrac{d^2y}{dt^2} = x-2$ ...(1)

$$\frac{d^2x}{dt^2} = y+2 \quad ...(2)$$

Differentiate equation (2) two times w.r.t. 't', we get

$$\frac{d^4x}{dt^4} = \frac{d^2y}{dt^2} \quad ...(3)$$

From (1) and (3), we get

$$\frac{d^4x}{dt^4} = x-2 \quad \Rightarrow \quad D^4x = x-2$$

$$\Rightarrow \quad (D^4-1)x = -2; \; D = \frac{d}{dt}$$

The A.E. is $\quad m^4 - 1 = 0 \quad \Rightarrow \quad (m^2-1)(m^2+1) = 0$

$$\Rightarrow \quad (m-1)(m+1)(m^2+1) = 0 \quad \Rightarrow \quad m = \pm 1, \; m = \pm i$$

$$\text{C.F.} = C_1e^t + C_2e^{-t} + C_3\cos t + C_4\sin t$$

$$\text{P.I.} = \frac{(-2)}{(D^4-1)}e^{0.t} = 2$$

$\therefore \quad x = C_1e^t + C_2e^{-t} + C_3\cos t + C_4\sin t + 2.$ **Ans.**

Differentiate w.r.t. 't', we get

$$\frac{dx}{dt} = C_1e^t - C_2e^{-t} - C_3\sin t + C_4\cos t$$

Again differentiate with respect to 't', we get

$$\frac{d^2x}{dt^2} = C_1e^t + C_2e^{-t} - C_3\cos t - C_4\sin t$$

Putting the value of $\dfrac{d^2x}{dt^2}$ in equation (2), we get

$$C_1e^t + C_2e^{-t} - C_3\cos t - C_4\sin t = y+2$$

$\Rightarrow \quad y = C_1e^t + C_2e^{-t} - C_3\cos t - C_4\sin t - 2.$ **Ans.**

Example 61: Solve $\dfrac{d^2x}{dt^2}+\dfrac{dy}{dt}+3x = e^{-t}$ *(U.P.T.U. 2007)*

$$\frac{d^2y}{dt^2}-4\frac{dx}{dt}+3y = \sin 2t$$

Solution: The given equations can be written as

$$(D^2+3)x+Dy = e^{-t} \quad \text{...(1)}$$

$$-4Dx+(D^2+3)y = \sin 2t \quad \text{...(2)}$$

Operating $4D$ on equation (1) and $(D^2 + 3)$ on equation (2), we get

$$4D(D^2+3)x+4D^2y = -\ 4e^{-t} \quad \text{...(3)}$$

$$-4D(D^2+3)x+(D^2+3)^2y = -4\sin 2t+3\sin 2t$$

$$\Rightarrow \quad -4D(D^2+3)x+(D^2+3)^2y = -\sin 2t \quad \text{...(4)}$$

Adding (3) and (4), we get

$$\left\{(D^2+3)^2+4D^2\right\}y = -4e^{-t}-\sin 2t$$

$$\Rightarrow \quad (D^4+10D^2+9)\,y = -4e^{-t}-\sin 2t$$

The A.E. is $m^4+10m^2+9 = 0$

$$\Rightarrow \quad (m^2+9)(m^2+1) = 0$$

$$\Rightarrow \quad m = \pm\, 3i,\ m = \pm\, i\text{C.F.} = C_1\cos 3t+C_2\sin 3t+C_3\cos t+C_4\sin t$$

$$\text{P.I.} = \frac{1}{(D^2+9)(D^2+1)}(-4e^{-t}-\sin 2t) = \frac{-4e^{-t}}{(1+9)(1+1)}-\frac{1}{(-4+9)(-4+1)}\sin 2t$$

$$\Rightarrow \quad \text{P.I.} = \frac{-4e^{-t}}{20}+\frac{\sin 2t}{15} = \frac{1}{5}\left(-e^{-t}+\frac{1}{3}\sin 2t\right)$$

$$\therefore \quad y = C_1\cos 3t+C_2\sin 3t+C_3\cos t+C_4\sin t+\frac{1}{5}\left(-e^{-t}+\frac{1}{3}\sin 2t\right) \quad \textbf{Ans.}$$

$$\Rightarrow \quad Dy = -3C_1\sin 3t+3C_2\cos 3t-C_3\sin t+C_4\cos t+\frac{1}{5}e^{-t}+\frac{2}{15}\cos 2t$$

$$\Rightarrow \quad D^2y = -9C_1\cos 3t-9C_2\sin 3t-C_3\cos t-C_4\sin t-\frac{1}{5}e^{-t}-\frac{4}{15}\sin 2t$$

$$\Rightarrow \quad (D^2+3)y = -9C_1\cos 3t-9C_2\sin 3t-C_3\cos t-C_4\sin t-\frac{1}{5}e^{-t}-\frac{4}{15}\sin 2t$$

$$+3C_1\cos 3t+3C_2\sin 3t+3C_3\cos t+3C_4\sin t-\frac{3}{5}e^{-t}+\frac{1}{5}\sin 2t$$

$$\Rightarrow \quad (D^2+3)y = -6C_1\cos 3t - 6C_2\sin 3t + 2C_3\cos t + 2C_4\sin t - \frac{4}{5}e^{-t} - \frac{1}{15}\sin 2t$$

Putting the value of $(D^2+3)y$ in equation (2), we get

$$-4Dx - 6C_1\cos 3t - 6C_2\sin 3t + 2C_3\cos t + 2C_4\sin t - \frac{4}{5}e^{-t} - \frac{1}{15}\sin 2t = \sin 2t$$

$$\Rightarrow -4Dx = 6C_1\cos 3t + 6C_2\sin 3t - 2C_3\cos t - 2C_4\sin t + \frac{4}{5}e^{-t} + \frac{16}{5}\sin 2t$$

$$\Rightarrow \quad -4x = 6C_1\int\cos 3t\,dt + 6C_2\int\sin 3t\,dt - 2C_3\int\cos t\,dt - 2C_4\int\sin t\,dt$$

$$+\frac{4}{5}\int e^{-t}dt + \frac{16}{15}\int\sin 2t\,dt + C_5$$

$$\Rightarrow \quad x = -\frac{1}{4}\left[2C_1\sin 3t - 2C_2\cos 3t - 2C_3\sin t + 2C_4\cos t - \frac{4}{5}e^{-t} - \frac{16}{15\times 2}\cos 2t\right] - \frac{C_5}{4}$$

$$\Rightarrow \quad x = -\frac{1}{2}C_1\sin 3t + \frac{C_2}{2}\cos 3t + \frac{1}{2}C_3\sin t - \frac{C_4}{2}\cos t + \frac{1}{5}e^{-t} + \frac{2}{15}\cos 2t - \frac{C_5}{4}$$

$$\Rightarrow \quad x = -\frac{C_1}{2}\sin 3t + \frac{C_2}{2}\cos 3t + \frac{C_3}{2}\sin t - \frac{C_4}{2}\cos t + \frac{1}{5}e^{-t} + \frac{2}{15}\cos 2t + A \text{ where } A = -\frac{C_5}{4}.$$

Ans.

Example 62: Solve the equations

$$\frac{dx}{dt} + \frac{dy}{dt} - 2y = 2\cos t - 7\sin t$$ *(U.P.T.U. 2001)*

$$\frac{dx}{dt} - \frac{dy}{dt} + 2x = 4\cos t - 3\sin t.$$

Solution: The given equations can be written as

$$Dx + (D-2)y = 2\cos t - 7\sin t \qquad \text{...(1)}$$

$$(D+2)x - Dy = 4\cos t - 3\sin t \qquad \text{...(2)}$$

Operating $(D + 2)$ on equation (1) and D on equation (2), we get

$$D(D+2)x + (D^2-4)y = (D+2)2\cos t - (D+2)7\sin t \qquad \text{...(3)}$$

$$D(D+2)x - D^2y = D(4\cos t) - D(3\sin t) \qquad \text{...(4)}$$

Subtracting equation (3) and (4), we get

$$(2D^2-4)y = -2\sin t + 4\cos t - 7\cos t - 14\sin t + 4\sin t + 3\cos t$$

$$\Rightarrow \quad (2D^2-4)y = -12\sin t$$

The A.E. is $2m^2 - 4 = 0 \Rightarrow m^2 - 2 = 0 \Rightarrow m = \pm\sqrt{2}$

$$\text{C.F.} = C_1 e^{\sqrt{2}t} + C_2 e^{-\sqrt{2}t}$$

$$\text{P.I.} = \left(\frac{1}{2D^2 - 4}\right)(-12\sin t) = \frac{-12}{-6}\sin t = 2\sin t$$

$$\therefore \quad y = C_1 e^{\sqrt{2}t} + C_2 e^{-\sqrt{2}t} + 2\sin t. \quad \textbf{Ans.}$$

$$\Rightarrow \quad Dy = C_1\sqrt{2}\, e^{\sqrt{2}t} - C_2\sqrt{2}e^{-\sqrt{2}t} + 2\cos t$$

$$\Rightarrow \quad (D-2)y = C_1\sqrt{2}e^{\sqrt{2}t} - C_2\sqrt{2}e^{-\sqrt{2}t} + 2\cos t - 2C_1 e^{\sqrt{2}t} - 2C_2 e^{-\sqrt{2}t} - 4\sin t$$

$$\Rightarrow \quad (D-2)y = \left(\sqrt{2}-2\right)C_1 e^{\sqrt{2}t} - \left(\sqrt{2}+2\right)C_2 e^{-\sqrt{2}t} + 2\cos t - 4\sin t$$

Using this value in equation (1), we get

$$Dx + (\sqrt{2}-2)C_1 e^{\sqrt{2}t} - (\sqrt{2}+2)C_2 e^{-\sqrt{2}t} + 2\cos t - 4\sin t = 2\cos t - 7\sin t$$

$$\Rightarrow \quad Dx = -(\sqrt{2}-2)C_1 e^{\sqrt{2}t} + (\sqrt{2}+2)C_2{}^{-\sqrt{2}t} - 3\sin t$$

$$\Rightarrow \quad x = -\frac{\left(\sqrt{2}-2\right)}{\sqrt{2}}C_1 e^{\sqrt{2}t} - \frac{\sqrt{2}+2}{\sqrt{2}}C_2 e^{-\sqrt{2}t} + 3\cos t + C_3$$

$$\Rightarrow \quad x = -(1-\sqrt{2})C_1 e^{\sqrt{2}t} - (1+\sqrt{2})C_2 e^{-\sqrt{2}t} + 3\cos t + C_3. \quad \textbf{Ans.}$$

Example 63: Solve $\frac{dx}{dt} + 4x + 3y = t.$ *(U.P.T.U. 2006)*

$$\frac{dy}{dt} + 2x + 5y = e^t$$

Solution: The given equations can be written as

$$(D+4)x + 3y = t \quad \text{...(1)}$$

$$2x + (D+5)y = e^t \quad \text{...(2)}$$

Operating $(D + 5)$ on equation (1) and multiplied equation (2) by 3, we get

$$(D+5)(D+4)x + 3(D+5)y = (D+5)t \quad \text{...(3)}$$

$$6x + 3(D+5)y = 3e^t \quad \text{...(4)}$$

Subtracting (3) and (4), we get

$$(D^2 + 9D + 20 - 6)x = 5t + 1 - 3e^t$$

$$\Rightarrow \quad (D^2 + 9D + 14)x = 5t - 3e^t + 1$$

The A.E. is $m^2 + 9m + 14 = 0 \Rightarrow (m+7)(m+2) = 0 \Rightarrow m = -7, -2$

$$\text{C.F.} = C_1 e^{-7t} + C_2 e^{-2t}$$

and $$\text{P.I.} = \frac{1}{(D^2+9D+14)}(5t-3e^t+1)$$

$$= \frac{5}{14}\left(1+\frac{D^2+9D}{14}\right)^{-1} t - \frac{3e^t}{(1+9+14)} + \frac{1}{14}$$

$$= \frac{5}{14}\left(1-\frac{9D}{14}\right)t - \frac{3e^t}{24} + \frac{1}{14} = \frac{5}{14}\left(t-\frac{9}{14}\right) - \frac{e^t}{8} + \frac{1}{14}$$

$$\Rightarrow \quad \text{P.I.} = \frac{5t}{14} - \frac{e^t}{8} - \frac{31}{196}$$

$$\therefore \quad x = C_1e^{-7t} + C_2e^{-2t} + \frac{5t}{14} - \frac{e^t}{8} - \frac{31}{196} \quad \textbf{Ans.}$$

Now $$(D+4)x = -7C_1e^{-7t} - 2C_2e^{-2t} + \frac{5}{14} - \frac{e^t}{8} + 4C_1e^{-7t} + 4C_2e^{-2t} + \frac{10}{7}t - \frac{e^t}{2} - \frac{31}{49}$$

$$\Rightarrow \quad (D+4)x = -3C_1e^{-7t} + 2C_2e^{-2t} - \frac{5e^t}{8} + \frac{10}{7}t - \frac{27}{98}$$

Using this value in equation (1), we get

$$3y = t + 3C_1e^{-7t} - 2C_2e^{-2t} + \frac{5}{8}e^t - \frac{10}{7}t + \frac{27}{98}.$$

$$\Rightarrow \quad y = -\frac{1}{7}t + C_1e^{-7t} - \frac{2}{3}C_2e^{-2t} + \frac{5}{24}e^t + \frac{9}{98}. \quad \textbf{Ans.}$$

Example 64: Solve $(D^2 + 8)x + 2Dy = 32t$ ...(1)

$3Dx + (D^2 - 2)y = 60e^{-t}$. ...(2)

Solution: Operating $(D^2 - 2)$ on equation (1) and $2D$ on (2), we get

$$(D^2-2)(D^2+8)x + 2D(D^2-2)y = 32(D^2-2)t \quad ...(3)$$

$$6D^2x + 2D(D^2-2)y = -120e^{-t} \quad ...(4)$$

Subtracting equation (3) and (4), we get

$$(D^4-16)x = 120e^{-t} - 64t$$

$$\Rightarrow \quad m^4 - 16 = 0 \quad \Rightarrow \quad (m^2-4)(m^2+4) = 0 \quad \Rightarrow \quad (m-2)(m+2)(m^2+4) = 0$$

$$\Rightarrow \quad m = -2, 2, m = 0 \pm 2i$$

$$\text{C.F.} = C_1e^{-2t} + C_2e^{2t} + C_3\cos 2t + C_4 \sin 2t$$

$$\text{P.I.} = \frac{1}{(D^4-16)}(120e^{-t}-64t) = \frac{120e^{-t}}{(1-16)} + \frac{64}{16}\left(1-\frac{D^4}{16}\right)^{-1} t$$

$$= -8e^{-t} + 4(t-0) = 4t - 8e^{-t}$$

$$\therefore \qquad x = C_1e^{-2t} + C_2e^{2t} + C_3\cos 2t + C_4\sin 2t + 4t - 8e^{t} \quad \textbf{Ans.}$$

$$\Rightarrow \qquad Dx = -2C_1e^{-2t} + 2C_2e^{2t} - 2C_3\sin 2t + 2C_4\cos 2t + 4 - 8e^{t}$$

and

$$D^2x = 4C_1e^{-2t} + 4C_2e^{2t} - 4C_3\cos 2t - 4C_4\sin 2t - 8e^{t}$$

$$\Rightarrow (D^2+8)x = 12C_1e^{-2t} + 12C_2e^{2t} + 4C_3\cos 2t + 4C_4\sin 2t - 72e^{t} + 32t$$

Using this value in equation (1), we get

$$2Dy = -12C_1e^{-2t} - 12C_2e^{2t} - 4C_3\cos 2t - 4C_4\sin 2t + 72e^{t}$$

$$\Rightarrow \qquad y = \frac{1}{2}\left[6C_1e^{-2t} - 6C_2e^{2t} - 2C_3\sin 2t + 2C_4\cos 2t + 72e^{t}\right] + C_5$$

$$\Rightarrow \qquad y = 3C_1e^{-2t} - 3C_2e^{2t} - C_3\sin 2t + C_4\cos 2t + 36e^{t} + C_5. \quad \textbf{Ans.}$$

Example 65: Solve $\frac{dx}{dt} + y = \sin t, \frac{dy}{dt} + x = \cos t$ *(U.P.T.U. 2004)*

$x(0) = 2, \quad y(0) = 0.$

Solution: We have $\frac{dx}{dt} + y = \sin t$ ...(1)

$$\frac{dy}{dt} + x = \cos t \qquad \text{...(2)}$$

Differentiate equation (1) w.r.t. '*t*', we get

$$\frac{d^2x}{dt^2} + \frac{dy}{dt} = \cos t \qquad \text{...(3)}$$

Using equation (2) in (3)

$$\frac{d^2x}{dt^2} - x = 0 \quad \Rightarrow \quad (D^2-1)x = 0$$

The A.E. is $m^2 - 1 = 0 \quad \Rightarrow \quad m = \pm 1$

$$\therefore \qquad x = C_1e^{t} + C_2e^{-t} \qquad \text{...(4)}$$

$$\Rightarrow \qquad \frac{dx}{dt} = C_1e^{t} - C_2e^{-t}$$

Using this value in equation (1), we get

$$y = \sin t - C_1e^{t} + C_2e^{-t} \qquad \text{...(5)}$$

Using given conditions

$$\left.\begin{array}{ll}\text{From (4)} & C_1 + C_2 = 2 \\ \text{From (5)} & -C_1 + C_2 = 0\end{array}\right\} \Rightarrow C_1 = C_2 = 1$$

Hence, the solutions are

$$x = e^t + e^{-t}$$

and

$$y = \sin t - e^t + e^{-t}. \quad \textbf{Ans.}$$

Example 66: Solve $\dfrac{dx}{dt} + \dfrac{2}{t}(x - y) = 1, \ \dfrac{dy}{dt} + \dfrac{1}{t}(x + 5y) = t$

Solution: The given equation can be written as *[U.P.T.U.(C.O.) 2005]*

$$t\frac{dx}{dt} + 2x - 2y = t \quad \text{...(1)}$$

$$t\frac{dy}{dt} + x + 5y = t^2 \quad \text{...(2)}$$

From equation (1) $2y = t\dfrac{dx}{dt} + 2x - t$ ...(3)

Using equation (3) in equation (2), we get

$$\frac{t}{2}\left(t\frac{d^2x}{dt^2} + \frac{dx}{dt} + 2\frac{dx}{dt} - 1\right) + x + \frac{5}{2}\left(t\frac{dx}{dt} + 2x - t\right) = t^2$$

$$\Rightarrow t^2\frac{d^2x}{dt^2} + 3t\frac{dx}{dt} - t + 2x + 5t\frac{dx}{dt} + 10x - 5t = 2t^2$$

$$\Rightarrow \quad t^2\frac{d^2x}{dt^2} + 8t\frac{dx}{dt} + 12x = 2t^2 + 6t \quad \text{...(4)}$$

Let $t = e^z \Rightarrow z = \log_e t$

which gives $t\dfrac{dx}{dt} = Dx; D = \dfrac{d}{dz}$

$$t^2\frac{d^2x}{dt^2} = D(D-1)x; D = \frac{d}{dz}$$

Substituting these values in equation (4), we get

$$\{D(D-1) + 8D + 12\}x = 2e^{2z} + 6e^z$$

$$\Rightarrow \quad (D^2 + 7D + 12)x = 2e^{2z} + 6e^z$$

The A.E. is $m^2 + 7m + 12 = 0 \Rightarrow (m+4)(m+3) = 0 \Rightarrow m = -3, -4$

$$\text{C.F.} = C_1e^{-3z} + C_2e^{-4z} \Rightarrow \text{C.F.} = \frac{C_1}{t^3} + \frac{C_2}{t^4}$$

$$\text{P.I.} = \frac{1}{(D^2+7D+12)}(2e^{2z}+6e^z) = \frac{2e^{2z}}{30}+\frac{6e^z}{20} = \frac{t^2}{15}+\frac{3t}{10}$$

$$\therefore \qquad x = \frac{C_1}{t^3}+\frac{C_2}{t^4}+\frac{t^2}{15}+\frac{3t}{10} \quad \textbf{Ans.}$$

$$\Rightarrow \qquad \frac{dx}{dt} = -\frac{3C_1}{t^4}-\frac{4C_2}{t^5}+\frac{2}{15}t+\frac{3}{10}$$

Putting this value in equation (3), we get

$$2y = -\frac{3C_1}{t^3}-\frac{4C_2}{t^4}+\frac{2t^2}{15}+\frac{3t}{10}+\frac{2C_1}{t^3}+\frac{2C_2}{t^4}+\frac{2t^2}{15}+\frac{3t}{5}-t$$

$$\Rightarrow \qquad y = \frac{1}{2}\left[-\frac{C_1}{t^3}-\frac{2C_2}{t^4}+\frac{4t^2}{15}+\frac{9t}{10}\right]$$

$$\Rightarrow \qquad y = -\frac{C_1}{2t^3}-\frac{C_2}{t^4}+\frac{2t^2}{15}+\frac{9t}{20}. \quad \textbf{Ans.}$$

Example 67: Solve the following simultaneous equations

$$\frac{d^2x}{dt^2}-3x-4y=0, \quad \frac{d^2y}{dt^2}+x+y=0. \qquad \textit{(U.P.T.U. 2005)}$$

Solution: The given equation can be written as

$$(D^2-3)x-4y = 0 \qquad \text{...(1)}$$

$$x+(D^2+1)y = 0 \qquad \text{...(2)}$$

Operating equation (2) by (D^2-3), we get

$$(D^2-3)x+(D^2-3)(D^2+1)y = 0 \qquad \text{...(3)}$$

Subtracting (1) and (3), we get

$$(D^4-2D^2+1)y = 0$$

$$\Rightarrow (D^2-1)^2 y = 0$$

The A.E. is $(m^2-1)^2 = 0 \Rightarrow m = \pm 1$ (twice)

C.F. $= (C_1+C_2t)e^t+(C_3+C_4t)e^{-t}$, P.I. $= 0$

$$\Rightarrow \qquad y = (C_1+C_2t)e^t+(C_3+C_4t)e^{-t} \quad \textbf{Ans.}$$

$$\Rightarrow \qquad Dy = C_1e^t+C_2e^t+C_2te^t-C_3e^{-t}+C_4e^{-t}-C_4te^{-t}$$

$$\Rightarrow \qquad D^2y = C_1e^t+C_2e^t+C_2e^t+C_2te^t+C_3e^{-t}-C_4e^{-t}+C_4te^{-t}-C_4e^{-t}$$

$$\Rightarrow \qquad D^2y = C_1e^t+2C_2e^t+C_2te^t+C_3e^{-t}-2C_4e^{-t}+C_4te^{-t}$$

$$\Rightarrow \quad (D^2+1)\,y = 2C_1e^t + 2C_2e^t + 2C_2te^t + 2C_3e^{-t} - 2C_4e^{-t} + 2C_4te^{-t}$$

Using this value in equation (2), we get

$x = -\ [(2C_1 + 2C_2 + 2C_2t)e^t + (2C_3 - 2C_4 + 2C_4t)e^{-t}]$. **Ans.**

Example 68: Solve $\dfrac{dx}{dt} + 5x - 2y = t$

$$\frac{dy}{dt} + 2x + y = 0$$

Also show that $x = y = 0$ when $t = 0$ for some definite values of constants. *(U.P.T.U. 2008)*

Solution: The given equations can be written as

$$(D+5)x - 2y = t \qquad \ldots(i)$$

$$2x + (D+1)y = 0 \qquad \ldots(ii)$$

Operating $(D + 5)$ on equation (ii) and multiplying (i) by 2, we get

$$2(D+5)x - 4y = 2t$$

$$\underset{-}{}2(D+5)x \underset{-}{+} (D+1)(D+5)y = 0$$

$$\overline{(D^2+6D+9)y = -\,2t}$$

The A.E. is $m^2 + 6m + 9 = 0 \;\Rightarrow\; (m+3)^2 = 0 \;\Rightarrow\; m = -3, -3$

$$\therefore \qquad \text{C.F.} = (C_1 + C_2t)e^{-3t}$$

$$\text{P.I.} = \frac{1}{(D+3)^2}(-2t) = -\frac{2}{9}\left(1+\frac{D}{3}\right)^{-2} t$$

$$= -\frac{2}{9}\left(1-\frac{2D}{3}\right)t = -\frac{2}{9}\left(t-\frac{2}{3}\right) = -\frac{2t}{9} + \frac{4}{27}$$

Thus
$$y = (C_1 + C_{2t})e^{-3t} - \frac{2t}{9} + \frac{4}{27} \qquad \ldots(iii)$$

Differentiate w.r.t. 't', we get

$$\frac{dy}{dt} = -3C_1e^{-3t} + C_2e^{-3t} - 3C_2te^{-3t} - \frac{2}{9} \qquad \ldots(iv)$$

Using (iii) and (iv) in equation (ii), we get

$$2x = 3C_1e^{-3t} - C_2e^{-3t} + 3C_2te^{-3t} + \frac{2}{9} - C_1e^{-3t} - C_2te^{-3t} + \frac{2t}{9} - \frac{4}{27}$$

$$2x = 2C_1e^{-3t} + 2C_2te^{-3t} - C_2e^{-3t} - C_1e^{-3t} + \frac{2t}{9} + \frac{2}{27}$$

$$\Rightarrow \qquad x = C_1e^{-3t} + C_2te^{-3t} - \frac{1}{2}C_2e^{-3t} - \frac{1}{2}C_1e^{-3t} + \frac{t}{9} + \frac{1}{27} \qquad \ldots(v)$$

Hence, the equations (*iii*) and (*v*) are required solutions.

Deduction: Putting $t = 0$ in equation (*iii*), we get

$$y = C_1 + \frac{4}{27}, \text{ if we suppose } C_1 = -\frac{4}{27}, \text{ then } \boxed{y=0}$$

Putting $t = 0$ in (*v*), we get

$$x = C_1 - \frac{1}{2}C_2 - \frac{1}{2}C_1 + \frac{1}{27} = -\frac{4}{27} - \frac{1}{2}C_2 + \frac{2}{27} + \frac{1}{27} \quad \left(\text{As } C_1 = -\frac{4}{27}\right)$$

$$\Rightarrow \quad x = -\frac{1}{27} - \frac{1}{2}C_2, \text{ if we suppose } C_2 = -\frac{2}{27} \text{ then } \boxed{x=0}$$

EXERCISE 1.11

Solve the following differential equations:

1. $Dx + Dy + 3x = \sin t$ (*U.P.T.U. 2003*)

$Dx + y - x = \cos t$

$$\left[\textbf{Ans. } x = C_1 e^{3t} + C_2 e^{-t} + \frac{1}{5}(2\sin t - \cos t)\right.$$

$$\left. y = \frac{1}{5}(2\cos t + \sin t) - 2C_1 e^{3t} + 2C_2 e^{-t}\right]$$

2. $2\dfrac{dx}{dt} + 3\dfrac{dy}{dt} + x + y = \sin t, \quad x(0) = y(0) = 0$

$\dfrac{dx}{dt} + \dfrac{dy}{dt} + 5x + 7y = 2$

$$\left[\textbf{Ans. } x = \frac{4}{3}(e^t - e^{-2t}) - 1 + 2\sin t + \cos t\right.$$

$$\left. y = 1 + \frac{4}{5}e^{-2t} - e^t - \frac{4}{5}\cos t - \frac{7}{5}\sin t\right]$$

3. $\dfrac{dx}{dt} - \dfrac{dy}{dt} + 2y = \cos 2t$

$\dfrac{dx}{dt} + \dfrac{dy}{dt} - 2x = \sin 2t$

$$\left[\textbf{Ans. } x = e^t(C_1 \cos t + C_2 \sin t) - \frac{1}{2}\cos 2t\right.$$

$$\left. y = e^t(C_1 \sin t - C_2 \cos t) - \frac{1}{2}\sin 2t\right]$$

4. $(D+5)x + (D+3)y = e^{-t}$

$(2D+1)x + (D+1)y = 3$ $\quad \left[\textbf{Ans. } x = -\frac{2}{3}C_1 e^t - \frac{C_2}{3}e^{-2t} - 9,\ y = C_1 e^t + C_2 e^{-2t} + \frac{15}{2} + \frac{e^{-t}}{2}\right]$

5. $\dfrac{dx}{dt} - 3x - 6y = t^2$

$\dfrac{dy}{dt} + \dfrac{dx}{dt} - 3y = e^t$

$$\left[\textbf{Ans. } x = C_3 \cos 3t + C_4 \sin 3t + \frac{3}{5}e^t - \frac{t^2}{3} + \frac{2t}{9} + \frac{2}{27}\right.$$

$$\left. y = \left(\frac{1}{2}C_4 - \frac{1}{2}C_3\right)\cos 3t + \left(\frac{-C_3}{2} - \frac{C_4}{2}\right)\sin 3t - \frac{e^t}{5} - \frac{2t}{9}\right]$$

6. $\dfrac{dx}{dt} + 2y + \sin t = 0$

$\dfrac{dx}{dt} - 2x - \cos t = 0$ $\quad$ $[\textbf{Ans. } x = C_1 \cos 2t + C_2 \sin 2t - \cos t, y = C_1 \sin 2t - C_2 \cos 2t - \sin t]$

7. $\dfrac{dy}{dt} = x$ and $\dfrac{dx}{dt} = y + e^{2t}$ $\quad$ $\left[\textbf{Ans. } x = C_1e^t + C_2e^{-t} + \dfrac{2}{3}e^{2t}, y = C_1e^t - C_2e^{-t} + \dfrac{1}{3}e^{2t}\right]$

8. $t\dfrac{dx}{dt} + y = 0$ and $t\dfrac{dy}{dt} + x = 0$

given $x(1) = 1$ and $y(-1) = 0$ $\quad$ $\left[\textbf{Ans. } x = \dfrac{1}{2}\left(t + \dfrac{1}{t}\right), y = \dfrac{1}{2}\left(\dfrac{1}{t} - t\right)\right]$

9. $\dfrac{dx}{dt} = 2y, \dfrac{dy}{dt} = 2z, \dfrac{dz}{dt} = 2x$

$$\left[\begin{aligned} \textbf{Ans. } x &= C_1e^{2t} + C_2e^{-t}\cos(\sqrt{3}t + C_3) \\ y &= C_1e^{2t} + C_2e^{-t}\cos\left(\sqrt{3}t + C_3 + \frac{2\pi}{3}\right) \\ z &= C_1e^{2t} + C_2e^{-t}\cos\left(\sqrt{3}t + C_3 + \frac{4\pi}{3}\right) \end{aligned}\right]$$

10. $2\dfrac{d^2x}{dt^2} + 3\dfrac{dy}{dt} = 4$ and $2\dfrac{d^2y}{dt^2} - 3\dfrac{dx}{dt} = 0$ with $x = y = \dfrac{dx}{dt} = \dfrac{dy}{dt} = 0$ at $t = 0$

$$\left[\textbf{Ans. } x = \frac{8}{9}\left(1 - \cos\frac{3}{2}t\right), y = \frac{4}{3}t - \frac{8}{9}\sin\frac{3}{2}t\right]$$

11. $\dfrac{dx}{dt} + 5x - 2y = t$ and $\dfrac{dy}{dt} + 2x + y = 0$ with $x(0) = y(0) = 0$

$$\left[\textbf{Ans. } x = -\frac{1}{27}(1 + 6t)e^{-3t} + \frac{1}{27}(1 + 3t), y = -\frac{2}{27}(2 + 3t)e^{-3t} + \frac{2}{27}(2 - 3t)\right]$$

12. $\dfrac{dx}{dt} = 3x + 8y$ and $\dfrac{dy}{dt} = -x - 3y$ with $x(0) = 6, y(0) = -2$

$$\left[\textbf{Ans.}\, x = 4e^{t} + 2e^{-t}, y = -(e^{t} + e^{-t})\right]$$

13. $(D-3)x + 2(D+2)y = 2\sin t$ and $2(D+1)x + (D-1)y = \cos t$

$$\left[\begin{aligned} \textbf{Ans. } x &= C_1 e^{-5t} + C_2 e^{-t/3} + \frac{1}{65}(8\sin t + \cos t) \\ y &= -\frac{4}{3}C_1 e^{-5t} + C_2 e^{-t/3} + \frac{1}{130}(61\sin t - 33\cos t) \end{aligned}\right]$$

1.6 SECOND ORDER LINEAR DIFFERENTIAL EQUATIONS WITH VARIABLE COEFFICIENTS [Solution by Changing Dependent and Independent Variables]

A differential equation of the form

$$\frac{d^2y}{dx^2} + P(x)\frac{dy}{dx} + Q(x)y = R(x) \qquad ...(i)$$

where $P(x)$, $Q(x)$ and $R(x)$ are the function of x is called second order linear differential equations with variable coefficients.

Since no general method of solving such equations are given, some particular cases are explained in the following subsections.

1.6.1 Method I: Complete Solution in Terms of Known Integral Belonging to the Complementary Function (One Part of Solution *i.e.*, Part of C.F. is Known or One Solution is Known)

Let $y = u$ be a part of complementary function of equation (*i*) and v is remaining solution of differential equation (*i*).

Then the complete solution of equation (*i*) is $y = uv$ ...(*ii*)

$$\Rightarrow \qquad \frac{dy}{dx} = v\frac{du}{dx} + u\frac{dv}{dx} \text{ and } \frac{d^2y}{dx^2} = v\frac{d^2u}{dx^2} + 2\frac{du}{dx}\frac{dv}{dx} + u\frac{d^2v}{dx^2}$$

Putting these values in equation (*i*) then, we get

$$v\frac{d^2u}{dx^2} + 2\frac{du}{dx}\frac{dv}{dx} + u\frac{d^2v}{dx^2} + P\left(v\frac{du}{dx} + u\frac{dv}{dx}\right) + Quv = R$$

$$\Rightarrow \qquad v\left[\frac{d^2u}{dx^2} + P\frac{du}{dx} + Qu\right] + u\left[\frac{d^2v}{dx^2} + P\frac{dv}{dx}\right] + 2\frac{du}{dx}\frac{dv}{dx} = R \qquad ...(iii)$$

Since u is a part of C.F. *i.e.*, solution of (*i*)

$$\therefore \qquad \frac{d^2u}{dx^2} + P\frac{du}{dx} + Qu = 0$$

Hence, equation (*iii*) becomes

$$u\left(\frac{d^2v}{dx^2}+P\frac{dv}{dx}\right)+2\frac{du}{dx}\frac{dv}{dx}=R$$

$$\Rightarrow \qquad \frac{d^2v}{dx^2}+\left[P+\frac{2}{u}\frac{du}{dx}\right]\frac{dv}{dx}=\frac{R}{u} \qquad ...(iv)$$

Let $\frac{dv}{dx}=z,$ so that $\frac{d^2v}{dx^2}=\frac{dz}{dx}$

Equation (*iv*) becomes

$$\boxed{\frac{dz}{dx}+\left[P+\frac{2}{u}\frac{du}{dx}\right]z=\frac{R}{u}}$$, which is linear in z.

Hence, z can be determined.

We obtain v, by integrating the relation $\frac{dv}{dx}=z$

$$v=\int z\,dx+C_1$$

Therefore, the solution of (*i*) is $y = u\left[\int z\,dx+C_1\right]$ *i.e.,* $y=uv$

Remarks: To Solve by the above method, u can be determined by inspection of the following rules:

(1) If $P+Qx=0,$ then $u=x$

(2) If $1+P+Q=0,$ then $u=e^x$

(3) If $1-P+Q=0,$ then $u=e^{-x}$

(4) If $1+\frac{P}{a}+\frac{Q}{a^2}=0,$ then $u=e^{ax}$

(5) If $2+2Px+Qx^2=0,$ then $u=x^2$

(6) If $m(m-1)+Pmx+Qx^2=0,$ then $u=x^m$.

Example 69: Solve $(x\sin x+\cos x)\frac{d^2y}{dx^2}-x\cos x\frac{dy}{dx}+y\cos x=0$ of which $y=x$ is a solution.

Solution: We have $(x\sin x+\cos x)\frac{d^2y}{dx^2}-x\cos x\frac{dy}{dx}+y\cos x=0$

or
$$\frac{d^2y}{dx^2}-\frac{x\cos x}{(x\sin x+\cos x)}\frac{dy}{dx}+\frac{\cos x}{(x\sin+\cos x)}y=0$$

Here $\qquad P(x)=-\frac{x\cos x}{x\sin x+\cos x}, \qquad Q(x)=\frac{\cos x}{(x\sin+\cos x)}, \qquad R=0$

Given that $y = x \Rightarrow u = x$

Let $y = uv \Rightarrow y = xv$ ...(1)

The reduced form as in the sub article 1.6.1

$$\frac{d^2v}{dx^2} + \left[P + \frac{2}{u}\frac{du}{dx}\right]\frac{dv}{dx} = 0 \quad | \text{ as } R = 0$$

$$\Rightarrow \quad \frac{d^2v}{dx^2} + \left[-\frac{x\cos x}{x\sin x + \cos x} + \frac{2}{x}\right]\frac{dv}{dx} = 0 \quad ...(2)$$

Let $z = \frac{dv}{dx}$, so that $\frac{dz}{dx} = \frac{d^2v}{dx^2}$

From equation (2), we get

$$\frac{dz}{dx} + \left[\frac{-x\cos x}{x\sin x + \cos x} + \frac{2}{x}\right] z = 0$$

$$\Rightarrow \quad \frac{dz}{dx} = -\left[\frac{2}{x} - \frac{x\cos x}{x\sin x + \cos x}\right] z$$

$$\Rightarrow \quad \frac{dz}{z} = \frac{x\cos x}{x\sin x + \cos x}\, dx - \frac{2}{x}\, dx$$

On integrating, $\int \frac{dz}{z} = \int \frac{x\cos x}{x\sin x + \cos x}\, dx - 2\int \frac{1}{x}\, dx + \log C$

$$\Rightarrow \quad \log_e z = \log_e(x\sin x + \cos x) - 2\log_e x + \log_e C \quad \left| \begin{array}{l} \text{Put } x\sin x + \cos x = t \\ \Rightarrow x\cos x\, dx = dt \end{array}\right.$$

$$\Rightarrow \quad \log_e z = \log_e C\left(\frac{x\sin x + \cos x}{x^2}\right)$$

$$\Rightarrow \quad z = C\left(\frac{\sin x}{x} + \frac{\cos x}{x^2}\right)$$

or

$$\frac{dv}{dx} = C\left(\frac{\sin x}{x} + \frac{\cos x}{x^2}\right)$$

$$\Rightarrow \quad \int dv = C\int\left(\frac{\sin x}{x} + \frac{\cos x}{x^2}\right) dx + C_1$$

$$\Rightarrow \quad v = C\left[\int \frac{\sin x}{x}\, dx - \frac{1}{x}\cos x - \int \frac{\sin}{x}\, dx\right] + C_1$$

$$\Rightarrow \qquad v = -\frac{C}{x}\cos x + C_1$$

Hence, the complete solution is $y = xv$

$$\Rightarrow \qquad y = -C\cos x + C_1 x. \quad \textbf{Ans.}$$

Example 70: Solve: $y'' - 4xy' + (4x^2 - 2)y = 0$ given that $y = e^{x^2}$ is an integral included in the complementary function. (*U.P.T.U. 2004*)

Solution: The given equation may be written as

$$\frac{d^2y}{dx^2} - 4x\frac{dy}{dx} + (4x^2 - 2)y = 0$$

Here, $P = -4x,\qquad Q = 4x^2 - 2,\qquad R = 0$

and $\qquad u = e^{x^2}$, so that $\dfrac{du}{dx} = 2xe^{x^2}$

Let $y = uv \qquad \Rightarrow \qquad y = e^{x^2}v$ ...(1)

We know that

$$\frac{d^2v}{dx^2} + \left[P + \frac{2}{u}\frac{du}{dx}\right]\frac{dv}{dx} = 0 \qquad \Big| \text{As } R = 0$$

$$\Rightarrow \qquad \frac{d^2v}{dx^2} + \left[\frac{2}{e^{x^2}} \times 2xe^{x^2} - 4x\right]\frac{dv}{dx} = 0$$

$$\Rightarrow \qquad \frac{d^2v}{dx^2} = 0 \qquad \Rightarrow \qquad \frac{dv}{dx} = C \qquad \Rightarrow \qquad v = Cx + C_1$$

Hence, the complete solution is $y = e^{x^2} \cdot v$

$$\Rightarrow \qquad y = Cx \cdot e^{x^2} + C_1 e^{x^2}. \quad \textbf{Ans.}$$

Example 71: Solve $x^2\dfrac{d^2y}{dx^2} - (x^2 + 2x)\dfrac{dy}{dx} + (x + 2)y = x^3e^x$ of which $y = x$ is a solution.

Solution: We have $\dfrac{d^2y}{dx^2} - \left(1 + \dfrac{2}{x}\right)\dfrac{dy}{dx} + \left(\dfrac{1}{x} + \dfrac{2}{x^2}\right)y = x \cdot e^x$

Here, $P = -\left(1 + \dfrac{2}{x}\right), Q = \dfrac{1}{x} + \dfrac{2}{x^2}$ and $R = x \cdot e^x$

Also given that $y = x \qquad \Rightarrow \qquad u = x$ so that $\dfrac{du}{dx} = 1$

Let $y = uv \quad \Rightarrow \quad y = xv$ ...(1)

We know that

$$\frac{d^2v}{dx^2}+\left[P+\frac{2}{u}\frac{du}{dx}\right]\frac{dv}{dx}=R/u$$

$$\Rightarrow \quad \frac{d^2v}{dx^2}+\left[-1-\frac{2}{x}+\frac{2}{x}\right]\frac{dv}{dx}=\frac{x\cdot e^x}{x}$$

$$\Rightarrow \quad \frac{d^2v}{dx^2}-\frac{dv}{dx}=e^x \quad \Rightarrow \quad (D^2-D)v=e^x; \quad \left|\, D=\frac{d}{dx}\right.$$

The A.E. is $m^2-m=0 \quad \Rightarrow \quad m(m-1)=0 \quad \Rightarrow \quad m=0,1$

$$\therefore \quad \text{C.F.} = C_1 + C_2e^x$$

and

$$\text{P.I.} = \frac{1}{(D^2-D)}e^x = \frac{x}{(2D-1)}e^x = \frac{xe^x}{2-1} = xe^x$$

Therefore $\quad v = C_1 + C_2e^x + xe^x$

Hence, the complete solution of given equation is

$$y = xv$$

$$\Rightarrow \quad y = x(C_1+C_2e^x+xe^x). \quad \textbf{Ans.}$$

Example 72: Solve the differental equation

$$(x+2)\frac{d^2y}{dx^2}-(2x+5)\frac{dy}{dx}+2y = (x+1)e^x.$$

Solution: We have $\dfrac{d^2y}{dx^2}-\dfrac{2x+5}{(x+2)}\dfrac{dy}{dx}+\dfrac{2}{(x+2)}y=\dfrac{(x+1)}{(x+2)}e^x$

Here $\quad P = -\dfrac{2x+5}{x+2}, \quad Q=\dfrac{2}{(x+2)} \quad$ and $\quad R=\dfrac{x+1}{(x+2)}e^x.$

We have $\quad 1+\dfrac{P}{a}+\dfrac{Q}{a^2}=0 \quad \Rightarrow \quad 1-\dfrac{2x+5}{a(x+2)}+\dfrac{2}{a^2(x+2)}=0$

$\Rightarrow \quad a^2(x+2)-(2x+5)a+2=0,$ which is quadratic equation in 'a'.

So,

$$a=\frac{(2x+5)\pm\sqrt{4x^2+20x+25-8x-16}}{2(x+2)}=\frac{(2x+5)\pm\sqrt{(2x+3)^2}}{2(x+2)}$$

$$\Rightarrow \quad a=\frac{(2x+5)\pm(2x+3)}{2(x+2)} \quad \Rightarrow \quad a=\frac{2x+5+2x+3}{2(x+2)} \quad \left|\; \text{Take +ve sign}\right.$$

$$\Rightarrow \qquad a = \frac{4x+8}{2(x+2)} = \frac{4(x+2)}{2(x+2)} = 2$$

$$\therefore \qquad u = e^{ax} = e^{2x} \quad \Rightarrow \quad \frac{du}{dx} = 2e^{2x}$$

Let $y = uv \quad \Rightarrow \quad y = e^{2x} v$ ...(1)

We know that

$$\frac{d^2v}{dx^2} + \left[P + \frac{2}{u} \cdot \frac{du}{dx} \right] \frac{dv}{dx} = R/u$$

$$\Rightarrow \qquad \frac{d^2v}{dx^2} + \left[-\frac{2x+5}{x+2} + \frac{2}{e^{2x}} .2e^{2x} \right] \frac{dv}{dx} = \left(\frac{x+1}{x+2} \right) \frac{e^x}{e^{2x}}$$

$$\Rightarrow \qquad \frac{d^2v}{dx^2} + \frac{2x+3}{(x+2)} \frac{dv}{dx} = \left(\frac{x+1}{x+2} \right) e^{-x} \qquad ...(2)$$

Let $\dfrac{dv}{dx} = z \quad \Rightarrow \quad \dfrac{d^2v}{dx^2} = \dfrac{dz}{dx}$

From equation (2), we get

$$\frac{dz}{dx} + \frac{2x+3}{(x+2)} .z = \left(\frac{x+1}{x+2} \right) e^{-x} \qquad ...(3)$$

which is a linear differential equation

$$\text{I.F.} = e^{\int \left(\frac{2x+3}{x+2} \right) dx} = e^{\int \left(2 - \frac{1}{x+2} \right) dx} = e^{2x - \log(x+2)} = \frac{e^{2x}}{(x+2)}$$

$\therefore$ The solution of equation (3) is

$$z \cdot \frac{e^{2x}}{x+2} = \int \frac{e^{2x}}{x+2} \cdot \frac{x+1}{x+2} e^{-x} dx + C$$

$$= \int \frac{e^x (x+1)}{(x+2)^2} dx + C$$

$$= \int e^x \left[\frac{1}{x+2} - \frac{1}{(x+2)^2} dx \right] + C$$

$$= \int \frac{e^x}{x+2} dx - \int \frac{e^x}{(x+2)^2} dx + C$$

$$\Rightarrow \qquad z\cdot\frac{e^{2x}}{x+2} = \frac{e^x}{(x+2)}+\int\frac{e^x}{(x+2)^2}dx-\int\frac{e^x}{(x+2)^2}dx+C=\frac{e^x}{(x+2)}+C$$

$$\Rightarrow \qquad z = e^{-x}+C(x+2)e^{-2x}$$

$$\Rightarrow \qquad \frac{dv}{dx} = e^{-x}+C(x+2)e^{-2x} \quad\Rightarrow\quad v=\int e^{-x}dx+C\int(x+2)e^{-2x}dx+C_1$$

$$\Rightarrow \qquad v = -e^{-x}+C\left[\frac{(x+2)e^{-2x}}{-2}-\frac{e^{-2x}}{4}\right]+C_1$$

$$\Rightarrow \qquad v = -e^{-x}-\frac{Ce^{-2x}}{4}(2x+5)+C_1$$

Hence, the solution of given equation is $y = e^{2x}v$

$$\Rightarrow \qquad y = e^{2x}\left[-e^{-x}-\frac{Ce^{-2x}}{4}(2x+5)+C_1\right] \quad\Rightarrow\quad y=-e^x-\frac{C}{4}(2x+5)+C_1e^{2x}. \qquad \textbf{Ans.}$$

Example 73: Solve $\frac{d^2y}{dx^2}-ax\frac{dy}{dx}+a^2(x-1)y=0.$

Solution: Here $P=-ax, Q=a^2(x-1), R=0$

$$1+\frac{P}{a}+\frac{Q}{a^2}=1-\frac{ax}{a}+\frac{a^2(x-1)}{a^2}=1-x+x-1=0$$

$$\therefore \qquad u=e^{ax} \quad\Rightarrow\quad \frac{du}{dx}=ae^{ax}$$

Let $\quad y=uv \quad\Rightarrow\quad y=e^{ax}\cdot v$

We know that $\quad \frac{d^2v}{dx^2}+\left[P+\frac{2}{u}\frac{du}{dx}\right]\frac{dv}{dx} = R/u$

$$\Rightarrow \qquad \frac{d^2v}{dx^2}+\left[-ax+\frac{2}{e^{ax}}ae^{ax}\right]\frac{dv}{dx}=0 \qquad \text{As } R=0$$

$$\Rightarrow \qquad \frac{d^2v}{dx^2}+[-ax+2a]\frac{dv}{dx}=0$$

Let $\quad \frac{dv}{dx}=z,$ so that $\quad \frac{d^2v}{dx^2}=\frac{dz}{dx}$

$$\Rightarrow \quad \frac{dz}{dx} + (-ax + 2a)z = 0 \quad \Rightarrow \quad \frac{dz}{z} = (ax - 2a)dx$$

On integrating $\quad \log_e z = \frac{ax^2}{2} - 2ax + \log C$

$$\Rightarrow \quad \log_e \frac{z}{C} = \frac{ax^2}{2} - 2ax \quad \Rightarrow \quad \frac{z}{C} = e^{\frac{ax^2}{2}} . e^{-2ax}$$

$$\Rightarrow \quad z = Ce^{\frac{ax^2}{2}} . e^{-2ax}$$

$$\Rightarrow \quad \frac{dv}{dx} = Ce^{\frac{ax^2}{2}} \cdot e^{-2ax} \quad \Rightarrow \quad dv = Ce^{\frac{ax^2}{2} - 2ax} dx$$

On integrating, we get

$$v = C\int e^{ax\left(\frac{x}{2} - 2\right)} \cdot dx + C_1$$

Hence, the complete solution is $y = e^{ax}. v$

$$\Rightarrow \quad y = Ce^{ax}\int e^{ax\left(\frac{x}{2} - 2\right)} \cdot dx + C_1 e^{ax}. \quad \textbf{Ans.}$$

EXERCISE 1.12

Solve the following differential equations:

1. $\sin^2 x \dfrac{d^2y}{dx^2} = 2y$; given $y = \cot x$ is a solution $\quad$ [**Ans.** $y = C_1 - C_1 x \cot x + C_2 \cot x$]

2. $\dfrac{d^2y}{dx^2} - \cot x \dfrac{dy}{dx} - (1 - \cot x)y = e^x \sin x$

$\left[\textbf{Ans. } y = -\dfrac{1}{2}e^x \cos x - \dfrac{1}{5}C_1 e^{-x}(2\sin x + \cos x) + C_2 e^x\right]$

3. $x\dfrac{d^2y}{dx^2} - (2x-1)\dfrac{dy}{dx} + (x-1)y = 0$ $\quad$ $\left[\textbf{Ans. } y = e^x(C_1 \log x + C_2)\right]$

4. $x\dfrac{d^2y}{dx^2} + (1+x)\dfrac{dy}{dx} + y = e^x$ $\quad$ $\left[\textbf{Ans. } y = e^{-x}\log x + C_1 e^{-x} + C_2 e^{-x}\int \dfrac{e^{-x}}{x}dx\right]$

5. $(1-x^2)\dfrac{d^2y}{dx^2}+x\dfrac{dy}{dx}-y=x(1-x^2)^{3/2}$

$$\left[\textbf{Ans. } y=-x\left[\frac{1}{9}(1-x^2)^{3/2}+\frac{C(1-x^2)^{1/2}}{x}+C\sin^{-1}x\right]\right]$$

6. $\dfrac{d^2y}{dx^2}+\left(1+\dfrac{2}{x}\cot x-\dfrac{2}{x^2}\right)y=x\cos x$, given $y=\dfrac{\sin x}{x}$ is a solution

$$\left[\textbf{Ans. } y=\frac{\sin x}{x}\left[\frac{x^3}{6}+\left(C_1-\frac{1}{4}\right)\{-x^2\cot x+2x\log\sin x-2\int\log\sin x\,dx\}+C_2\right]\right]$$

7. $(1-x^2)\dfrac{d^2y}{dx^2}-x\dfrac{dy}{dx}-a^2y=0$, given $y=ce^{a\sin^{-1}x}$ is a solution

$$\left[\textbf{Ans. } y=\left(-\frac{C_1}{2a}e^{-2a\sin^{-1}x}+C_2\right)Ce^{a\sin^{-1}x}\right]$$

8. $\dfrac{d^2y}{dx^2}-x^2\dfrac{dy}{dx}+xy=x$, given $y=x$ is a solution $\quad\left[\textbf{Ans. } y=1+C_1x\int\dfrac{1}{x^2}e^{x^3/3}dx+C_2x\right]$

9. $x^2\dfrac{d^2y}{dx^2}-(x^2+2x)\dfrac{dy}{dx}+(x+2)y=x^3e^x$, given $y=x$ is a solution

$$\left[\textbf{Ans. } y=(x^2-x+Cx)e^x+C_1x\right]$$

10. $x^2y''-2x(1+x)y'+2(1+x)y=x^3$ $\quad\left[\textbf{Ans. } y=x\left(-\dfrac{x}{2}+\dfrac{C_1}{2}e^{2x}+C_2\right)\right]$

1.6.2 Method II: Removal of the First Order Derivative (Normal Form)

Let $\quad\dfrac{d^2y}{dx^2}+P\dfrac{dy}{dx}+Qy=R \qquad ...(i)$

Putting $y=uv$, we get

$$v\left[\frac{d^2u}{dx^2}+P\frac{du}{dx}+Qu\right]+u\left[\frac{d^2v}{dx^2}+P\frac{dv}{dx}\right]+2\frac{du}{dx}\frac{dv}{dx}=R \qquad \left[\begin{array}{c}\text{Art. 1.6.1}\\ \text{(equation } iii)\end{array}\right]$$

$$\Rightarrow \quad \frac{d^2v}{dx^2}+\left(\frac{2}{u}\frac{du}{dx}+P\right)\frac{dv}{dx}+v\left(\frac{1}{u}\frac{d^2u}{dx^2}+\frac{P}{u}\frac{du}{dx}+Q\right)=\frac{R}{u} \qquad ...(ii)$$

But the first order derivative must be remove

So $\quad\dfrac{2}{u}\dfrac{du}{dx}+P=0 \quad\Rightarrow\quad \dfrac{du}{u}=-\dfrac{1}{2}Pdx$

$$\Rightarrow \qquad \log u = -\int \frac{P}{2}\,dx \quad \Rightarrow \quad u = e^{-\int \frac{P}{2}dx}$$

Since
$$\frac{du}{dx} = -\frac{Pu}{2} \quad \Rightarrow \quad \frac{d^2u}{dx^2} = -\frac{1}{2}\left[P\frac{du}{dx} + u\frac{dP}{dx}\right]$$

$$\Rightarrow \qquad \frac{d^2u}{dx^2} = -\frac{1}{2}\left[P\left(\frac{-Pu}{2}\right) + u\frac{dP}{dx}\right] = \frac{P^2u}{4} - \frac{u}{2}\frac{dp}{dx}$$

From equation (*ii*)

$$\frac{d^2v}{dx^2} + v\left(\frac{P^2}{4} - \frac{1}{2}\frac{dP}{dx} - \frac{P^2}{2} + Q\right) = \frac{R}{u}$$

$$\Rightarrow \qquad \frac{d^2v}{dx^2} + v\left(Q - \frac{1}{2}\frac{dP}{dx} - \frac{P^2}{4}\right) = \frac{R}{u}$$

$$\Rightarrow \qquad \boxed{\frac{d^2v}{dx^2} + Q_1 v = R_1}$$

This equation is called **normal form** of equation (*i*).

where
$$Q_1 = Q - \frac{1}{2}\frac{dP}{dx} - \frac{P^2}{4} \quad \text{and} \quad R_1 = \frac{R}{u}.$$

Example 74: Solve $\frac{d^2y}{dx^2} - 2\tan x\frac{dy}{dx} + 5y = e^x.\sec x$

Solution: Here $\qquad P = -2\tan x, \quad Q = 5, \quad R = e^x.\sec x$

$$u = e^{-\frac{1}{2}\int P dx} = e^{+\frac{2\times 1}{2}\int \tan x\, dx} = e^{\log\sec x} = \sec x$$

$$Q_1 = Q - \frac{1}{2}\frac{dP}{dx} - \frac{P^2}{4} = 5 - \frac{1}{2}(-2\sec^2 x) - \frac{4\tan^2 x}{4}$$

$$\Rightarrow \qquad Q_1 = 5 + \sec^2 x - (\sec^2 x - 1) = 6$$

and
$$R_1 = \frac{R}{u} = \frac{e^x \sec x}{\sec x} = e^x$$

The normal from is $\frac{d^2v}{dx^2} + Q_1 v = R_1$

$$\Rightarrow \quad \frac{d^2v}{dx^2}+6v = e^x$$

$$\Rightarrow \quad (D^2+6)v = 0 \quad \Rightarrow \quad m^2+6=0$$

$$\Rightarrow \quad m = \pm\sqrt{6}\, i$$

$$\text{C.F.} = C_1\cos\sqrt{6}x + C_2\sin\sqrt{6}x$$

$$\text{P.I.} = \frac{1}{(D^2+6)}e^x = \frac{1}{7}e^x$$

$$v = C_1\cos\sqrt{6}x + C_2\sin\sqrt{6}x + \frac{1}{7}e^x$$

Hence the solution of given differential equation is given by

$$y = uv = \sec x\left(C_1\cos\sqrt{6}x + C_2\sin\sqrt{6}x + \frac{1}{7}e^x\right) \quad \textbf{Ans.}$$

Example 75: Solve $\frac{d^2y}{dx^2} - 4x\frac{dy}{dx} + (4x^2-1)y = -3e^{x^2}\sin 2x.$ [*U.P.T.U. (C.O.) 2004*]

Solution: Given that

$$P = -4x, \quad Q = (4x^2-1),\ R = -3e^{x^2}\sin 2x$$

Now $$u = e^{-\frac{1}{2}\int P dx} = e^{-\frac{1}{2}\int -4x dx} = e^{x^2}$$

$$Q_1 = Q - \frac{1}{2}\frac{dP}{dx} - \frac{P^2}{4} = 4x^2 - 1 + 2 - \frac{16x^2}{4} = 1$$

$$R_1 = \frac{R}{u} = \frac{-3e^{x^2}\sin 2x}{e^{x^2}} = -3\sin 2x$$

The normal equation is $\frac{d^2v}{dx^2} + Q_1 v = R_1$

$$\Rightarrow \quad \frac{d^2v}{dx^2} + v = -3\sin 2x$$

$$\Rightarrow \quad (D^2+1)v = -3\sin 2x$$

The A.E. is $m^2+1 = 0 \quad \Rightarrow \quad m = \pm i$

$\therefore$ $$\text{C.F.} = C_1\cos x + C_2\sin x$$

and $$\text{P.I.} = \frac{1}{(D^2+1)}(-3\sin 2x) = \frac{-3}{-4+1}\sin 2x = \sin 2x$$

$\Rightarrow$ $$v = C_1\cos x + C_2\sin x + \sin 2x$$

Hence, the solution of given equation is

$$y = uv = e^{x^2}(C_1\cos x + C_2\sin x + \sin 2x). \quad \textbf{Ans.}$$

EXERCISE 1.13

Solve the following differential equations:

1. $x^2\frac{d^2y}{dx^2} - 2(x^2+x)\frac{dy}{dx} + (x^2+2x+2)y = 0$ $\quad\left[\textbf{Ans.}\ y = (C_1x + C_2)xe^x\right]$

2. $\frac{d^2y}{dx^2} - 4x\frac{dy}{dx} + (4x^2-3)y = e^{x^2}$ $\quad\left[\textbf{Ans.}\ y = e^{x^2}(C_1e^x + C_2e^{-x} - 1)\right]$

3. $x\frac{d}{dx}\left(x\frac{dy}{dx} - y\right) - 2x\frac{dy}{dx} + 2y + x^2y = 0$ $\quad\left[\textbf{Ans.}\ y = x(C_1\cos x + C_2\sin x)\right]$

4. $\frac{d^2y}{dx^2} + \frac{2}{x}\frac{dy}{dx} = n^2y$ $\quad\left[\textbf{Ans.}\ y = \frac{1}{x}(C_1e^{nx} + C_2e^{-nx})\right]$

5. $\frac{d^2y}{dx^2} + \frac{1}{x^{1/3}}\frac{dy}{dx} + \left(\frac{1}{4x^{2/3}} - \frac{1}{6x^{4/3}} - \frac{6}{x^2}\right)y = 0$ $\quad\left[\textbf{Ans.}\ y = e^{-\frac{3}{4}x^{2/3}}\left(C_1x^2 + \frac{C_2}{x^2}\right)\right]$

6. $\frac{d^2y}{dx^2} - 2x\frac{dy}{dx} + (x^2+2)y = e^{\frac{1}{2}(x^2+2x)}$ $\quad\left[\textbf{Ans.}\ y = \left(C_1\cos\sqrt{3}x + C_2\sin\sqrt{3x}\right)e^{\frac{x^2}{2}} + \frac{1}{4}e^xe^{\frac{x^2}{2}}\right]$

7. $\frac{d}{dx}\left(\cos^2 x\frac{dy}{dx}\right) + y\cos^2 x = 0$ $\quad\left[\textbf{Ans.}\ y = \sec x\left\{C_1\cos x\sqrt{2} + C_2\sin x\sqrt{2}\right\}\right]$

8. $\frac{d^2y}{dx^2} - \frac{1}{\sqrt{x}}\frac{dy}{dx} + \frac{y}{4x^2}\left(x + \sqrt{x} - 8\right) = 0$ $\quad\left[\textbf{Ans.}\ y = e^{\sqrt{x}}\left(C_1x^2 + C_2x^{-1}\right)\right]$

1.6.3 Method III: Change of Independent Variable

$$\frac{d^2y}{dx^2} + P\frac{dy}{dx} + Qy = R \qquad ...(i)$$

Let $$z = f(x)$$

$$\therefore \quad \frac{dy}{dx} = \frac{dy}{dz}\cdot\frac{dz}{dx} \text{ and } \frac{d^2y}{dx^2} = \frac{d^2y}{dz^2}\left(\frac{dz}{dx}\right)^2 + \frac{dy}{dz}\frac{d^2z}{dx^2}$$

From equation (*i*), we get

$$\frac{d^2y}{dz^2}\left(\frac{dz}{dx}\right)^2 + \frac{dy}{dz}\frac{d^2z}{dx^2} + P\frac{dy}{dz}\cdot\frac{dz}{dx} + Qy = R$$

$$\Rightarrow \quad \frac{d^2y}{dz^2}\left(\frac{dz}{dx}\right)^2 + \left(P\frac{dz}{dx} + \frac{d^2z}{dx^2}\right)\frac{dy}{dz} + Qy = R$$

$$\Rightarrow \quad \frac{d^2y}{dz^2} + \frac{\left(P\frac{dz}{dx} + \frac{d^2z}{dx^2}\right)}{\left(\frac{dz}{dx}\right)^2}\cdot\frac{dy}{dz} + \frac{Q}{\left(\frac{dz}{dx}\right)^2}y = \frac{R}{\left(\frac{dz}{dx}\right)^2}$$

$$\Rightarrow \quad \boxed{\frac{d^2y}{dz^2} + P_1\frac{dy}{dz} + Q_1 y = R_1}$$

where $$P_1 = \frac{\left(P\frac{dz}{dx} + \frac{d^2z}{dx^2}\right)}{\left(\frac{dz}{dx}\right)^2},\ Q_1 = \frac{Q}{\left(\frac{dz}{dx}\right)^2} \text{ and } R_1 = \frac{R}{\left(\frac{dz}{dx}\right)^2}.$$

Note: Choose z such that $\left(\frac{dz}{dx}\right)^2 = Q$ [Take only whole square term of Q and neglect negative sign].

Example 76: $\frac{d^2y}{dx^2} + (3\sin x - \cot x)\frac{dy}{dx} + 2y\sin^2 x = e^{-\cos x}\cdot\sin^2 x.$

Solution: Here $P = 3\sin x - \cot x, Q = 2\sin^2 x, R = e^{-\cos x}\cdot\sin^2 x$

Let $$\left(\frac{dz}{dx}\right)^2 = \sin^2 x \quad \Rightarrow \quad \frac{dz}{dx} = \sin x \quad \Rightarrow \quad z = -\cos x$$

Now the transformed equation is

$$\frac{d^2y}{dz^2} + P_1\frac{dy}{dz} + Q_1 y = R_1 \qquad \text{...(1)}$$

where $$P_1 = \frac{Pdz/dx + \dfrac{d^2z}{dx^2}}{\left(\dfrac{dz}{dx}\right)^2} = \frac{(3\sin x - \cot x)\sin x + \cos x}{\sin^2 x} = 3$$

$$Q_1 = Q\bigg/\left(\frac{dz}{dx}\right)^2 = \frac{2\sin^2 x}{\sin^2 x} = 2 \text{ and } R_1 = \frac{e^{-\cos x}\sin^2 x}{\sin^2 x} = e^{-\cos x}$$

Using these values in equation (1), we get

$$\frac{d^2y}{dz^2} + 3\frac{dy}{dz} + 2y = e^{-\cos x} = e^z \qquad \Big| As\ z = -\cos x$$

$$\Rightarrow \qquad (D^2 + 3D + 2)y = e^z$$

The A.E. is $m^2 + 3m + 2 = 0 \quad \Rightarrow \quad m = -1, -2$

$$\therefore \qquad \text{C.F.} = C_1e^{-z} + C_2e^{-2z} = C_1e^{\cos x} + C_2e^{2\cos x}$$

and $$\text{P.I.} = \frac{1}{(D^2 + 3D + 2)}e^z = \frac{1}{6}e^z = \frac{1}{6}e^{-\cos x}$$

Hence the required solution is

$$y = C_1e^{\cos x} + C_2e^{2\cos x} + \frac{1}{6}e^{-\cos x}. \quad \textbf{Ans.}$$

Example 77: Solve $\dfrac{d^2y}{dx^2} - \dfrac{1}{x}\dfrac{dy}{dx} + 4x^2y = x^4$. *(U.P.T.U. 2003, 2002)*

Solution: Here $P = -\dfrac{1}{x}$, $\quad Q = 4x^2 \quad$ and $\quad R = x^4$

Let $$\left(\frac{dz}{dx}\right)^2 = 4x^2 \quad \Rightarrow \quad \frac{dz}{dx} = 2x \quad \Rightarrow \quad z = x^2 + C$$

The transformed equation is

$$\frac{d^2y}{dz^2} + P_1\frac{dy}{dz} + Q_1y = R_1 \qquad \text{...(1)}$$

$$P_1 = \frac{P\left(\dfrac{dz}{dx}\right) + \dfrac{d^2z}{dx^2}}{\left(\dfrac{dz}{dx}\right)^2} = \frac{-\dfrac{1}{x}\dfrac{dz}{dx} + \dfrac{d^2z}{dx^2}}{\left(\dfrac{dz}{dx}\right)^2} = \frac{-\dfrac{1}{x}(2x) + 2}{(2x)^2} = 0$$

$$Q_1 = \frac{Q}{\left(\frac{dz}{dx}\right)^2} = \frac{4x^2}{4x^2} = 1, \quad R_1 = \frac{R}{\left(\frac{dz}{dx}\right)^2} = \frac{x^4}{4x^2} = \frac{x^2}{4} = \frac{z-C}{4}$$

From equation (1), we get

$$\frac{d^2y}{dz^2} + y = \frac{z-C}{4} \quad \Rightarrow \quad (D^2+1)y = \frac{z-C}{4}$$

The A.E. is $\quad m^2 + 1 = 0 \quad \Rightarrow \quad m = \pm i$

$\therefore \quad$ C.F. $= C_1 \cos z + C_2 \sin z$

C.F. $= C_1 \cos(x^2 + C) + C_2 \sin(x^2 + C)$

$$\text{P.I.} = \frac{1}{(D^2+1)}\left(\frac{z-C}{4}\right) = (1+D^2)^{-1}\left(\frac{z-C}{4}\right) = (1-D^2)\left(\frac{z-C}{4}\right)$$

$$= \frac{z-C}{4} = \frac{x^2}{4}$$

Hence the required solution is

$$y = C_1 \cos(x^2 + C) + C_2 \sin(x^2 + C) + \frac{x^2}{4}. \quad \textbf{Ans.}$$

Example 78: Solve $(1+x^2)^2 \frac{d^2y}{dx^2} + 2x(1+x^2)\frac{dy}{dx} + 4y = 0.$

Solution: We have $\frac{d^2y}{dx^2} + \frac{2x}{(1+x^2)}\frac{dy}{dx} + \frac{4}{(1+x^2)^2}y = 0$

Choosen z, such that $\left(\frac{dz}{dx}\right)^2 = \frac{4}{(1+x^2)^2} \quad \Rightarrow \quad \frac{dz}{dx} = \frac{2}{(1+x^2)}$

$\Rightarrow \quad z = 2\tan^{-1} x \quad \Rightarrow \quad \frac{z}{2} = \tan^{-1} x \quad \Rightarrow \quad x = \tan\frac{z}{2}$

The transformed equation is

$$\frac{d^2y}{dz^2} + P_1 \frac{dy}{dz} + Q_1 y = R_1 \qquad \text{...(1)}$$

Now, $\quad P_1 = \frac{(P\,dz/dx + d^2z/dx^2)}{\left(\frac{dz}{dx}\right)^2} = \frac{\left(\frac{2x}{1+x^2}\cdot\frac{2}{1+x^2} - \frac{4x}{(1+x^2)^2}\right)}{\frac{4}{(1+x^2)^2}} = 0$

$$Q_1 = \frac{Q}{\left(\frac{dz}{dx}\right)^2} = \frac{\frac{4}{(1+x^2)^2}}{\frac{4}{(1+x^2)^2}} = 1, \quad R_1 = \frac{0}{\left(\frac{dz}{dx}\right)^2} = 0$$

From equation (1), we get

$$\frac{d^2y}{dz^2} + y = 0 \quad \Rightarrow \quad (D^2+1)y = 0$$

The A.E. is $m^2 + 1 = 0 \quad \Rightarrow \quad m = \pm i$

$\therefore$ C.F. $= C_1 \cos z + C_1 \sin z$

$\Rightarrow$ C.F. $= C_1 \cos(2\tan^{-1} x) + C_2 \sin(2\tan^{-1} x)$

P.I. = 0.

Hence the required solution is

$$y = C_1 \cos(2\tan^{-1} x) + C_2 \sin(2\tan^{-1} x).$$

Example 79: Solve the following differential equation by changing the independent variable

$$x\frac{d^2y}{dx^2} + (4x^2-1)\frac{dy}{dx} + 4x^3y = 2x^3. \qquad \textit{(U.P.T.U. 2006)}$$

Solution: The given equation may be written as

$$\frac{d^2y}{dx^2} + \left(\frac{4x^2-1}{x}\right)\frac{dy}{dx} + 4x^2y = 2x^2$$

Here $P = \frac{4x^2-1}{x}, \; Q = 4x^2, \; R = 2x^2$

Choosen z such that $\left(\frac{dz}{dx}\right)^2 = 4x^2 \quad \Rightarrow \quad \frac{dz}{dx} = 2x \quad \Rightarrow \quad z = x^2$

The transformed equation is

$$\frac{d^2y}{dx^2} + P_1\frac{dy}{dz} + Q_1 y = R_1 \qquad ...(1)$$

Now $$P_1 = \frac{P\left(\frac{dz}{dx}\right) + \frac{d^2z}{dx^2}}{\left(\frac{dz}{dx}\right)^2} = \frac{\left(\frac{4x^2-1}{x}\right)(2x) + 2}{(2x)^2} = 2$$

$$Q_1 = \frac{4x^2}{4x^2} = 1 \text{ and } R_1 = \frac{2x^2}{4x^2} = \frac{1}{2}$$

Using these values in equation (1), we get

$$\frac{d^2y}{dz^2} + 2\frac{dy}{dz} + y = \frac{1}{2} \quad \Rightarrow \quad (D^2 + 2D + 1)y = \frac{1}{2}$$

The A.E. is $\quad m^2 + 2m + 1 = 0 \Rightarrow (m+1)^2 = 0 \Rightarrow m = -1, -1$

$$\text{C.F.} = (C_1 + C_2 z)e^{-z} \quad \Rightarrow \quad \text{C.F.} = (C_1 + C_2 x^2)e^{-x^2}$$

$$\text{P.I.} = \frac{1}{(D^2 + 2D + 1)}\left(\frac{1}{2}\right) = \frac{1}{0+0+1}\left(\frac{1}{2}\right) = \frac{1}{2}$$

Hence the required solution is

$$y = (C_1 + C_2 x^2)e^{-x^2} + \frac{1}{2}. \quad \textbf{Ans.}$$

Example 80: $x\dfrac{d^2y}{dx^2} - \dfrac{dy}{dx} + 4x^3 y = 8x^3 \sin x^2.$

Solution: The given equation can be written as

$$\frac{d^2y}{dx^2} - \frac{1}{x}\frac{dy}{dx} - 4x^2 y = 8x^2 \sin x^2$$

Here $\qquad P = -\dfrac{1}{x}, \quad Q = 4x^2, \quad R = 8x^2 \sin x^2$

Choosen z such that $\left(\dfrac{dz}{dx}\right)^2 = Q = 4x^2$

$$\Rightarrow \qquad \frac{dz}{dx} = 2x \quad \Rightarrow \quad z = x^2$$

The transformed equation is $\dfrac{d^2y}{dz^2} + P_1\dfrac{dy}{dz} + Q_1 y = R_1$ ...(1)

Now
$$P_1 = \frac{P\dfrac{dz}{dx} + \dfrac{d^2z}{dx^2}}{\left(\dfrac{dz}{dx}\right)^2} = \frac{-\dfrac{1}{x} \times 2x + 2}{4x^2} = 0$$

$$Q_1 = \frac{Q}{\left(\dfrac{dz}{dx}\right)^2} = \frac{4x^2}{4x^2} = 1 \text{ and } R_1 = \frac{R}{\left(\dfrac{dz}{dx}\right)^2} = \frac{8x^2 \sin x^2}{4x^2} = 2\sin x^2$$

From equation (1)

$$\frac{d^2y}{dz^2}+y = 2\sin z \qquad (\text{As } z = x^2)$$

$$\Rightarrow \qquad (D^2+1)y = 2\sin z$$

The A.E. is $m^2+1=0 \quad \Rightarrow \quad m=\pm i$

$$\therefore \qquad \text{C.F.} = (C_1\cos z + C_2\sin z) = C_1\cos x^2 + C_2\sin x^2$$

and $$\text{P.I.} = \frac{1}{(D^2+1)}2\sin z = \frac{2}{-1+1}\sin z \qquad \text{(Case fail)}$$

$$\Rightarrow \qquad \text{P.I.} = \frac{z}{2D}2\sin z = -z\cos z = -x^2\cos x^2$$

Hence the required solution is

$$y = C_1\cos x^2 + C_2\sin x^2 - x^2\cos x^2. \quad \textbf{Ans.}$$

Example 81: Solve $\dfrac{d^2y}{dx^2}+(\tan x-1)^2\dfrac{dy}{dx}-n(n-1)y\sec^4 x=0.$

Solution: Here $P=(\tan x-1)^2, Q=-n(n-1)\sec^4 x$ and $R=0$

Choosen z such that $\left(\dfrac{dz}{dx}\right)^2 = \sec^4 x \quad \Rightarrow \dfrac{dz}{dx}=\sec^2 x \quad \Rightarrow \quad z=\tan x$

The transformed equation is $\dfrac{d^2y}{dz^2}+P_1\dfrac{dy}{dz}+Q_1y=R_1$ ...(1)

where $$P_1 = \frac{P\dfrac{dz}{dx}+\dfrac{d^2z}{dx^2}}{\left(\dfrac{dz}{dx}\right)^2} = \frac{(\tan x-1)^2\sec^2 x + 2\sec^2 x\tan x}{\sec^4 x}$$

$$\Rightarrow \qquad P_1 = \frac{\sec^2 x(1+\tan^2 x)}{\sec^4 x} = 1$$

$$Q_1 = \frac{Q}{\left(\dfrac{dz}{dx}\right)^2} = \frac{-n(n-1)\sec^4 x}{\sec^4 x} = -n(n-1)$$

and $$R_1 = 0$$

From equation (1), we get

$$\frac{d^2y}{dz^2}+\frac{dy}{dz}+\{-n(n-1)\}y = 0$$

$$\Rightarrow \quad \frac{d^2y}{dz^2}+\frac{dy}{dz}-n(n-1)y = 0$$

$$\Rightarrow \quad \{D^2+D-n(n-1)\}y = 0$$

The A.E. is

$$m^2+m-n(n-1) = 0$$

$$m^2+m-n^2+n = 0$$

$$\Rightarrow \quad (m+n)(m-n+1) = 0$$

$$\Rightarrow \quad m = -n, (n-1)$$

$$\therefore \quad y = C_1e^{-nz}+C_2e^{-(n-1)z}$$

Hence the required soution is

$$y = C_1e^{-n\tan x}+C_2e^{-(n-1)\tan x}. \quad \textbf{Ans.}$$

EXERCISE 1.14

Solve the following differential equations:

1. $x^4\dfrac{d^2y}{dx^2}+2x^3\dfrac{dy}{dx}+n^2y=0$ $\left[\textbf{Ans. } y=C_1\cos\left(\dfrac{n}{x}+C_2\right)\right]$

2. $\dfrac{d^2y}{dx^2}+\left(1-\dfrac{1}{x}\right)\dfrac{dy}{dx}+4x^2e^{-2x}\,y=4(x^2+x^3)e^{-3x}$

$\left[\textbf{Ans. } y=C_1\cos\{2(x+1)e^{-x}\}-C_2\sin\{2(x+1)e^{-x}\}+(x+1)e^{-x}\right]$

3. $\dfrac{d^2y}{dx^2}+(\tan x-3\cos x)\dfrac{dy}{dx}+(2\cos^2 x)y=\cos^4 x$

$\left[\textbf{Ans. } y=C_1e^{\sin x}+C_2e^{2\sin x}-\dfrac{1}{2}\sin^2 x-\dfrac{3}{2}\sin x-\dfrac{5}{4}\right]$

4. $x^6\dfrac{d^2y}{dx^2}+3x^5\dfrac{dy}{dx}+a^2y=\dfrac{1}{x^2}$ $\left[\textbf{Ans. } y=C_1\cos\left(C_2-\dfrac{a}{2x^2}\right)+\dfrac{1}{a^2x^2}\right]$

5. $(1+x)^2\dfrac{d^2y}{dx^2}+(1+x)\dfrac{dy}{dx}+y=4\cos\log(1+x)$

$\left[\textbf{Ans. } y=C_1\cos\{\log(1+x)+C_2\}+2\log(1+x)\sin\{\log(1+x)\}\right]$

6. Change the independent variable x to z by the relation $z = \sin x$ in the following differential equation and solve it.

$$\cos x\frac{d^2y}{dx^2}+\sin x\frac{dy}{dx}-2y\cos^3 x=2\cos^5 x \qquad \left[\textbf{Ans. } y=C_1e^{\sqrt{2}\sin x}+C_2e^{-\sqrt{2}\sin x}+\sin^2 x\right]$$

7. Solve $\dfrac{d^2y}{dx^2}-(8e^{2x}+2)\dfrac{dy}{dx}+4e^{4x}\cdot y=e^{6x}$ $\left[\textbf{Ans. } y=C_1e^{(2+\sqrt{3})e^{2x}}+C_2e^{(2-\sqrt{3})e^{2x}}+\frac{1}{4}e^{2x}+1\right]$

1.6.4 Method IV: Method of Variation of Parameters

Consider the differential equation

$$\frac{d^2y}{dx^2}+P\frac{dy}{dx}+Qy = R \qquad ...(i)$$

Let the complementary function of equation (*i*) is

$$y= C_1u+C_2v \qquad ...(ii)$$

where u and v are the parts of C.F.

Therefore, we have

$$\frac{d^2u}{dx^2}+P\frac{du}{dx}+Qu = 0 \qquad ...(iii)$$

and

$$\frac{d^2v}{dx^2}+P\frac{dv}{dx}+Qv = 0 \qquad ...(iv)$$

Let us assume that particular solution (P.I.) of equation (*i*) is

$$y= Au+Bv \qquad ...(v)$$

where A and B are unknown function of x

Differentiate equation (*v*) w.r.t. 'x' we have

$$y' = Au'+Bv'+A'u+B'v \qquad ...(vi)$$

Let A and B be choosen such that they satisfy the condition

$$A'u+B'v = 0 \qquad ...(vii)$$

Using (*vii*), (*vi*) becomes $y'= Au'+Bv'$ ...(*viii*)

Differentiating (*viii*) w.r.t. 'x' we have

$$y'' = Au''+Bv''+A'u'+B'v' \qquad ...(ix)$$

Substituting the values of y, y' and y'' in (*i*), we get

$$Au''+Bv''+A'u'+B'v'+P(Au'+Bv')+Q(Au+Bv) = R$$

$$\Rightarrow A(u''+Pu'+Qu)+B(v''+Pv'+Qv)+A'u'+B'v' = R$$

$$\Rightarrow A\left(\frac{d^2u}{dx^2}+P\frac{du}{dx}+Qu\right)+B\left(\frac{d^2v}{dx^2}+P\frac{dv}{dx}+Qv\right)+A'u'+B'v' = R$$

Using (*iii*) and (*iv*), we get

$$A'u' + B'v' = R \qquad ...(x)$$

From (*vii*) and (*x*), we get

$$A'u + B'v = 0$$

and

$$A'u' + B'v' - R = 0$$

$$\Rightarrow \quad \frac{A'}{-Rv} = -\frac{B'}{-Ru} = \frac{1}{uv' - u'v}$$

$$\Rightarrow \quad A' = \frac{-Rv}{uv' - u'v}, B' = \frac{Ru}{uv' - u'v}$$

$$\Rightarrow \quad \boxed{A = \int \frac{-Rv}{uv' - u'v} dx \quad and \quad B = \int \frac{Ru}{uv' - u'v} dx}$$

Hence, the complete solution is

$$\boxed{y = \text{C.F.} + \text{P.I.} = (C_1 u + C_2 v) + (Au + Bv)}$$

where A and B are given above.

Example 82: Use variation of parameters method to solve

$$y'' + y = \sec x. \qquad (U.P.T.U.\ 2002,\ A.M.I.E.\ 2004)$$

Solution: We have

$$\frac{d^2y}{dx^2} + y = \sec x \quad \Rightarrow \quad (D^2 + 1)y = \sec x$$

The A.E. is $\quad m^2 + 1 = 0 \quad \Rightarrow \quad m^2 = i^2 \Rightarrow m = \pm i$

$$\text{C.F.} = C_1 \cos x + C_2 \sin x$$

Here $\quad u = \cos x, \quad v = \sin x$

$\Rightarrow \quad u' = -\sin x, \quad v' = \cos x$

$$\text{P.I.} = Au + Bv \qquad ...(1)$$

Where

$$A = \int \frac{-Rv}{uv' - u'v} dx, \quad B = \int \frac{Ru}{uv' - u'v} dx$$

$$\Rightarrow \quad A = \int \frac{-\sec x \sin x}{\cos^2 x + \sin^2 x} dx = -\int \tan x dx = \log \cos x$$

$$B = \int \frac{Ru}{uv' - u'v} dx = \int \frac{\sec x . \cos x}{1} dx = \int dx = x$$

Putting the values of A and B in equation (1), we get

$$\text{P.I.} = (\log \cos x). \cos x + x \sin x$$

Hence, the complete solution of given equation is

$$y = \text{C.F.} + \text{P.I.} = C_1 \cos x + C_2 \sin x + (\log \cos x) \cdot \cos x + x \sin x. \quad \textbf{Ans.}$$

Example 83: Obtain general solution of the differential equation

$$x^2y'' + xy' - y = x^5e^x. \qquad \textit{(U.P.T.U. Summer, 2002)}$$

Solution: We have $x^2\dfrac{d^2y}{dx^2} + x\dfrac{dy}{dx} - y = x^5e^x$ ...(1)

Let $\quad x = e^z \quad \Rightarrow \quad x\dfrac{dy}{dx} = Dy; D = \dfrac{d}{dz}$

$$x^2\frac{d^2y}{dx^2} = D(D-1)y; D = \frac{d}{dz}$$

Using these values in equation (1), we get

$$\{D(D-1) + D - 1\}y = e^{5z}e^{e^z}$$

$$\Rightarrow \quad (D^2 - 1)y = e^{5z}.e^{e^z}$$

The A.E. is $\quad m^2 - 1 = 0 \quad \Rightarrow \quad m = \pm 1$

$$\therefore \quad \text{C.F.} = C_1e^z + C_2e^{-z} = C_1x + \frac{C_2}{x}$$

Here $\quad u = x, \quad v = \dfrac{1}{x} \quad \Rightarrow \quad u' = 1, \quad v' = -\dfrac{1}{x^2}$

$$\text{P.I.} = Au + Bv = Ax + \frac{B}{x} \qquad \text{...(2)}$$

Now $\quad A = \displaystyle\int \frac{-Rv}{uv' - u'v}dx = -\int \frac{x^3e^x.\frac{1}{x}}{-\frac{x}{x^2} - \frac{1}{x}}dx = \int \frac{x^2e^x}{\frac{2}{x}}dx$

$$\Rightarrow \quad A = \frac{1}{2}\int x^3e^x dx = \frac{1}{2}\left[x^3e^x - 3x^2e^x + 6xe^x - 6e^x\right]$$

$$\Rightarrow \quad A = \frac{1}{2}[x^3 - 3x^2 + 6x - 6]e^x$$

$$B = \int \frac{Ru}{uv' - u'v}dx = \int \frac{x.x^3e^x}{x\left(-\frac{1}{x^2}\right) - \frac{1}{x}(1)}dx = \int \frac{x^4e^x}{-\frac{2}{x}}dx = -\frac{1}{2}\int x^5e^x dx$$

$$B = -\frac{1}{2}[x^5e^x - 5x^4e^x + 20x^3e^x - 60x^2e^x + 120xe^x - 120e^x]$$

From equation (2)

$$\text{P.I.} = \frac{1}{2}(x^3 - 3x^2 + 6x - 6)e^x \cdot x - \frac{1}{2}(x^5 - 5x^4 + 20x^3 - 60x^2 + 120x - 120)e^x.\frac{1}{x}$$

$$\Rightarrow \qquad \text{P.I.} = \frac{e^x}{2}\left[x^4 - 3x^3 + 6x^2 - 6x - x^4 + 5x^3 - 20x^2 + 60x - 120 + \frac{120}{x}\right]$$

$$= \frac{e^x}{2}\left[2x^3 - 14x^2 + 54x - 120 + \frac{120}{x}\right]$$

$$y = \text{C.F.} + \text{P.I.}$$

$$\Rightarrow \qquad y = C_1 x + \frac{C_2}{x} + \left(x^3 - 7x^2 + 27x - 60 + \frac{60}{x}\right)e^x. \quad \textbf{Ans.}$$

Example 84: Solve by the method of variation of parameters

$$\frac{d^2y}{dx^2} - y = \frac{2}{1+e^x}. \qquad (\textit{A.M.I.E.T.E. 2001, U.P.T.U. 2001})$$

Solution: Given that $\dfrac{d^2y}{dx^2} - y = \dfrac{2}{1+e^x}$

$$\Rightarrow \qquad (D^2 - 1)y = \frac{2}{1+e^x}$$

The A.E. is $\quad m^2 - 1 = 0 \quad \Rightarrow \quad m = \pm 1$

$$\text{C.F.} = C_1 e^x + C_2 e^{-x}$$

$$u = e^x, \quad v = e^{-x} \quad \Rightarrow \quad u' = e^x, \quad v' = -e^{-x}$$

$$\text{P.I.} = Au + Bv \qquad \ldots(1)$$

Now

$$A = \int \frac{-Rv}{uv' - u'v}\,dx = -\int \frac{e^{-x}}{-2} \times \frac{2}{1+e^x}\,dx = \int \frac{e^{-x}}{1+e^x}\,dx$$

$$= \int \frac{dx}{e^x(1+e^x)} = \int\left(\frac{1}{e^x} - \frac{1}{1+e^x}\right)dx$$

$$\Rightarrow \qquad A = \int e^{-x}dx - \int \frac{e^{-x}}{e^{-x}+1}\,dx = -e^{-x} + \log(e^{-x}+1)$$

$$B = \int \frac{Ru}{uv' - u'v}\,dx = \int \frac{e^x}{-2}\frac{2}{(1+e^x)}\,dx = -\int \frac{e^x}{1+e^x}\,dx$$

$$\Rightarrow \qquad B = -\log(1+e^x)$$

$$\text{P.I.} = \left[-e^{-x} + \log(e^{-x}+1)\right]e^x - e^{-x}\log(1+e^x)$$

$$= -1 + e^x \log(e^{-x}+1) - e^{-x}\log(e^x+1)$$

$$\therefore \qquad y = \text{C.F.} + \text{P.I.} = C_1 e^x + C_2 e^{-x} - 1 + e^x \log(e^{-x}+1) - e^{-x}\log(e^x+1). \quad \textbf{Ans.}$$

Example 85: Solve by the method of variation of parameters

$$\frac{d^2y}{dx^2} - 2\frac{dy}{dx} = e^x \sin x. \qquad (U.P.T.U.\ 2003)$$

Solution: We have $\frac{d^2y}{dx^2} - 2\frac{dy}{dx} = e^x \sin x$

$$\Rightarrow \qquad (D^2 - 2D)y = e^x \sin x$$

The A.E. is $m^2 - 2m = 0 \quad \Rightarrow \quad m(m-2) = 0 \quad \Rightarrow \quad m = 0, 2$

$$\therefore \qquad \text{C.F.} = C_1 + C_2 e^{2x}$$

So, $u = 1, \quad v = e^{2x} \quad \Rightarrow \quad u' = 0, v' = 2e^{2x}$

$$\text{P.I.} = Au + Bv = A + Be^{2x} \qquad \ldots(1)$$

Now $$A = \int \frac{-Rv}{uv' - u'v} dx = \int \frac{-e^x \sin x \times e^{2x}}{2e^{2x} - 0(e^{2x})} dx = -\frac{1}{2}\int e^x \sin x\, dx$$

$$= -\frac{1}{2}\frac{e^x}{(1+1)}(\sin x - \cos x) = -\frac{e^x}{4}(\sin x - \cos x)$$

and $$B = \int \frac{Ru}{uv' - u'v} dx = \int \frac{1.e^x \sin x}{1(2e^{2x}) - 0(e^{2x})} dx = \frac{1}{2}\int e^{-x}.\sin x\, dx$$

$$= \frac{1}{2}\frac{e^{-x}}{\{(-1)^2 + (1)^2\}}[-\sin x - \cos x] = -\frac{e^{-x}}{4}(\sin x + \cos x)$$

From equation (1) $$\text{P.I.} = -\frac{e^x}{4}(\sin x - \cos x) - \frac{e^{-x}}{4}(\sin x + \cos x)e^{2x}$$

$$= -\frac{e^x}{4}[\sin x - \cos x + \sin x + \cos x] = -\frac{e^x}{2}\sin x$$

Hence, the complete solution is

$$y = C_1 + C_2 e^{2x} - \frac{e^x}{2}\sin x. \quad \textbf{Ans.}$$

Example 86: Apply the method of variation of parameters to solve

$$\frac{d^2y}{dx^2} - 3\frac{dy}{dx} + 2y = \frac{e^x}{1+e^x}. \qquad (U.P.T.U.\ 2005)$$

Solution: We have $\frac{d^2y}{dx^2} - 3\frac{dy}{dx} + 2y = \frac{e^x}{1+e^x}$

$$\Rightarrow \qquad (D^2 - 3D + 2)y = \frac{e^x}{1+e^x}$$

The A.E. is $\quad m^2 - 3m + 2 = 0 \quad \Rightarrow \quad (m-1)(m-2) = 0 \quad \Rightarrow \quad m = 1, 2$

$$\text{C.F.} = C_1 e^x + C_2 e^{2x}$$

Here $\quad u = e^x, \quad v = e^{2x} \quad \Rightarrow \quad u' = e^x, v' = 2e^{2x}$

$$\text{P.I.} = Au + Bv = Ae^x + Be^{2x} \qquad \ldots(1)$$

where
$$A = \int \frac{-Rv}{uv' - u'v} dx = \int \frac{-e^{2x} \cdot \dfrac{e^x}{1+e^x}}{e^x(2e^{2x}) - e^{2x}(e^x)} dx = \int \frac{\dfrac{-e^{3x}}{1+e^x}}{2e^{3x} - e^{3x}} dx$$

$$\Rightarrow \qquad A = \int \frac{-e^{3x}}{e^{3x}(1+e^x)} dx = -\int \frac{1}{1+e^x} dx = -\int \frac{e^{-x}}{e^{-x}+1} dx = \log(e^{-x}+1)$$

and
$$B = \int \frac{Ru}{uv' - u'v} dx = \int \frac{e^x\left(\dfrac{e^x}{1+e^x}\right)}{e^{3x}} dx = \int \frac{1}{e^x(1+e^x)} dx$$

$$\Rightarrow \qquad B = \int\left(\frac{1}{e^x} - \frac{1}{1+e^x}\right) dx = \int\left(e^{-x} - \frac{e^{-x}}{e^{-x}+1}\right) dx = -e^{-x} + \log(e^{-x}+1)$$

$$\text{P.I.} = e^x \log(e^{-x}+1) + e^{2x}\left\{-e^{-x} + \log(e^{-x}+1)\right\}$$

$$\text{P.I.} = e^x \log(e^{-x}+1) - e^x + e^{2x} \log(e^{-x}+1)$$

The required solution of given differential equation is

$$y = C_1 e^x + C_2 e^{2x} + e^x \log(e^{-x}+1) - e^x + e^{2x} \log(e^{-x}+1). \quad \textbf{Ans.}$$

Example 87: Apply the method of variation of parameters to solve

$$\frac{d^2y}{dx^2} + y = \tan x. \qquad \textit{(U.P.T.U. 2000, 2009)}$$

Solution: The given equation can be written as

$$(D^2 + 1)y = \tan x$$

The A.E. is $\quad m^2 + 1 = 0 \quad \Rightarrow \quad m^2 = -1 \quad \Rightarrow \quad m = \pm i$

$$\text{C.F.} = C_1 \cos x + C_2 \sin x$$

Here $\quad u = \cos x, \quad v = \sin x \quad \Rightarrow \quad u' = -\sin x, \quad v' = \cos x$

$$\text{P.I.} = Au + Bv = A\cos x + B\sin x \qquad \ldots(1)$$

Now $$A = \int \frac{-Rv}{uv' - u'v}\,dx = \int \frac{-\tan x \times \sin x}{1}\,dx = -\int \frac{\sin^2 x}{\cos x}\,dx$$

$$= -\int \frac{1-\cos^2 x}{\cos x}\,dx = -\int \sec x\,dx + \int \cos x\,dx$$

$\Rightarrow$ $$A = -\log(\sec x + \tan x) + \sin x$$

$$B = \int \frac{Ru}{uv' - u'v}\,dx = \int \cos x \tan x\,dx = \int \sin x\,dx = -\cos x$$

From (1), $$\text{P.I.} = -\cos x \log(\sec x + \tan x)$$

Hence, the complete solution is

$$y = C_1 \cos x + C_2 \sin x - \cos x \cdot \log(\sec x + \tan x).$$ **Ans.**

Example 88: By the method of variation of parameters, solve the differential equation

$$\frac{d^2y}{dx^2} + (1 - \cot x)\frac{dy}{dx} - y\cot x = \sin^2 x.$$ *(U.P.T.U. Special Exam., 2001)*

Solution: We find C.F. *i.e.*, solution of

$$\frac{d^2y}{dx^2} + (1 - \cot x)\frac{dy}{dx} - y\cot x = 0 \qquad \text{...(1)}$$

Here $P = 1 - \cot x$, $Q = -\cot x$ and $1 - P + Q = 1 - 1 + \cot x - \cot x = 0$

$\therefore$ $u = e^{-x}$ is part of C.F.

Putting $y = e^{-x}.\ v$, equation (1) reduces to

$$\frac{d^2v}{dx^2} - (1 + \cot x)\frac{dv}{dx} = 0$$

$\Rightarrow$ $$\frac{dP}{dx} - (1 + \cot x)P = 0, \text{ where } P = \frac{dv}{dx}$$

$\Rightarrow$ $$\frac{dP}{P} = (1 + \cot x)dx$$

On integrating, $$\log P = x + \log\sin x + \log C$$

$\Rightarrow$ $$P = Ce^x \sin x$$

$\Rightarrow$ $$\frac{dv}{dx} = Ce^x \sin x$$

On integrating again, we get

$$v = \frac{C}{2}e^x(\sin x - \cos x) + C_2$$

$\therefore$ C.F. of given equation is $y = e^{-x} \cdot v$

or $$y = \frac{C}{2}(\sin x - \cos x) + C_2 e^{-x}$$

or $$y = C_1(\sin x - \cos x) + C_2 e^{-x}, \text{ where } C_1 = \frac{C}{2}$$

$$\Rightarrow \quad \text{C.F.} = C_1(\sin x - \cos x) + C_2 e^{-x}$$

Here $$U = \sin x - \cos x, \quad V = e^{-x}$$

$$\Rightarrow \quad U' = \cos x + \sin x, \quad V' = -e^{-x}$$

$$\text{P.I.} = AU + BV = A(\sin x - \cos x) + Be^{-x} \quad \ldots(1)$$

$$A = \int \frac{-RV}{UV' - U'V} dx = \int \frac{-\sin^2 x \times (e^{-x})}{-(\sin x - \cos x)e^{-x} - (\cos x + \sin x)e^{-x}} dx$$

$$= \int \frac{e^{-x} \cdot \sin^2 x}{e^{-x}\sin x - e^{-x}\cos x + e^{-x}\cos x + e^{-x}\sin x} dx = \frac{1}{2}\int \sin x \, dx$$

$$\Rightarrow \quad A = -\frac{1}{2}\cos x$$

$$B = \int \frac{RU}{UV' - U'V} dx = \int \frac{(\sin x - \cos x)\sin^2 x}{-2e^{-x}\sin x} dx$$

$$= \frac{1}{2}\int e^x (\sin x \cos x - \sin^2 x) dx$$

$$= \frac{1}{4}\int e^x [\sin 2x - (1 - \cos 2x)] dx$$

$$= \frac{1}{4}\int e^x (\sin 2x - 1 + \cos 2x) dx$$

$$= \frac{1}{4} \cdot \frac{e^x}{5}(\sin 2x - 2\cos 2x) - \frac{e^x}{4} + \frac{1}{4} \cdot \frac{e^x}{5}(\cos 2x + 2\sin 2x)$$

$$B = \frac{e^x}{20}(3\sin 2x - \cos 2x) - \frac{e^x}{4}$$

$$\text{P.I.} = -\frac{1}{2}\cos x(\sin x - \cos x) + \frac{1}{20}(3\sin 2x - \cos 2x) - \frac{1}{4}$$

$$\text{P.I.} = -\frac{1}{10}\sin 2x + \frac{1}{5}\cos 2x$$

Complete solution of given equation is

$$y = \text{C.F.} + \text{P.I.}$$

$$\Rightarrow \qquad y = C_1(\sin x - \cos x) + C_2 e^{-x} - \frac{1}{10}\sin 2x + \frac{1}{5}\cos 2x. \quad \textbf{Ans.}$$

Example 89: Using variation of parameters method, solve:

$$x^2\frac{d^2y}{dx^2} + 2x\frac{dy}{dx} - 12y = x^3 \log x. \qquad (U.P.T.U.\ 2004)$$

Solution: We have $x^2\dfrac{d^2y}{dx^2} + 2x\dfrac{dy}{dx} - 12y = x^3 \log x$ ...(1)

Let $\quad x = e^z \quad \Rightarrow \quad x\dfrac{dy}{dx} = Dy, \quad x^2\dfrac{d^2y}{dx^2} = D(D-1)y\,;\, D = \dfrac{d}{dz}$

From (1) $\{D(D-1) + 2D - 12\}y = ze^{3z}$

$$\Rightarrow \qquad (D^2 + D - 12)y = ze^{3z}$$

The auxiliary equation is $m^2 + m - 12 = 0 \quad \Rightarrow \quad m = 3, -4$

$$\therefore \qquad \text{C.F.} = C_1 e^{3z} + C_2 e^{-4z} = C_1 x^3 + C_2 \cdot \frac{1}{x^4}$$

Here $\quad u = x^3, \quad v = \dfrac{1}{x^4} \quad \Rightarrow \quad u' = 3x^2; \quad v' = -\dfrac{4}{x^5}$

$$\text{P.I.} = Au + Bv = Ax^3 + \frac{B}{x^4} \qquad \text{...(2)}$$

where $\quad A = \displaystyle\int \frac{-Rv}{uv' - u'v}dx = -\int \frac{x\log x \cdot x^{-4}}{x^3(-4x^{-5}) - 3x^2(x^{-4})}dx = -\int \frac{x^{-3}\log x}{-7x^{-2}}dx$

$$\Rightarrow \qquad A = \frac{1}{7}\int \frac{\log x}{x}dx = \frac{1}{7}\frac{(\log x)^2}{2} = \frac{1}{14}(\log x)^2$$

and $\quad B = \displaystyle\int \frac{Ru}{uv' - u'v}dx = \int \frac{x\log x \cdot x^3}{-7x^{-2}}dx = -\frac{1}{7}\int x^6 \log x\, dx$

$$= -\frac{1}{7}\left[\log x \cdot \frac{x^7}{7} - \int \frac{1}{x}\cdot\frac{x^7}{7}dx\right]$$

$$\Rightarrow \qquad B = -\frac{1}{7}\left[\frac{x^7 \log x}{7} - \frac{1}{7}\left(\frac{x^7}{7}\right)\right] = \frac{x^7}{49}\left(\frac{1}{7} - \log x\right)$$

From equation (2) $$\text{P.I.} = \frac{1}{14}(\log x)^2 \times x^3 + \frac{1}{x^4}\cdot\frac{x^7}{49}\left(\frac{1}{7} - \log x\right)$$

$$\text{P.I.} = \frac{x^3}{14}(\log x)^2 + \frac{x^3}{49}\left(\frac{1}{7} - \log x\right)$$

The required solution is $y = \text{C.F.} + \text{P.I.}$

$$y = C_1 x^3 + \frac{C_2}{x^4} + \frac{x^3}{14}(\log x)^2 + \frac{x^3}{343} - \frac{x^3}{49}\log x$$

$$\Rightarrow \quad y = \left(C_1 + \frac{1}{343}\right)x^3 + \frac{C_2}{x^4} + \frac{x^3}{98}\log x(7\log x - 2). \quad \textbf{Ans.}$$

Example 90: Apply the method of variation of parameters to solve $(D^2 - 1)y = 2(1 - e^{-2x})^{-1/2}$

(*U.P.T.U. Summer, 2009*)

Solution: We have

$$(D^2 - 1)y = 2(1 - e^{-2x})^{-1/2}$$

The A.E. is $$m^2 - 1 = 0 \Rightarrow m = \pm 1$$

$$\text{C.F.} = C_1 e^x + C_2 e^{-x}$$

Here $u = e^x, \; v = e^{-x} \Rightarrow u' = e^x, \; v' = -e^{-x}$

$$\text{P.I.} = Au + Bv = Ae^x + Be^{-x} \quad \text{...}(i)$$

Now $$A = \int \frac{-Rv}{uv' - u'v}dx = \int \frac{-2(1 - e^{-2x})^{-1/2}\cdot e^{-x}}{-1-1}dx$$

$$A = \int (1 - e^{-2x})^{-1/2}\cdot e^{-x}dx$$

Let $$e^{-x} = \sin\theta \Rightarrow e^{-x}dx = -\cos\theta\, d\theta$$

$$A = -\int (1 - \sin^2\theta)^{-1/2}\cdot\cos\theta\, d\theta = \int (\cos^2\theta)^{-1/2}\cos\theta\, d\theta$$

$$= -\int \frac{\cos\theta}{\cos\theta}d\theta = -\int d\theta = -\theta$$

$$\Rightarrow \quad A = -\sin^{-1}(e^{-x}) \qquad \Big| \text{ As } \theta = \sin^{-1}(e^{-x})$$

$$B = \int \frac{Ru}{uv' - u'v}dx = \int \frac{2(1 - e^{-2x})^{-1/2}\cdot e^x}{-2}dx$$

$$B = -\int (1-e^{-2x})^{-1/2} e^{x} dx$$
$$= -\int (e^{2x}-1)^{-1/2} \cdot e^{2x} dx$$

Let $\qquad e^{2x} - 1 = t \;\Rightarrow\; e^{2x} dx = \dfrac{dt}{2}$

$$B = \frac{-1}{2}\int t^{-1/2} dt = -t^{1/2}$$

$\Rightarrow \qquad B = -(e^{2x}-1)^{1/2}$

From equation (*i*) P.I. $= -\sin^{-1}(e^{-x}) \cdot e^{x} - (e^{2x}-1)^{1/2} e^{-x}$

Hence the complete solution is

$$y = C_1 e^{x} + C_2 e^{-x} - \sin^{-1}(e^{-x}) \cdot e^{x} - (e^{2x}-1)^{1/2} \cdot e^{-x} \quad \textbf{Ans.}$$

Example 91: Apply the method of variation of parameters to solve the differential equation

$$(1-x)\frac{d^2y}{dx^2} + x\frac{dy}{dx} - y = (1-x)^2$$

Solution: The given equation may be written as

$$\frac{d^2y}{dx^2} + \frac{x}{1-x}\frac{dy}{dx} - \frac{1}{1-x}y = (1-x) \qquad \text{...(1)}$$

Here $\qquad P = \dfrac{x}{1-x}, \quad Q = -\dfrac{1}{1-x}$ and $R = (1-x)$

and $\qquad P + Qx = \dfrac{x}{1-x} - \dfrac{x}{1-x} = 0$

Part of C.F. $\qquad u = x$

Now we find the complementary function of differential equation (1), consider the equation

$$\frac{d^2y}{dx^2} + \frac{x}{1-x}\frac{dy}{dx} - \frac{1}{1-x}y = 0 \qquad \text{...(2)}$$

Let $\qquad y = xv \;\Rightarrow\; \dfrac{dy}{dx} = v + x\dfrac{dv}{dx}, \; \dfrac{d^2y}{dx^2} = 2\dfrac{dv}{dx} + \dfrac{xd^2v}{dx^2}$

From equation (2), we have

$$\frac{d^2v}{dx^2} + \left(\frac{x}{1-x} + \frac{2}{x}\right)\frac{dv}{dx} = 0 \;\Rightarrow\; \frac{d^2v}{dx^2} + \left(-1 - \frac{1}{x-1} + \frac{2}{x}\right)\frac{dv}{dx} = 0$$

$$\Rightarrow \qquad \frac{dp}{dx} + \left(-1 - \frac{1}{x-1} + \frac{2}{x}\right)p = 0, \text{ where } p = \frac{dv}{dx}$$

$$\Rightarrow \qquad \frac{dp}{p}+\left(-1-\frac{1}{x-1}+\frac{2}{x}\right)dx = 0$$

On integrating

$$\log p - x - \log(x-1) + 2\log x = \log C_1$$

$$\Rightarrow \qquad \log\frac{px^2}{C_1(x-1)} = x \quad \Rightarrow \quad px^2 = C_1(x-1)e^x$$

$$\Rightarrow \qquad p = \frac{C_1(x-1)e^x}{x^2} \quad \Rightarrow \quad \frac{dv}{dx} = C_1\left(\frac{1}{x}-\frac{1}{x^2}\right)e^x$$

Integrating again $\quad v = C_1\dfrac{e^x}{x}+C_2$

$\therefore$ C.F. of given equation is $y = xv$

or $\qquad y = C_1e^x + C_2x \quad \Rightarrow \quad \text{C.F.} = C_1e^x + C_2x$

Here $\qquad U = e^x, \quad V = x \quad \Rightarrow U' = e^x, \quad V' = 1$

$$\text{P.I.} = AU + BV = Ae^x + Bx \qquad \text{...(3)}$$

where $\qquad A = \displaystyle\int\frac{-RV}{UV'-U'V}dx = \int\frac{-(1-x)x}{e^x - e^x.x}dx = -\int xe^{-x}dx$

$$\Rightarrow \qquad A = -\int xe^{-x}dx = xe^{-x} - \int e^{-x}dx = xe^{-x} + e^{-x} = (x+1)e^{-x}$$

and $\qquad B = \displaystyle\int\frac{RU}{UV'-U'V}dx = \int\frac{(1-x)e^x}{e^x(1-x)}dx = x$

From (3) $\qquad \text{P.I.} = x + 1 + x^2$

Hence, the complete solution of given equation is

$$y = \text{C.F.} + \text{P.I.}$$

or $\qquad y = C_1e^x + C_2x + x^2 + x + 1.$ **Ans.**

Example 92: Solve by method of variation of parameters

$$\frac{d^2y}{dx^2}+2\frac{dy}{dx}+y = e^{-x}\log x \qquad \textit{(U.P.T.U. 2008)}$$

Solution: We have $\dfrac{d^2y}{dx^2}+2\dfrac{dy}{dx}+y = e^{-x}\log x$

$$\Rightarrow \qquad (D^2+2D+1)y = e^{-x}\log x$$

The A.E. is $\quad m^2+2m+1=0 \;\Rightarrow\; (m+1)^2=0 \;\Rightarrow\; m=-1,-1$

$$\therefore \qquad \text{C.F.} = (C_1+C_2x)e^{-x} = C_1e^{-x}+C_2xe^{-x}$$

Here $\qquad u = e^{-x}, \quad v = xe^{-x}$

$$\therefore \qquad u' = -e^{-x}, \quad v' = -xe^{-x}+e^{-x}$$

$$\text{P.I.} = Au + Bv \qquad \ldots(i)$$

where
$$A = \int \frac{-Rv}{uv'-u'v}\,dx = \int \frac{-e^{-x}\cdot \log x \cdot xe^{-x}}{-xe^{-2x}+e^{-2x}+xe^{-2x}}\,dx$$

$$A = \int \frac{-xe^{-2x}\log x}{e^{-2x}}\,dx = -\int x\log x\,dx$$

$$= -\frac{x^2}{2}\cdot \log x + \int \frac{x^2}{2}\cdot\frac{1}{x}\,dx$$

$$A = -\frac{x^2}{2}\log x + \frac{x^2}{4}$$

And
$$B = \int \frac{Ru}{uv'-u'v}\,dx = \int \frac{e^{-x}\log x\cdot e^{-x}}{e^{-2x}}\,dx$$

$$= \int \underset{II}{1}\cdot \underset{I}{\log x}\,dx = x\cdot\log x - \int x\cdot\frac{1}{x}\,dx$$

$$B = x\log x - x$$

From (i)
$$\text{P.I.} = -\frac{x^2e^{-x}}{2}\log x + e^{-x}\cdot\frac{x^2}{4} + x^2e^{-x}\log x - x^2e^{-x}$$

$$\Rightarrow \qquad \text{P.I.} = \frac{x^2e^{-x}}{2}\log x - \frac{3}{4}x^2e^{-x}$$

Hence
$$y = (C_1+C_2x)e^{-x} + \frac{x^2e^{-x}}{2}\log x - \frac{3}{4}x^2e^{-x} \qquad \textbf{Ans.}$$

EXERCISE 1.15

Solve the following differential equations by variation of parameters method:

1. $\dfrac{d^2y}{dx^2} - y = e^{-x}\sin(e^{-x}) + \cos(e^{-x})$ $\left[\textbf{Ans. } y = C_1e^x + C_2e^{-x} - \dfrac{3}{2}\cos(e^{-x})\right]$

2. $\dfrac{d^2y}{dx^2} + y = \operatorname{cosec} x$ $[\textbf{Ans. } y = A\cos x + B\sin x - x\cos x + \sin x \log\sin x]$

3. $x^2\dfrac{d^2y}{dx^2} - 2x(1+x)\dfrac{dy}{dx} + 2(1+x)y = x^3$ $[\textbf{Ans. } y = (C_1 + C_2e^{2x} - x)x]$

4. $\dfrac{d^3y}{dx^3} - 6\dfrac{d^2y}{dx^2} + 11\dfrac{dy}{dx} - 6y = e^{2x}$ $[\textbf{Ans. } y = C_1e^x + C_2e^{2x} + C_3e^{3x} - xe^{2x}]$

5. $\dfrac{d^2y}{dx^2} + a^2y = \sec ax$ $\left[\textbf{Ans. } y = C_1\cos ax + C_2\sin ax + \dfrac{1}{a^2}\cos ax \log\cos ax + \dfrac{x}{a}\sin ax\right]$

6. $x^2\dfrac{d^2y}{dx^2} - 2x(1+x)\dfrac{dy}{dx} + 2(1+x)y = x^3$ $\left[\textbf{Ans. } y = C_1x + C_2xe^{2x} - \dfrac{1}{2}x^2 - \dfrac{1}{4}x\right]$

7. $x^2\dfrac{d^2y}{dx^2} + x\dfrac{dy}{dx} - y = x^2e^x$ $\left[\textbf{Ans. } y = C_1x + C_2\dfrac{1}{x} + e^x - \dfrac{1}{x}e^x\right]$

8. $\dfrac{d^2y}{dx^2} - 2\dfrac{dy}{dx} = e^x\sin x$ $\left[\textbf{Ans. } y = C_1 + C_2e^{2x} - \dfrac{1}{2}e^x\sin x\right]$

1.7 EQUATION OF THE TYPE $\dfrac{d^ny}{dx^n} = f(x)$

These type of differential equations are solved by successive integration.

Example 93: Solve $\dfrac{d^4y}{dx^4} = x + e^{-x} - \cos x$

Solution: We have $\dfrac{d^4y}{dx^4} = x + e^{-x} - \cos x$ ...(1)

Integrating the differential equation (1), we get

$$\frac{d^3y}{dx^3} = \frac{x^2}{2} - e^{-x} - \sin x + C \quad ...(2)$$

Again integrating (2), we have

$$\frac{d^2y}{dx^2} = \frac{x^3}{6} + e^{-x} + \cos x + Cx + C_1 \quad ...(3)$$

Again integrating (3), we get

$$\frac{dy}{dx} = \frac{x^4}{24} - e^{-x} + \sin x + \frac{Cx^2}{2} + C_1x + C_2 \quad ...(4)$$

Again integrating

$$y = \frac{x^5}{120} + e^{-x} - \cos x + \frac{C_1x^3}{6} + \frac{C_1x^2}{2} + C_2x + C_3. \quad \textbf{Ans.}$$

Example 94: Solve $x^2 \dfrac{d^2y}{dx^2} = \log x$

Solution: We have $x^2 \dfrac{d^2y}{dx^2} = \log x$

$$\Rightarrow \quad \frac{d^2y}{dx^2} = \frac{1}{x^2}\log x \quad ...(1)$$

Integrating equation (1), we get

$$\frac{dy}{dx} = \int \frac{1}{x^2} \log x \, dx + C$$

$$\frac{dy}{dx} = -\frac{1}{x}\log x + \int \frac{1}{x^2} dx + C$$

$$\frac{dy}{dx} = -\frac{1}{x}\log x - \frac{1}{x} + C \quad ...(2)$$

Again integrating, we get

$$y = -\int \frac{1}{x}\log x \, dx - \int \frac{1}{x} dx + Cx + C_1$$

$$\Rightarrow \quad y = -\frac{(\log x)^2}{2} - \log x + Cx + C_1. \quad \textbf{Ans.}$$

EXERCISE 1.16

Solve the following differential equations:

1. $\dfrac{d^2y}{dx^2} = x + \sin x$ $\left[\textbf{Ans. } y = \dfrac{x^3}{6} - \sin x + C_1x + C_2\right]$

2. $\dfrac{d^3y}{dx^3} = x + \log x$ $\left[\textbf{Ans. } y = \dfrac{x^4}{24} + \dfrac{x^3}{6}\log x - \dfrac{11}{36}x^3 + C_1\dfrac{x^2}{2} + C_2x + C_3\right]$

3. $\dfrac{d^2y}{dx^2} = x^2 \cdot \sin x$ $\left[\textbf{Ans. } y = -x^2 \sin x - 4x\cos x + 6\sin x + C_1x + C_2\right]$

4. $\dfrac{d^3y}{dx^3} = \sin^2 x$ $\left[\textbf{Ans. } y = \dfrac{x^3}{12} + \dfrac{\sin 2x}{16} + \dfrac{C_1x^2}{2} + C_2x + C_3\right]$

5. $\dfrac{d^3y}{dx^3} = x^2e^x$ $\left[\textbf{Ans. } y = x^2e^x - 6xe^x + 12e^x + \dfrac{C_1x^2}{2} + C_2x + C_3\right]$

1.8 EQUATION OF THE TYPE $\dfrac{d^2y}{dx^2} = f(y)$

Multiplying given equation by $2\dfrac{dy}{dx}$, we get

$$2\frac{dy}{dx}\frac{d^2y}{dx^2} = 2f(y)\frac{dy}{dx}$$

$$\Rightarrow \quad \frac{d}{dx}\left(\frac{dy}{dx}\right)^2 = 2f(y)\frac{dy}{dx} \qquad ...(i)$$

Integrate equation (*i*), w.r.t. '*x*' we get

$$\left(\frac{dy}{dx}\right)^2 = 2\int f(y)dy + C = \psi(y)$$

$$\Rightarrow \quad \frac{dy}{dx} = \sqrt{\psi(y)} \quad \Rightarrow \frac{dy}{\sqrt{\psi(y)}} = dx$$

$$\Rightarrow \quad \boxed{\int \frac{dy}{\sqrt{\psi(y)}} = x + C}$$

Example 95: Solve $\frac{d^2y}{dx^2} = 2(y^3 + y)$, under the conditions $y = 0, \frac{dy}{dx} = 1$ when $x = 0$ *(U.P.T.U. 2003)*

Solution: We have
$$\frac{d^2y}{dx^2} = 2(y^3 + y) \quad ...(1)$$

Multiplying equation (1) by $2\frac{dy}{dx}$, we get

$$2\frac{dy}{dx}\frac{d^2y}{dx^2} = 4(y^3 + y)\frac{dy}{dx}$$

$$\Rightarrow \quad \frac{d}{dx}\left(\frac{dy}{dx}\right)^2 = 4(y^3 + y)\frac{dy}{dx}$$

Integrating, we get

$$\left(\frac{dy}{dx}\right)^2 = y^4 + 2y^2 + C \quad ...(2)$$

On putting $y = 0$ and $\frac{dy}{dx} = 1$, in (2), we get $1 = C$

Equation (2) reduces to

$$\left(\frac{dy}{dx}\right)^2 = y^4 + 2y^2 + 1 = (y^2 + 1)^2$$

$$\Rightarrow \quad \frac{dy}{dx} = 1 + y^2 \quad \Rightarrow \quad \frac{dy}{1 + y^2} = dx$$

Again integrating, we get $\tan^{-1}y = x + C_1$ ...(3)

Again on putting $y = 0$ and $x = 0$ in (3), we get $0 = C_1$

The equation (3) reduce in the form.

$$\tan^{-1} y = x \quad \Rightarrow \quad y = \tan x. \quad \textbf{Ans.}$$

EXERCISE 1.17

Solve the following differential equations:

1. $e^{2y}\frac{d^2y}{dx^2} = 1$ $\left[\textbf{Ans. } C_1 e^y = \cosh(C_1 x + C_2)\right]$

2. $\sin^3 y \dfrac{d^2y}{dx^2} = \cos y$ $\left[\textbf{Ans. } \sin\left[(x+C_2)\sqrt{1+C_1}\right] + \dfrac{\sqrt{1+C_1}}{C_1}\cos y = 0\right]$

3. $\dfrac{d^2y}{dx^2} = \sec^2 y \tan y$ under the condition $y = 0$ and $\dfrac{dy}{dx} = 1$, when $x = 0$ $\left[\textbf{Ans. } y = \sin^{-1} x\right]$

4. $\dfrac{d^2y}{dx^2} = \dfrac{1}{\sqrt{ay}}$ $\left[\textbf{Ans. } 3x + C_2 = 2a^{1/4}(\sqrt{y} - 2C_1)\sqrt{(\sqrt{y} + C_1)}\right]$

1.9 DIFFERENTIAL EQUATIONS WHICH DO NOT CONTAIN 'y' DIRECTLY

Such type differential equation may be solved by putting

$$\frac{dy}{dx} = p, \frac{d^2y}{dx^2} = \frac{dp}{dx} \text{ and so on}$$

Example 96: Solve $\dfrac{d^2y}{dx^2} = \left[1 - \left(\dfrac{dy}{dx}\right)^2\right]^{1/2}$ *(U.P.T.U. 2002)*

Solution: We have $\dfrac{d^2y}{dx^2} = \left[1 - \left(\dfrac{dy}{dx}\right)^2\right]^{1/2}$ ...(1)

Putting $\dfrac{dy}{dx} = p \Rightarrow \dfrac{d^2y}{dx^2} = \dfrac{dp}{dx}$ in equation (1), we get

$$\frac{dp}{dx} = (1 - p^2)^{1/2} \quad \Rightarrow \quad \frac{dp}{\sqrt{1 - p^2}} = dx$$

On integrating

$$\sin^{-1} p = x + C \quad \Rightarrow \quad p = \sin(x + C)$$

$$\Rightarrow \quad \frac{dy}{dx} = \sin(x + C)$$

Again integrating, we get

$$y = -\cos(x + C) + C_1. \textbf{ Ans.}$$

Example 97: Solve $\dfrac{d^4y}{dx^4} + \tan x \dfrac{d^3y}{dx^3} = 0$

Solution: We have $\dfrac{d^4y}{dx^4} + \tan x \dfrac{d^3y}{dx^3} = 0$ ...(1)

Putting $\frac{d^3y}{dx^3} = p \Rightarrow \frac{d^4y}{dx^4} = \frac{dp}{dx}$ in equation (1), we get

$$\frac{dp}{dx} + p\tan x = 0$$

$$\Rightarrow \qquad \frac{dp}{p} = -\tan x\, dx$$

On integrating $\quad \log p = \log \cos x + \log C$

$$\Rightarrow \qquad \log p = \log C \cdot \cos x$$

$$\Rightarrow \qquad p = C\cos x \quad \Rightarrow \quad \frac{d^3y}{dx^3} = C\cos x$$

Integrating w.r.t. 'x', we get

$$\frac{d^2y}{dx^2} = C\sin x + C_1$$

Again integrating

$$\frac{dy}{dx} = -C\cos x + C_1x + C_2$$

Again integrating

$$y = -C\sin x + \frac{C_1x^2}{2} + C_2x + C_3. \quad \textbf{Ans.}$$

1.10 DIFFERENTIAL EQUATIONS WHICH DO NOT CONTAIN 'x' DIRECTLY

In this situation put also $\quad \frac{dy}{dx} = p$

and then get $\quad \frac{d^2y}{dx^2} = \frac{dp}{dx} = \frac{dp}{dy}\frac{dy}{dx} = p\frac{dp}{dy}$

and so on.

Example 98: Solve $2y\frac{d^2y}{dx^2} - 3\left(\frac{dy}{dx}\right)^2 - 4y^2 = 0$

Solution: We have $2y\frac{d^2y}{dx^2} - 3\left(\frac{dy}{dx}\right)^2 - 4y^2 = 0 \qquad \text{...(1)}$

Let $\quad \frac{dy}{dx} = p \quad \Rightarrow \quad \frac{d^2y}{dx^2} = \frac{dp}{dx} = \frac{dp}{dy}\frac{dy}{dx} = p\frac{dp}{dy}$

From (1), we get

$$2py\frac{dp}{dy} - 3p^2 - 4y^2 = 0$$

$$\Rightarrow \qquad \frac{dp}{dy} - \frac{3}{2y}p = \frac{2y}{p}$$

Dividing on both side by $\frac{1}{p}$, we get

$$p\frac{dp}{dy} - \frac{3}{2y}p^2 = 2y \qquad ...(2)$$

Putting $\qquad p^2 = z \quad \Rightarrow \quad 2p\frac{dp}{dy} = \frac{dz}{dy} \quad \Rightarrow \quad p\frac{dp}{dy} = \frac{1}{2}\frac{dz}{dy}$

From (2), we get

$$\frac{1}{2}\frac{dz}{dy} - \frac{3}{2y}z = 2y$$

$$\Rightarrow \qquad \frac{dz}{dy} - \frac{3}{y}z = 4y \qquad ...(3)$$

which is the linear differential equation

$$\text{I.F.} = e^{-3\int \frac{1}{y}dy} = e^{-3\log y} = e^{\log \frac{1}{y^3}} = \frac{1}{y^3}$$

The solution of equation (3) is

$$z \cdot \frac{1}{y^3} = \int \frac{1}{y^3}.4y\, dy + C$$

$$\frac{z}{y^3} = 4\int \frac{1}{y^2}dy + C$$

$$\Rightarrow \qquad \frac{z}{y^3} = -\frac{4}{y} + C \quad \Rightarrow \quad z = -4y^2 + Cy^3 = y^2(Cy - 4)$$

$$\Rightarrow \qquad p^2 = y^2(Cy - 4) \quad \Rightarrow \quad p = y\sqrt{Cy - 4}$$

$$\Rightarrow \qquad \frac{dy}{dx} = y\sqrt{C}\sqrt{y - \left(\frac{2}{\sqrt{C}}\right)^2}$$

$$\Rightarrow \qquad \int \frac{dy}{y\sqrt{C}\sqrt{y - \left(\frac{2}{\sqrt{C}}\right)^2}} = \int dx + a$$

Putting $y = \left(\frac{2}{\sqrt{C}}\right)^2 \sec^2\theta \quad \Rightarrow \quad dy = 2\left(\frac{2}{\sqrt{C}}\right)^2 \sec^2\theta \tan\theta \, d\theta$

$$\Rightarrow \int \frac{\frac{8}{C}\sec^2\theta\tan\theta}{\frac{4}{\sqrt{C}}\sec^2\theta \frac{2}{\sqrt{C}}\tan\theta} d\theta = x + a$$

$$\Rightarrow \qquad \int d\theta = x + a \quad \Rightarrow \quad \theta = x + a$$

$$\Rightarrow \qquad \sec^{-1}\sqrt{y}\left(\frac{\sqrt{C}}{2}\right) = x + a \quad \Rightarrow \quad \frac{\sqrt{yC}}{2} = \sec(x+a)$$

$$\Rightarrow \qquad \sqrt{y} = \frac{2}{\sqrt{C}}\sec(x+a)$$

$$\Rightarrow \qquad y = \frac{4}{C}\sec^2(x+a)$$

$$\Rightarrow \qquad y = b\sec^2(x+a) \text{ where } b = \frac{4}{C}.$$ **Ans.**

Example 99: Solve $y\dfrac{d^2y}{dx^2} = \left\{\dfrac{dy}{dx}\right\}^2\left[1 - \dfrac{dy}{dx}\cos y + y\dfrac{dy}{dx}\sin y\right]$

Solution: The given equation can be written as

$$\frac{y\frac{d^2y}{dx^2}}{\left(\frac{dy}{dx}\right)^2} - 1 = -\frac{dy}{dx}\cos y + y\frac{dy}{dx}\sin y$$

$$\Rightarrow \qquad \frac{y\frac{d^2y}{dx^2} - \left(\frac{dy}{dx}\right)^2}{\left(\frac{dy}{dx}\right)^2} = -\frac{dy}{dx}\cos y + y\frac{dy}{dx}\sin y$$

$$\Rightarrow \qquad -\frac{d}{dx}\left\{\frac{y}{\left(\frac{dy}{dx}\right)}\right\} = -\frac{d}{dx}(y\cos y) \quad \Rightarrow \quad \frac{d}{dx}\left\{\frac{y}{\frac{dy}{dx}}\right\} = \frac{d}{dx}(y\cos y)$$

$\Rightarrow$ On integrating, we get

$$\frac{y}{\frac{dy}{dx}} = y\cos y + C$$

$$\Rightarrow \qquad \frac{dx}{dy} = \cos y + \frac{C}{y}$$

Again integrating, we get the required solution

$$x = \sin y + C\log y + C_1. \quad \textbf{Ans.}$$

Example 100: Solve $y\dfrac{d^2y}{dx^2} - \left(\dfrac{dy}{dx}\right)^2 = y^2 \log y$

Solution: We have $y\dfrac{d^2y}{dx^2} - \left(\dfrac{dy}{dx}\right)^2 = y^2 \log y$ ...(1)

Putting $\dfrac{dy}{dx} = p, \dfrac{d^2y}{dx^2} = p\dfrac{dp}{dy}$ in equation (1), we get

$$py\frac{dp}{dy} - p^2 = y^2 \log y$$

or $$p\frac{dp}{dy} - \frac{1}{y}p^2 = y\log y \qquad ...(2)$$

Let $\qquad p^2 = z \quad \Rightarrow \quad 2p\dfrac{dp}{dy} = \dfrac{dz}{dy}$

From equation (2)

$$\frac{dz}{dy} - \frac{2}{y}z = 2y\log y \qquad ...(3)$$

which is a linear differential equation of first order

Here $\qquad P = -\dfrac{2}{y}, \quad Q = 2y\log y$

$$I.F. = e^{-\int \frac{2}{y}dy} = e^{-2\log y} = e^{\log \frac{1}{y^2}} = \frac{1}{y^2}$$

The solution of equation (3) is

$$z \cdot \frac{1}{y^2} = 2\int \frac{1}{y}\log y\, dy + C$$

or $$z \cdot \frac{1}{y^2} = 2\frac{(\log y)^2}{2} + C$$

or $$z = (y \log y)^2 + Cy^2$$

or $$p^2 = (y \log y)^2 + Cy^2 \Rightarrow \frac{dy}{dx} = y\sqrt{(\log y)^2 + C}$$

or $$\int \frac{dy}{y\sqrt{(\log y)^2 + C}} = \int dx + C_1$$

Let $$\log y = t \Rightarrow \frac{1}{y} dy = dt$$

$$\Rightarrow \int \frac{dt}{\sqrt{t^2 + (\sqrt{C})^2}} = x + C_1 \Rightarrow \sin h^{-1} \frac{t}{\sqrt{C}} = x + C_1 \quad \left| \text{As} \int \frac{dx}{\sqrt{x^2 + a^2}} = \sin h^{-1} \frac{x}{a} \right.$$

$$\Rightarrow \frac{t}{\sqrt{C}} = \sinh(x + C_1) \Rightarrow \log y = \sqrt{C}\left\{\frac{e^{(x+C_1)} - e^{-(x+C_1)}}{2}\right\}$$

$$\Rightarrow \log y = \frac{\sqrt{C} \cdot e^{C_1}}{2} e^x - \frac{\sqrt{C} e^{-C_1} e^{-x}}{2}$$

$$\Rightarrow \log y = Ae^x + Be^{-x}, \text{ where } \left[A = \frac{\sqrt{C} e^{C_1}}{2},\ B = -\frac{\sqrt{C} e^{-C_1}}{2}\right]. \quad \textbf{Ans.}$$

EXERCISE 1.18

Solve the following differential equations:

1. $x\frac{d^2y}{dx^2} + x\left(\frac{dy}{dx}\right)^2 - \frac{dy}{dx} = 0$ $\left[\textbf{Ans. } y = \log(x^2 + 2C_1) + C_2\right]$

2. $e^{x^2/2}\left[x\frac{d^2y}{dx^2} - \frac{dy}{dx}\right] = x^3$ $\left[\textbf{Ans. } y = e^{-x^2/2} + C_1\frac{x^2}{2} + C_2\right]$

3. $\frac{d^2y}{dx^2} + \left(\frac{dy}{dx}\right)^2 + 1 = 0$ $\left[\textbf{Ans. } y = \log\sin(x - a) + b\right]$

4. $(1 + x^2)\frac{d^2y}{dx^2} + 1 + \left(\frac{dy}{dx}\right)^2 = 0$ $\left[\textbf{Ans. } y = -\frac{x}{C} + \frac{1 + C^2}{C^2}\log(1 + Cx) + a\right]$

5. $y(1-\log y)\dfrac{d^2y}{dx^2}+(1+\log y)\left(\dfrac{dy}{dx}\right)^2=0$ $\left[\textbf{Ans. } (1-\log y)=\dfrac{1}{Cx+d}\right]$

6. $y\dfrac{d^2y}{dx^2}+\left(\dfrac{dy}{dx}\right)^2=y^2$ $\left[\textbf{Ans. } y^2=k\sin h(\sqrt{2}x+C)\right]$

7. $y\dfrac{d^2y}{dx^2}+\dfrac{dy}{dx}+\left(\dfrac{dy}{dx}\right)^3=0$ $\left[\textbf{Ans. } y=a-\sin^{-1}(be^{-x})\right]$

1.11 APPLICATIONS OF DIFFERENTIAL EQUATIONS

1.11.1 Problems Related to Electric Circuit

There are some formulae which are useful to solve such type of problems.

(*a*) $i=\dfrac{dq}{dt}$

(*b*) Voltage drop across resistance R is $V_R=Ri$

(*c*) Voltage drop across inductance L is $V_L=L.\dfrac{di}{dt}$

(*d*) Voltage drop across capacitance C is $V_C=\dfrac{q}{C}$

(*e*) Equation of *L-R* circuit is $L\dfrac{di}{dt}+Ri=E$

Example 101: The current i in a circuit containing an inductance L, resistance R and voltage $E\sin\omega t$ is given by $L\dfrac{di}{dt}+Ri=E\sin\omega t$. If $i=0$ at $t=0$, show that at any time t,

$$i=\frac{E}{\sqrt{R^2+L^2\omega^2}}\left[\sin(\omega t-\phi)+\sin\phi\cdot e^{-Rt/L}\right]\quad\text{where } \tan\phi=\frac{L\omega}{R}.$$

Solution: The given equation can be written as

$$\frac{di}{dt}+\frac{R}{L}i=\frac{E}{L}\sin\omega t\quad\text{which is linear in } i \qquad ...(1)$$

$$\text{I.F.}=e^{\int\frac{R}{L}dt}=e^{\frac{Rt}{L}}$$

The solution of equation (1) is

$$i \cdot e^{\frac{Rt}{L}} = \int \frac{E}{L} \sin \omega t \times e^{\frac{Rt}{L}} dt + C$$

$$= \frac{E}{L} \frac{e^{\frac{Rt}{L}}}{\omega^2 + \frac{R^2}{L^2}} \left[\frac{R}{L} \sin \omega t - \omega \cos \omega t \right] + C$$

$$\Rightarrow \qquad i = Ce^{-\frac{Rt}{L}} + \frac{E}{\omega^2 L^2 + R^2} [R \sin \omega L - \omega L \cos \omega t] \qquad \text{...(2)}$$

If we write here $R = k \cos \phi$ and $\omega L = k \sin \phi$, then

$$k^2 = R^2 + \omega^2 L^2 \text{ and } \tan \phi = \frac{\omega L}{R}$$

Then
$$i = Ce^{-\frac{Rt}{L}} + \frac{E}{\sqrt{\omega^2 L^2 + R^2}} \sin(\omega t - \phi)$$

Putting $i = 0$ at $t = 0$ and get

$$C = \frac{E \sin \phi}{\sqrt{\omega^2 L^2 + R^2}}$$

$$\Rightarrow \qquad i = \frac{E \sin \phi . e^{-\frac{Rt}{L}}}{\sqrt{\omega^2 L^2 + R^2}} + \frac{E}{\sqrt{L^2 \omega^2 + R^2}} \sin(\omega t - \phi)$$

$$\Rightarrow \qquad i = \frac{E}{\sqrt{\omega^2 L^2 + R^2}} \left[\sin \phi . e^{-\frac{Rt}{L}} + \sin(\omega t - \phi) \right]. \quad \textbf{Proved.}$$

Example 102: The equation of electromotive force in terms of current i for an electrical circuit having resistance R and a condenser of capacity C, in series, is $E = Ri + \int \frac{i}{C} dt$. Find the current i at any time t, when $E = E_0 \sin \omega t$. *(U.P.T.U. 2006)*

Solution: The given equation is

$$Ri + \int \frac{i}{C} dt = E_0 \sin \omega t \qquad \Big| As \ E = E_0 \sin \omega t$$

Differentiate w.r.t. 't', we get

$$R\frac{di}{dt} + \frac{i}{C} = E_0 \omega \cos \omega t$$

$$\Rightarrow \qquad \frac{di}{dt}+\frac{i}{RC} = \frac{E_0}{R}\omega\cos\omega t \qquad ...(1)$$

which is the linear differential equation

$$\Rightarrow \qquad \text{I.F.} = e^{\int \frac{1}{RC}dt} = e^{\frac{t}{RC}}$$

The solution of equation (1) is

$$i\cdot e^{\frac{t}{RC}} = \int \frac{E_0}{R}\omega\cos\omega t.e^{t/RC}\,dt + C_1$$

$$i\cdot e^{t/RC} = \frac{E_0\omega}{R}\frac{e^{\frac{t}{RC}}}{\frac{1}{R^2C^2}+\omega^2}\left[\frac{1}{RC}\cos\omega t+\omega\sin\omega t\right]+C_1$$

$$= \frac{E_0\omega}{R}\frac{R^2C^2\,e^{t/RC}}{(1+\omega^2R^2C^2)}\left[\frac{1}{RC}\sin\omega t+\omega\cos\omega t\right]+C_1$$

$$\Rightarrow \qquad i = E_0\omega\cdot\frac{RC^2}{1+\omega^2R^2C^2}\left[\frac{1}{RC}\sin\omega t+\omega\cos\omega t\right]+C_1e^{-t/RC}. \qquad \textbf{Ans.}$$

Example 103: The damped L-C-R circuit is governed by the equation

$$L\frac{d^2Q}{dt^2}+R\frac{dQ}{dt}+\left(\frac{1}{C}\right)Q = 0$$

where L, C, R are positive constants. Find the conditions under which the circuit is overdamped, underdamped and critically damped. Find also the critical resistance. (*U.P.T.U. 2005*)

Solution: We have

$$L\frac{d^2Q}{dt^2}+R\frac{dQ}{dt}+\left(\frac{1}{C}\right)Q = 0$$

$$\Rightarrow \qquad \frac{d^2Q}{dt^2}+\frac{R}{L}\frac{dQ}{dt}+\left(\frac{1}{LC}\right)Q = 0$$

$$\Rightarrow \qquad \frac{d^2Q}{dt^2}+2P\frac{dQ}{dt}+\omega^2Q = 0$$

For simplification Let $\frac{R}{L} = 2P$ and $\frac{1}{LC} = \omega^2$

$$\Rightarrow \qquad (D^2+2PD+\omega^2)Q = 0 \qquad \left(D=\frac{d}{dt}\right)$$

The A.E. is

$$m^2+2Pm+\omega^2 = 0$$

$$m = -P\pm\sqrt{P^2-\omega^2}$$

Case I: When $P > \omega$, roots are real and distinct therefore

$$Q = C_1 e^{(-P+\sqrt{P^2-\omega^2})t} + C_2 e^{(-P-\sqrt{P^2-\omega^2})t}$$

In this case Q is always positive, this is a condition of over damping.

Thus $$P > \omega$$

$$\frac{R}{2L} > \frac{1}{\sqrt{LC}}$$

or $$R > 2\sqrt{\frac{L}{C}}$$

Case II: When $P < \omega$, roots are imaginary

$$Q = e^{-P}\left\{C_1 \cos(\sqrt{\omega^2 - P^2})t + C_2 \sin(\sqrt{\omega^2 - P^2})t\right\}$$

Period of oscillation decreases and this condition is under damping.

Case III: Where $P = \omega$, roots are equal

$$Q = (C_1 + C_2 t)e^{-Pt}$$

This is condition for critically damped

Critical resistance is given by

$$P = \omega$$

$$\frac{R}{2L} = \frac{1}{\sqrt{LC}}$$

$\Rightarrow$ $$R = 2\sqrt{\frac{L}{C}}$$

Example 104: An inductance of 2 henries and resistance of 20 ohms are connected in series with an e.m.f. 100 sin150*t*. If the current is zero when $t = 0$, find the current at the end of 0.01 sec.

Solution: We know that the equation of *L-R* circuit is

$$L\frac{di}{dt} + Ri = E$$

$\Rightarrow$ $$L\frac{di}{dt} + Ri = 100 \sin 150t \qquad \left|\; E = 100 \sin 150t\right.$$

or $$\frac{di}{dt} + \frac{R}{L}i = \frac{100}{L}\sin 150t$$

$$\text{I.F.} = e^{\frac{R}{L}\int dt} = e^{\frac{R}{L}t}$$

The general solution is

$$i \cdot e^{\frac{R}{L}t} = \int \frac{100}{L}(\sin 150t)e^{\frac{R}{L}t}\,.\,dt + a$$

$$= \frac{100}{L}\left[\frac{e^{\frac{R}{L}t}}{(150)^2 + \frac{R^2}{L^2}}\left(\frac{R}{L}\sin 150t - 150\cos 150t\right)\right] + a$$

$$\Rightarrow \qquad i = \frac{100L}{(150L)^2 + R^2}\left(\frac{R}{L}\sin 150t - 150\cos 150t\right) + ae^{-\frac{R}{L}t} \qquad \text{...(1)}$$

Putting the value $i = 0$ at $t = 0$ in equation (1), we obtain

$$0 = a - \frac{100 \times 150L}{R^2 + (150)^2 L^2} \quad \text{or} \quad a = \frac{100 \times 150L}{R^2 + (150)^2 L^2}$$

From equation (1)

$$i = \frac{100L}{R^2 + (150)^2 L^2}\left[\frac{R}{L}\sin 150t - 150\cos 150t + 150e^{-\frac{R}{L}t}\right]$$

When $R = 20\ \Omega$, $L = 2$ henries, $t = 0.01$ sec, then, we get

$$i = \frac{100 \times 2}{400 + 90000}\left[\frac{20}{2}\sin 150(0.01) - 150\cos 150(0.01) + 150e^{-0.1}\right]$$

$$\frac{200}{90400}[10\sin 1.5 - 150\cos 1.5 + 150e^{-0.1}]$$

or

$$= \frac{1}{452}[135.3] \approx 0.299 \text{ Amp.} \quad \textbf{Ans.}$$

1.11.2 Problems of Newton's Law of Cooling

The Newton's law of cooling states that:

"The temperature of a body changes at a rate proportional to the difference in temperatures of the surrounding medium and that of the body itself".

Thus if T is the temperature of the body at a time t and T_0 is the temperature of the surrounding medium, then

$$\frac{dT}{dt} \propto -(T - T_0) \quad \text{or} \quad \frac{dT}{dt} = -k(T - T_0)$$

where k is the constant of proportionality.

Example 105: If the temperature of a body drops from 100°C to 60°C in one minute when the surrounding temperature is 20°C, find the time when the temperature will come down to 30°C.

Solution: We know that

$$\frac{dT}{dt} = -k(T - T_0) = -k(T - 20) \qquad \Big| \text{ As } T_0 = 20°C$$

$$\Rightarrow \qquad \frac{dT}{(T-20)} = -kdt$$

On integrating, we get

$$\log(T-20) = -kt + C \qquad ...(1)$$

Initially at $t = 0, T = 100°C$ here $C = \log 80°C$

Therefore (1) becomes $\log\frac{T-20}{80} = -kt$ ...(2)

Next it is given that, $T = 60°C$ when $t = 1$ min. therefore from (2)

$$\log\frac{60-20}{80} = -k \quad \Rightarrow \quad k = \log 2$$

From (2)

$$\log\frac{T-20}{80} = -t\log 2$$

$$\Rightarrow \qquad \log\frac{30-20}{80} = -t\log 2 \qquad \text{We are to find } t \text{ when } T \text{ becomes } 30°C$$

$$\Rightarrow \qquad -3\log 2 = -t\log 2 \quad \Rightarrow \quad t = 3 \text{ minutes}$$

Therefore after 3 minutes the body temperature will come down to 30°C. **Ans.**

Example 106: The rate at which a body cools is proportional to the difference between the temperature of the body and that of the surrounding air. If a body in air at 25°C will cool from 100°C to 75°C in one minute, find its temperature at the end of the three minutes.

Solution: We know that

$$\frac{dT}{dt} = -K(T-25) \quad \Rightarrow \quad \frac{dT}{T-25} = -Kdt$$

On integrating $\log(T-25) = -Kt + \log A$

$$\Rightarrow \qquad \log\frac{(T-25)}{A} = -Kt \quad \Rightarrow \quad T - 25 = Ae^{-Kt} \qquad ...(1)$$

When $t = 0$, then $T = 100$, from equation (1)

$$100 - 25 = A \quad \Rightarrow \quad A = 75$$

and when $t = 1$, then $T = 75$, $A = 75$, from (1)

$$75 - 25 = 75\, e^{-K}$$

$$\Rightarrow \qquad \frac{50}{75} = e^{-K} \quad \Rightarrow \quad \frac{2}{3} = e^{-K} \qquad ...(2)$$

When $t = 3$ minutes, then temperature is, from equation (1)

$$T = 25 + 75e^{-3K} = 25 + 75(e^{-K})^3 = 25 + 75\left(\frac{2}{3}\right)^3 \qquad \text{Using (2)}$$

$$\Rightarrow \qquad T = 25 + 75 \times \frac{8}{27} = 47.22°\text{C.} \quad \textbf{Ans.}$$

Example 107: An inductance (L) of 2.0 H and a resistance (R) of 20 Ω are connected in series with an e.m.f. E volt. If current (i) is zero, when $t = 0$, find current (i) at the end of 0.01 second if $E = 100$ V, using the following differential equation:

$$L\frac{di}{dt} + iR = E. \qquad \textit{(U.P.T.U. 2008)}$$

Solution: We have $L\dfrac{di}{dt} + iR = E$

$$\Rightarrow \qquad L\frac{di}{dt} + iR = 100 \qquad \Big| \text{As } E = 100 \text{ V}$$

or
$$\frac{di}{dt} + \frac{R}{L}i = \frac{100}{L}$$

which is linear

$$\therefore \qquad \text{I.F.} = e^{\frac{R}{L}\int dt} = e^{\frac{R}{L}t} \qquad \Big| \text{I.F.} = e^{\int P dx}$$

The general solution is

$$i.e^{\frac{R}{L}t} = \int \frac{100}{L} \times e^{\frac{R}{L}t}\, dt + a$$

$$= \frac{100}{L} \times \frac{e^{\frac{R}{L}t}}{\frac{R}{L}} + a = \frac{100}{R} e^{\frac{R}{L}t} + a$$

$$\Rightarrow \qquad i = \frac{100}{R} + ae^{-\frac{R}{L}t} \qquad \ldots(1)$$

Putting $i = 0$ at $t = 0$ in equation (1), we get

$$0 = \frac{100}{R} + a \Rightarrow a = -\frac{100}{R}$$

From equation (1)

$$i = \frac{100}{R}\left(1 - e^{-\frac{R}{L}t}\right)$$

when $R = 20$ Ω, $L = 2.0$ H, $t = 0.01$ second then, we get

$$i = \frac{100}{20}\left(1 - e^{-\frac{20}{2}\times 0.01}\right)$$

$$i = 5(1 - e^{-0.1}) = 5(0.095162581)$$

$$\Rightarrow \quad i = 0.475812 \text{ amp.} \qquad \textbf{Ans.}$$

1.11.3 Heat Conduction Problems

The amount of heat in a body becomes proportional to its mass and temperature.

Let Q is the amount of heat flowing per second through a slab of surface area A and thickness δx whose faces are at temperatures t and $t + \delta t$, is given by

$$Q = -KA\frac{dt}{dx}$$

where K is the coefficient of thermal conductivity of the material of the body.

Example 108: A pipe 20 cm in diameter contains steam at 160°C and is protected with a covering 5 cm thick for which K (the coefficient of thermal conductivity) = 10025. If the temperature of the outer surface of the covering is 50°C, find the temperature at half way through the covering under steady conditions.

Solution: We know that

$$Q = -KA\frac{dt}{dx} = -K(2\pi x)\frac{dt}{dx} \qquad \left| \begin{array}{l} A \text{ is the surface} \\ \text{area of the pipe} \\ \text{of length 1 cm} \\ i.e., A = 2\pi x \end{array} \right.$$

$$\Rightarrow \quad dt = -\frac{Q}{2K\pi}\left(\frac{dx}{x}\right) \qquad \text{...(1)}$$

We are given that $t = 160°C$ at $x = 10$ cm and $t = 50°C$ at $x = 15$ cm

From equation (1), $$\int_{160}^{50} dt = -\frac{Q}{2\pi K}\int_{10}^{15}\frac{dx}{x}$$

$$\Rightarrow \quad -110°C = -\frac{Q}{2\pi K}\log\frac{15}{10}$$

$$\Rightarrow \quad \frac{Q}{2\pi K} = \frac{110}{\log 1.5} \qquad \text{...(2)}$$

Let $t = T$ when $x = 12.5$ (that is half way through the covering)

We have $$\int_{160}^{T} dt = -\frac{Q}{2\pi K}\int_{10}^{12.5}\frac{dx}{x} \Rightarrow T - 160°C = -\frac{Q}{2\pi K}\left[\log x\right]_{10}^{12.5}$$

Using (2), we get

$$T = 160°C - 110°C\frac{\log 1.25}{\log 1.5} = 99.5°C. \qquad \textbf{Ans.}$$

1.11.4 Problems on Chemical Reactions

Example 109: Uranium disintegrates at a rate proportional to the amount present at the instant. If m_1 and m_2 grams of uranium are present at times t_1 and t_2 respectively, find the half-life of uranium.

Solution: The differential equation of disintegration of uranium is

$$\frac{dm}{dt} = -Km \qquad ...(1)$$

where m is the mass of uranium at any time t and K is any constant.

From equation (1)

$$\int_{m_1}^{m_2} \frac{dm}{m} = -\int_{t_1}^{t_2} K\,dt \quad \Rightarrow \quad \log m_2 - \log m_1 = -K(t_2 - t_1)$$

$$\Rightarrow \qquad K = \frac{\log\left(\frac{m_1}{m_2}\right)}{(t_2 - t_1)}$$

Now, we want to compute time T in which the uranium will disintegrate to half its original amount M (say). As such

$$\int_{M}^{M/2} \frac{dm}{m} = -K\int_0^T dt \quad \Rightarrow \quad \log\left(\frac{M/2}{M}\right) = -KT$$

$$KT = \log 2 \quad \Rightarrow \quad T = \frac{\log 2}{K} = \frac{(t_2 - t_1)\log 2}{\log(m_1/m_2)}.$$ **Ans.**

EXERCISE 1.19

1. A condenser of capacitance C is charged through a resistance R by a steady voltage V. The charge Q on the plate is given by $R\frac{dQ}{dt} + \frac{Q}{C} = V.$

If initially the plate is chargeless find the charge and the current at time t.

$$\left[\textbf{Ans.}\, Q = VC(1 - e^{-\frac{t}{RC}}), i = \frac{V}{R}e^{-\frac{t}{RC}}\right]$$

2. An inductance of 2 henries and resistance of 20 ohms are connected in series with an e.m.f. E volts. If the current is zero when $t = 0$, find the current at the end of 0.01 sec if $E = 100$ volts. $[\textbf{Ans.}\, i = 0.475\ (Approx.)]$

3. Solve $L\frac{di}{dt} + Ri = 200\cos 300t,$ when $R = 100$, $L = 0.05$ and find i given that $i = 0$ when $t = 0$. What value does i approach after a long time.

$$\left[\textbf{Ans.}\ i = \frac{40}{409}(20\cos 300t + 3\sin 300t) - \frac{809}{409}e^{-200t}; \frac{40}{\sqrt{409}}\right]$$

4. When a resistance R ohms and a capacitance C farads are connected in series with an e.m.f. E volts, the current i amperes is given by $R\frac{di}{dt}+\frac{i}{C}=\frac{dE}{dt}$.

   If $R = 1000$ ohms, $C = 50 \times 10^{-1}$ farads, $i = 10$ amperes and $t = 0$. Find the current for $t = 1$ sec and $E = 100 \sin 120\,\pi t$ volts. **[Ans.** 8.187amperes**]**

5. A cup of tea at temperature 100°C is placed on a room whose temperature is 15°C and it cools to 60°C in 5 minutes. Find its temperature after a further interval of 5 minutes. **[Ans.** 38.9°C**]**

6. Water at temperature 100°C cools in 10 minutes to 80°C in a room of temperature 25°C. Find (*i*) the temperature of water after 20 minutes (*ii*) the time when the temperature is 40°C. **[Ans.** (*i*) 65.5°C, (*ii*) 51.9 minutes**]**

7. If the air is maintained at 30°C and the substance cools from 80°C to 60°C in 12 minutes, find the temperature of the body after 24 minutes. **[Ans.** 48°C**]**

8. Heat flows through the wall of a hollow sphere of inner and outer radii r_1 and r_2 respectively in a steady state. The temperature u at a distance r $(r_1 < r < r_2)$ from the centre of the sphere, is given by $r\frac{d^2u}{dr^2}+2\frac{du}{dr}=0$. If u_1 and u_2 are the temperatures of the inner and the outer surfaces of the sphere, find u in terms of r. $\left[\textbf{Ans. } u=\frac{1}{r_1-r_2}\left[u_1r_1-u_2r_2\right]-\frac{r_1r_2(u_1-u_2)}{r}\right]$

9. The rate of which certain mass of radium decomposes is at any instant proportional to the mass and the present mass decreases to half its original values in 1500 years. If the initial mass be 100 mg in how many years will it be 80 mg? $\left[\textbf{Ans.}\ 1500\frac{\log\frac{4}{5}}{\log\frac{2}{5}}\text{ years}\right]$

10. In a certain chemical reaction, the rate of conversion of a substance at time t is proportional to the quantity of the substance still untransformed at the instant. At the end of one hour 60 gms remain and at the end of four hour 21 gms. How many grams of the first substance were there initially? **[Ans.** 85 grams.**]**

11. The temperature T of a body satisfies the equation $\frac{dT}{dt}=-k(T-T_0)$ where T_0 is the temperature of the surrounding medium. A thermometer which reads 75°F indoors is taken outdoors, after 5 minutes it reads 65°F and after another 5 minutes, it reads 60°F. What is the outside temperature ? **[Ans.** 55°F**]**

1.11.5 Problems on Vertical Motion

Example 110: A particle falls under gravity in resisting medium whose resistance varies with velocity v. Find the relation between distance and velocity if initially the particle starts from rest. *(U.P.T.U. 2003)*

Solution: Air resistance $\propto v = kv$

$\therefore$ Upthrust $= mkv$

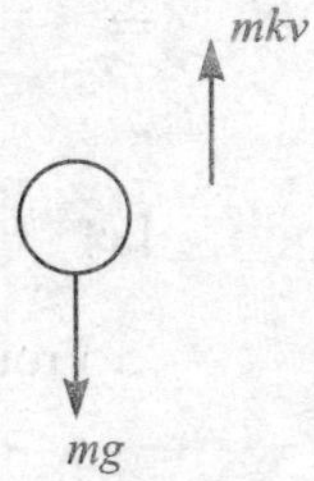

Now by Newton's second law of motion the equation of motion of the body

$$m\frac{d^2x}{dt^2} = mg - mkv \quad \Rightarrow \quad mv\frac{dv}{dx} = mg - mkv$$

$$\Rightarrow \quad \frac{vdv}{dx} = g - kv \quad \Rightarrow \quad \frac{vdv}{g - kv} = dx$$

$$\Rightarrow \quad \frac{1}{k}\left(\frac{-g + kv + g}{g - kv}\right)dv = dx$$

$$-\frac{dv}{k} + \frac{g}{k(g - kv)}dv = dx$$

Integrating, we get

$$-\frac{v}{k} + \frac{g}{k}\left(-\frac{1}{k}\right)\log(g - kv) = x + c$$

$$-\frac{v}{k} - \frac{g}{k^2}\log(g - kv) = x + c \qquad \ldots(1)$$

Initially $\quad x = 0,\ v = 0$

$$\Rightarrow \quad -\frac{g}{k^2}\log g = c$$

Equation (1) becomes

$$-\frac{v}{k} - \frac{g}{k^2}\log(g - kv) = x - \frac{g}{k^2}\log g$$

$$\Rightarrow \quad -\frac{v}{k} - \frac{g}{k^2}\log\frac{g - kv}{g} = x. \quad \textbf{Ans.}$$

Example 111: A 4 kg object falls from rest of time $t = 0$ in a medium offering a resistance in kg numerically equal to twice its instantaneous velocity in m/sec. Find the velocity and distance travelled at any time $t > 0$, and also the limiting velocity. *(U.P.T.U. 2007)*

Solution: Air resistance $= 2v$

$\therefore$ Upthrust $= 2 \times 4v = 8v$ | As $m = 4$ kg

By Newton's second law of motion the equation of motion of body

$$4\frac{d^2x}{dt^2} = 4g - 8v \Rightarrow \frac{d^2x}{dt^2} = g - 2\frac{dx}{dt}$$

$$\Rightarrow \quad \frac{d^2x}{dt^2} + 2\frac{dx}{dt} = g \qquad ...(1)$$

Let $\quad \frac{dx}{dt} = p \Rightarrow \frac{d^2x}{dt^2} = \frac{dp}{dt}$

$\therefore$ From equation (1), we get

$$\frac{dp}{dt} + 2p = g \qquad ...(2)$$

which is linear differential equation of first order.

$$\therefore \quad I.F. = e^{\int 2dt} = e^{2t}$$

Solution of equation (2) is $p \cdot e^{2t} = \int ge^{2t} \cdot dt + c = \frac{g}{2}e^{2t} + c$

$$\Rightarrow \quad p = \frac{g}{2} + ce^{-2t} \Rightarrow \frac{dx}{dt} = \frac{g}{2} + ce^{-2t} \qquad ...(3)$$

At $t = 0$, $\frac{dx}{dt} = 0$ which gives $c = -\frac{g}{2}$

From equation (3) $\quad \frac{dx}{dt} = \frac{g}{2}(1 - e^{-2t}) \Rightarrow$ Velocity $= \frac{g}{2}(1 - e^{-2t})$. **Ans.**

Again integrating above equation, we get

$$x = \frac{gt}{2} + \frac{g}{4}e^{-2t} + c_1 \qquad ...(4)$$

At $t = 0$, $x = 0 \Rightarrow 0 = g/4 + c_1 \Rightarrow c_1 = -g/4$

From equation (4) $\quad x = \frac{gt}{2} + \frac{g}{4}(e^{-2t} - 1) \Rightarrow$ Distance $= \frac{gt}{2} + \frac{g}{4}(e^{-2t} - 1)$. **Ans.**

and Limiting velocity $= \left(\lim_{t\to\infty} \frac{dx}{dt}\right) = \lim_{t\to\infty} \frac{g}{2}(1 - e^{-2t}) = \frac{g}{2}$. **Ans.**

Example 112: If $\frac{d^2r}{dt^2} = -\frac{gR^2}{r^2}$ where R is the radius of the earth and $(r \geq R)$ is the distance of an object at any time t, projected vertically upwards from the centre of the earth with velocity $v_0 = \sqrt{2gR}$, show that the object will never return to the earth.

Solution: We have $\frac{d^2r}{dt^2} = \frac{d}{dt}\left(\frac{dr}{dt}\right) = \frac{d}{dt}v$, where $v = \frac{dr}{dt}$

$$= \frac{dv}{dr}\cdot\frac{dr}{dt} = v\frac{dv}{dr}$$

$\therefore$ The given equation becomes $v\frac{dv}{dr} = -\frac{gR^2}{r^2} \quad \Rightarrow \quad 2vdv = \frac{-2gR^2 dr}{r^2}$

On integrating, we get

$$v^2 = \frac{2gR^2}{r} + C$$

We are given that, $\quad v = v_0 = \sqrt{2gR}$, when $r = R$.

$$\therefore \quad 2gR = \frac{2gR^2}{R} + C \Rightarrow C = 0$$

$$\therefore \quad v^2 = \frac{2gR^2}{r} \quad \Rightarrow \quad v = R\sqrt{\frac{2g}{r}}$$

Thus, the object will not come to rest at $v = 0$ only when $r = \infty$, meaning thereby that the object will never come back to earth.

EXERCISE 1.20

1. An object of mass m is falling under gravity and facing air resistance proportional to the square of the velocity. If after falling through a distance x it possess a velocity v at that instant prove that $\frac{2Kx}{m} = \log\frac{a^2}{a^2 - v^2}$, where K is the constant of proportionality and $\frac{mg}{K} = a^2$.
2. A body of mass of m falls from rest under gravity in a field whose resistance is mK times the velocity of the body. Find the terminal velocity of the body and also the time taken to acquire one-half of its limiting speed. $\left[\textbf{Ans.}\ \frac{g}{K}, \frac{\log_e^2}{K}\right]$
3. Two particles fall freely, one in a medium whose resistance is equal to K times the velocity and the other in medium whose resistance is equal to K times the square of the velocity. If V_1 and V_2 are there maximum velocities respectively show that $V_1 = V_2^2$.
4. A particle falls in a vertical line gravity (supposed constant) and the force of air resistance to its motion is proportional to its velocity. Show that its velocity cannot exceed a particular limit. $\left[\textbf{Ans.}\ V = \frac{g}{k}\right]$

1.11.6 Simple Harmonic Motion (S.H.M.)

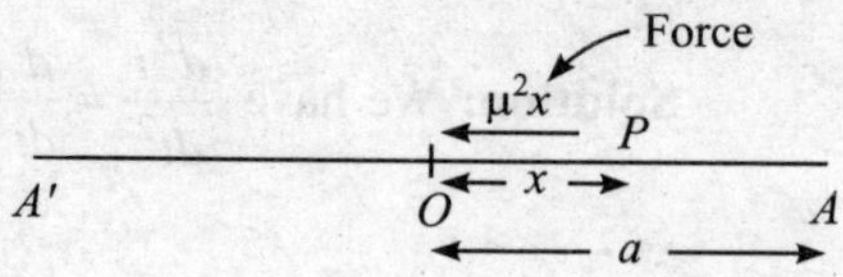

When the acceleration of a moving particle is proportional to its displacement from a fixed point O and is always directed towards it, it is called a simple harmonic motion.

The equation of motion is

$$\frac{d^2x}{dt^2} = -\mu^2 x \qquad ...(i)$$

which gives,

$$x = C_1 \cos\mu t + C_2 \sin\mu t \qquad ...(ii)$$

The velocity of particle at P *is* $\frac{dx}{dt} = -C_1\mu\sin\mu t + C_2\mu\cos\mu t \qquad ...(iii)$

At A, velocity $\frac{dx}{dt} = 0, x = (OA = a), t = 0$

Putting $x = a$, $t = 0$ in (*ii*), we get $C_1 = a$

and putting $\frac{dx}{dt} = 0$, $t = 0$ in (*iii*), we get $C_2 = 0$

From equations (*ii*) and (*iii*), we get

$$x = a\cos\mu t \text{ and } \frac{dx}{dt} = -a\mu\sin\mu t$$

$$\Rightarrow \quad \boxed{\text{Velocity} = -\mu\sqrt{a^2 - x^2}}$$

The maximum distance from the centre is a, is called the amplitude.

The time of one complete oscillation is $\frac{2\pi}{\mu}$ called the period and the frequency $= \frac{1}{\text{Period}} = \frac{\mu}{2\pi}$.

Example 113: A particle moves with S.H.M. in a straight line under the action of a force which is proportional to the distance of the particle from $x = 0$. If it starts at $x = 5$ cm with a velocity 10 cm/sec and it reaches an extreme position $x = 10$ cm, at what speed does it pass through the origin?

Solution: We know that

$$V^2 = \mu(a^2 - x^2) \qquad ...(1)$$

$$x = 5 \text{ cm}, V = 10 \text{ cm/sec}$$

$$\Rightarrow \quad 100 = \mu(a^2 - 25) \qquad ...(2)$$

Since the maximum distance from the centre is 10 cm

$\therefore$ Amplitude $a = 10$ cm

$$\Rightarrow \quad V^2 = \mu(100 - x^2) \qquad ...(3)$$

Dividing equation (3) by (2), we obtain

$$\frac{V^2}{100} = \frac{100 - x^2}{100 - 25}$$

$$\Rightarrow \qquad V^2 = \frac{4}{3}(100 - x^2)$$

The speed through the origin ($x = 0$) is

$$V^2 = \frac{4}{3}(100 - 0) = \frac{400}{3} = 133.33$$

$$V = \sqrt{133.33} = 11.546 \text{ cm/sec.} \qquad \textbf{Ans.}$$

Example 114: A point moves in a straight line towards a centre of force $\mu/(\text{distance})^3$, starting from rest at a distance 'a' from the centre of force; show that the time of reaching a point distant b from the centre of force is $\left(\dfrac{a}{\sqrt{u}}\right)\sqrt{a^2 - b^2}$ and that its velocity is $\dfrac{\sqrt{\mu}}{ab}\sqrt{a^2 - b^2}$. *(U.P.T.U. 2001)*

Solution: Let O is the centre of force and let a point moves from P towards the centre of force O.

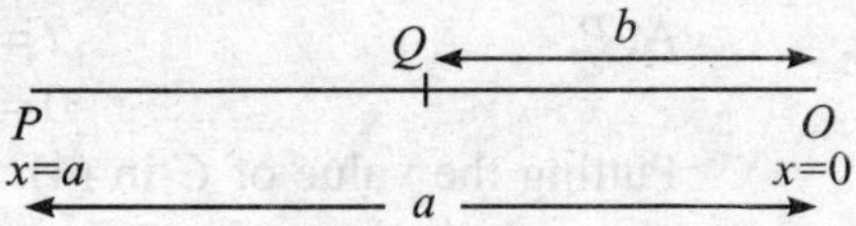

The equation of motion is

$$\frac{d^2x}{dt^2} = -\frac{\mu}{x^3} \qquad \ldots(1)$$

$$\Rightarrow \qquad 2\frac{dx}{dt}\frac{d^2x}{dt^2} = -2\frac{\mu}{x^3}\frac{dx}{dt}$$

On integrating, we get

$$\left(\frac{dx}{dt}\right)^2 = -2\mu\frac{x^{-2}}{-2} + C = \frac{\mu}{x^2} + C$$

$$\Rightarrow \qquad \frac{dx}{dt} = \pm\sqrt{\frac{\mu}{x^2} + C} \qquad \ldots(2)$$

At P $\dfrac{dx}{dt} = 0$ and $x = a$

$$\Rightarrow \qquad 0 = \sqrt{\frac{\mu}{a^2} + C} \quad \Rightarrow \quad C = -\frac{\mu}{a^2}$$

From equation (2) $$\frac{dx}{dt} = \pm\sqrt{\frac{\mu}{x^2} - \frac{\mu}{a^2}} = \pm\sqrt{\mu}\,\frac{\sqrt{a^2 - x^2}}{ax} \qquad \ldots(3)$$

The velocity at $x = b$ is

$$V = \pm\sqrt{\mu}\,\frac{\sqrt{a^2 - b^2}}{ab} \qquad \textbf{Ans.}$$

or $$V = -\sqrt{\mu}\frac{\sqrt{a^2 - b^2}}{ab} \quad \text{(As the point 'P' is moving towards } O)$$

From equation (3) $$\frac{dx}{dt} = -\sqrt{\mu}\frac{\sqrt{a^2 - x^2}}{ax}$$

$$\Rightarrow \quad dt = -\frac{1}{\sqrt{\mu}} \cdot \frac{xa}{\sqrt{a^2 - x^2}} dx$$

On integrating, we get

$$t = \frac{a}{\sqrt{\mu}}\sqrt{a^2 - x^2} + C \quad \text{....(4)}$$

At P, $t = 0$, $x = a$, in (4), we get

$$0 = C$$

Putting the value of C in (4), we have

$$t = \frac{a}{\sqrt{\mu}}\sqrt{a^2 - x^2}$$

At $$x = b$$

$$t = \frac{a}{\sqrt{\mu}}\sqrt{a^2 - b^2}. \quad \textbf{Proved.}$$

Example 115: A particle moves with S.H.M. of period 6 secs, travels 6 cm from the position of rest in 1 sec. Find the amplitude, the maximum velocity and the velocity at the end of 1 sec.

Solution: $T = \frac{2\pi}{\sqrt{\mu}} \Rightarrow 6 = \frac{2\pi}{\sqrt{\mu}} \Rightarrow \sqrt{\mu} = \frac{\pi}{3}$

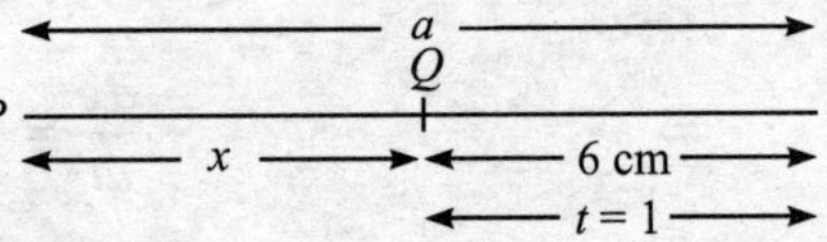

Let a be the amplitude OP

$OQ = 6$ cm, $PQ = x = a - 6$, $t = 1$ sec.

We have $$x = a\cos\sqrt{\mu}\, t$$

$$a - 6 = a\cos\sqrt{\mu} = a\cos\frac{\pi}{3} = \frac{a}{2} \quad \Big| \text{ Taken } \mu \text{ as } \sqrt{\mu}$$

$$\Rightarrow \quad a - \frac{a}{2} = 6 \Rightarrow \frac{a}{2} = 6 \Rightarrow a = 12$$

Maximum velocity = $\sqrt{\mu}\, a = \frac{\pi}{3} \times 12 = 4\pi = 12.56$ cm/sec.

Velocity V at the end of one second = $\sqrt{\mu}\sqrt{a^2 - x^2} = \frac{\pi}{3}\sqrt{144 - 36}$

$$= \frac{\pi}{3}\sqrt{108} = \frac{6\pi\sqrt{3}}{3} \text{ cm/sec.} \quad \textbf{Ans.}$$

EXERCISE 1.21

1. A particle moving in a straight line with S.H.M. with amplitude 1.60 meters and the time period 4 seconds. Find the time required by the particle in passing between points which are at distances 1.2 and 0.6 meters from the centre of force and are on the same side of it.

[**Ans.** 0.30 sec]

2. A moving body is opposed by a force per unit mass of value Cx and the resistance per unit mass of value bv^2 where x and v are the displacement and velocity of the particle at that instant. Find the velocity of the particle in terms of x, if it starts from rest.

$$\left[\textbf{Ans. } V^2 = \frac{-Cx}{b} + \frac{C}{2b^2} - \frac{C}{2b^2} e^{-2bx}\right]$$

3. A particle of mass m is oscillating in a straight line about a centre of force O towards which the force is mn^2r when at a distance r. The particle recives a blow in the direction of motion which generates a velocity na when it is at a distance $\frac{a\sqrt{3}}{2}$ from O. a is the amplitude of oscillation. If this velocity is away from O show that the new amplitude is $a\sqrt{3}$.

4. A point executes S.H.M. such that in two of its positions; the velocity are u, v and the corresponding accelerations α, β. Show that the distance between the positions is $(v^2 - u^2)/(\alpha + \beta)$, and find the amplitude of motion.

5. A particle is performing a simple harmonic motion of period T about a centre O and it passes through a point P, where $OP = b$ with velocity v in the direction OP. Prove that the time which elapses before it returns to P is $\frac{T}{\pi}\tan^{-1}\left(\frac{VT}{2\pi b}\right)$.

1.11.7 Oscillations of a Spring

(*i*) Free Oscillation of a Spring

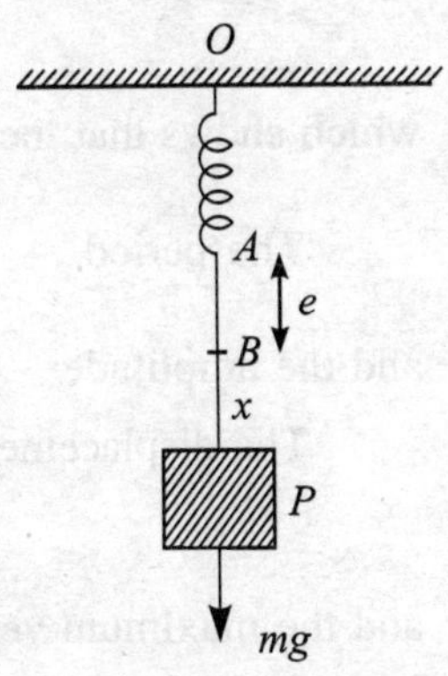

Let a mass m suspended from the end A of a spring which is fixed at the point O.

Let ($e = AB$) be the elongation produced in the spring by the mass m and suppose that K is the restoring force per unit stretch of the spring due to elasticity. Then, for the equilibrium at B, we have

$$mg = Ke \qquad \ldots(i)$$

Next, let the mass be displaced and the motion starts. Let P is the position of the mass at any time t where $BP = x$. The acceleration of the mass at P is $\frac{d^2x}{dt^2}$ and the forces acting upon it are its own weight mg and the restoring force $K(e + x)$ upwards.

By Newton's second law, the equation of motion is

$$m\frac{d^2x}{dt^2} = mg - K(e+x) = Ke - Ke - Kx = -Kx$$

or
$$\frac{d^2x}{dt^2} + \frac{K}{m}x = 0 \quad \Rightarrow \quad \frac{d^2x}{dt^2} + \mu^2 x, \text{ where } \mu^2 = \frac{K}{m}.$$

This equation represents free oscillations of the spring, which are of simple harmonic from centre having at *B*. The period *T* of oscillation is given by

$$T = \frac{2\pi}{\sqrt{\mu}} = 2\pi\sqrt{\frac{m}{K}} = 2\sqrt{\frac{e}{g}}.$$

Example 116: A spring is such that a pull 20 kg weight can stretch it by 10 cm. Then a body weighing 10 kg hangs from the spring. The body is further pulled down by 20 cm below the static equilibrium position and then released. Find the displacement of the body from its equilibrium position at any *t* and the period of oscillation.

Solution : Since a pull of 20 kg weight can stretch the spring by 10 cm (= 0.1 m) $\Rightarrow e = 0.1$

We have $$mg = Ke \quad \Rightarrow \quad 20 = K \times 0.1 \quad \Rightarrow \quad \text{K} = 200 \text{ kg/meter}.$$

Next when a body of 10 kg weight hangs from the spring it would stretch it by a length e_1, (say), then in the static equilibrium state, we have

$$10 = Ke_1 = 200\, e_1. \quad \therefore \quad e_1 = 0.05 \text{ meter}$$

Further, when the body is pulled down to 20 cm (= 0.2m) below the static equilibrium, the equation of motion is

$$m\frac{d^2x}{dt^2} = mg - K(x + e_1) = Ke_1 - Kx - Ke_1 = -Kx$$

$$\therefore \quad \frac{d^2x}{dt^2} + \mu^2 x = 0 \text{ where } \mu^2 = \frac{K}{m} = \frac{200}{\frac{10}{g}} = 20g = 20(9.8)$$

$$\Rightarrow \quad \mu^2 = 2 \times 98 \quad \Rightarrow \quad \mu = 14.0$$

which shows that the motion of the body is simple harmonic.

The period $$= \frac{2\pi}{\mu} = \frac{2\times 3.1415}{14} = 0.45 \text{sec}.$$

and the amplitude $= 0.2$ m $= 20$ cm.

The displacement of the body at any time *t* is given by

$$x = 0.2\cos\mu t = 0.2\cos 14t$$

and the maximum velocity $= 0.2\,\mu = 2.8$ m/sec. **Ans.**

Example 117: In a spring if T_1 and T_2 be the periods corresponding to two different attached weights and s_1 and s_2 the statical extensions due to these weights prove that, $g = \dfrac{4\pi^2(s_1 - s_2)}{T_1^2 - T_2^2}$.

Solution: Let the stiffness of the spring be K. Let T_1 and T_2 be the periods of masses m_1 and m_2 respectively when attached to spring, then we have

$$T_1 = 2\pi\sqrt{\left(\frac{m_1}{K}\right)} \quad \Rightarrow \quad T_1^2 = 4\pi^2\frac{m_1}{K} \qquad \text{...(1)}$$

$$T_2 = 2\pi\sqrt{\left(\frac{m_2}{K}\right)} \quad \Rightarrow \quad T_2^2 = 4\pi^2\frac{m_2}{K} \qquad \text{...(2)}$$

Subtracting equation (2) from equation (1), we get

$$T_1^2 - T_2^2 = \frac{4\pi^2}{K}(m_1 - m_2) \qquad \text{...(3)}$$

During static equilibrium spring force equal to gravitational force *i.e.*, $Ks = mg$

$$\therefore \quad Ks_1 = m_1 g \qquad \text{...(4)}$$

$$Ks_2 = m_2 g \qquad \text{...(5)}$$

Subtracting equation (4) from equation (5), we get

$$K(s_1 - s_2) = (m_1 - m_2)g$$

$$\Rightarrow \quad \frac{(m_1 - m_2)}{K} = \frac{1}{g}(s_1 - s_2)$$

Substituting the value of $\left(\dfrac{m_1 - m_2}{K}\right)$ in equation (3), we get

$$T_1^2 - T_2^2 = 4\pi^2 \cdot \frac{1}{g}(s_1 - s_2)$$

$$\Rightarrow \quad g = \frac{4\pi^2(s_1 - s_2)}{T_1^2 - T_2^2}. \quad \textbf{Proved.}$$

(*ii*) Damped Free Oscillations in Mechanical System

If the motion of the mass m be subjected to a resisting force then the oscillation are called damped.

The damping force may be either constant or proportional to the velocity $= r\dfrac{dx}{dt}$, say, where r is a constant.

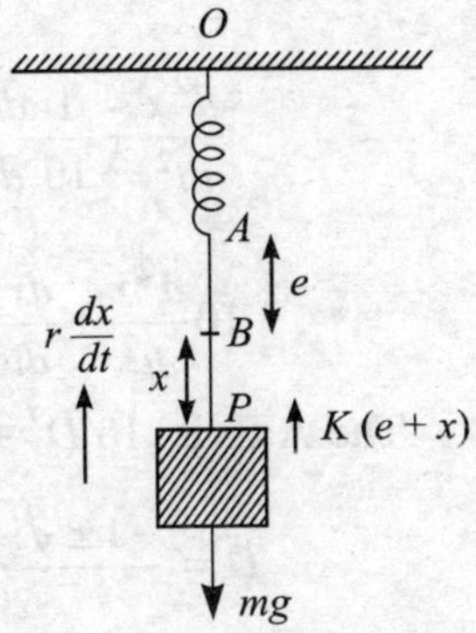

Then the equation of motion becomes

$$m\frac{d^2x}{dt^2} = mg - K(e + x) - r\frac{dx}{dt}$$

$$= -Kx - r\frac{dx}{dt} \qquad \text{As } mg = ke$$

or $$\frac{d^2x}{dt^2} + 2\lambda\frac{dx}{dt} + \mu^2 x = 0 \qquad ...(i)$$

where $\mu^2 = \frac{K}{m}$ and $\lambda = \frac{r}{2m}$.

Example 118: A spring stretches by 0.5 cm under a load of 10 gm. The spring fixed at the upper end supports a weight 980 gm at the lower end. The resistance (in gm wt) to the motion of the weight is one-tenth of the velocity of the weight in cm/sec. The weight is now pulled down by $\frac{1}{4}$ cm below its equillibrium position and then released. Express the distance of the weight from its equillibrium position at any time t during its first upward motion.

Solution: Let O be the fixed upper end of the spring.

Since load of 10 gm attached at the lower end of the spring stretches by 0.5 cm, we have

$$10 = K \cdot \frac{1}{2} \quad \Rightarrow \quad K = 20 \text{ gm/cm}.$$

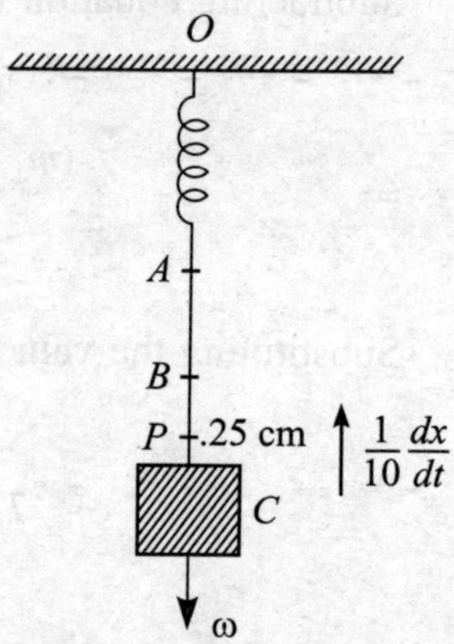

Let B be the equilibrium after attaching a weight 980 gm at A. After the release of the spring at C, at any time t let the weight be at P so that $BP = x$, then the tension is given by

$$T = K.AP = 20\ (49 + x) = 980 + 20x$$

The resistance to the motion $= \frac{1}{10}\frac{dx}{dt}$.

Therefore, the equation of the motion of the weight is

$$m\frac{d^2x}{dt^2} = W - T - \frac{1}{10}\frac{dx}{dt}$$

or $$\frac{980}{g}\frac{d^2x}{dt^2} = 980 - (980 + 20x) - \frac{1}{10}\frac{dx}{dt} \qquad \text{since } g = 980 \text{ cm/s}^2$$

or $$\frac{d^2x}{dt^2} + \frac{1}{10}\frac{dx}{dt} + 20x = 0$$

or $$10\frac{d^2x}{dt^2} + \frac{dx}{dt} + 200x = 0 \qquad ...(1)$$

The A.E. is $10\,D^2 + D + 200 = 0$

$$D = \frac{-1 \pm \sqrt{1 - 8000}}{20} = -0.05 \pm 4.5\,i$$

$\therefore$ The solution of equation (1) is

$$x = e^{-0.05t}(C_1 \cos 4.5t + C_2 \sin 4.5t) \qquad ...(2)$$

Initially at $t = 0$, $\quad x = \frac{1}{4}$ cm.

From equation (2), $\quad \frac{1}{4} = C_1 \Rightarrow C_1 = 0.25$

Now $$\frac{dx}{dt} = -0.05e^{-0.05t}(C_1 \cos 4.5t + C_2 \sin 4.5t)$$

$$+ e^{-0.05t}(-4.5C_1 \sin 4.5t + 4.5C_2 \cos 4.5t) \qquad ...(3)$$

At $t = 0$, $\quad \frac{dx}{dt} = 0$

From equation (3), we get $\quad 0 = -0.05\, C_1 + 4.5\, C_2 \Rightarrow C_2 = 0.003$

Hence $\quad x = e^{-0.05t}(0.25 \cos 4.5t + 0.003 \sin 4.5t)$. **Ans.**

(*iii*) Forced Oscillation (without Damping)

Let an external force is impressed upon the system for instance, the point of support of the spring is also vibrating with an external force, say $Q \cos nt$, the resulting motion is called forced oscillatory system (without damping). Then the equation of the motion is

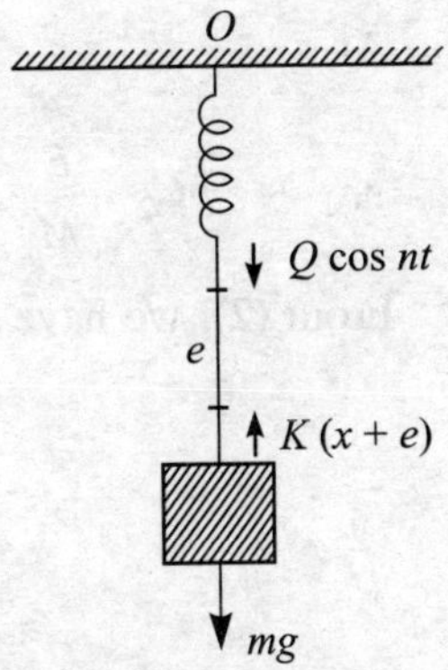

$$m\frac{d^2x}{dt^2} = mg - K(x+e) + Q\cos nt$$

$$\Rightarrow \quad \frac{d^2x}{dt^2} + \frac{K}{m}x = \frac{Q}{m}\cos nt \quad \boxed{\text{As } mg = Ke}$$

$$\Rightarrow \quad \frac{d^2x}{dt^2} + \mu^2 x = E\cos nt, \text{ where } \mu^2 = \frac{K}{m}, E = \frac{Q}{m}.$$

Example 119: A mass M suspended from the end of a helical spring is subjected to a periodic force $f = F \sin \omega t$ in the direction of its length. The force f is measured positive vertically downwards and at zero time M is at rest. If the spring stiffness is S, prove that the displacement of M at time t from the commencement of motion is given by $x = \frac{F}{M(p^2 - \omega^2)}\left(\sin \omega t - \frac{\omega}{p}\sin pt\right)$, where $p^2 = \frac{S}{M}$ and damping effects are neglected. *(U.P.T.U. 2000)*

Solution: Let x be the displacement from the equilibrium position at any time t then the equation of the motion is

$$M\frac{d^2x}{dt^2} = -Sx + F\sin \omega t$$

or $$\frac{d^2x}{dt^2} + \frac{S}{M}x = \frac{F}{M}\sin \omega t$$

or $$\frac{d^2x}{dt^2} + p^2 x = \frac{F}{M}\sin \omega t \qquad \left(\text{As } \frac{S}{M} = p^2\right) \qquad ...(1)$$

The A.E. is $m^2 + p^2 = 0 \quad \Rightarrow \quad m = \pm\, ip$

$$\text{C.F.} = C_1 \cos pt + C_2 \sin pt$$

$$\text{P.I.} = \frac{1}{(D^2 + p^2)}\left(\frac{F}{M}\sin \omega t\right) = \frac{F}{M}\cdot\frac{1}{(-\omega^2 + p^2)}\sin \omega t$$

$\therefore$ $$x = C_1 \cos pt + C_2 \sin pt + \frac{F}{M}\cdot\frac{1}{(p^2 - \omega^2)}\sin \omega t \qquad ...(2)$$

Initially, at $t = 0, \quad x = 0 \quad \therefore \quad C_1 = 0.$

Differentiating equation (2) w.r.t. 't', we get

$$\frac{dx}{dt} = -pC_1 \sin pt + pC_2 \cos pt + \frac{F}{M}\cdot\frac{\omega}{p^2 - \omega^2}\cos \omega t$$

At, $t = 0, \quad \frac{dx}{dt} = 0,$

$\therefore$ $$pC_2 + \frac{F}{M}\cdot\frac{\omega}{p^2 - \omega^2} = 0 \quad \text{or} \quad C_2 = -\frac{\omega}{p}\cdot\frac{F}{M(p^2 - \omega^2)}$$

From (2), we have

$$x = -\frac{\omega}{p}\frac{F}{M(p^2 - \omega^2)}\sin pt + \frac{F}{M}\cdot\frac{1}{p^2 - \omega^2}\sin \omega t$$

or $$x = \frac{F}{M(p^2 - \omega^2)}\left(\sin \omega t - \frac{\omega}{p}\sin pt\right) \quad \textbf{Ans.}$$

EXERCISE 1.22

1. A spring which stretches by an amount e under a force mk^2e is suspended from a support P and has a mass m at its lower end. At time $t = 0$, the mass is at rest in its equilibrium position at a point below P. A vertical oscillation is now given to the support P such that ($t > 0$) its displacement below its initial position is $a \sin nt$. Show that the displacement x of the mass below A is given by

$$\frac{d^2x}{dt^2} + k^2 x = k^2 a \sin nt$$

Hence, show that if $n \neq k$, the displacement is given by $x = \dfrac{Ka}{K^2 - n^2}(K \sin nt - n \sin Kt)$ what happens, when $n = K$.

2. A mass of 200 gm is tied at the end of a spring which extends to 4 cm under a force 196,000 dynes. The spring is pulled 5 cm and released. Find the displacement, *t* seconds after release, if there be a damping force of 2000 dynes per cm per second. What should be the damping force for the dead beat motion?

$$\left[\textbf{Ans. } x = e^{-5t}\left(5\cos\sqrt{220}\,t + \frac{25}{\sqrt{220}}\sin\sqrt{220}\,t\right),\ p = 6261\right]$$

3. A spring is such that it would stretch 72 mm by a mass of 15 kg. A mass of 30 kg is attached and brought to rest. The resistance of the medium is numerically equal to $20\dfrac{dr}{dt}$ Newtons. Find the equation of the motion of the weight if it is pulled down 140 mm and given an upward velocity 3 m/sec.

$$\left[\textbf{Ans. } x = e^{-2t}\ (0.14\cos 8t - 0.34\sin 8t)\right]$$

4. A particle of mass *m* is attached to one end of a light spring of modulus λ, the other end being fixed and the spring verticle. Prove that the velocity of the particle when it has travelled a distance '*a*' is $\sqrt{2ag - \dfrac{\lambda}{m}a^2}$.

1.11.8 Electric Circuits

The equation of vibration of charge in the condenser, is

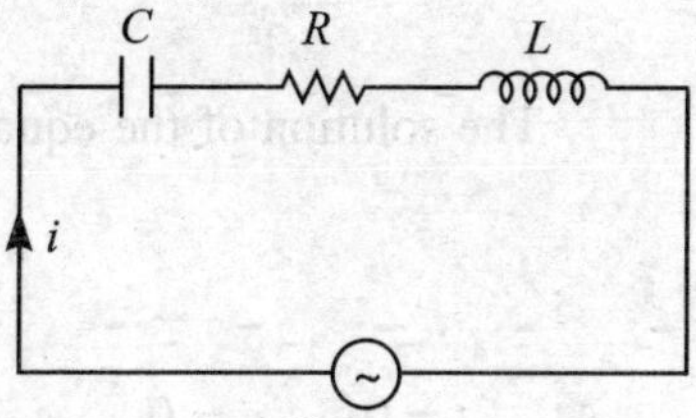

$$L\frac{di}{dt} + Ri + \frac{q}{C} = Q\cos nt$$

or
$$L\frac{d^2q}{dt^2} + R\frac{dq}{dt} + \frac{q}{C} = Q\cos nt$$

or
$$\frac{d^2q}{dt^2} + \frac{R}{L}\frac{dq}{dt} + \frac{q}{LC} = \frac{Q}{L}\cos nt$$

or
$$\frac{d^2q}{dt^2} + 2\lambda\frac{dq}{dt} + \mu^2 q = E\cos nt$$

where
$$2\lambda = \frac{R}{L},\ \mu^2 = \frac{1}{LC} \text{ and } E = \frac{Q}{L}.$$

Example 120: An uncharged condenser of capacity *C* is charged by applying an e.m.f. $E\sin\dfrac{t}{\sqrt{LC}}$, through leads of self-inductance *L* and negligible resistance. Prove that at time *t*, the charge on one of the plates is

$$\frac{EC}{2}\left[\sin\frac{t}{\sqrt{LC}} - \frac{t}{\sqrt{LC}}\cos\frac{t}{\sqrt{LC}}\right].$$ (*U.P.T.U. 2003*)

Solution: The differential equation for the circuit is

$$L\frac{d^2q}{dt^2}+\frac{q}{C}=E_0$$

or $$L\frac{d^2q}{dt^2}+\frac{q}{C}=E\sin\frac{t}{\sqrt{LC}} \qquad \left|\ \text{As } E_0=E\sin\frac{t}{\sqrt{LC}}\right.$$

or $$\frac{d^2q}{dt^2}+\frac{q}{LC}=\frac{E}{L}\sin\frac{t}{\sqrt{LC}} \qquad \text{...(1)}$$

The A.E. is $$m^2+\frac{1}{LC}=0 \Rightarrow m=\pm\frac{i}{\sqrt{LC}}$$

$$\text{C.F.}=C_1\cos\frac{t}{\sqrt{LC}}+C_2\sin\frac{t}{\sqrt{LC}}$$

$$\text{P.I.}=\frac{1}{\left(D^2+\frac{1}{LC}\right)}\cdot\frac{E}{L}\sin\frac{t}{\sqrt{LC}} \qquad \text{(case of failure)}$$

$$=\frac{Et}{L}\cdot\frac{1}{(2D)}\sin\frac{t}{\sqrt{LC}}=-\frac{Et}{2L}\left(\cos\frac{t}{\sqrt{LC}}\right)\sqrt{LC}$$

$$=-\frac{Et}{2}\sqrt{\frac{C}{L}}\cos\frac{t}{\sqrt{LC}}$$

∴ The solution of the equation is

$$q=C_1\cos\frac{t}{\sqrt{LC}}+C_2\sin\frac{t}{\sqrt{LC}}-\frac{Et}{2}\sqrt{\frac{C}{L}}\cos\frac{t}{\sqrt{LC}} \qquad \text{...(2)}$$

At $t=0,\quad q=0 \quad \therefore C_1=0$

Differentiating equation (2) w.r.t. 't', we get

$$\frac{dq}{dt}=-\frac{C_1}{\sqrt{LC}}\sin\frac{t}{\sqrt{LC}}+\frac{C_2}{\sqrt{LC}}\cos\frac{t}{\sqrt{LC}}+\frac{Et}{2}\sqrt{\frac{C}{L}}\cdot\sin\left(\frac{t}{\sqrt{LC}}\right)\cdot\frac{1}{\sqrt{LC}}-\frac{E}{2}\sqrt{\frac{C}{L}}\cos\frac{t}{\sqrt{LC}}$$

Initially, $$\frac{dq}{dt}=i=0, \text{ when } t=0$$

$$\therefore \quad \frac{C_2}{\sqrt{LC}}-\frac{E}{2}\sqrt{\frac{C}{L}}=0 \Rightarrow C_2=\frac{EC}{2}$$

From equation (2), we get

$$q=\frac{EC}{2}\sin\frac{t}{\sqrt{LC}}-\frac{Et}{2}\sqrt{\frac{C}{L}}\cos\frac{t}{\sqrt{LC}}$$

or $$q = \frac{EC}{2}\left(\sin\frac{t}{\sqrt{LC}} - \frac{t}{\sqrt{LC}}\cos\frac{t}{\sqrt{LC}}\right)$$ **Proved.**

Example 121: In the *L-C-R* circuit, the charge q on a plate of a condenser is given by $L\frac{d^2q}{dt^2} + R\frac{dq}{dt} + \frac{q}{C} = E \sin pt$. The circuit is turned to resonance so that $p^2 = \frac{1}{LC}$. If initially the current i and the charge q be zero, show that for small values of $\frac{R}{L}$, the current in the circuit at time t is given by $\left(\frac{Et}{2L}\right)\sin pt$. [*U.P.T.U. (SUM) 2004*]

Solution: We have

$$L\frac{d^2q}{dt^2} + R\frac{dq}{dt} + \frac{q}{C} = E\sin pt$$

$$\Rightarrow \quad \left(LD^2 + RD + \frac{1}{C}\right)q = E\sin pt \qquad ...(1)$$

The auxiliary equation is

$$Lm^2 + Rm + \frac{1}{C} = 0$$

$$m = -\frac{R}{2L} \pm \frac{1}{2}\sqrt{\frac{R^2}{L^2} - \frac{4}{LC}}$$

$$m = -\frac{R}{2L} \pm \frac{1}{2}\sqrt{\frac{-4}{LC}}$$

$$= -\frac{R}{2L} \pm \frac{2i}{2}\sqrt{\frac{1}{LC}}$$

Since $\frac{R^2}{L^2}$ is very small quantity, so we neglect this quantity.

$$\Rightarrow \quad m = -\frac{R}{2L} \pm i\sqrt{\frac{1}{LC}} = -\frac{R}{2L} \pm ip \qquad \left(\because P = \frac{1}{\sqrt{LC}}\right)$$

$$\therefore \quad \text{C.F.} = e^{-\frac{Rt}{2L}}(C_1 \cos pt + C_2 \sin pt)$$

$$= \left(1 - \frac{Rt}{2L}\right)(C_1 \cos pt + C_2 \sin pt)$$

(Neglect the higher power of $\frac{R}{L}$ in the expansion of $e^{-\frac{Rt}{2L}}$)

$$\text{P.I.} = \frac{1}{\left(LD^2 + RD + \frac{1}{C}\right)}(E\sin pt)$$

$$= E\frac{1}{\left(-Lp^2 + RD + \frac{1}{C}\right)}\sin pt$$

$$= E\cdot\frac{1}{RD}\sin pt = -\frac{E}{RP}\cos pt \qquad \left(\because p^2 = \frac{1}{LC}\right)$$

Complete solution of equation (1) is

$$q = \left(1 - \frac{Rt}{2L}\right)(C_1 \cos pt + C_2 \sin pt) - \frac{E}{Rp}\cos pt \qquad \text{...(2)}$$

$$\therefore \qquad i = \frac{dq}{dt} = \left(1 - \frac{Rt}{2L}\right)(-pC_1 \sin pt + pC_2 \cos pt)$$

$$-\frac{R}{2L}(C_1 \cos pt + C_2 \sin pt) + \frac{E}{R}\sin pt \qquad \text{...(3)}$$

At $t = 0$, $q = 0$, and $i = 0$

From equation (2) $\qquad 0 = C_1 - \frac{E}{Rp} \quad \Rightarrow \quad C_1 = \frac{E}{Rp}$

and from equation (3), $\qquad 0 = C_2 p - \frac{R}{2L}C_1 \quad \Rightarrow \quad C_2 = \frac{E}{2Lp^2}$

Putting the values of C_1 and C_2 in (3), we get

$$i = \left(1 - \frac{Rt}{2L}\right)\left(-\frac{E}{R}\sin pt + \frac{E}{2Lp}\cos pt\right)$$

$$-\frac{R}{2L}\left(\frac{E}{Rp}\cos pt + \frac{E}{2Lp^2}\sin pt\right) + \frac{E}{R}\sin pt$$

$$= \frac{Et}{2L}\sin pt - \frac{ERt}{4pL^2}\cos pt - \frac{ER}{4L^2p^2}\sin pt$$

$$= \frac{Et}{2L}\sin pt; \left(\text{Neglecting the remain terms as } \frac{R}{L^2} \text{ is very small}\right).$$ **Proved.**

1.11.9 Applications to Deflection of Beams, Struts, Columns and Whirling of Shafts

A bar whose length is much greater than its cross-section and its thickness is called a beam.

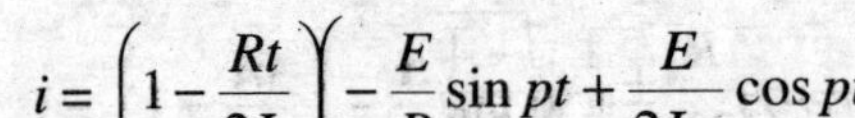

Supported Beam: If a beam may just rest on a support like a knife edge is called a supported beam.

Fixed Beam: If one or both ends of a beam are firmly fixed then it is called fixed beam.

Cantilever: If one end of a beam is fixed and the other end is loaded, it is called a cantilever.

Bending Moment : If M is the bending moment of the forces acting on it, then

$$M = \frac{EI}{R} \quad ...(i)$$

where, E = Modulus of elasticity of the beam.

I = Moment of inertia of the cross-section of beam.

R = Radius of curvature of the curved beam.

$$R = \frac{\left[1+\left(\frac{dy}{dx}\right)^2\right]^{3/2}}{\frac{d^2y}{dx^2}} = \frac{1}{\frac{d^2y}{dx^2}} \quad \left[\text{neglecting } \frac{dy}{dx}\right]$$

Thus equation (i) reduces in the form

$$\boxed{M = EI\frac{d^2y}{dx^2}}$$

Boundary Conditions

(i) At the freely supported end there will be no deflection and no bending moment.

$$y = 0, \ \frac{d^2y}{dx^2} = 0.$$

(ii) If the end is fixed horizontally then deflection and slope of the beam are zero.

$$y = 0, \ \frac{dy}{dx} = 0.$$

(iii) If the end is perfectly free then there is no bending moment.

$$\frac{d^2y}{dx^2} = 0, \ \frac{d^3y}{dx^3} = 0.$$

Convention of Signs

The sign of the moment about NN' on the left NN' is positive if anticlockwise and negative if clockwise.

The downward deflection is positive and length x on right-side is also positive. Slope $\frac{dy}{dx}$ is positive if downward.

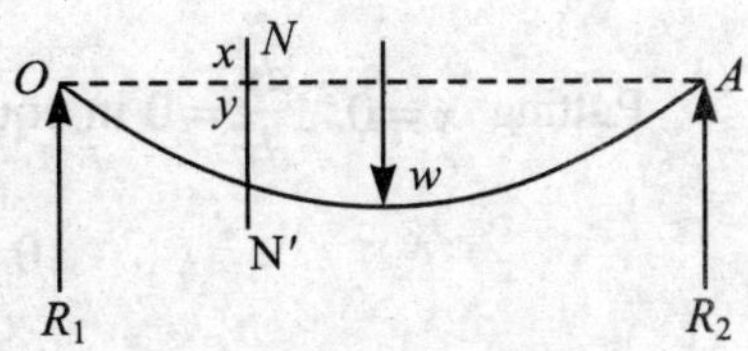

Example 122: The differential equation satisfied by a beam uniformly loaded (W kg/metre), with one end fixed and the second end subjected to tensile for P, is given by

$$EI\frac{d^2y}{dx^2} = Py - \frac{1}{2}Wx^2$$

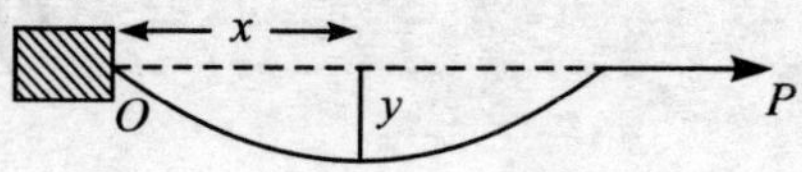

Show that the elastic curve for the beam with conditions

$$y = 0 = \frac{dy}{dx} \text{ at } x = 0, \text{ is given by}$$

$$y = \frac{W}{Pn^2}(1 - \cosh nx) + \frac{Wx^2}{2P} \quad \text{where } n^2 = \frac{P}{EI}.$$

Solution: We have $EI\dfrac{d^2y}{dx^2} = Py - \dfrac{1}{2}W \cdot x^2$ ...(1)

$$\Rightarrow \quad \frac{d^2y}{dx^2} - \frac{P}{EI}y = -\frac{W}{2EI}x^2 \quad \text{or} \quad \left(D^2 - \frac{P}{EI}\right)y = -\frac{W}{2EI}x^2$$

A.E. is $\quad D^2 - \dfrac{P}{EI} = 0 \quad$ or $\quad D^2 = \dfrac{P}{EI} = n^2 \quad$ or $\quad D = \pm n \qquad \left(n^2 = \dfrac{P}{EI}\right)$

$$\text{C.F.} = c_1 e^{nx} + c_2 e^{-nx}$$

$$\text{P.I.} = \frac{1}{D^2 - \frac{P}{EI}}\left(-\frac{W}{2EI}\right)x^2 = -\frac{W}{2EI}\,\frac{1}{D^2 - n^2} \cdot x^2$$

$$= \frac{W}{2n^2EI}\left(1 - \frac{D^2}{n^2}\right)^{-1} \cdot x^2 = \frac{W}{2n^2EI}\left(1 + \frac{D^2}{n^2}\right) \cdot x^2$$

$$= \frac{W}{2n^2EI}\left(x^2 + \frac{2}{n^2}\right)$$

$$\therefore \quad y = c_1e^{nx} + c_2e^{-nx} + \frac{W}{2n^2EI}\left(x^2 + \frac{2}{n^2}\right) \quad \text{...(2)}$$

Differentiating equation (2) w.r.t., x, we get

$$\frac{dy}{dx} = n\,c_1e^{nx} - n\,c_2e^{-nx} + \frac{W}{2n^2EI}(2x) \quad \text{...(3)}$$

Putting $x = 0$, $\dfrac{dy}{dx} = 0$ in equation (3), we get

$$0 = nc_1 - nc_2 \quad \text{or} \quad c_1 = c_2$$

Putting $x = 0$, $y = 0$ in equation (2), wet get

$$0 = c_1 + c_2 + \frac{W}{2n^2EI}\frac{2}{n^2} \quad \text{or} \quad 0 = c_1 + c_2 + \frac{W}{n^4EI} \qquad ...(4)$$

Putting $c_1 = c_2$ in equation (4), we get

$$0 = 2c_1 + \frac{W}{n^4EI} \quad \text{or} \quad c_1 = -\frac{W}{2n^4EI}$$

but $$n^2 = \frac{P}{EI} \quad \text{or} \quad n^2EI = P$$

so $$c_1 = c_2 = -\frac{W}{2n^2P}$$

Putting the values of c_1 and c_2 in equation (2), we get

$$y = \frac{-W}{2n^2P}(e^{nx} + e^{-nx}) + \frac{W}{2P}\left(x^2 + \frac{2}{n^2}\right)$$

$$y = \frac{-W}{n^2P}\cosh nx + \frac{W}{2P}x^2 + \frac{W}{Pn^2}$$

$$y = \frac{W}{Pn^2}(1 - \cosh nx) + \frac{Wx^2}{2P}. \quad \textbf{Ans.}$$

Example 123: For a strut of length l, freely hinged at each end, prove that the bending moment at the centre is $-\frac{Wl^2}{8}\frac{Q}{Q-P}$, where $Q = \frac{EI\pi^2}{l^2}$. The differential equation for a strut is

$$EI\frac{d^2y}{dx^2} + Py = -\frac{Wl^2}{8}\sin\frac{\pi x}{l}.$$

Solution: We have $EI\frac{d^2y}{dx^2} + Py = -\frac{Wl^2}{8}\sin\frac{\pi x}{l}$ ...(1)

$$\Rightarrow \quad \frac{d^2y}{dx^2} + \frac{P}{EI}y = -\frac{W\pi^2}{8Q}\sin\frac{\pi x}{l} \qquad \left[Q = EI\frac{\pi^2}{l^2} \Rightarrow \frac{1}{EI} = \frac{\pi^2}{l^2Q}\right]$$

$$\Rightarrow \quad \frac{d^2y}{dx^2} + \frac{P\pi^2}{Ql^2}y = -\frac{W\pi^2}{8Q}\sin\frac{\pi x}{l} \quad \Rightarrow \quad \left(D^2 + \frac{P\pi^2}{Ql^2}\right)y = -\frac{W\pi^2}{8Q}\sin\frac{\pi x}{l}$$

The A.E. is $D^2 + \frac{P\pi^2}{Ql^2} = 0$ or $D = \pm i\frac{\pi}{l}\sqrt{\frac{P}{Q}}$

$$\text{C.F.} = C_1 \cos\left(\frac{\pi}{l}\sqrt{\frac{P}{Q}}\right)x + C_2 \sin\left(\frac{\pi}{l}\sqrt{\frac{P}{Q}}\right)x$$

$$\text{P.I.} = \frac{1}{D^2 + \frac{P\pi^2}{Ql^2}}\left(-\frac{W\pi^2}{8Q}\sin\frac{\pi x}{l}\right)$$

$$= -\frac{W\pi^2}{8Q}\frac{1}{-\frac{\pi^2}{l^2}+\frac{p}{Q}\frac{\pi^2}{l^2}}\sin\frac{\pi x}{l} = \frac{W\pi^2}{8Q}\frac{l^2}{\pi^2}\frac{1}{1-\frac{P}{Q}}\sin\frac{\pi x}{l}$$

$$= \frac{Wl^2}{8}\frac{1}{Q-P}\sin\frac{\pi x}{l}$$

The general solution of equation (1) is

$$y = C_1 \cos\left(\frac{\pi}{l}\sqrt{P/Q}\right)x + C_2 \sin\left(\frac{\pi}{l}\sqrt{P/Q}\right)x + \frac{Wl^2}{8}\frac{1}{Q-P}\sin\frac{\pi x}{l} \quad \text{...(2)}$$

Since the ends freely hinged,

$$y = 0, \ \frac{d^2y}{dx^2} = 0 \ \text{ at } x = 0 \quad \text{and} \quad y = 0, \frac{d^2y}{dx^2} = 0 \text{ at } x = l$$

On putting $x = 0$, $y = 0$ in equation (2), we get $0 = c_1$

Equation (2) becomes $$y = c_2 \sin\left(\frac{\pi}{l}\sqrt{P/Q}\right)x + \frac{Wl^2}{8}\frac{1}{Q-P}\sin\frac{\pi x}{l} \quad \text{...(3)}$$

On putting $x = l$ and $y = 0$ in (3), we get

$$0 = c_2 \sin\pi\sqrt{P/Q} \quad \Rightarrow \quad c_2 = 0$$

Equation (3) reduces to $$y = \frac{Wl^2}{8}\frac{1}{Q-P}\sin\frac{\pi x}{l}$$

But bending moment is given by

$$M = EI\frac{d^2y}{dx^2}$$

$$M = EI\left[-\frac{Wl^2}{8}\frac{1}{Q-P}\frac{\pi^2}{l^2}\sin\frac{\pi x}{l}\right] = -EI\frac{W\pi^2}{8}\frac{1}{Q-P}\sin\frac{\pi x}{l}$$

or $$M = -\frac{Wl^2}{8}\frac{Q}{Q-P}\sin\frac{\pi x}{l} \qquad \left(Q = \frac{EI\pi^2}{l^2}\right)$$

M at the centre $\left(\text{at } x = \frac{l}{2}\right) = -\frac{Wl^2}{8}\frac{Q}{Q-P}\sin\frac{\pi}{2} = -\frac{Wl^2}{8}\frac{Q}{Q-P}$. **Ans.**

Example 124: A beam of length 'l' is clamped horizontally at its end $x = 0$ and is free at the end $x = l$. A point load W is applied at the end $x = l$, in addition to a uniform load w per unit length from $x = 0$ to $x = \frac{l}{2}$. Find the deflection at any point. *(U.P.T.U. 2002)*

Solution: Let us consider the load diagram as shown below. From balanced equation, we get

$$R = W + \frac{wl}{2}$$

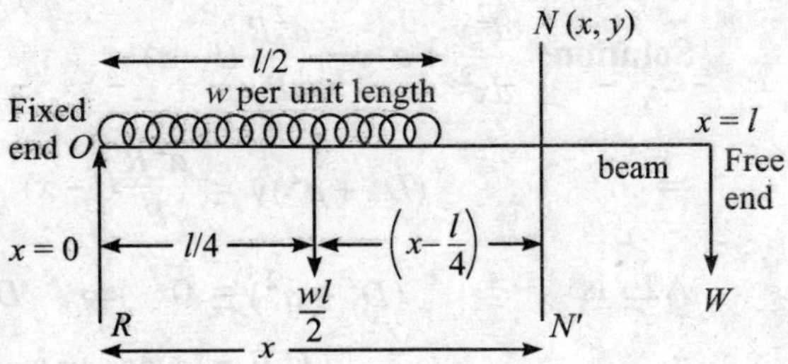

Choose a random axis NN', let the coordinates of N are x and y as shown in the figure. Taking moment about N, we get

$$EI\frac{d^2y}{dx^2} = \frac{wl}{2}\times\left(x - \frac{l}{4}\right) - R\times x$$

$$= \frac{wl}{2}\left(x - \frac{l}{4}\right) - \left(W + \frac{wl}{2}\right)x \qquad \left(\because R = W + \frac{wl}{2}\right)$$

$$EI\frac{d^2y}{dx^2} = -\frac{wl^2}{8} - Wx \qquad \text{...(1)}$$

Integrating w.r.t. x, we get

$$EI\frac{dy}{dx} = -\frac{wl^2}{8}x - \frac{Wx^2}{2} + C$$

Using boundary condition at $x = 0 \quad \Rightarrow \quad$ slope $= \frac{dy}{dx} = 0, \quad \Rightarrow \quad C = 0$

$$\Rightarrow \qquad EI\frac{dy}{dx} = -\frac{wl^2}{8}x - \frac{Wx^2}{2} \qquad \text{...(2)}$$

Again integrating w.r.t. x, we get

$$EI\,y = \frac{-wl^2x^2}{16} - \frac{Wx^3}{6} + C_1$$

Again at $x = 0,\ y = 0 \quad \Rightarrow \quad C_1 = 0$

$$\Rightarrow \qquad EI\,y = -\frac{wl^2x^2}{16} - \frac{Wx^3}{6}$$

$$y = \frac{-1}{EI}\left[\frac{Wx^3}{6} + \frac{wl^2x^2}{16}\right]. \qquad \textbf{Ans.}$$

Example 125: The deflection of a strut of length l with one end $(x = 0)$ built in and the other supported and subjected to end thrust P, satisfies the equation

$$\frac{d^2y}{dx^2}+a^2y = \frac{a^2R}{P}(l-x).$$

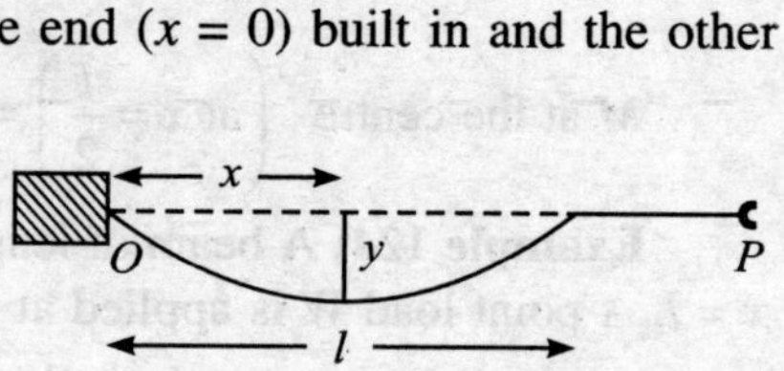

Prove that the deflection curve is

$$y = \frac{R}{P}\left(\frac{1}{a}\sin ax - l\cos ax + l - x\right)$$ (*U.P.T.U. 2001*)

Solution: $$\frac{d^2y}{dx^2}+a^2y=\frac{a^2R}{P}(l-x) \quad ...(1)$$

$$\Rightarrow \quad (D^2+a^2)y = \frac{a^2R}{P}(l-x)$$

A.E. is $(D^2+a^2) = 0 \quad \Rightarrow \quad D = \pm ia$

$$\therefore \quad \text{C.F.} = c_1\cos ax + c_2\sin ax$$

$$\text{P.I.} = \frac{1}{D^2+a^2}\frac{a^2R}{P}(l-x) = \frac{1}{a^2}\frac{1}{\left(1+\frac{D^2}{a^2}\right)}\frac{a^2R}{P}(l-x)$$

$$= \frac{R}{P}\left[1+\frac{D^2}{a^2}\right]^{-1}(l-x) = \frac{R}{P}\left[1-\frac{D^2}{a^2}\right](l-x) = \frac{R}{P}(l-x)$$

The complete solution is C.F. + P.I. $\Rightarrow \; y = c_1\cos ax + c_2\sin ax + \frac{R}{P}(l-x) \quad ...(2)$

On differentiating equation (2), we get

$$\frac{dy}{dx} = -ac_1\sin ax + ac_2\cos ax - \frac{R}{P} \quad ...(3)$$

Since the end O is built in, at $x = 0, \quad y = \frac{dy}{dx} = 0$

On putting $x = 0$, $y = 0$ in equation (2), we get $0 = c_1 + \frac{Rl}{P}$ or $c_1 = -\frac{Rl}{P}$

On putting $\frac{dy}{dx} = 0,\ x = 0$ in equation (3), we have $0 = ac_2 - \frac{R}{p}$ or $c_2 = \frac{R}{aP}$

On putting the values of c_1 and c_2 in (2), we get

$$y = -\frac{Rl}{P}\cos ax + \frac{R}{aP}\sin ax + \frac{R}{P}(l-x)$$

$$y = \frac{R}{P}\left[-l\cos ax + \frac{1}{a}\sin ax + l - x\right].$$ **Hence Proved**

Whirling of Shafts

All rotating shafts deflect during rotation. The magnitude of the deflection depends upon the stiffness of shaft, its support, mass of the shaft and attached parts, imbalanceness and the amount of damping in the system. At a certain speed the rotating shaft will fracture. The dangerous speeds are called *whirling* speed or *critical* speeds of the shafts.

Differential Equation of a Rotating Shaft

Let W be the weight of the shaft per unit length and ω be its angular velocity at the whirling speed. Let y be the maximum deflection of an element of length δx at P at a distance x from one end O.

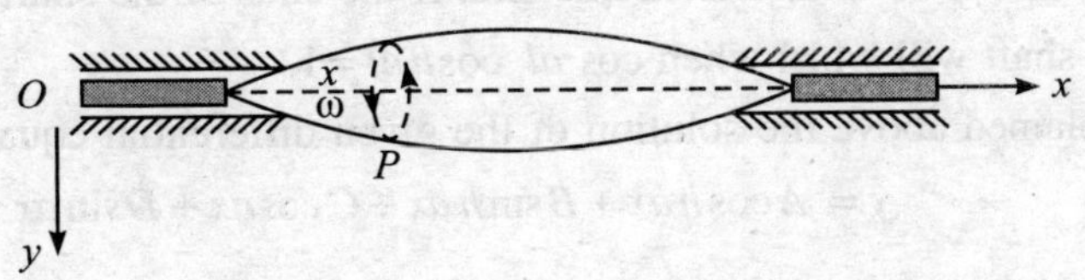

Centrifugal force per unit length at $P = \dfrac{mv^2}{r} = (mr\omega^2) = \dfrac{Wy}{g}\omega^2$. We know that for a uniformly loaded beam, the intensity of loading at $P(x, y) = EI\dfrac{d^4y}{dx^4}$.

This is the restoring force due to stiffness of the shaft, balances the centrifugal force.

Equating the centrifugal and restoring forces, we get

$$EI\frac{d^4y}{dx^4} = \frac{W}{g}y\omega^2 \quad \Rightarrow \quad \frac{d^4y}{dx^4} - a^4y = 0, \quad \text{where } a^4 = \frac{W\omega^2}{gEI}$$

$$D^4y - a^4y = 0 \text{ or } (D^4 - a^4)y = 0$$

$$\text{A.E. is } D^4 - a^4 = 0 \text{ or } (D-a)(D+a)(D^2+a^2) = 0 \quad \Rightarrow \quad D = \pm a, \ \pm ia$$

$\therefore$ $$\text{C.F. is } y = c_1e^{ax} + c_2e^{-ax} + c_3\cos ax + c_4\sin ax$$

$\Rightarrow$ $$y = A\cosh ax + B\sinh ax + C\cos ax + D\sin ax \qquad \ldots(i)$$

The arbitrary constants A, B, C and D in (i) are to be determined from the boundary conditions. The possible boundary conditions are as follows:

(*i*) *In case of short or flexible bearings*: No deflection and no bending moment.

$$y = 0 \Rightarrow \frac{d^2y}{dx^2} = 0.$$

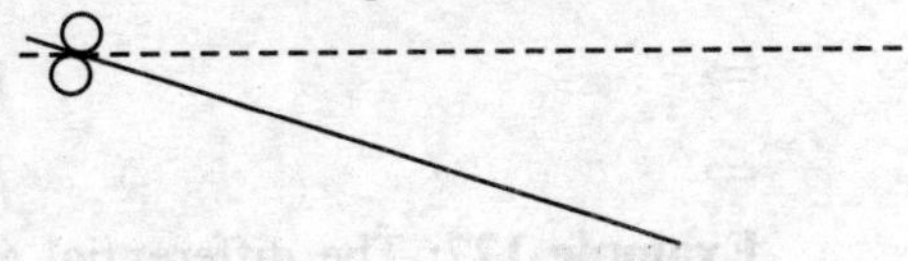

(*ii*) *If the shaft has long bearing*: The deflection and the slope of the shaft being both zero.

$$y = 0 \Rightarrow \frac{dy}{dx} = 0.$$

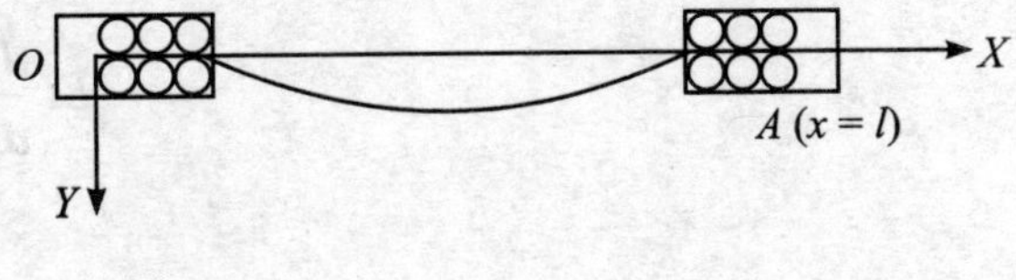

(*iii*) In a prefectly free end there being no shear force and no bending moment.

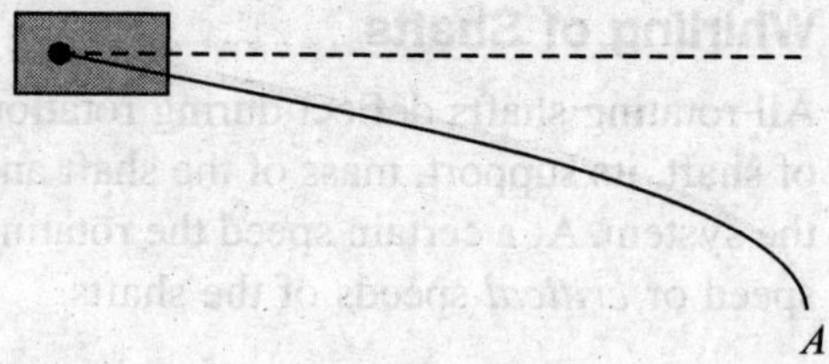

$$\frac{d^2y}{dx^2} = 0 \Rightarrow \frac{d^3y}{dx^3} = 0.$$

Example 126: The whirling speed of a saft of length l is given by $\frac{d^4y}{dx^4} - a^4 y = 0$, where $a^4 = \frac{W\omega^2}{gEI}$ and y is the displacement at a distance x from one end. If the ends of the shaft are constrained in long bearings show that the shaft will whirl when $\cos al \cosh al = 1$.

Solution: As explained above the solution of the given differential equation is

$$y = A\cosh ax + B\sinh ax + C\cos ax + D\sin ax \quad ..(1)$$

On differentiation it gives $\frac{dy}{dx} = Aa\sinh ax + Ba\cosh ax - Ca\sin ax + Da\cos ax$ ...(2)

The boundary conditions are

when $x = 0$, $y = 0$, $\frac{dy}{dx} = 0$ and also at $x = l$.

Therefore, we have $A + C = 0 \Rightarrow C = -A$ and $0 = Ba + Da$

Hence $D = -B$

and $A\cosh al + B\sinh al + C\cos al + D\sin al = 0$ ...(3)

$$A\sinh al + B\cosh al - C\sin al + D\cos al = 0 \quad ...(4)$$

Since $C = -A$ and $D = -B$, equation (3) and (4) becomes

$$A(\cosh al - \cos al) + B(\sinh al - \sin al) = 0$$

and $$A(\sinh al + \sin al) + B(\cosh al - \cos al) = 0$$

Eliminating A and B, we get

$$\begin{vmatrix} \cosh al - \cos al & \sinh al - \sin al \\ \sinh al + \sin al & \cosh al - \cos al \end{vmatrix} = 0$$

$$\Rightarrow \cosh^2 al + \cos^2 al - 2\cosh al \cos al - \sinh^2 al + \sin^2 al = 0$$

$$\Rightarrow 1 + 1 - 2\cosh al \cos al = 0$$

$$\Rightarrow \cosh al \cdot \cos al = 1.$$ **Ans.**

Example 127: The differential equation for the displacement y of a whirling shaft when the weight of the shaft is taken into account is

$$EI\frac{d^4y}{dx^4} - \frac{W\omega^2}{g}y = W.$$

Taking the shaft of length $2l$ with the origin at the centre and short bearings at both ends, show that the maximum deflection of the shaft is

$$\frac{g}{2\omega^2}(\sec hal + \sec al - 2).$$

Solution: The given differential equation may be written as

$$\frac{d^4y}{dx^4} - a^4 y = \frac{W}{EI}, \text{ where } a^4 = \frac{W\omega^2}{EIg} \qquad \text{...(1)}$$

$$\text{C.F.} = c_1 \cos hax + c_2 \sin hax + c_3 \cos ax + c_4 \sin ax$$

$$\text{P.I.} = \frac{1}{(D^4 - a^4)} \cdot \frac{W}{EI} = -\frac{W}{EIa^4} = -\frac{g}{\omega^2}$$

Complete solution is,

$$y = c_1 \cos hax + c_2 \sin hax + c_3 \cos ax + c_4 \sin ax - \frac{g}{\omega^2} \qquad \text{...(2)}$$

Differentiating, we get

$$\frac{1}{a}\frac{dy}{dx} = ac_1 \sin hax + ac_2 \cos hax - ac_3 \sin ax + ac_4 \cos ax$$

and

$$\frac{1}{a^2}\frac{d^2y}{dx^2} = c_1 \cos hax + c_2 \sin hax - c_3 \cos ax - c_4 \sin ax \qquad \text{...(3)}$$

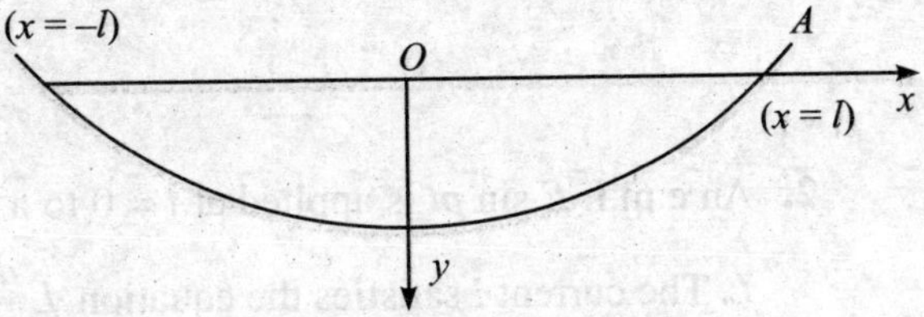

Using boundary conditions

At A, $x = l$, $y = 0$, $\dfrac{d^2y}{dx^2} = 0$.

From equation (2) and (3)

$$0 = c_1 \cos hal + c_2 \sin hal + c_3 \cos al + c_4 \sin al - \frac{g}{\omega^2} \qquad \text{...(4)}$$

$$0 = c_1 \cos hal + c_2 \sin hal - c_3 \cos al - c_4 \sin al \qquad \text{...(5)}$$

Similarly, at end B, $\quad x = -l$, $y = 0$ and $\dfrac{d^2y}{dx^2} = 0$.

Again from equation (2) and (3), we get

$$0 = c_1 \cos hal - c_2 \sin hal + c_3 \cos al - c_4 \sin al - \frac{g}{\omega^2} \qquad \text{...(6)}$$

$$0 = c_1 \cos hal - c_2 \sin hal + c_3 \cos al - c_4 \sin al \qquad \text{...(7)}$$

Adding equations (4), (6), (5) and (7), we get

$$c_1 \cos hal + c_3 \cos al = \frac{g}{\omega^2}$$

and $\quad c_1 \cos hal - c_3 \cos al = 0$

$$\therefore \qquad c_1 = \frac{g}{2\omega^2 \cosh al} \text{ and } c_3 = \frac{g}{2\omega^2 \cos al}$$

Again subtracting equations (6) from (4) and (7) from (5), we get

$$c_1 \sinh al + c_4 \sin al = 0$$

and $$c_2 \sinh al - c_4 \sin al = 0$$

$$\Rightarrow \qquad c_1 = 0,\ c_4 = 0$$

$$\therefore \qquad y = \frac{g}{2\omega^2}\left[\frac{\cosh ax}{\cosh al} + \frac{\cos ax}{\cos al} - 2\right]$$

Thus maximum deflection = Value of y at the centre (at $x = 0$)

$$= \frac{g}{2\omega^2}(\operatorname{sech} al + \sec al - 2).$$

EXERCISE 1.23

1. An electric circuit consists of an inductance 0.1 henry, a resistance of 20 ohms and a condenser of capacitance 25 micro farads. Find the charge q and the current i at time t, given the initial conditions $q = 0.05$ coulombs, $i = 0$ when $t = 0$.

$$\left[\textbf{Ans. } q = e^{-100t}[0.05 \cos 624.5t + 0.008 \sin 624.5t; \quad i = -32e^{-100t} \sin 624.5t (\text{approximately})\right]$$

2. An e.m.f. $E \sin pt$ is applied at $t = 0$ to a circuit containing a capacitance C and inductance L. The current i satisfies the equation $L\frac{di}{dt} + \frac{1}{C}\int i\,dt = E \sin pt$. If $p^2 = 1/LC$ and initially the current i and the charge q are zero, show that the current at time t is $\frac{Et}{2L}$ sin pt, where $i = dq/dt$.

3. A condenser of capacity C is discharged through an inductance L and resistance R in series and the charge q at time t satisfies the equation $L\frac{d^2q}{dt^2} + R\frac{dq}{dt} + \frac{q}{C} = 0$ give that $L = 0.25$ henries, $R = 250$ ohms, $C = 2 \times 10^{-6}$ farads and that when $t = 0$, charge q is 0.002 coulombs and current $i = 0$, obtain the value of q in terms of t.

$$\left[\textbf{Ans. } q = e^{-500t}[0.002 \cos 1323t + 0.0008 \sin 1323t]\right]$$

4. An L-C-R circuit has $R = 180$ ohm, $C = \frac{1}{280}$ farad, $L = 20$ henries and an applied voltage $E(t) = 10 \sin t$. Assuming that no charge is present but an initial current of i ampere is flowing at $t = 0$ when the voltage is first applied find q and $i = \frac{dq}{dt}$ at any time t, q is given

by the differential equation $L\frac{d^2q}{dt^2}+R\frac{dq}{dt}+\frac{q}{C}=E(t).$

$$\left[\textbf{Ans. } \frac{11}{50}e^{-2t}-\frac{101}{500}e^{-7t}+\frac{1}{1000}(26\sin t-18\cos t)\right]$$

5. A horizontal tie-rod of length l freely pinned at its ends carries a uniformaly distributed load w per unit length and is subjected to a horizontal tension T. Find the central deflection and show that maximum bending moment is $\frac{w}{n^2}\left(\sec h\frac{nl}{2}-1\right)$ where $n^2=\frac{T}{EI}$.

6. A beam of length l meters is fixed horizontally at one end. Find the equation of the elastic curve and the maximum deflection, if there is uniform load W Newtons/m along the length and load W Newtons at the free end.

$$\left[\textbf{Ans. } y=\frac{W}{24EI}(-3l^4-8l^3)=\frac{Wl^3}{24EI}(3l+8)\right]$$

7. The differential equation for the displacement of a heavy whirling shaft is $\frac{d^4y}{dx^4}=a^4\left(y+\frac{g}{\omega^2}\right)$ where $a^4=\frac{W\omega^2}{gEI}$. If both ends are short bearings, then ends being $x = 0$ and $x = l$, find the bending moment of the center of the shaft.

$$\left[\textbf{Ans. } \frac{W}{2a^2}\left(\sec h\frac{al}{2}-\sec al\right)\right]$$

OBJECTIVE TYPE QUESTIONS

A. Pick the correct answer of the choices given below:

1. The differential equation $y''-y=0$ with $y(0) = 4$, $y'(0) = -2$ has the solution

   (*i*) $y=e^{-x}-3e^{x}$ (*ii*) $y=e^{x}-3e^{-x}$ (*iii*) C_1+C_2 (*iv*) $y=e^{x}+2e^{-x}$

2. Solution of the differential equation $\frac{d^2y}{dx^2}-5\frac{dy}{dx}+6y=0$ is

   (*i*) $y=(C_1+C_2x)e^{2x}$ (*ii*) $y=C_1e^{2x}-C_2e^{3x}$

   (*iii*) $y=C_1e^{-2x}+C_2e^{3x}$ (*iv*) $y=C_1e^{2x}+C_2e^{3x}$

3. If $(D^2+1)y=\sin(3x+5)$ then P.I. is

   (*i*) $-\frac{1}{7}\sin(3x+5)$ (*ii*) $-\frac{1}{8}\sin(3x+5)$ (*iii*) $-\frac{1}{6}\sin(3x+5)$ (*iv*) zero

4. P.I. of $(D^2+3D+2)y = e^{e^x}$ is

(i) $e^{2x} \cdot e^{e^x}$ (ii) $e^{-2x} \cdot e^{e^x}$ (iii) e^{e^x} (iv) e^{-2x}

5. If $y = x$ is a solution of $x^2 \dfrac{d^2y}{dx^2} + x\dfrac{dy}{dx} - y = 0$ then second part of the solution is

(i) $\dfrac{1}{x}$ (ii) x^2 (iii) x^{-2} (iv) x^n

6. P.I. of the differential equation $\dfrac{d^2y}{dx^2} - \dfrac{dy}{dx} - 2y = \cos x + 3\sin x$ is

(i) sin x (ii) cos x (iii) – sin x (iv) – cos x

7. Solution of differential equation $\dfrac{d^2y}{dx^2} - y = k$; $y(0) = 0$ and $\lim_{x\to\infty} y(x) =$ finite is

(i) $y = k(e^{-x} - 1)$ (ii) $y = k(e^x - 1)$ (iii) $y = (e^x - 1)/k$ (iv) None of these

8. The solution of $\dfrac{d^2s}{dt^2} = g$; $s(0) = 0$, $s'(0) = u$ is

(i) $s = \dfrac{1}{2}gt^2$ (ii) $s = at$ (iii) $s = g/t$ (iv) $s = ut + \dfrac{1}{2}gt^2$

9. The C.F. of the differential equation $x^2y'' + xy' - y = x^5e^x$ is

(i) $C_1x + \dfrac{C_2}{x}$ (ii) $C_1x - \dfrac{C_2}{x}$ (iii) $(C_1 + C_2x)e^x$ (iv) $C_1x + C_2$

10. If $\dfrac{d^2y}{dx^2} - 4x\dfrac{dy}{dx} + (4x^2 - 1)y = -3e^{x^2}\sin 2x$ then integral solution *i.e.* u is

(i) e^{x^2} (ii) e^{-x^2} (iii) e^x (iv) e^{-x}

B. Fill in the blanks:

1. The differential equation $\left(\dfrac{d^3y}{dx^3}\right)^4 - 6x^2\left(\dfrac{dy}{dx}\right)^2 + e^x = \sin xy$ is of __________ order and __________ degree. *(U.P.T.U. 2009)*

2. A particular solution of the differential equation $\dfrac{dy}{dt} = \dfrac{1}{2}(y^2 - 1)$ that satisfies the initial condition $y(0) = 2$ is __________ . *(U.P.T.U. 2009)*

3. The general solution of the differential equation $\frac{d^5y}{dx^5}-\frac{d^3y}{dx^3}=0$ is given by ________ .

(U.P.T.U. 2009)

4. Under the boundary conditions $y = 0$ and $\frac{dy}{dx}=1$ when $x = 0$ the solution of the equation

$\frac{d^2y}{dx^2}=\sec^2 y \tan y$ is given by ________ . [U.P.T.U. (Special) – 2009]

5. Solution of $(D^2-1)y=e^{-x}$ is ________ .

6. $\frac{1}{(D^2-1)}x\cdot e^x$ is ________ .

7. P.I. of $y''-y'-2y=-6\sin 2x-18\cos 2x$ is ________ .

8. P.I. of $(D^2+2D+1)y=\frac{e^{-x}}{x^2}$ is ________ .

9. Solution of the equation $\frac{d^2y}{dx^2}=\left[1-\left(\frac{dy}{dx}\right)^2\right]^{1/2}$ is ________ .

10. P.I. of $(D^2-12D+32)y=(e^x+e^{-2x})^2$ is ________ .

C. Indicate True or False for the following statements:

1. Two functions $y_1(x)$ and $y_2(x)$ are said to be linearly independent if $C_1y_1+C_2y_2=0$ where $C_1 = C_2 = 0$.
2. A symbol, which retains the same value during a set of mathematical operations is said to be constant for these operations.
3. If $R = e^{ax}$ and $f(a) = 0$ then P.I. $=\frac{1}{f(a)}e^{ax}$.
4. If $R = \cos ax$ and $f(-a^2)$ then P.I. $=\frac{x}{f'(-a^2)}\cos ax$.
5. $x\frac{dy}{dx}+y=x^3y^8$ is a homogeneous differential equation.
6. If $1+\frac{P}{a}+\frac{Q}{a^2}=0,$ then $u = e^{-ax}$.
7. The particular integral is also known as particular solution.

8. $\frac{1}{(D+a)}\phi(x) = e^{-ax}\int e^{-ax}\phi(x)dx.$

9. Order of differential equation $\left(\frac{d^3y}{dx^3}\right)^2 + \left(\frac{dy}{dx}\right) - 4y = 0$ is 3.

10. $x^3\frac{d^3y}{dx^3} = D(D-1)(D-2)y.$

ANSWERS TO OBJECTIVE TYPE QUESTIONS

A. Pick the correct answers:

1. (*iii*)	**2.** (*iv*)	**3.** (*ii*)	**4.** (*ii*)	**5.** (*i*)
6. (*iii*)	**7.** (*i*)	**8.** (*iv*)	**9.** (*i*)	**10.** (*i*)

B. Fill in the blanks:

1. 3, 4

2. $\frac{y-1}{y+1} = \frac{1}{3}e^t$

3. $y = (C_1 + C_2x + C_3x^2) + C_4e^x + C_5e^{-x}$

4. $y = \sin^{-1}x$

5. $y = (c_1 + c_2x)e^x - \frac{x}{2}e^{-x}$

6. $\frac{e^x}{2}\left(\frac{x^2}{2} - \frac{x}{2} + \frac{1}{4}\right)$

7. $3\cos 2x$

8. $-e^{-x}\log x$

9. $y = -\cos(x + C) + C_1$

10. $\frac{1}{12}e^{2x} + \frac{2}{45}e^{-x} + \frac{1}{96}e^{-4x}$

C. True or False:

1. T	**2.** T	**3.** F	**4.** T	**5.** F
6. F	**7.** T	**8.** F	**9.** T	**10.** T.

UNIT II

Laplace Transform

2.1 INTRODUCTION

Laplace* transform is a powerful technique to replace the operations of calculus by operation of algebra. For example, with the application of Laplace transform to an initial value problem, consisting of an ordinary (or partial) differential equation together with initial condition is reduced to problem of solving an algebraic equation.

2.2 INTEGRAL TRANSFORM

Let $k(s, t)$ be a function of two variables s and t, where s is a parameter (may be real or complex) independent of t. The function $f(s)$ defined by the integral

$$f(s) = \int_{-\infty}^{\infty} k(s,t) \cdot F(t)dt \qquad ...(i)$$

is called the **'Integral transform'** of the function $F(t)$ and is denoted by $T\{F(t)\}$. The function $k(s, t)$ is called the Kernel of transformation.

2.3 LAPLACE TRANSFORM

Definition: Let $F(t)$ be a given function defined for all $t \geq 0$ and if the Kernel $k(s, t)$ is defined as

$$k(s, t) = \begin{cases} 0, & t < 0 \\ e^{-st}, & t \geq 0 \end{cases}$$

then

$$f(s) = \int_{0}^{\infty} e^{-st} F(t)\, dt \qquad ...(i)$$

---

*Pierre Simon Marquis De Laplace (1749–1827), French mathematician, known as the Newton of France and teacher to Napoleon Bonaparte.

The function $f(s)$ defined by the integral (i) is called the **'Laplace transform'** of $F(t)$ and denoted by $L\{F(t)\}$ or $f(s)$.

Therefore $\boxed{L\{F(t)\} = f(s) = \int_0^\infty e^{-st} F(t)\,dt}$, where L is an operator.

2.4 LINEARITY PROPERTY OF LAPLACE TRANSFORM

If c_1 and c_2 are two constants then

$$L\{c_1F_1(t) + c_2F_2(t)\} = c_1L\{F_1(t)\} + c_2L\{F_2(t)\}$$

Proof: We have

$$L\{F(t)\} = \int_0^\infty e^{-st} F(t)dt$$

$$\Rightarrow \quad L\{c_1F_1(t) + c_2F_2(t)\} = \int_0^\infty e^{-st}\{c_1F_1(t) + c_2F_2(t)\}dt$$

$$= c_1\int_0^\infty e^{-st}F_1(t)dt + c_2\int_0^\infty e^{-st}F_2(t)\,dt$$

$$\Rightarrow \quad \boxed{L\{c_1F_1(t) + c_2F_2(t)\} = c_1L\{F_1(t)\} + c_2L\{F_2(t)\}}$$

2.5 LAPLACE TRANSFORMS OF SOME FUNCTIONS

1. Find Laplace transform of '1' *i.e.*, $L\{1\}$

Proof: We know that $L\{F(t)\} = \int_0^\infty e^{-st} \cdot F(t)dt$

$$\Rightarrow \quad L\{1\} = \int_0^\infty e^{-st} \cdot dt = \left[\frac{e^{-st}}{-s}\right]_0^\infty = -\frac{1}{s}\left[\lim_{t\to\infty} e^{-st} - 1\right]$$

$$\Rightarrow \quad \boxed{\mathbf{L\{1\} = \frac{1}{s}, s > 0}} \qquad \Big|\text{As } \lim_{t\to\infty} e^{-st} \to 0$$

2. To show $L\{e^{-at}\} = \dfrac{1}{s+a}$ and $L\{e^{at}\} = \dfrac{1}{s-a}, s > a$

Proof: $$L\{e^{-at}\} = \int_0^\infty e^{-st} \cdot e^{-at}dt = \int_0^\infty e^{-(s+a)t}\,dt = -\frac{1}{(s+a)}\left[e^{-(s+a)t}\right]_0^\infty$$

$$= -\frac{1}{(s+a)}\left[\lim_{t\to\infty} e^{-(s+a)t} - 1\right] = \frac{1}{s+a}$$

$\Rightarrow$ $\boxed{L\{e^{-at}\} = \frac{1}{s+a}}$, similarly we can show

$$\boxed{L\{e^{at}\} = \frac{1}{s-a}, s > a.}$$

2. To show $L\{\sin at\} = \frac{a}{s^2+a^2}, s > 0.$

Proof: $L\{\sin at\} = L\left\{\frac{e^{ait} - e^{-ait}}{2i}\right\} = \frac{1}{2i}\left[L\{e^{iat}\} - L\{e^{-iat}\}\right]$

$$= \frac{1}{2i}\left[\frac{1}{s-ia} - \frac{1}{s+ia}\right] = \frac{1}{2i}\left[\frac{s+ia-s+ia}{s^2+a^2}\right]$$

$$= \frac{a}{s^2+a^2} \quad \Rightarrow \quad \boxed{L\{\sin at\} = \frac{a}{s^2+a^2}, s > 0.}$$

4. To show $L\{\cos at\} = \frac{s}{s^2+a^2}, s > 0.$

Proof: $L\{\cos at\} = L\left\{\frac{e^{iat} + e^{-iat}}{2}\right\} = \frac{1}{2}L\{e^{iat} + e^{-iat}\}$

$$= \frac{1}{2}\left[\frac{1}{s-ia} + \frac{1}{s+ia}\right] = \frac{1}{2}\left[\frac{2s}{(s^2+a^2)}\right]$$

$\Rightarrow$ $\boxed{L\{\cos at\} = \frac{s}{(s^2+a^2)}, s > 0.}$

5. To show $L\{\sinh at\} = \frac{a}{s^2-a^2}, s > |a|.$

Proof: $L\{\sinh at\} = L\left\{\frac{e^{at} - e^{-at}}{2}\right\} = \frac{1}{2}\left[L\{e^{at}\} - L\{e^{-at}\}\right]$

$$= \frac{1}{2}\left[\frac{1}{s-a} - \frac{1}{s+a}\right] = \frac{1}{2}\left[\frac{2a}{(s^2-a^2)}\right]$$

$$\Rightarrow \quad \boxed{L\{\sinh at\} = \frac{a}{s^2 - a^2}, s > |a|.}$$

6. To show $L\{\cosh at\} = \dfrac{s}{s^2 - a^2}, s > |a|.$

Proof:
$$L\{\cosh at\} = L\left\{\frac{e^{at} + e^{-at}}{2}\right\} = \frac{1}{2}\left[L\{e^{at}\} + L\{e^{-at}\}\right]$$

$$= \frac{1}{2}\left[\frac{1}{s-a} + \frac{1}{s+a}\right] = \frac{1}{2}\left[\frac{2s}{(s^2 - a^2)}\right]$$

$$\Rightarrow \quad \boxed{L\{\cosh at\} = \frac{s}{s^2 - a^2}, s > |a|.}$$

7. To show $L\{t^n\} = \dfrac{\lfloor n}{s^{n+1}}$, where n is a positive integer.

Proof:
$$L\{t^n\} = \int_0^\infty e^{-st} \cdot t^n dt$$

$$= \left[-\frac{1}{s} \cdot t^n e^{-st}\right]_0^\infty + \frac{1}{s}\int_0^\infty n e^{-st} \cdot t^{n-1} dt \qquad \text{(Integrating by parts)}$$

$$= -\frac{1}{s} \cdot \lim_{t \to \infty} (t^n e^{-st}) + \frac{n}{s}\int_0^\infty e^{-st} \cdot t^{n-1} dt$$

$$= 0 + \frac{n}{s}\int_0^\infty e^{-st} \cdot t^{n-1} dt \qquad \Big| \lim_{t \to \infty} (t^n e^{-st}) \to 0$$

$$= 0 + \frac{n(n-1)}{s^2}\int_0^\infty e^{-st} \cdot t^{n-2} dt \qquad \text{(Integrate again by parts)}$$

....

....

$$L\{t^n\} = \frac{n(n-1)(n-2)......3 \cdot 2 \cdot 1}{s^n}\int_0^\infty e^{-st} dt \qquad \Big| \text{As } t^{n-n} = t^0 = 1$$

$$= \frac{\lfloor n}{s^n}\left[\frac{e^{-st}}{-s}\right]_0^\infty = \frac{\lfloor n}{s^{n+1}}, s > 0$$

$$\Rightarrow \quad \boxed{\mathbf{L\{t^n\} = \frac{\lfloor n}{s^{n+1}}, s > 0}} \quad \text{or} \quad \boxed{\mathbf{L\{t^n\} = \frac{\overline{|n+1}}{s^{n+1}}}.}$$

Example 1: Find $L\{\sin\sqrt{t}\}$.

Solution: $L\{\sin\sqrt{t}\} = L\left[t^{1/2} - \frac{t^{3/2}}{\lfloor 3} + \frac{t^{5/2}}{\lfloor 5} - \ldots\right]$

$$= L\{t^{1/2}\} - \frac{1}{\lfloor 3}L[t^{3/2}] + \frac{1}{\lfloor 5}L\{t^{5/2}\} - \ldots\ldots \qquad \Big| \text{ As } \sin\theta = \theta - \frac{\theta^3}{\lfloor 3} + \ldots$$

We know $\quad L\{t^n\} = \frac{\lfloor n}{s^{n+1}} = \frac{\overline{|n+1}}{s^{n+1}}$

$$\therefore \quad L\{\sin\sqrt{t}\} = \frac{\overline{|3/2}}{s^{3/2}} - \frac{1}{3\cdot 2}\frac{\overline{|5/2}}{s^{5/2}} + \frac{1}{5\cdot 4\cdot 3\cdot 2}\frac{\overline{|7/2}}{s^{7/2}} - \ldots\ldots$$

$$= \frac{\frac{1}{2}\overline{|\frac{1}{2}}}{s^{3/2}} - \frac{1}{3\cdot 2}\frac{3/2\cdot\frac{1}{2}\overline{|1/2}}{s^{5/2}} + \frac{5/2}{5\cdot 4\cdot 3\cdot 2}\frac{3/2\cdot 1/2\overline{|1/2}}{s^{7/2}} - \ldots\ldots \qquad \boxed{\overline{|n} = (n-1)\overline{|n-1}}$$

$$= \frac{\sqrt{\pi}}{2s^{3/2}}\left[1 - \frac{1}{4s} + \frac{1}{\lfloor 2}\left[\frac{1}{4s}\right]^2 - \ldots\ldots\right] \qquad \Big| \text{As } \overline{|1/2} = \sqrt{\pi}$$

$$= \frac{\sqrt{\pi}}{2s^{3/2}}\left[e^{-\frac{1}{4s}}\right]. \quad \textbf{Ans.}$$

Example 2: Find $L\{F(t)\}$ where $F(t) = \begin{cases}\sin t, & 0 < t < \pi\\ 0, & t > \pi\end{cases}$

Solution: We know that $L\{F(t)\} = \int_0^\infty e^{-st}\cdot F(t)\,dt$

$$\Rightarrow \quad L\{F(t)\} = \int_0^\pi e^{-st}\cdot \sin t\,dt + \int_\pi^\infty e^{-st}\cdot 0\,dt \qquad \ldots(1)$$

Let $$I = \int_0^{\pi} \underset{I}{e^{-st}} \cdot \underset{II}{\sin t}\, dt = [-e^{-st} \cos t]_0^{\pi} - s\int_0^{\pi} e^{-st} \cdot \cos t\, dt$$

$$\Rightarrow \quad I = e^{-s\pi} + 1 - s[e^{-st} \cdot \sin t]_0^{\pi} - s^2 \int_0^{\pi} e^{-st} \cdot \sin t\, dt$$

$$\Rightarrow \quad I = 1 + e^{-s\pi} - s^2 I$$

$$\Rightarrow \quad I(1+s^2) = 1 + e^{-s\pi} \quad \Rightarrow \quad I = \frac{1+e^{-s\pi}}{1+s^2}$$

Hence, from equation (1)

$$L\{F(t)\} = \frac{1+e^{-s\pi}}{1+s^2}. \quad \textbf{Ans.}$$

Example 3: Find the Laplace transform of the function $F(t)$, where

$$F(t) = \{7e^{2t} + 9e^{-2t} + 5\cos t + 7t^3 + 5\sin 3t + 2\}.$$

Solution: $L\{F(t)\} = 7L\{e^{2t}\} + 9L\{e^{-2t}\} + 5L\{\cos t\} + 7L\{t^3\} + 5L\{\sin 3t\} + 2L\{1\}$

$$= 7\cdot\frac{1}{s-2} + 9\cdot\frac{1}{s+2} + 5\cdot\frac{s}{s^2+1} + 7\cdot\frac{\lfloor 3}{s^4} + 5\cdot\frac{3}{s^2+9} + 2.\frac{1}{s}$$

$$= \frac{16s-4}{s^2-4} + \frac{5s}{s^2+1} + \frac{2s^3+42}{s^4} + \frac{15}{s^2+9}. \quad \textbf{Ans.}$$

Example 4: Find $L\left\{\frac{1}{\sqrt{\pi t}}\right\}$.

Solution: $$L\left\{\frac{1}{\sqrt{\pi t}}\right\} = \frac{1}{\sqrt{\pi}} L\left\{t^{-1/2}\right\} = \frac{1}{\sqrt{\pi}} \frac{\overline{\left|-\frac{1}{2}+1\right.}}{s^{-\frac{1}{2}+1}} \qquad \left[\text{As } L\{t^n\} = \frac{\overline{|n+1}}{s^{n+1}}\right]$$

$$= \frac{1}{\sqrt{\pi}} \cdot \frac{\overline{\left|\frac{1}{2}\right.}}{\sqrt{s}} = \frac{1}{\sqrt{\pi}} \cdot \frac{\sqrt{\pi}}{\sqrt{s}} = \frac{1}{\sqrt{s}} \qquad \left(\because \overline{\left|\frac{1}{2}\right.} = \sqrt{\pi}\right)$$

$$\Rightarrow \quad L\left\{\frac{1}{\sqrt{\pi t}}\right\} = \frac{1}{\sqrt{s}}. \quad \textbf{Ans.}$$

Example 5: Find the Laplace transform of

$$F(t) = \begin{cases} t^2, & 0<t<2 \\ t-1, & 2<t<3. \\ 7, & t>3 \end{cases} \qquad (U.P.T.U.\ 2007)$$

Solution: $L\{F(t)\} = \int_0^2 e^{-st}\cdot t^2 dt + \int_2^3 e^{-st}\cdot(t-1)\,dt + \int_3^\infty 7e^{-st}dt$

$$= \left[t^2\frac{e^{-st}}{-s}\right]_0^2 + \frac{2}{s}\int_0^2 e^{-st}\cdot t\,dt + \left[(t-1)\frac{e^{-st}}{-s}\right]_2^3$$

$$+\frac{1}{s}\int_2^3 e^{-st}dt + \frac{7}{-s}\left[e^{-st}\right]_3^\infty$$

$$= -\frac{4}{s}e^{-2s} + \frac{2}{s}\left[\frac{e^{-st}\cdot t}{-s}\right]_0^2 + \frac{2}{s^2}\int_0^2 e^{-st}dt$$

$$+\left[-\frac{2}{s}e^{-3s} + \frac{e^{-2s}}{s}\right] - \frac{1}{s^2}\left[e^{-3s} - e^{-2s}\right] + \frac{7}{s}e^{-3s} \qquad \left|e^{-\infty}\to 0\right.$$

$$= -\frac{4}{s}e^{-2s} - \frac{2}{s^2}\times 2e^{-2s} - \frac{2}{s^3}e^{-2s} + \frac{2}{s^3} - \frac{2}{s}e^{-3s} + \frac{e^{-2s}}{s}$$

$$-\frac{e^{-3s}}{s^2} + \frac{e^{-2s}}{s^2} + \frac{7}{s}e^{-3s}$$

$$= -\frac{3}{s}e^{-2s} - \frac{3}{s^2}e^{-2s} - \frac{2}{s^3}e^{-2s} + \frac{2}{s^3} + \frac{5}{s}e^{-3s} - \frac{e^{-3s}}{s^2}$$

$$= \frac{2}{s^3} + \frac{1}{s^2}(5s-1)e^{-3s} - \frac{1}{s^3}(3s^2+3s+2)e^{-2s}. \quad \textbf{Ans.}$$

2.6 FIRST TRANSLATION PROPERTY OR SHIFTING THEOREM

If $L\{F(t)\} = f(s)$, then $L\{e^{at}\cdot F(t)\} = f(s-a)$.

Proof: We know that

$$f(s) = L\{F(t)\} = \int_0^\infty e^{-st}\cdot F(t)\,dt$$

$$\Rightarrow \qquad L\{e^{at}\cdot F(t)\} = \int_0^\infty e^{at}\cdot e^{-st}\cdot F(t)\,dt$$

$$= \int_0^\infty e^{-(s-a)t}\cdot F(t)dt = f(s-a)$$

$$\Rightarrow \qquad \boxed{L\{e^{at}\cdot F(t)\} = f(s-a).}$$

2.7 SECOND TRANSLATION PROPERTY OR SHIFTING THEOREM

If $L\{F(t)\} = f(s)$ and $G(t) = \begin{cases} F(t-a), t > a \\ 0, \quad t < a \end{cases}$

then $L\{G(t)\} = e^{-as}\cdot f(s).$ *(U.P.T.U. 2006, 2008)*

Proof: We know that $L\{F(t)\} = \int_0^\infty e^{-st}\cdot F(t)\,dt$

$$\Rightarrow \qquad L\{G(t)\} = \int_0^a e^{-st}\cdot 0\,dt + \int_a^\infty e^{-st}\cdot F(t-a)\,dt$$

$$\Rightarrow \qquad L\{G(t)\} = \int_a^\infty e^{-st}\cdot F(t-a)\,dt$$

Let $t - a = x \quad \Rightarrow \quad dt = dx$

$$\therefore \qquad L\{G(t)\} = \int_0^\infty e^{-s(x+a)}\cdot F(x)\,dx$$

$$= e^{-as}\int_0^\infty e^{-sx}\cdot F(x)\,dx$$

$$= e^{-as}\int_0^\infty e^{-st}\cdot F(t)\,dt \qquad \left|\ \int_0^a f(x)\,dx = \int_0^a f(t)\,dt\right.$$

$$\Rightarrow \qquad \boxed{L\{G(t)\} = e^{-as}f(s)}.$$

2.8 CHANGE OF SCALE PROPERTY

If $L\{F(t)\} = f(s)$ then $L\{F(at)\} = \dfrac{1}{a}f(s/a).$

Proof. We have $L\{F(t)\} = \int_0^\infty e^{-st} \cdot F(t)\, dt$

$$\therefore \qquad L\{F(at)\} = \int_0^\infty e^{-st} \cdot F(at)\, dt$$

Let $at = x \quad \Rightarrow \quad dt = \dfrac{dx}{a}$

$$\Rightarrow \qquad L\{F(at)\} = \frac{1}{a}\int_0^\infty e^{-\frac{s}{a}x} \cdot F(x)\, dx = \frac{1}{a}\int_0^\infty e^{-\frac{s}{a}t} \cdot F(t)\, dt \qquad \text{(By definite integral)}$$

$$\Rightarrow \qquad \boxed{L\{F(at)\} = \frac{1}{a} f\left(\frac{s}{a}\right)}$$

Example 6: Find the Laplace transform of the following function (*U.P.T.U. 2008*)

$$F(t) = \begin{cases} e^{t-a}, t > a \\ 0 \quad, t < a \end{cases}$$

Solution: Since $L\left\{e^t\right\} = \dfrac{1}{s-1}$

By second shifting theorem

$$L\{F(t)\} = e^{-as} \cdot \frac{1}{s-1} \quad \textbf{Ans.}$$

Example 7: If $L\{\cos^2 t\} = \dfrac{s^2+2}{s(s^2+4)}$, find $L\{\cos^2 at\}$. (*U.P.T.U. 2006*)

Solution: We have

$$L\{\cos^2 t\} = f(s) = \frac{s^2+2}{s(s^2+4)}$$

By change of scale property, we have

$$L\{\cos^2 at\} = \frac{1}{a} f\left(\frac{s}{a}\right) = \frac{1}{a} \frac{\left(\frac{s}{a}\right)^2 + 2}{\left(\frac{s}{a}\right)\left(\frac{s^2}{a^2} + 4\right)} = \frac{1}{a}\left[\frac{s^2+2a^2}{\frac{s}{a}(s^2+4a^2)}\right]$$

$$= \frac{s^2 + 2a^2}{s(s^2 + 4a^2)}. \quad \textbf{Ans.}$$

Example 8: If $L\{J_0(\sqrt{t})\} = \frac{e^{-\frac{1}{4s}}}{s}$, find $L\{J_0(2\sqrt{t})\}$.

Solution: Let $F(t) = J_0(\sqrt{t})$

$$\therefore \quad F(4t) = J_0(\sqrt{4t}) = J_0(2\sqrt{t})$$

$$L\{F(t)\} = f(s) = \frac{e^{-\frac{1}{4s}}}{s}$$

By change of scale property

$$L\{F(4t)\} = L\{J_0(2\sqrt{t})\} = \frac{1}{4} f(s/4) = \frac{1}{4} \cdot \frac{e^{-\frac{1}{4 \times s/4}}}{\frac{s}{4}}$$

$$\Rightarrow \quad L\{J_0(2\sqrt{t})\} = \frac{1}{s} e^{-\frac{1}{s}}. \quad \textbf{Ans.}$$

Example 9: Find $L\{F(t)\}$, where

$$F(t) = \begin{cases} \cos\left(t - \frac{2\pi}{3}\right), & t > \frac{2\pi}{3} \\ 0, & t < \frac{2\pi}{3} \end{cases}$$

Solution: Let $G(t) = \cos t$, $\therefore \; G\left(t - \frac{2\pi}{3}\right) = \cos\left(t - \frac{2\pi}{3}\right)$

$$\Rightarrow \quad F(t) = \begin{cases} G\left(t - \frac{2\pi}{3}\right), & t > \frac{2\pi}{3} \\ 0, & t < \frac{2\pi}{3} \end{cases}$$

Now $L\{G(t)\} = L\{\cos t\} = \frac{s}{s^2 + 1} = f(s)$

From second shifting property

$$L\{F(t)\} = e^{-as} \cdot f(s) = e^{-\frac{2\pi s}{3}} \cdot \frac{s}{s^2+1}. \quad \textbf{Ans.} \quad \Big| \text{As } a = \frac{2\pi}{3}$$

Example 10: If $L\{F(t)\} = \frac{1}{s} e^{-1/s}$, prove that

$$L\{e^{-t} F(3t)\} = \frac{e^{-3/(s+1)}}{s+1}.$$

Solution: We have $L\{F(t)\} = \frac{1}{s} e^{-1/s} = f(s)$

By change of scale property, we have

$$L\{F(3t)\} = \frac{1}{3} f(s/3) = \frac{1}{3} \cdot \frac{1}{s/3} e^{-\frac{1}{s/3}} = \frac{1}{s} e^{-3/s} = f_1(s)$$

Now by first shifting property

$$L\left\{e^{(-t)} \cdot F(3t)\right\} = f_1(s+1) = \frac{1}{s+1} e^{-\frac{3}{(s+1)}}. \quad \textbf{Proved.}$$

Example 11: If $L\left\{\frac{\sin t}{t}\right\} = \tan^{-1}\left(\frac{1}{s}\right)$, find $L\left\{\frac{\sin at}{t}\right\}$.

Solution: We have $L\left\{\frac{\sin t}{t}\right\} = f(s) = \tan^{-1}\left(\frac{1}{s}\right)$

Let $\quad F(t) = \frac{\sin t}{t} \Rightarrow F(at) = \frac{\sin at}{at}$

By change of scale property, we have

$$L\{F(at)\} = L\left\{\frac{\sin at}{at}\right\} = \frac{1}{a} f(s/a) = \frac{1}{a} \tan^{-1} \frac{a}{s}$$

$$\Rightarrow \frac{1}{a} L\left\{\frac{\sin at}{t}\right\} = \frac{1}{a} \tan^{-1} \frac{a}{s}$$

$$\Rightarrow L\left\{\frac{\sin at}{t}\right\} = \tan^{-1} \frac{a}{s}. \quad \textbf{Ans.}$$

EXERCISE 2.1

Find the Laplace transform of the following:

1. $3\cos h5t - 4\sin h5t.$ $\left[\textbf{Ans.}\ \dfrac{3s-20}{s^2-25}, s>5\right]$

2. $3t^4 - 2t^3 + 4e^{-3t} - 2\sin 5t + 3\cos 2t.$ $\left[\textbf{Ans.}\ \dfrac{72}{s^6} - \dfrac{12}{s^4} + \dfrac{4}{s+3} - \dfrac{10}{s^2+25} + \dfrac{3s}{s^2+4}\right]$

3. $\cos at \sin h\, at.$ $\left[\textbf{Ans.}\ \dfrac{1}{2}\dfrac{s-a}{(s-a)^2+a^2} - \dfrac{s+a}{(s+a)^2+a^2}\right]$

4. $e^{-t} \cdot \cos^2 t.$ $\left[\textbf{Ans.}\ \dfrac{1}{2s+2} + \dfrac{s+1}{2s^2+4s+10}\right]$

5. $\sin h^3\, t.$ $[\textbf{Ans.}\ 6/(s^2-1)(s^2-9)]$

6. Find $L\{F(t)\}$, if $F(t) = \begin{cases} e^t, 0 < t \le 1 \\ 0, t > 1 \end{cases}$ $\left[\textbf{Ans.}\ \dfrac{1}{s-1}\{1 - e^{-(s-1)}\}\right]$

7. Find $L\{F(t)\}$, if $F(t) = \begin{cases} 0, 0 < t < 2 \\ 4, t > 2 \end{cases}$ $\left[\textbf{Ans.}\ \dfrac{4e^{-2s}}{s}\right]$

8. Find $L\left\{\dfrac{e^{-at} \cdot t^{n-1}}{\lfloor n-1}\right\}.$ $\left[\textbf{Ans.}\ \dfrac{1}{(s+a)^n}\right]$

9. Find $L\{e^t \cdot \sin^2 t\}.$ $[\textbf{Ans.}\ 2/(s-1)(s^2-2s+5)]$

10. Find $L\{e^{-t}(3\sin 2t - 5\cos h2t)\}.$ $[\textbf{Ans.}\ 6/(s^2+2s+5) - (5s+5)/(s^2+2s-3)]$

11. If $L\{F(t)\} = f(s)$, find $L\{F(t) \cdot \cos \omega t\}.$ $\left[\textbf{Ans.}\ \dfrac{1}{2}[f(s-i\omega) + f(s+i\omega)]\right]$

12. Find $L\{F(t)\}$ where $F(t) = \begin{cases} \sin\left(t - \dfrac{\pi}{3}\right), & t > \dfrac{\pi}{3} \\ 0, & t < \dfrac{\pi}{3} \end{cases}$ $\left[\textbf{Ans.}\ e^{-s\pi/3} \cdot \dfrac{1}{s^2+1}, s>0\right]$

13. Apply change of scale property, obtain the Laplace transform of $\sin h\ 3t$.

$\left[\textbf{Ans.}\ \dfrac{3}{s^2-9}\right]$

14. Find the Laplace transform of $\left(\sqrt{t}+\frac{1}{\sqrt{t}}\right)^3$. $\left[\textbf{Ans. } \sqrt{\pi}\left\{\frac{3}{4s^{5/2}}-2\sqrt{s}+\frac{3}{2s^{3/2}}+\frac{3}{\sqrt{s}}\right\}\right]$

15. Find $L\{1 + te^{-t}\}^3$. $\left[\textbf{Ans. } \frac{1}{s}+\frac{3}{(s+1)^2}+\frac{6}{(s+2)^3}+\frac{6}{(s+3)^4}\right]$

2.9 PIECEWISE (OR SECTIONALLY) CONTINUOUS FUNCTIONS

A function $F(t)$ is said to be piecewise (or sectionally) continuous on a closed interval $a \le t \le b$ if that interval can be divided into a finite number of subintervals, in each of which $F(t)$ is continuous and has finite right and left hand limits.

The figure shown in the adjoining diagram is piecewise continuous over the interval $(0, t_3)$.

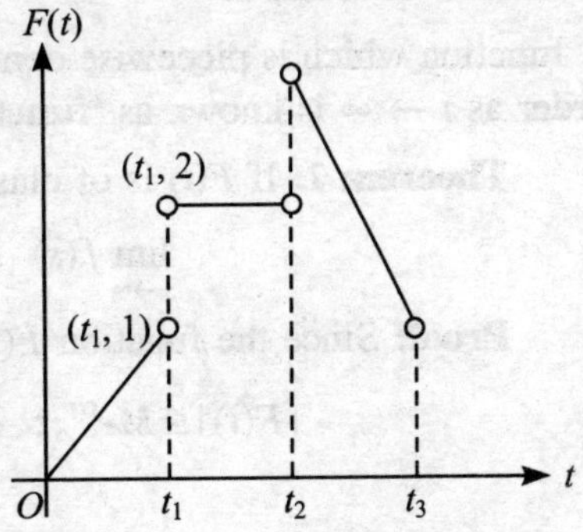

Since $\lim_{t \to t_1-} \{F(t)\} = 1$

and $\lim_{t \to t_1+} \{F(t)\} = 2,$

(where $t \to t_1-$ and $t \to t_1+$ mean that t tends to t_1 from the left and right respectively), it follows that $F(t)$ is discontinuous at $t = t_1$ and has a finite jump. A similar jump occurs at $t = t_2$.

2.10 FUNCTIONS OF EXPONENTIAL ORDER

A function $F(t)$ is said to of exponential order as $t \to \infty$ if there exist a positive constant M and a number α and a fixed value t_0 of t such that

$$|F(t)| < Me^{\alpha t} \text{ for } t \ge t_0 \qquad ...(i)$$

If this inequality is satisfied and α is to emphasized, we say that $F(t)$ is of the order of $e^{\alpha t}$ as $t \to \infty$.

we also write

$$F(t) = 0(e^{\alpha t}), t \to \infty \qquad ...(ii)$$

it means that $F(t)$ is of exponential order, the exponential being $e^{\alpha t}$, as $t \to \infty$. Thus equation (*ii*) is another way of expressing equation (*i*).

Example 12: Show that t^2 is of exponential order as $t \to \infty$.

Solution: $F(t) = t^2$

Now $$\lim_{t\to\infty} e^{-\alpha t} \cdot t^2 = \lim_{t\to\infty} \frac{t^2}{e^{\alpha t}} = \lim_{t\to\infty} \frac{2t}{\alpha e^{\alpha t}} = 0$$

Hence $F(t) = t^2$ is of exponential order,

$$F(t) = t^2 = 0(e^{\alpha t}), t \to \infty$$

for any fixed positive α.

Similarly we can show that t^n $(n \geq 0)$ is of exponential order with $\alpha > 0$.

Example 13: Show that e^{t^2} is not of exponential order as $t \to \infty$.

Solution: Let $F(t) = e^{t^2}$

Now $$\lim_{t\to\infty} e^{-\alpha t} \cdot e^{t^2} = \lim_{t\to\infty} e^{t(t-\alpha)} = \infty \quad ...(1)$$

Thus for any fixed value α, the limit in equation (1) is infinite. Hence $F(t) = e^{t^2}$ is not an exponential order.

Function of Class *A*

A function which is piecewise continuous on every finite interval in the range $t \geq 0$ and is exponential order as $t \to \infty$ is known as **'function of class A'**.

Theorem 1: If $F(t)$ is of class A and if $L\{F(t)\} = f(s)$ then

$$\lim_{s\to\infty} f(s) = 0.$$

Proof: Since the function $F(t)$ is class A function so

$$|F(t)| \leq Me^{\alpha t}, t \geq 0$$

Since $$f(s) = \int_0^\infty e^{-st} \cdot F(t)\,dt$$

So $$\left|\int_0^\infty e^{-st} \cdot F(t)\,dt\right| \leq M\int_0^\infty e^{-st} \cdot e^{\alpha t}\,dt$$

$$= \frac{M}{s-\alpha}, \text{for } s > \alpha$$

$$= 0 \text{ as } s \to \infty$$

Hence $$\boxed{\lim_{s\to\infty} f(s) = 0.}$$

Existence of Laplace Transforms (Existence Theorem)

Theorem 2: If a function $F(t)$ is piecewise continuous on every finite interval in the range $t \geq 0$ and is of exponential order, $|F(t)| \leq Me^{\alpha t}$ or $F(t) = 0(e^{\alpha t})$, as $t \to \infty$, then the Laplace transform

$$L\{F(t)\} = \int_0^\infty e^{-st} \cdot F(t)\,dt = f(s) \quad ...(1)$$

exists for $s > \alpha$.

Proof: The integral in equation (1) may be split into two parts as follows:

$$\int_0^{\infty} e^{-st} \cdot F(t)\,dt = \int_0^{T} e^{-st} \cdot F(t)\,dt + \int_T^{\infty} e^{-st} \cdot F(t)\,dt \qquad \text{...(2)}$$

The integral $\int_0^{T} e^{-st} \cdot F(t)\,dt$ exists since $F(t)$ is piecewise continuous on every finite interval $0 \le t \le T$.

Now $$\left|\int_T^{\infty} e^{-st} \cdot F(t)\,dt\right| \le \int_T^{\infty} e^{-st}\,|F(t)|\,dt$$

$$\le M\int_T^{\infty} e^{-st} e^{\alpha t}\,dt \qquad \Big|\text{As } |F(t)| \le Me^{\alpha t}$$

$$\le M\int_T^{\infty} e^{-(s-\alpha)t}\,dt = \frac{Me^{-(s-\alpha)T}}{(s-\alpha)}, \text{ for } s > \alpha$$

The term $\frac{Me^{-(s-\alpha)T}}{(s-\alpha)} \to 0$ as $T \to \infty$.

Therefore, the second integral of equation (2) exists. Hence the Laplace transform of $F(t)$ exists and the theorem is established.

Remark: Conditions in the above theorem are sufficient but not necessary for the existence of the Laplace transform. If these conditions are satisfied, the Laplace transform must exist. If these conditions are not satisfied, the Laplace transform may or may not exist. We can show this by the following example.

Example 14: Let $F(t) = \frac{1}{\sqrt{t}}$

Here $F(t) \to \infty$ as $t \to 0$

Thus the function $F(t)$ is not piecewise continuous on every finite interval in the range $t \ge 0$.

But $$L\{t^{-1/2}\} = \frac{\left|\overline{\frac{-1}{2}+1}\right.}{s^{-\frac{1}{2}+1}} = \frac{\left|\overline{1/2}\right.}{s^{1/2}} = \sqrt{\frac{\pi}{s}}, s > 0$$

Hence $L\left\{\frac{1}{\sqrt{t}}\right\}$ exists for $s > 0$ even $\frac{1}{\sqrt{t}}$ is not piecewise continuous in the range $t \ge 0$.

2.11 LAPLACE TRANSFORM OF THE DERIVATIVE OF $F(t)$

If $L\{F(t)\} = f(s)$; then $L\{F'(t)\} = sL\{F(t)\} - F(0)$

Proof: We know that $L\{F(t)\} = \int_0^\infty e^{-st} \cdot F(t)\, dt$

$$\therefore \quad L\{F'(t)\} = \int_0^\infty \underset{I}{e^{-st}} \cdot \underset{II}{F'(t)}\, dt = [e^{-st} \cdot F(t)]_0^\infty + s\int_0^\infty e^{-st} \cdot F(t)\, dt$$

$$\Rightarrow \quad L\{F'(t)\} = \lim_{t\to\infty} e^{-st} \cdot F(t) - F(0) + sL\{F(t)\}$$

$$\Rightarrow \quad \boxed{L\{F'(t)\} = sL\{F(t)\} - F(0).}$$

Since $|F(t)| \le Me^{\alpha t}, t \ge 0$

$\therefore \lim_{t\to\infty} e^{-st} \cdot F(t) = 0$

2.12 LAPLACE TRANSFORM OF *n*th DERIVATIVE

$$L\{F^n(t)\} = s^n L\{F(t)\} - s^{n-1}F(0)\} - s^{n-2}F'(0) - \cdots\cdots - F^{n-1}(0)$$

Proof: We have

$$L\{F'(t)\} = sL\{F(t)\} - F(0) \qquad \text{...(1)}$$

$$\Rightarrow \quad L\{F''(t)\} = sL\{F'(t)\} - F'(0)$$

$$\Rightarrow \quad L\{F''(t)\} = s[sL\{F(t)\} - F(0)] - F'(0)$$

$$\Rightarrow \quad \boxed{L\{F''(t)\} = s^2L\{F(t)\} - sF(0) - F'(0)} \qquad \text{...(2)}$$

Again applying (1) to third order derivative $F'''(t)$, we get

$$L\{F'''(t)\} = sL\{f''(t)\} - F''(0)$$

$$= s[s^2L\{F(t)\} - sF(0) - F'(0)] - F''(0)$$

$$\Rightarrow \quad \boxed{L\{F'''(t)\} = s^3L\{F(t)\} - s^2F(0) - sF'(0) - F''(0)}$$

Proceeding n times, we have

$$\boxed{L\{F^n(t)\} = s^nL\{F(t)\} - s^{n-1}F(0) - s^{n-2}F'(0) - \cdots - F^{n-1}(0).}$$

2.13 INITIAL VALUE THEOREM

If $F(t)$ is continuous for all $t \ge 0$ and is of exponential order as $t \to \infty$ and if $F'(t)$ is of class A, then

$$\lim_{t\to 0} F(t) = \lim_{s\to\infty} sL\{F(t)\}.$$

Proof: We have

$$L\{F'(t)\} = sL\{F(t)\} - F(0)$$

$$\Rightarrow \quad \int_0^\infty e^{-st}F'(t)\,dt = sL\{F(t)\} - F(0) \qquad ...(1)$$

Taking $\lim\limits_{s\to\infty}$, we get

$$\lim_{s\to\infty}\int_0^\infty e^{-st}F'(t)\,dt = \lim_{s\to\infty} sL\{F(t)\} - F(0)$$

$$\Rightarrow \quad 0 = \lim_{s\to\infty} sL\{F(t)\} - F(0)$$

$$\Rightarrow \quad F(0) = \lim_{s\to\infty} sL\{F(t)\}$$

or

$$\boxed{\lim_{t\to 0} F(t) = \lim_{s\to\infty} sL\{F(t)\}.}$$

2.14 FINAL VALUE THEOREM

If $F(t)$ is continuous for all $t \geq 0$ and is of exponential order as $t \to \infty$ and if $F'(t)$ is of class A, then

$$\lim_{t\to\infty} F(t) = \lim_{s\to 0} sL\{F(t)\}$$

Proof: From Section 2.13, equation (1), we have

$$L\{F'(t)\} = \int_0^\infty e^{-st}\cdot F'(t)\,dt = sL\{F(t)\} - F(0)$$

Taking limit as $s \to 0$, we get

$$\lim_{s\to 0}\int_0^\infty e^{-st}F'(t)\,dt = \lim_{s\to 0} sL\{F(t)\} - F(0)$$

$$\Rightarrow \quad \int_0^\infty F'(t)\,dt = \lim_{s\to 0} sL\{F(t)\} - F(0)$$

$$\Rightarrow \quad [F(t)]_0^\infty = \lim_{s\to 0} sL\{F(t)\} - F(0)$$

or

$$\lim_{t\to\infty} F(t) - F(0) = \lim_{s\to 0} sL\{F(t)\} - F(0)$$

$$\Rightarrow \quad \boxed{\lim_{t\to\infty} F(t) = \lim_{s\to 0} sL\{F(t)\}.}$$

2.15 LAPLACE TRANSFORM OF INTEGRALS

If $F(t)$ is piecewise continuous and satisfies $|F(t)| \leq Me^{\alpha t}, t \geq 0$

then $$L\left\{\int_0^t F(t)dt\right\} = \frac{1}{s}L\{F(t)\} = \frac{1}{s}f(s).$$

Proof: Let $$G(t) = \int_0^t F(t)\,dt \qquad \text{...(1)}$$

We know that

$$L\{G'(t)\} = sL\{G(t)\} - G(0) \qquad \text{...(2)}$$

From (1) $$|\{G(t)| \leq \int_0^t |F(t)|\,dt \leq M\int_0^t e^{\alpha t}dt = M\frac{e^{\alpha t}-1}{\alpha}$$

At $t = 0$, $$|G(0)| = M\frac{e^0 - 1}{\alpha} = 0$$

$$\Rightarrow \qquad G(0) = 0 \qquad \text{...(3)}$$

Using (3) in (2), we get

$$L\{G'(t)\} = sL\{G(t)\} \quad \Rightarrow \quad L\{G(t)\} = \frac{1}{s}L\{G'(t)\}$$

$$\Rightarrow \quad \boxed{L\left\{\int_0^t F(t)dt\right\} = \frac{1}{s}L\{F(t)\} = \frac{1}{s}f(s).} \qquad \left|\begin{array}{l}\text{from (1)}\\ G'(t) = F(t)\end{array}\right.$$

2.16 LAPLACE TRANSFORMATION OF *t.F* (*t*), (MULTIPLICATION BY *t*)

If $L\{F(t)\} = f(s)$, then $L\{tF(t)\} = -\dfrac{d}{ds}f(s)$.

Proof: We have $f(s) = \displaystyle\int_0^\infty e^{-st}\cdot F(t)\,dt$

Differentiating w.r.t. 's', we get

$$\frac{d}{ds}f(s) = \frac{d}{ds}\int_0^\infty e^{-st}\cdot F(t)\,dt$$

$$= \int_0^\infty \frac{\partial}{\partial s} e^{-st} \cdot F(t)\, dt$$

By Leibnitz's rule for differentiating under the sign of integral.

$$= -\int_0^\infty e^{-st} \cdot tF(t)\, dt$$

$$\Rightarrow \qquad \frac{d}{ds} f(s) = -L\{tF(t)\}$$

Hence, $\boxed{L\{tF(t) = -\frac{d}{ds} f(s).}$

2.17 MULTIPLICATION BY t^n

If $L\{F(t)\} = f(s)$, then $L\{t^n F(t)\} = (-1)^n \frac{d^n}{ds^n}[f(s)]$. *(U.P.T.U. 2005)*

Proof: We shall prove this theorem by Mathematical induction.
We have already shown that

$$L\{tF(t)\} = (-1)^1 \frac{d}{ds} f(s)$$

This shows that the given result is true for $n = 1$.
Now assume that the theorem is true for $n = k$, then we have

$$L\{t^k F(t)\} = (-1)^k \frac{d^k}{ds^k}\left[f(s)\right]$$

$$\Rightarrow \qquad \int_0^\infty e^{-st} \cdot t^k \cdot F(t) dt = (-1)^k \cdot \frac{d^k}{ds^k} f(s)$$

Differentiating both sides w.r.t. 's', we get

$$\frac{d}{ds}\int_0^\infty e^{-st} t^k F(t)\, dt = (-1)^k \frac{d^{k+1}}{ds^{k+1}} f(s)$$

$$\Rightarrow \qquad \int_0^\infty \frac{\partial}{\partial s} e^{-st} \cdot t^k \cdot F(t)\, dt = (-1)^k \frac{d^{k+1}}{ds^{k+1}} f(s)$$

$$\Rightarrow \qquad -\int_0^\infty e^{-st} \cdot t^{k+1} \cdot F(t)\, dt = (-1)^k \cdot \frac{d^{k+1}}{ds^{k+1}} f(s)$$

$$\Rightarrow \qquad L\{t^{k+1} F(t)\} = (-1)^{k+1} \cdot \frac{d^{k+1}}{ds^{k+1}} f(s)$$

Thus the result is true for $n = k + 1$

The above relation shows that if the result is true for any particular value of n, it is true for the next value of n. But we have already seen that the result is true for $n = 1$. Hence it is true for $n = 1 + 1 = 2$ and $n = 2 + 1 = 3$, etc.

Hence, $\boxed{L\{t^n F(t)\} = (-1)^n \frac{d^n}{ds^n} f(s).}$

2.18 DIVISION BY *t*

If $L\{F(t)\} = f(s)$, then $L\left\{\frac{F(t)}{t}\right\} = \int_s^\infty f(s)ds$ provided $\lim_{t\to 0}\left\{\frac{1}{t}F(t)\right\}$ exists. (*U.P.T.U. 2005, 2007*)

Proof: Let

$$G(t) = \frac{F(t)}{t} \qquad \Rightarrow \qquad F(t) = tG(t)$$

$$\therefore \qquad L\{F(t)\} = L\{tG(t)\} = -\frac{d}{ds}g(s)$$

or

$$f(s) = -\frac{d}{ds}L\{G(t)\} \qquad ...(1)$$

Since $\lim_{t\to 0}\left\{\frac{1}{t}F(t)\right\}$ exists it follows that $G(t) = \frac{F(t)}{t}$ is piecewise continuous as well as of exponential order, and so $L\{G(t)\}$ exists.

Integrating equation (1) between s and ∞, we get

$$L\{G(t)\} = -\int_\infty^s f(s)ds = \int_s^\infty f(s)ds$$

$$\Rightarrow \qquad \boxed{L\left\{\frac{F(t)}{t}\right\} = \int_s^\infty f(s)ds.}$$

Example 15: Find the Laplace transform of the function

$$F(t) = \int_0^t \frac{\sin\tau}{\tau}d\tau. \qquad (U.P.T.U.\ 2007)$$

Solution: We have $F(t) = \int_0^t \frac{\sin\tau}{\tau}\,d\tau = \int_0^t \frac{\sin t}{t}\,dt$ | By definite integral

Now $$L\{\sin t\} = \frac{1}{s^2+1}$$

$$\Rightarrow \quad L\left\{\frac{\sin t}{t}\right\} = \int_s^\infty \frac{1}{s^2+1}\,ds \quad \text{(By division formula)}$$

$$= [\tan^{-1} s]_s^\infty = \tan^{-1}\infty - \tan^{-1} s = \frac{\pi}{2} - \tan^{-1} s$$

$$\Rightarrow \quad L\left\{\frac{\sin t}{t}\right\} = \cot^{-1} s \quad \left(\frac{\pi}{2} - \tan^{-1} s = \cot^{-1} s\right)$$

Here $$f_1(s) = \cot^{-1} s$$

$$\therefore \quad L\{F(t)\} = L\left\{\int_0^t \frac{\sin t}{t}\,dt\right\} = \frac{1}{s}\cot^{-1} s. \quad \textbf{Ans.} \quad \left[\text{By } L\left\{\int_0^t F(t)\right\} = \frac{1}{s} f(s)\right]$$

Example 16: Show that $L\left\{\frac{\cos\sqrt{t}}{\sqrt{t}}\right\} = \sqrt{\frac{\pi}{s}}\,e^{-1/4s}$.

Solution: Let $F(t) = \sin\sqrt{t}$

then $$F'(t) = \frac{\cos\sqrt{t}}{2\sqrt{t}} \text{ and } F(0) = \sin 0 = 0$$

$\therefore$ From $$L\{F'(t)\} = sL\{F(t)\} - F(0)$$

$$L\left\{\frac{\cos\sqrt{t}}{2\sqrt{t}}\right\} = sL\{\sin\sqrt{t}\}$$

$$\Rightarrow \quad L\left\{\frac{\cos\sqrt{t}}{\sqrt{t}}\right\} = 2sL\{\sin\sqrt{t}\} = s\cdot\frac{\sqrt{\pi}}{s^{3/2}}e^{-1/4s} \cdot$$ See Ex. 1 Page 253

$$= \sqrt{\frac{\pi}{s}}\,e^{-1/4s}. \quad \textbf{Proved.}$$

Example 17: Find the Laplace transform of $\dfrac{e^{at}-\cos bt}{t}$ *(U.P.T.U. 2003)*

Solution: Let $F(t) = \dfrac{e^{at}-\cos bt}{t}$

$$\therefore \quad L\{e^{at}-\cos bt\} = L\{e^{at}\}-L\{\cos bt\} = \frac{1}{s-a}-\frac{s}{s^2+b^2}$$

Now,
$$L\{F(t)\} = L\left\{\frac{e^{at}-\cos bt}{t}\right\} = \int_s^\infty \left(\frac{1}{s-a}-\frac{s}{s^2+b^2}\right)ds \quad \text{(By division formula)}$$

$$= \left[\log(s-a)-\frac{1}{2}\log(s^2+b^2)\right]_s^\infty$$

$$= \frac{1}{2}[\log(s-a)^2-\log(s^2+b^2)]_s^\infty$$

$$= \frac{1}{2}\left[\log\frac{(s-a)^2}{(s^2+b^2)}\right]_s^\infty = \frac{1}{2}\left[\log\frac{\left(1-\frac{a}{s}\right)^2}{\left(1+\frac{b^2}{s^2}\right)}\right]_s^\infty$$

$$= \frac{1}{2}\left[\log 1-\log\frac{\left(1-\frac{a}{s}\right)^2}{\left(1+\frac{b^2}{s^2}\right)}\right]$$

$$= \frac{1}{2}\left[0-\log\frac{(s-a)^2}{(s^2+b^2)}\right] = \frac{1}{2}\log\left[\frac{(s^2+b^2)}{(s-a)^2}\right]. \quad \textbf{Ans.}$$

Example 18: If $F(t)$ is continuous except for an ordinary discontinuity at $t = a$ $(a > 0)$ as given in the figure.

Then show that

$$L\{F'(t)\} = sL\{F(t)\}-F(0)-e^{-as}[F(a+0)-F(a-0)]. \quad \textit{(U.P.T.U. 2003)}$$

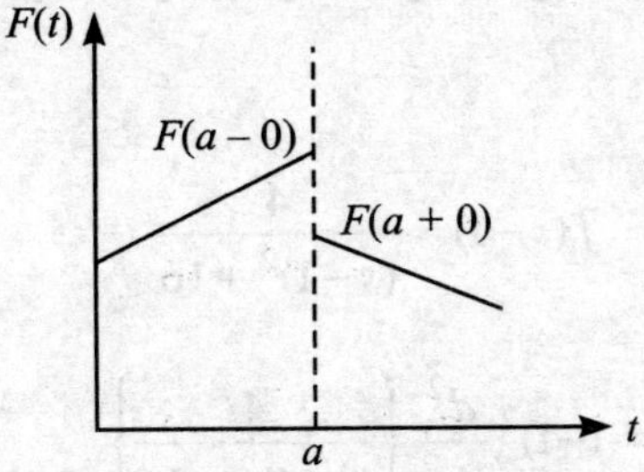

Solution: $L\{F'(t)\} = \int_0^\infty e^{-st} F'(t)\,dt$

$$= \int_0^a e^{-st} \cdot F'(t)\,dt + \int_a^\infty e^{-st} \cdot F'(t)\,dt$$

$$= [e^{-st}F(t)]_0^a + s\int_0^a e^{-st}F(t)\,dt + [e^{-st} \cdot F(t)]_a^\infty + s\int_a^\infty e^{-st}F(t)\,dt \quad \text{[Integrate by part]}$$

$$= e^{-as}F(a-0) - F(0) + s\int_0^\infty e^{-st} \cdot F(t)\,dt + \lim_{t\to\infty}\{e^{-st} \cdot F(t)\} - e^{-as}F(a+0)$$

$$\Rightarrow \quad L\{F'(t)\} = sL\{F(t)\} - F(0) - e^{-as}[F(a+0) - F(a-0)]. \quad \textbf{Proved.}$$

| As $\lim_{t\to\infty} e^{-st}F(t) = 0$

Example 19: Find the Laplace transform of

(*i*) $te^{-t}\sin 2t$

(*ii*) $t^2 e^t \sin 4t$. *(U.P.T.U. 2002)*

Solution: (*i*) $L\{\sin 2t\} = \dfrac{2}{s^2+4} = f(s)$

$$\therefore \quad L\{e^{-t}\sin 2t\} = f(s+1) = \frac{2}{(s+1)^2+4}$$ | By first shifting theorem

$$\Rightarrow \quad L\{t(e^{-t}\sin 2t)\} = -\frac{d}{ds}\left\{\frac{2}{(s+1)^2+4}\right\} \quad \left[By\ L\{tF(t)\} = -\frac{d}{ds}f(s)\right]$$

$$= \frac{2\{2(s+1)\}}{\{(s+1)^2+4\}^2} = \frac{4(s+1)}{\{(s+1)^2+4\}^2}$$

$$\Rightarrow \quad L\{te^{-t}\sin 2t\} = \frac{4(s+1)}{\{s^2+2s+5\}^2}. \quad \textbf{Ans.}$$

(*ii*) $L\{\sin 4t\} = \dfrac{4}{s^2+16} = f(s)$

$$\therefore \quad L\{e^t \sin 4t\} = f(s-1) = \frac{4}{(s-1)^2+16}$$

$$\Rightarrow \quad L\{t^2(e^t \sin 4t)\} = (-1)^2 \frac{d^2}{ds^2}\left\{\frac{4}{(s-1)^2+16}\right\} \qquad \left| L\{t^n F(t)\} = (-1)^n \frac{d^n}{ds^n} f(s) \right.$$

$$= 4\frac{d}{ds}\left[-\frac{2(s-1)}{\{(s-1)^2+16\}^2}\right]$$

$$= -8\frac{d}{ds}\left[\frac{s-1}{\{(s-1)^2+16\}^2}\right]$$

$$= -8\left[\frac{\{(s-1)^2+16\}^2-(s-1)^2 4\{(s-1)^2+16\}}{\{(s-1)^2+16\}^4}\right]$$

$$= -\frac{8\{(s-1)^2+16\}[(s-1)^2+16-4(s-1)^2]}{\{(s+1)^2+16\}^4}$$

$$= \frac{8[3s^2-6s-13]}{\{(s-1)^2+16\}^3}. \qquad \textbf{Ans.}$$

Example 20: Evaluate the integral $\displaystyle\int_0^\infty \frac{\sin^2 t}{t^2}\, dt.$

Solution: $L\{\sin^2 t\} = L\left\{\dfrac{1-\cos 2t}{2}\right\} = \dfrac{1}{2}L\{1\} - \dfrac{1}{2}L\{\cos 2t\}$

$$= \frac{1}{2s} - \frac{1}{2}\frac{s}{s^2+4}$$

$$\therefore \quad L\left\{\frac{\sin^2 t}{t}\right\} = \frac{1}{2}\int_s^\infty \left(\frac{1}{s} - \frac{s}{s^2+4}\right) ds \qquad \text{(By division formula)}$$

$$= \frac{1}{2}\left[\log s - \log\sqrt{s^2+4}\right]_s^\infty = \frac{1}{2}\left[\log\frac{s}{\sqrt{s^2+4}}\right]_s^\infty$$

$$= \frac{1}{2}\left[\log \frac{1}{\sqrt{1+\frac{4}{s^2}}}\right]_s^\infty = \frac{1}{2}\log 1 - \frac{1}{2}\log \frac{s}{\sqrt{s^2+4}}$$

$$= \frac{1}{2}\log\left(\frac{s^2+4}{s^2}\right)^{1/2} \qquad \Big| \text{ As } \log 1 = 0$$

$$\Rightarrow \qquad L\left\{\frac{\sin^2 t}{t}\right\} = \frac{1}{4}\log\left(\frac{s^2+4}{s^2}\right)$$

Again apply the division formula, we have

$$L\left\{\frac{\left(\frac{\sin^2 t}{t}\right)}{t}\right\} = \frac{1}{4}\int_s^\infty 1 \cdot \log\left(\frac{s^2+4}{s^2}\right) ds$$

$$= \frac{1}{4}\left[\left(s \cdot \log \frac{s^2+4}{s^2}\right)_s^\infty + \int_s^\infty \frac{8s^3}{s^3(s^2+4)} ds\right]$$

$$= \frac{1}{4}\left[\left\{s \log\left(1+\frac{4}{s^2}\right)\right\}_s^\infty + 4\left\{\tan^{-1}\frac{s}{2}\right\}_s^\infty\right]$$

$$= \frac{1}{4}\left[0 + s\log\frac{s^2+4}{s^2} + 4\tan^{-1}\infty - 4\tan^{-1}\frac{s}{2}\right]$$

$$\Rightarrow \qquad L\left\{\frac{\sin^2 t}{t^2}\right\} = \frac{s}{4}\log\frac{s^2+4}{s^2} + \left(\frac{\pi}{2} - \tan^{-1}\frac{s}{2}\right) = \frac{s}{4}\log\frac{s^2+4}{s^2} + \cot^{-1}\frac{s}{2}$$

or

$$\int_0^\infty e^{-st} \cdot \left(\frac{\sin^2 t}{t^2}\right) dt = \frac{s}{4}\log\frac{s^2+4}{s^2} + \cot^{-1}\frac{s}{2}$$

Taking $\lim\limits_{s \to 0}$, we get

$$\int_0^\infty \frac{\sin^2 t}{t^2} dt = \frac{\pi}{2}. \qquad \textbf{Ans.} \qquad \Big| \text{ As } \cot^{-1} 0 = \frac{\pi}{2}$$

Example 21: Find the Laplace transform of $\int_0^t \frac{e^{-at}-e^{-bt}}{t}\,dt$.

Solution: $L\{e^{-at}-e^{-bt}\}=L\{e^{-at}\}-L\{e^{-bt}\}$

$$= \frac{1}{s+a}-\frac{1}{s+b}$$

$$\therefore \quad L\left\{\frac{e^{-at}-e^{-bt}}{t}\right\} = \int_s^{\infty}\left[\frac{1}{(s+a)}-\frac{1}{(s+b)}\right]ds \qquad \text{(By division form)}$$

$$= [\log(s+a)-\log(s+b)]_s^{\infty} = \left[\log\frac{(s+a)}{(s+b)}\right]_s^{\infty}$$

$$= \left[\log\frac{1+\frac{a}{s}}{1+\frac{b}{s}}\right]_s^{\infty} = 0-\log\frac{(s+a)}{(s+b)} = \log\frac{(s+b)}{(s+a)}$$

Now apply the formula $L\left\{\int_0^t F(t)\,dt\right\} = \frac{1}{s}f(s) = \frac{1}{s}L\{F(t)\}$

$$\therefore \quad L\left\{\int_0^t \frac{e^{-at}-e^{-bt}}{t}\,dt\right\} = \frac{1}{s}L\left\{\frac{e^{-at}-e^{-bt}}{t}\right\} = \frac{1}{s}\log\frac{(s+b)}{(s+a)}. \qquad \textbf{Ans.}$$

Example 22: Show that $\int_0^{\infty} te^{-3t}\sin t\,dt = \frac{3}{50}$.

Solution: $\because \quad L\{\sin t\} = \frac{1}{s^2+1}$

$$\therefore \quad L\{t\sin t\} = -\frac{d}{ds}\left\{\frac{1}{s^2+1}\right\} = \frac{2s}{(s^2+1)^2}$$

$$\Rightarrow \quad \int_0^{\infty} e^{-st}\cdot t\sin t\,dt = \frac{2s}{(s^2+1)^2} \qquad \left|\ \text{As } L\{F(t)\} = \int_0^{\infty} e^{-st}\cdot F(t)\,dt\right.$$

Putting $s = 3$, we get

$$\int_0^\infty te^{-3t}\sin t\,dt = \frac{3}{50}. \quad \textbf{Proved.}$$

Example 23: Evaluate $\int_0^\infty e^{-t}\frac{\sin t}{t}dt.$

Solution: $\because \quad L\{\sin t\} = \frac{1}{s^2+1}$

$$\therefore \quad L\left\{\frac{\sin t}{t}\right\} = \int_s^\infty \frac{1}{s^2+1}ds \qquad \left| L\left\{\frac{F(t)}{t}\right\} = \int_s^\infty f(s)\,ds \right.$$

$$= [\tan^{-1} s]_s^\infty = \frac{\pi}{2} - \tan^{-1} s = \cot^{-1} s$$

$$\Rightarrow \quad L\left\{\frac{\sin t}{t}\right\} = \cot^{-1} s$$

or $$\int_0^\infty e^{-st}\cdot\frac{\sin t}{t}dt = \cot^{-1} s \qquad \left| L\{F(t)\} = \int_0^\infty e^{-st}\cdot F(t)\,dt \right.$$

Now putting $s = 1$, we have

$$\int_0^\infty e^{-t}\frac{\sin t}{t}\cdot dt = \cot^{-1} 1 = \frac{\pi}{4}. \quad \textbf{Ans.} \qquad \left| \cot^{-1} 1 = \frac{\pi}{4} \right.$$

Example 24: Using Laplace transform, evaluate $\int_0^\infty e^{-t}\frac{\sin^2 t}{t}dt$ *(U.P.T.U. 2008)*

Solution: $L\{\sin^2 t\} = L\left\{\frac{1-\cos 2t}{2}\right\} = \frac{1}{2}\left\{\frac{1}{s} - \frac{s}{s^2+4}\right\}$

$$\therefore \quad L\left\{\frac{\sin^2 t}{t}\right\} = \frac{1}{2}\int_s^\infty \left(\frac{1}{s} - \frac{s}{s^2+4}\right)ds \qquad |\text{Apply division formula}$$

$$= \frac{1}{2}\left[\log s - \log\sqrt{s^2+4}\right]_s^\infty$$

$$= \frac{1}{2}\left[\log \frac{s}{\sqrt{s^2+4}}\right]_s^\infty$$

$$= \frac{1}{2}\left[\log \frac{1}{\sqrt{1+\frac{4}{s^2}}}\right]_s^\infty = \frac{1}{2}\left[0 - \log \frac{1}{\sqrt{1+\frac{4}{s^2}}}\right]$$

$$\Rightarrow \quad L\left\{\frac{\sin^2 t}{t}\right\} = \frac{1}{2}\log\left(\frac{\sqrt{s^2+4}}{s}\right)$$

$$\Rightarrow \quad \int_0^\infty e^{-st} \cdot \frac{\sin^2 t}{t}\, dt = \frac{1}{2}\log\left(\frac{\sqrt{s^2+4}}{s}\right)$$

Putting $s = 1$, we get

$$\int_0^\infty e^{-t} \cdot \frac{\sin^2 t}{t}\, dt = \frac{1}{2}\log\left(\frac{\sqrt{5}}{1}\right) = \frac{1}{4}\log 5. \quad \textbf{Ans.}$$

EXERCISE 2.2

Find the Laplace transform of the following:

1. $t^2 \sin at.$ $\left[\textbf{Ans. } \frac{2a(3s^2-a^2)}{(s^2+a^2)^3}, s > a\right]$

2. $t^n e^{at}.$ $\left[\textbf{Ans. } \frac{\underline{|n}}{(s-a)^{n+1}}, s > a\right]$

3. $(\sin at - at \cos at).$ $\left[\textbf{Ans. } \frac{2a^3}{(s^2+a^2)^2}\right]$

4. (*i*) $t \cos h\, 3t$ (*ii*) $t \sin h\, at.$ $\left[\textbf{Ans. } (i) \frac{s^2+9}{(s^2-9)^2} \; (ii) \frac{2as}{(s^2-a^2)^2}\right]$

5. $\left\{\frac{\sin ht}{t}\right\}.$ $\left[\textbf{Ans. } \frac{1}{2}\log\frac{s-1}{s+1}\right]$

6. $te^{-t}\sin 2t$. $\left[\textbf{Ans.}\ \dfrac{4(s+1)}{[(s+1)^2+4]^2}\right]$ (*U.P.T.U. 2002*)

7. $\displaystyle\int_0^t \frac{1-\cos 2t}{t}\,dt$. $\left[\textbf{Ans.}\ \dfrac{1}{2s}\log\left(1+\dfrac{4}{s^2}\right)\right]$

8. $\displaystyle\int_0^t e^{-2t}\cdot t\sin^3 t\,dt$. $\left[\textbf{Ans.}\ \dfrac{3(s+2)}{2s}\left[\dfrac{1}{[(s+2)^2+9]^2}-\dfrac{1}{[(s+2)^2+1]^2}\right]\right]$

9. $\displaystyle\int_0^t \frac{\cos at-\cos bt}{t}\,dt$. $\left[\textbf{Ans.}\ \dfrac{1}{2s}\log\left(\dfrac{s^2+b^2}{s^2+a^2}\right)\right]$

10. $\displaystyle\int_0^t \frac{1-e^{-t}}{t}\,dt$. $\left[\textbf{Ans.}\ \dfrac{1}{s}\log\left(1+\dfrac{1}{s}\right)\right]$

11. $\displaystyle\int_0^{t/2} \frac{1-e^{-2x}}{x}\,dx$. $\left[\textbf{Ans.}\ \dfrac{1}{s}\log\left(1+\dfrac{1}{s}\right)\right]$

12. Prove that $\displaystyle\int_0^\infty \frac{\sin t}{t}\,dt=\frac{\pi}{2}$. (*U.P.T.U. 2009*)

13. Prove that $\displaystyle\int_0^\infty \frac{e^{-t}\cdot\sin t}{t}\,dt=\frac{\pi}{4}$.

14. Evaluate $\displaystyle\int_0^\infty \frac{e^{-3t}-e^{-6t}}{t}\,dt$. $[\textbf{Ans.}\ \log 2]$

15. Evaluate $\displaystyle\int_0^\infty te^{-2t}\cos t\,dt$. $\left[\textbf{Ans.}\ \dfrac{3}{25}\right]$

2.19 PERIODIC FUNCTIONS

A function $F(t)$ is said to be periodic with period $T > 0$, if for all t, $F(t + T) = F(t)$, $F(t + 2T) = F(t)$, etc. where T is positive constant, the least value of $T > 0$ is called the period of $F(t)$.

Theorem: If $F(t)$ is a periodic function with period T, when $T > 0$ and piecewise continuous, then

$$L\{F(t)\} = \frac{\int_0^T e^{-st} \cdot F(t)\,dt}{1-e^{-sT}}$$

Proof: By the definition, we have

$$L\{F(t)\} = \int_0^\infty e^{-st} \cdot F(t)\,dt$$

$$= \int_0^T e^{-st} \cdot F(t)\,dt + \int_T^{2T} e^{-st} \cdot F(t)\,dt + \int_{2T}^{3T} e^{-st} \cdot F(t)\,dt + \ldots\ldots$$

Now putting $t = x + T$, $t = x + 2T$, etc., in 2nd, 3rd, integrals respectively, we get

$$L\{F(t)\} = \int_0^T e^{-st} \cdot F(t)\,dt + \int_0^T e^{-s(x+T)} \cdot F(x+T)\,dx + \int_0^T e^{-s(x+2T)} \cdot F(x+2T)\,dx + \ldots\ldots$$

$$= \int_0^T e^{-st} \cdot F(t)\,dt + e^{-sT}\int_0^T e^{-st} F(t+T)\,dt + e^{-2sT}\int_0^T e^{-st} \cdot F(t)\,dt + \ldots\ldots \quad \Big|\ \text{By definite integral}$$

$$= (1+e^{-sT}+e^{-2sT}+\ldots)\int_0^T e^{-st} F(t)\,dt$$

$$= \frac{1}{(1-e^{-sT})}\int_0^T e^{-st} \cdot F(t)\,dt \quad \Bigg|\ \text{Sum of infinite (G.P.) series} = \frac{a}{1-r},\ r<1$$

Hence $$\boxed{L\{F(t)\} = \frac{\int_0^T e^{-st} \cdot F(t)\,dt}{1-e^{-sT}}}.$$

2.20 UNIT STEP FUNCTION

The unit step function is defined as

$$U(t-a) = \begin{cases} 0, & t<a \\ 1, & t \geq a, \end{cases} \quad a \geq 0$$

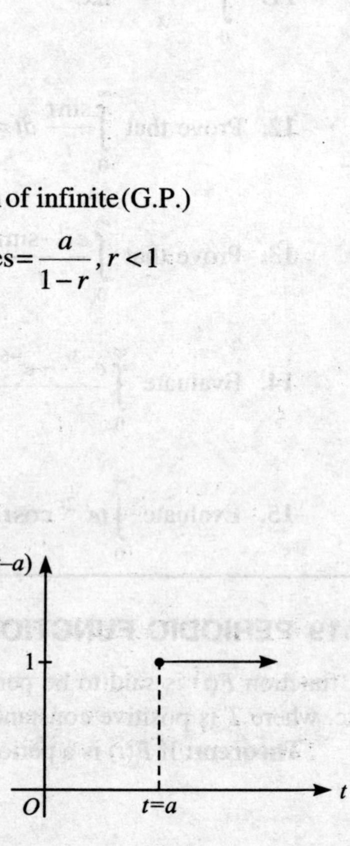

It is also called Heaviside's unit function.

2.20.1 Laplace Transform of Unit Step Function

$$L\{U(t-a)\} = \int_0^\infty e^{-st} \cdot U(t-a)\,dt = \int_0^a e^{-st} \cdot 0\,dt + \int_a^\infty e^{-st}\,dt$$

$$= -\frac{1}{s}[e^{-st}]_a^\infty = \frac{1}{s}e^{-as}$$

$$\Rightarrow \boxed{L\{U(t-a)\} = \frac{1}{s}e^{-as}.}$$

2.20.2 Second Shifting Theorem in Unit Step's Term

If $\quad L\{F(t)\} = f(s)$ then $L\{F(t-a) \cdot U(t-a)\} = e^{-as} \cdot f(s)$.

Proof: $L\{F(t-a) \cdot U(t-a)\} = \int_0^\infty e^{-st} \cdot F(t-a) \cdot U(t-a)\,dt$

$$= \int_0^a e^{-st} \cdot F(t-a) \cdot 0\,dt + \int_a^\infty e^{-st} \cdot F(t-a)\,dt = \int_a^\infty e^{-st} \cdot F(t-a)\,dt$$

Putting $t-a=x \quad \Rightarrow \quad t = x+a$, we get

$$L\{F(t-a) \cdot U(t-a)\} = \int_0^\infty e^{-s(x+a)} \cdot F(x)\,dx = e^{-sa}\int_0^\infty e^{-st} \cdot F(t)\,dt \qquad \text{| By definite integral}$$

$$\Rightarrow \quad \boxed{L\{F(t-a) \cdot U(t-a)\} = e^{-sa} \cdot f(s).}$$

Example 25: Using unit step function. Find the Laplace transform of

(*i*) $(t-1)^2 \cdot u(t-1)$ (*U.P.T.U. 2008*)

(*ii*) $\sin t \cdot u(t-\pi)$

Solution: (*i*) Using second shifting theorem

$$L\left\{(t-1)^2 u(t-1)\right\} = e^{-s} \cdot L\{t^2\}$$

$$= e^{-s} \cdot \frac{\lfloor 2}{s^{2+1}} = \frac{2e^{-s}}{s^3} \qquad \textbf{Ans.}$$

(*ii*) $\sin t \cdot u(t-\pi) = \sin(t-\pi+\pi) \cdot u(t-\pi)$

$$= \left\{\sin(t-\pi)\cos\pi + \sin\pi\cos(t-\pi)\right\} u(t-\pi)$$

$$= -\sin(t-\pi) \cdot u(t-\pi) \qquad \text{| As } \sin\pi = 0$$

$$\therefore \quad L\{\sin t \cdot u(t-\pi)\} = -L\{\sin(t-\pi)\cdot u(t-\pi)\}$$

$$= -e^{-\pi s} \cdot L\{\sin t\} \qquad \text{|By IInd shifting theorem}$$

$$= \frac{-e^{-\pi s}}{s^2+1} \qquad \textbf{Ans.}$$

Example 26: Find the Laplace transform of the following periodic function

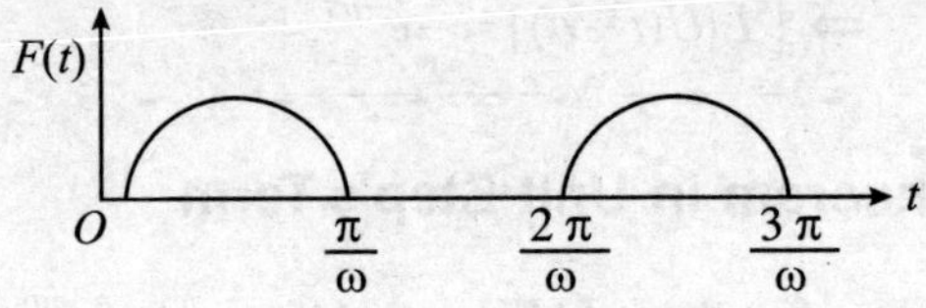

Half wave rectifier

Solution: From the figure, we have

$$F(t) = \begin{bmatrix} \sin \omega t & \text{for } 0 < t < \dfrac{\pi}{\omega} \\ 0 & \text{for } \dfrac{\pi}{\omega} < t < \dfrac{2\pi}{\omega} \end{bmatrix}$$ [*U.P.T.U. 2002, (Summer) 2009*]

The given function is a periodic function with the period $\dfrac{\text{'}2\pi\text{'}}{\omega}\cdot$ *i.e.,* $T = \dfrac{2\pi}{\omega}$

$$\therefore \quad L\{F(t)\} = \frac{1}{(1-e^{-sT})}\int_0^T e^{-st}\cdot F(t)dt$$

$$= \frac{1}{1-e^{-s\frac{2\pi}{\omega}}}\int_0^{2\pi/\omega} e^{-st}\cdot F(t)\,dt$$

$$= \frac{1}{1-e^{-\frac{2\pi s}{\omega}}}\left[\int_0^{\pi/\omega} e^{-st}\cdot \sin\omega t\,dt + \int_{\pi/\omega}^{2\pi/\omega} e^{-st}\cdot 0\,dt\right]$$

$$L\{F(t)\} = \frac{1}{1-e^{-\frac{2\pi s}{\omega}}}\int_0^{\pi/\omega} e^{-st}\cdot \sin\omega t\,dt \qquad \text{...(1)}$$

Now let
$$I = \int_0^{\pi/\omega} \underset{I}{e^{-st}}\cdot \underset{II}{\sin\omega t}\,dt = -\frac{1}{\omega}[e^{-st}\cos\omega t]_0^{\pi/\omega} - \frac{s}{\omega}\int_0^{\pi/\omega} \underset{I}{e^{-st}}\cdot \underset{II}{\cos\omega t}\,dt$$

$$\Rightarrow \quad I = -\frac{1}{\omega}[e^{-\frac{s\pi}{\omega}}\cdot\cos\pi - 1] - \frac{s}{\omega}\left(\frac{1}{\omega}\right)[e^{-st}\cdot\sin\omega t]_0^{\pi/\omega} - \frac{s^2}{\omega^2}\int_0^{\pi/\omega} e^{-st}\cdot\sin\omega t\,dt$$

$$\Rightarrow \qquad I\left(1+\frac{s^2}{\omega^2}\right) = \frac{1}{\omega}(1+e^{-\frac{s\pi}{\omega}}) \quad \Rightarrow \quad I\left(\frac{s^2+\omega^2}{\omega^2}\right) = \frac{1}{\omega}\left(1+e^{-\frac{s\pi}{\omega}}\right)$$

$$\Rightarrow \qquad I = \frac{\omega(1+e^{-s\pi/\omega})}{(s^2+\omega^2)} \qquad \text{...(2)}$$

Using (2) in (1), we get

$$L\{F(t)\} = \frac{\omega(1+e^{-s\pi/\omega})}{(s^2+\omega^2)(1-e^{-2\pi s/\omega})} = \frac{\omega(1+e^{-s\pi/\omega})}{(s^2+\omega^2)(1+e^{-s\pi/\omega})(1-e^{-s\pi/\omega})} = \frac{\omega}{(s^2+\omega^2)(1-e^{-s\pi/\omega})}.$$ **Ans.**

Example 27: A periodic square wave function *F(t)* in space, terms of unit step function is written as

$$F(t) = k\{u_0(t) - 2u_a(t) + 2u_{2a}(t) - 2u_{3a}(t) + \cdots\}$$

Show that $\qquad L\{F(t)\} = \frac{k}{s}\tan h\left(\frac{as}{2}\right).$

Solution: $L\{F(t)\} = kL\{u_0(t)\} - 2kL\{u_a(t)\} + 2kL\{u_{2a}(t)\} \cdots + \cdots$

$$= kL\{u(t+0)\} - 2kL\{u(t+a)\} + 2kL\{u(t+2a)\} - .....$$

$$= k\left[\frac{1}{s} - 2\frac{e^{-as}}{s} + 2\frac{e^{-2as}}{s} - \frac{2e^{-3as}}{s} + \cdots\right] = \frac{k}{s}\left[1 - 2e^{-as} + 2e^{-2as} - 2e^{-3as} + \cdots\right]$$

$$= \frac{k}{s}[1 - 2(e^{-as} - e^{-2as} + e^{-3as} - \cdots)] = \frac{k}{s}\left[1 - 2\frac{e^{-as}}{1+e^{-as}}\right]$$

$$= \frac{k}{s}\left[\frac{1+e^{-as}-2e^{-as}}{1+e^{-as}}\right] = \frac{k}{s}\left[\frac{1-e^{-as}}{1+e^{-as}}\right] = \frac{k}{s}\left[\frac{e^{as/2}-e^{-as/2}}{e^{as/2}+e^{-as/2}}\right] = \frac{k}{s}\tan h\left(\frac{as}{2}\right).$$ **Proved.**

Example 28: Obtain Laplace transform of rectangular wave function given by

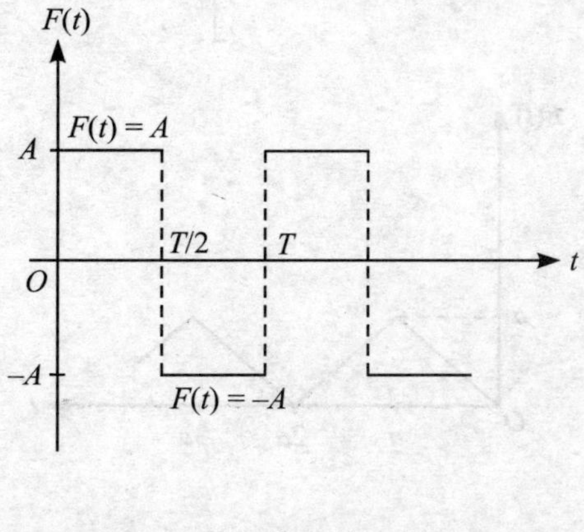

Solution: From the figure, we have

$$F(t) = \begin{bmatrix} A, & 0<t<T/2 \\ -A, & \frac{T}{2}<t<T \end{bmatrix}$$

The given function is a periodic function with period T.

$$\therefore \quad L\{F(t)\} = \frac{1}{1-e^{-sT}}\left[\int_0^{T/2} e^{-st}\cdot A\,dt + \int_{T/2}^{T} e^{-st}(-A)\,dt\right]$$

$$= \frac{A}{1-e^{-sT}}\left[\left\{\frac{e^{-st}}{-s}\right\}_0^{T/2} - \left\{\frac{e^{-st}}{-s}\right\}_{T/2}^{T}\right]$$

$$= \frac{A}{1-e^{-sT}}\left[-\frac{e^{-sT/2}}{s}+\frac{1}{s}+\frac{e^{-sT}}{s}-\frac{e^{-sT/2}}{s}\right]$$

$$= \frac{A}{s(1-e^{-sT})}[e^{-sT}-2e^{-sT/2}+1] = \frac{A}{s(1-e^{-sT})}(1-e^{-\frac{sT}{2}})^2. \quad \textbf{Ans.}$$

Example 29: Draw the graph of the following periodic function and find its Laplace transform:

$$F(t) = \begin{cases} t & \text{for } 0<t\le a \\ 2a-t & \text{for } a<t<2a \end{cases} \qquad (U.P.T.U.\ 2002)$$

Solution: The function $F(t)$ is a periodic function with period $2a$, we know that

$$L\{F(t)\} = \frac{1}{1-e^{-sT}}\int_0^{\infty} e^{-st}\cdot F(t)\,dt$$

$$= \frac{1}{1-e^{-2as}}\left[\int_0^{a} e^{-st}\cdot t\,dt + \int_a^{2a} e^{-st}\cdot(2a-t)\,dt\right]$$

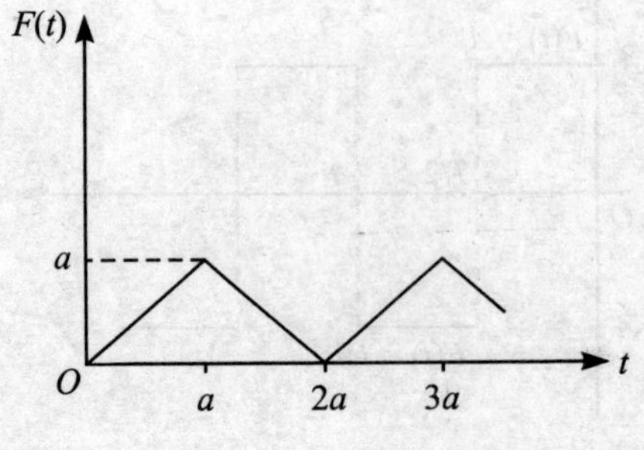

$$= \frac{1}{1-e^{-2as}}\left[\left\{t\cdot\frac{e^{-st}}{-s}-\frac{e^{-st}}{s^2}\right\}_0^a+\left\{(2a-t)\frac{e^{-st}}{-s}+\frac{e^{-st}}{s^2}\right\}_a^{2a}\right]$$

$$= \frac{1}{1-e^{-2as}}\left[-a\frac{e^{-as}}{s}-\frac{e^{-as}}{s^2}+\frac{1}{s^2}+\frac{e^{-2as}}{s^2}+a\frac{e^{-as}}{s}-\frac{e^{-as}}{s^2}\right]$$

$$= \frac{1}{1-e^{-2as}}\left[\frac{1}{s^2}+\frac{e^{-2as}}{s^2}-2\frac{e^{-as}}{s^2}\right]=\frac{1}{s^2(1-e^{-2as})}[1+e^{-2as}-2e^{-as}]$$

$$= \frac{(1-e^{-as})^2}{s^2(1-e^{-as})(1+e^{-as})}=\frac{1-e^{-as}}{s^2(1+e^{-as})}=\frac{1}{s^2}\frac{e^{\frac{as}{2}}-e^{-\frac{as}{2}}}{e^{as/2}+e^{-as/2}}=\frac{1}{s^2}\tan h\frac{as}{2}.$$ **Ans.**

Example 30: Draw the graph of the periodic function

$$F(t) = \begin{cases} t, & 0<t<\pi \\ \pi-t, & \pi<t<2\pi \end{cases}$$ (*U.P.T.U. 2003*)

and find its Laplace transform.

Solution: Here period $T = 2\pi$

Laplace transform of periodic functions

$$L\{F(t)\} = \frac{\int_0^T e^{-st}F(t)\,dt}{1-e^{-sT}}$$

$$L\{F(t)\} = \frac{1}{1-e^{-2\pi s}}\int_0^{2\pi} e^{-st}F(t)\,dt = \frac{1}{1-e^{-2\pi s}}\left[\int_0^{\pi} e^{-st}t\,dt+\int_\pi^{2\pi} e^{-st}(\pi-t)\,dt\right]$$

$$= \frac{1}{1-e^{-2\pi s}}\cdot\left\{\frac{t\cdot e^{-st}}{-s}-1\cdot\frac{e^{-st}}{(-s)^2}\right\}_0^{\pi}+\left\{(\pi-t)\frac{e^{-st}}{(-s)}-(-1)\frac{e^{-st}}{(-s)^2}\right\}_\pi^{2\pi}$$

$$= \frac{1}{1-e^{-2\pi s}}\left[\left\{\frac{\pi\cdot e^{-\pi s}}{-s}-\frac{e^{-\pi s}}{(-s)^2}-0+\frac{1}{s^2}\right\}+\left\{(\pi-2\pi)\frac{e^{-2\pi s}}{-s}+\frac{e^{-2\pi s}}{s^2}-\left((\pi-\pi)\frac{e^{-\pi s}}{-s}+\frac{e^{-\pi s}}{s^2}\right)\right\}\right]$$

$$= \frac{1}{1-e^{-2\pi s}}\left\{-\frac{\pi e^{-\pi s}}{s}-\frac{e^{-\pi s}}{s^2}+\frac{1}{s^2}+\pi\frac{e^{-2\pi s}}{s}+\frac{e^{-2\pi s}}{s^2}-0-\frac{e^{-\pi s}}{s^2}\right\}$$

$$= \frac{1}{1-e^{-2\pi s}}\left\{-\frac{\pi}{s}e^{-\pi s} + \frac{\pi}{s}e^{-2\pi s} + \frac{1}{s^2} - \frac{1}{s^2}\cdot e^{-\pi s} + \frac{1}{s^2}e^{-2\pi s} - \frac{e^{-\pi s}}{s^2}\right\}$$

$$= \frac{1}{1-e^{-2\pi s}}\left[\frac{\pi}{s}(e^{-2\pi s} - e^{-\pi s}) + \frac{1}{s^2}(1 + e^{-2\pi s} - 2e^{-\pi s})\right]. \quad \textbf{Ans.}$$

Graph of the function

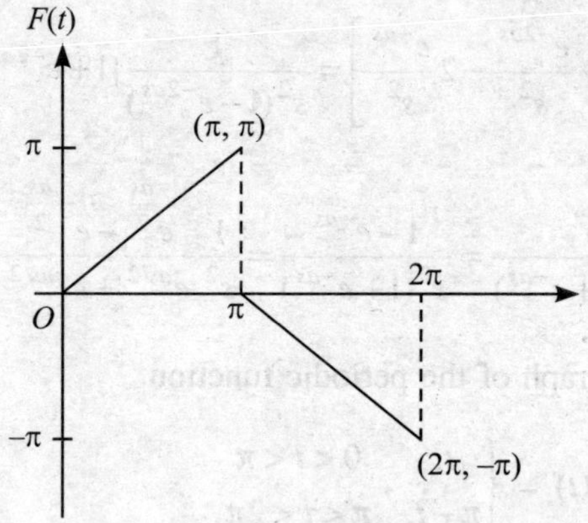

Example 31: Express the function shown in the diagram in terms of unit step function and obtain its Laplace transform. (*U.P.T.U. 2009*)

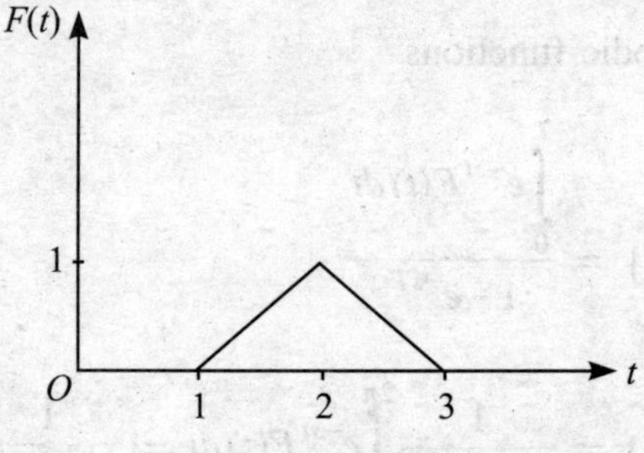

Solution: Here,

$$F(t) = \begin{cases} t-1, & 1 < t < 2 \\ 3-t, & 2 < t < 3 \end{cases}$$

$$\therefore \quad F(t) = (t-1)\{U(t-1) - U(t-2)\} + (3-t)\{U(t-2) - U(t-3)\}$$

$$= (t-3)U(t-3) - 2(t-2)U(t-2) + (t-1)U(t-1)$$

Hence, $$L\{F(t)\} = L\{(t-3)U(t-3)\} - 2L\{(t-2)U(t-2) + L\{(t-1)U(t-1)\}$$

$$= \frac{e^{-3s}}{s^2} - \frac{2e^{-2s}}{s^2} + \frac{e^{-s}}{s^2} \quad \text{(by second shifting theorem)}$$

$$= \frac{e^{-s}(1-e^{-s})^2}{s^2}. \quad \textbf{Ans.}$$

Example 32: Express the following function in terms of unit step function and find its Laplace transform: (*U.P.T.U. 2002*)

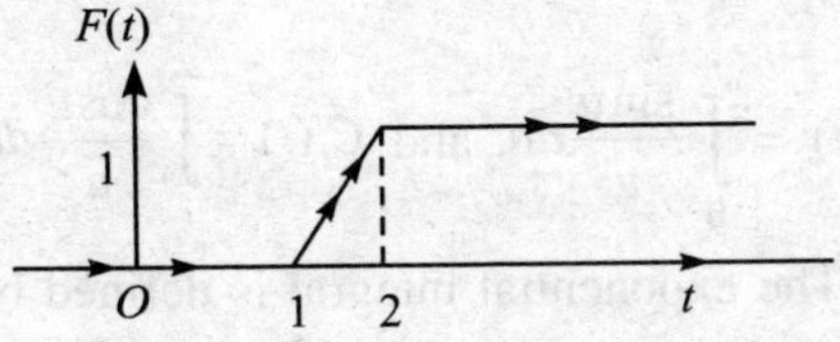

Solution: The above function shown in the figure is expressed in algebraic form.

$$F(t) = \begin{cases} 0, & 0 < t < 1 \\ t-1, & 1 < t < 2 \\ 1, & 2 < t \end{cases} \quad ...(1)$$

$$F(t) = (t-1)[U(t-1) - U(t-2)] + U(t-2)$$

$$= (t-1)U(t-1) - U(t-2)\{t-1-1\} = (t-1)U(t-1) - (t-2)U(t-2)$$

$$\therefore \quad L\{F(t)\} = L(t-1)U(t-1) - L(t-2)U(t-2)$$

$$= \frac{e^{-s}}{s^2} - \frac{e^{-2s}}{s^2}. \quad \textbf{Ans.}$$

Example 33: Find the Laplace transform of $t^2\ U\ (t-3)$.

Solution: $t^2 \cdot U(t-3) = [(t-3)^2 + 6(t-3) + 9]U(t-3)$

$$= (t-3)^2 \cdot U(t-3) + 6(t-3) \cdot U(t-3) + 9U(t-3)$$

$$L\{t^2 U(t-3)\} = L\{(t-3)^2 \cdot U(t-3)\} + 6L\{(t-3) \cdot U(t-3)\} + 9L\{U(t-3)\}$$

$$= e^{-3s}\left[\frac{2}{s^3} + \frac{6}{s^2} + \frac{9}{s}\right]. \quad \textbf{Ans.}$$

2.21 SOME SPECIAL FUNCTIONS

1. **The error function:** The error function denoted by $erf(t)$ is defined by

$$erf(t) = \frac{2}{\sqrt{\pi}} \int_0^t e^{-u^2}\, du$$

or

$$erf(\sqrt{t}) = \frac{2}{\sqrt{\pi}} \int_0^{\sqrt{t}} e^{-u^2}\, du.$$

2. **The sine and consine integrals:** The sine and cosine integrals denoted by $S_i(t)$ and $C_i(t)$ respectively are defined by

$$S_i(t) = \int_0^t \frac{\sin u}{u} du, \text{ and } C_i(t) = \int_t^\infty \frac{\cos u}{u} du.$$

3. **The exponential integral:** The exponential integral is defined by

$$E_i(t) = \int_t^\infty \frac{e^{-u}}{u} du.$$

4. **The complementary error function:** The complementary error function is defined as

$$erf\,c(t) = 1 - erf(t) = \frac{2}{\sqrt{\pi}} \int_t^\infty e^{-u^2} du.$$

Example 34: Find $L\{erf\sqrt{t}\}$ and hence prove that

$$L\{t \cdot erf(2\sqrt{t})\} = \frac{3s+8}{s^2(s+4)^{3/2}} \qquad (U.P.T.U.\ 2001)$$

Solution: We know that

$$erf(\sqrt{t}) = \frac{2}{\sqrt{\pi}} \int_0^{\sqrt{t}} e^{-u^2} du$$

$$\therefore \quad L\{erf(\sqrt{t})\} = \frac{2}{\sqrt{\pi}} L\left\{ \int_0^{\sqrt{t}} e^{-u^2} du \right\}$$

$$= \frac{2}{\sqrt{\pi}} L\left\{ \int_0^{\sqrt{t}} \left(1 - \frac{u^2}{\underline{|1}} + \frac{u^4}{\underline{|2}} - \ldots + \ldots du \right) \right\}$$

$$= \frac{2}{\sqrt{\pi}} L\left\{ \left(u - \frac{u^3}{3\underline{|1}} + \frac{u^5}{5\underline{|2}} + \ldots\ldots \right)_0^{\sqrt{t}} \right\}$$

$$= \frac{2}{\sqrt{\pi}} L\left\{ t^{1/2} - \frac{t^{3/2}}{3} + \frac{t^{5/2}}{10} - \ldots\ldots \right\}$$

$$= \frac{2}{\sqrt{\pi}} \left[L\{t^{1/2}\} - \frac{1}{3} L\{t^{3/2}\} + \frac{1}{10} L\{t^{5/2}\} - \ldots\ldots \right]$$

$$= \frac{2}{\sqrt{\pi}}\left[\frac{\overline{|1/2+1}}{s^{3/2}} - \frac{1}{3}\frac{\overline{|3/2+1}}{s^{5/2}} + \frac{1}{10}\frac{\overline{|5/2+1}}{s^{7/2}} - \ldots\ldots\right]$$

$$= \frac{2}{\sqrt{\pi}}\left[\frac{\overline{|3/2}}{s^{3/2}} - \frac{\overline{|5/2}}{3s^{5/2}} + \frac{\overline{|7/2}}{10s^{7/2}} - \ldots\ldots\right]$$

$$= \frac{2}{\sqrt{\pi}}\left[\frac{\frac{1}{2}\overline{|\frac{1}{2}}}{s^{3/2}} - \frac{\frac{3}{2}.\frac{1}{2}.\overline{|\frac{1}{2}}}{3\cdot s^{5/2}} + \frac{\frac{5}{2}.\frac{3}{2}.\frac{1}{2}.\overline{|\frac{1}{2}}}{10\cdot s^{7/2}} - \ldots\ldots\right]$$

$$= \frac{1}{s^{3/2}}\left[1 - \frac{1}{2}\cdot\frac{1}{s} + \frac{1\cdot 3}{2\cdot 4}\cdot\frac{1}{s^2} - \ldots\ldots\right] \qquad \left|\text{As } \overline{|1/2} = \sqrt{\pi}\right.$$

or $$L\{erf(\sqrt{t})\} = \frac{1}{s^{3/2}}\left(1+\frac{1}{s}\right)^{-1/2} = \frac{1}{s\sqrt{s+1}}.$$

Now apply change of scale property, we have

$$L\{erf 2\sqrt{t}\} = L\{erf(\sqrt{4t})\} = \frac{1}{4}\cdot\frac{1}{\frac{s}{4}\sqrt{\frac{s}{4}+1}} = \frac{2}{s\sqrt{s+4}}$$

or $$L\{t.erf 2\sqrt{t}\} = -\frac{d}{ds}\left\{\frac{2}{s(\sqrt{s+4})}\right\} \qquad \left|\begin{array}{l}\text{As}\\ L\{tF(t)\} = -\dfrac{d}{ds}f(s)\end{array}\right.$$

$$= \frac{s(3s+8)}{s^3(s+4)^{3/2}} = \frac{3s+8}{s^2(s+4)^{3/2}}. \quad \textbf{Ans.}$$

Example 35: Prove that $L\{J_0(t)\} = \dfrac{1}{\sqrt{s^2+1}}$, where $J_0(t)$ is Bessel's function of order zero.

Solution: We know that

$$J_n(t) = \sum_{r=0}^{\infty}\frac{(-1)^r}{\underline{|r}\,\overline{|n+r+1}}\left(\frac{t}{2}\right)^{n+2r}$$

$\therefore$ $$J_0(t) = \sum_{r=0}^{\infty}\frac{(-1)^r}{(\underline{|r})^2}\left(\frac{t}{2}\right)^{2r}$$

or $$J_0(t) = 1 - \frac{1}{(\lfloor 1)^2}\left(\frac{t}{2}\right)^2 + \frac{1}{(\lfloor 2)^2}\left(\frac{t}{2}\right)^4 - \frac{1}{(\lfloor 3)^2}\left(\frac{t}{2}\right)^6 + \ldots$$

$$\therefore \quad L\{J_0(t)\} = L\{1\} - \frac{1}{1\cdot 2^2}\cdot L\{t^2\} + \frac{1}{2^2 . 4^2} L\{t^4\} - \frac{1}{2^2 . 4^2 . 6^2} L\{t^6\} + \ldots$$

$$= \frac{1}{s} - \frac{1}{2^2}\frac{\lfloor 2}{s^3} + \frac{1}{2^2 . 4^2}\frac{\lfloor 4}{s^5} - \frac{1}{2^2 . 4^2 . 6^2}\frac{\lfloor 6}{s^7} + \ldots.$$

$$= \frac{1}{s}\left[1 - \frac{1}{2}\left(\frac{1}{s^2}\right) + \frac{1\cdot 3}{2\cdot 4}\left(\frac{1}{s^2}\right)^2 - \frac{1\cdot 3\cdot 5}{2\cdot 4\cdot 6}\left(\frac{1}{s^2}\right)^3 + \ldots.\right]$$

$$\Rightarrow \quad L\{J_0(t)\} = \frac{1}{s}\left(1 + \frac{1}{s^2}\right)^{-1/2} = \frac{1}{\sqrt{s^2+1}}.$$ **Proved.**

Example 36: Show that $L\{E_i(t)\} = \frac{\log(s+1)}{s}$.

Solution: We have $E_i(t) = \int_t^\infty \frac{e^{-u}}{u}\,du$

$$\therefore \quad L\{E_i(t)\} = L\left\{\int_t^\infty \frac{e^{-u}}{u}\,du\right\}$$

Let $u = tv$, $du = t\,dv$

$$= L\left\{\int_1^\infty \frac{e^{-tv}}{v}\,dv\right\} = \int_0^\infty e^{-st}\left\{\int_1^\infty \frac{e^{-tv}}{v}\,dv\right\}dt$$

$$= \int_1^\infty \frac{1}{v}\left\{\int_0^\infty e^{-st}\cdot e^{-tv}\,dt\right\}dv$$ (Changing the order of integration)

$$= \int_1^\infty \frac{1}{v}\cdot\frac{1}{(s+v)}\,dv = \int_1^\infty \frac{1}{s}\left(\frac{1}{v} - \frac{1}{s+v}\right)dv$$ (By partial fraction)

$$= \frac{1}{s}\left[\log v - \log(s+v)\right]_1^\infty = \frac{1}{s}\left[-\log\left(\frac{s}{v}+1\right)\right]_1^\infty$$

$$= \frac{1}{s}\log(s+1). \qquad \textbf{Proved.}$$

2.22 DIRAC'S* DELTA FUNCTION (or Unit Impulse Function)

The forces (like earthquake) that produce large effects on a system when applied for a short time interval, this short time interval is called impulse and can be represented by an impulse function which is a discontinuous function and is highly irregular from mathematical point of view.

$\therefore$ The impulse of a force $f(t)$ in the interval $(a, a+\varepsilon) = \int_a^{a+\varepsilon} f(t)\,dt.$

Now define the function

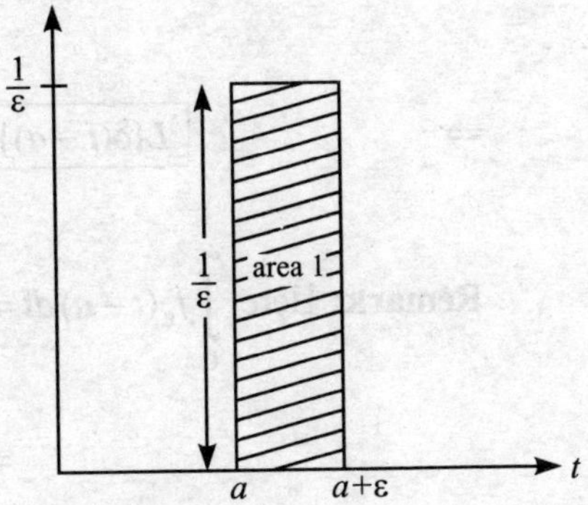

$$f_\varepsilon(t-a) = \begin{cases} 0 & \text{for } t < a \\ \dfrac{1}{\varepsilon} & \text{for } a \le t \le a+\varepsilon \\ 0 & \text{for } t > a \end{cases}$$

or
$$f_\varepsilon(t-a) = \frac{1}{\varepsilon}[U(t-a) - U(t-(a+\varepsilon))]$$

Taking Laplace transform, we have

$$L\{f_\varepsilon(t-a)\} = \frac{1}{\varepsilon}L[U(t-a) - U(t-(a+\varepsilon))] = \frac{1}{\varepsilon s}[e^{-as} - e^{-(a+\varepsilon)s}]$$

$$\boxed{L\{f_\varepsilon(t-a)\} = e^{-as}\cdot\frac{(1-e^{-\varepsilon s})}{\varepsilon s}.} \qquad \ldots(A)$$

Dirac's delta function or unit impulse function denoted by $\delta(t-a)$ is defined as the limit of $f_\varepsilon(t-a)$ as $\varepsilon \to 0$ *i.e.*,

$$\delta(t-a) = \lim_{\varepsilon\to 0} f_\varepsilon(t-a)$$

$$= \lim_{\varepsilon\to 0}\begin{cases} 0, & \text{for } \quad t < a \\ \dfrac{1}{\varepsilon}, & \text{for } \quad a \le t \le a+\varepsilon \\ 0, & \text{for } \quad t > a \end{cases} = \begin{cases} 0, & \text{for } \quad t < a \\ \infty & \text{for } \quad a = t = a \\ 0, & \text{for } \quad t > a \end{cases}$$

*Paul Dirac (1902–1984), English Physicist, Nobel prize winner.

$$\therefore \quad \boxed{\delta(t-a) = \begin{cases} \infty & \text{for } t = a \\ 0 & \text{for } t \neq a \end{cases}}, \text{ Subject to } \int_0^\infty \delta(t-a)\,dt = 1$$

Laplace Transform of Unit Impulse Function

$$L\{\delta(t-a)\} = \lim_{\varepsilon \to 0} L\{f_\varepsilon(t-a)\}$$

$$= \lim_{\varepsilon \to 0} \left\{ e^{-as} \cdot \frac{(1-e^{-\varepsilon s})}{\varepsilon s} \right\} \qquad \Big| \text{ from } (A)$$

$$= e^{-as} \lim_{\varepsilon \to 0} \left[\frac{1}{\varepsilon s} \left\{ 1 - 1 + \frac{\varepsilon s}{\underline{|1}} - \frac{(\varepsilon s)^2}{\underline{|2}} + \ldots\ldots \right\} \right]$$

$$\Rightarrow \qquad \boxed{L\{\delta(t-a)\} = e^{-as}.}$$

Remark: Here $\int_0^\infty f_\varepsilon(t-a)\,dt = \int_0^a 0\,dt + \int_a^{a+\varepsilon} \frac{1}{\varepsilon}\,dt + \int_{a+\varepsilon}^\infty 0 = \frac{1}{\varepsilon}[t]_a^{a+\varepsilon} = 1$

$$\Rightarrow \text{The impulse } \boxed{I_\varepsilon = 1.}$$

Example 37: Find the value of $\int_{-\infty}^{\infty} e^{-3t} \delta'(t-2)\,dt$

$$= [e^{-3t} \cdot \delta(t-2)]_{-\infty}^{\infty} + 3\int_{-\infty}^{\infty} e^{-3t} \cdot \delta(t-2)\,dt$$

$$= 0 + 3\int_{-\infty}^{0} e^{-3t} \cdot \delta(t-2)\,dt + 3\int_0^\infty e^{-3t} \cdot \delta(t-2)\,dt$$

$$= 3\int_0^\infty e^{-3t} \cdot \delta(t-2)\,dt = 3e^{-3\times 2} = 3e^{-6}. \quad \textbf{Ans.}$$

As $\int_0^\infty e^{-st} \cdot \delta(t-a) = e^{-as}$

For $s = 3, a = 2$

$\int_0^\infty e^{-3t} \cdot \delta(t-2) = e^{-6}$.

Example 38: Evaluate $\int_0^\infty e^{-3t} \cdot \delta(t-4)\,dt$.

Solution: We know that $L\{\delta(t-a)\} = e^{-as}$

For $a = 4$, we get

$$L\{\delta(t-4) = e^{-4s}$$

$$\Rightarrow \quad \int_0^\infty e^{-st}\cdot\delta(t-4)dt = e^{-4s}. \qquad \left|\text{As } L\{F(t)\} = \int_0^\infty e^{-st}\cdot F(t)\,dt\right.$$

Putting $s = 3$, we get

$$\int_0^\infty e^{-3t}\cdot\delta(t-4)dt = e^{-12}. \qquad \textbf{Ans.}$$

EXERCISE 2.3

Find the Laplace transform of the following functions:

1. $F(t) = \begin{cases} 1, 0 < t < 1 \\ -1, 0 < t < 2 \end{cases}$ and $F(t+2) = F(t)$. $\left[\textbf{Ans.} \frac{1}{s}\tanh\frac{s}{2}\right]$

2. $F(t) = \begin{cases} t, 0 < t < 1 \\ 0, 1 < t < 2 \end{cases}$ where $F(t)$ has period 2. $\left[\textbf{Ans.} \frac{1-e^{s}(s+1)}{s^2(1-e^{-2s})}\right]$

3. $F(t) = \begin{cases} \sin t, 0 < t < \pi \\ 0, \quad \pi < t < 2\pi \end{cases}$ where $F(t)$ has period 2π. $\left[\textbf{Ans.} \frac{1+e^{-s\pi}}{1+s^2}\right]$

4. $F(t) = \int_0^\infty \cos tx^2\,dx.$ $[\textbf{Ans.} \, \pi/2\sqrt{2s}]$

5. Prove that $L\{J_0(a\sqrt{t})\} = \frac{1}{s}e^{-(a^2/4s)}$.

6. Evaluate $\int_0^\infty te^{-3t}J_0(4t)\,dt.$ $\left[\textbf{Ans.} \frac{3}{125}\right]$

7. Evaluate $\int_0^\infty te^{-t}\cdot E_i(t)\,dt.$ $\left[\textbf{Ans.} \log 2 - \frac{1}{2}\right]$

8. Given that $L\left\{2\sqrt{\frac{t}{\pi}}\right\} = \frac{1}{s^{3/2}}$, show that $\frac{1}{s^{1/2}} = L\left\{\frac{1}{\sqrt{\pi t}}\right\}$.

9. Express the following functions in terms of unit step function and find Laplace transform.

(*i*) $\begin{cases}(t-1), 1 < t < 2\\(3-t), 2 < t < 3\end{cases}$ $\left[\textbf{Ans. } \dfrac{e^{-s} - 2e^{-2s} + e^{-3s}}{s^2}\right]$

(*ii*) $\begin{cases}1, 0 \le t \le a\\2, a \le t \le 2a\\3, 2a < t < 3a\end{cases}$ $\left[\textbf{Ans. } \dfrac{1}{s}(1 - e^{-as})\right]$

10. Find the Laplace transform of the periodic function $F(t) = \dfrac{kt}{T}$ for $0 < t < T$, $f(t + T) = f(t)$.

$\left[\textbf{Ans. } \dfrac{k}{s^2 T} - \dfrac{ke^{-sT}}{s(1 - e^{-aT})}\right]$ [*U.P.T.U. (C.O.) 2004*]

11. Find (*i*) $L\{e^{t-3} \cdot U(t-3)\}$ (*ii*) $L\{t^2 U(t-1) + \delta(t-1)\}$.

$\left[\textbf{Ans. } (i)\ \dfrac{e^{-3s}}{s-3}\quad (ii)\ e^{-s}\dfrac{(2 + 2s + s^2 + s^3)}{s^3}\right]$

12. Find the Laplace transform of the wave form $F(t) = \left(\dfrac{2t}{3}\right), 0 \le t \le 3.$

$\left[\textbf{Ans. } \dfrac{2e^{-3s}}{-s(1 - e^{-3s})} + \dfrac{2}{3s^2}\right]$

2.23 INVERSE LAPLACE TRANSFORM

If $L\{F(t)\} = f(s)$ then $F(t)$ is known as inverse Laplace transform or inverse transform or simply inverse of $f(s)$ and is denoted by $L^{-1}\{f(s)\}$.

Thus $$F(t) = L^{-1}\{f(s)\}$$

L^{-1} is called the 'inverse Laplace transform operator' and is such that

$$LL^{-1} = L^{-1}L = 1.$$

2.24 LINEARITY PROPERTY

If $L\{F_1(t)\} = f_1(s), L\{F_2(t)\} = f_2(s)$ then $L^{-1}\{c_1 f_1(s) + c_2 f_2(s)\} = c_1 L^{-1}\{f_1(s)\} + c_2 L^{-1}\{f_2(s)\}$.

Proof: From the Linearity property for Laplace transform, we have

$$L\{c_1 F_1(t) + c_2 F_2(t)\} = c_1 L\{F_2(t)\} + c_1 L\{F_2(t)\}$$
$$= c_1 f_1(s) + c_2 f_2(s)$$

$$\therefore \quad L^{-1}\{c_1 f_1(s) + c_2 f_2(s)\} = c_1 F_1(t) + c_2 F_1(t)$$

$$\Rightarrow \quad \boxed{L^{-1}\{c_1 f_1(s) + c_2 f_2(s)\} = c_1 L^{-1}\{f_1(s)\} + c_2 L^{-1}\{f_2(s)\}.}$$

Example 39: Find (*i*) $L^{-1}\left\{\dfrac{1}{s^2+a^2}\right\}$ (*ii*) $L^{-1}\left\{\dfrac{s}{s^2+a^2}\right\}$

(*iii*) $L^{-1}\left\{\dfrac{s}{s^2-a^2}\right\}$ (*iv*) $L^{-1}\left\{\dfrac{s}{s^2-a^2}\right\}$

Solution: (*i*) We know that $L\{\sin at\} = \dfrac{a}{s^2+a^2} \quad \Rightarrow \quad \dfrac{1}{a} L\{\sin at\} = \dfrac{1}{s^2+a^2}$

or

$$L\left\{\frac{\sin at}{a}\right\} = \frac{1}{s^2+a^2} \quad \Rightarrow \quad \boxed{L^{-1}\left\{\frac{1}{s^2+a^2}\right\} = \frac{\sin at}{a}.}$$

(*ii*) $L\{\cos at\} = \dfrac{s}{s^2+a^2} \quad \Rightarrow \quad \boxed{L^{-1}\left\{\dfrac{s}{s^2+a^2}\right\} = \cos at.}$

(*iii*) $L\{\sinh at\} = \dfrac{a}{s^2-a^2} \quad \Rightarrow \quad \boxed{L^{-1}\left\{\dfrac{1}{s^2-a^2}\right\} = \dfrac{\sinh at}{a}.}$

(*iv*) $L\{\cosh at\} = \dfrac{s}{s^2-a^2} \quad \Rightarrow \quad \boxed{L^{-1}\left\{\dfrac{s}{s^2-a^2}\right\} = \cosh at.}$

Example 40: Find (*i*) $L^{-1}\left\{\dfrac{1}{s}\right\}$ (*ii*) $L^{-1}\left\{\dfrac{1}{s^{n+1}}\right\}$ (*iii*) $L^{-1}\left\{\dfrac{1}{s+a}\right\}$.

Solution: (*i*) We have $L\{1\} = \dfrac{1}{s} \quad \Rightarrow \quad \boxed{L^{-1}\left\{\dfrac{1}{s}\right\} = 1.}$

(*ii*) $L\{t^n\} = \dfrac{\Gamma(n+1)}{s^{n+1}} \quad \Rightarrow \quad \boxed{L^{-1}\left\{\dfrac{1}{s^{n+1}}\right\} = \dfrac{t^n}{\Gamma(n+1)} = \dfrac{t^n}{\underline{|n}}}$

(*iii*) $L\{e^{-at}\} = \dfrac{1}{s+a} \quad \Rightarrow \quad \boxed{L^{-1}\left\{\dfrac{1}{s+a}\right\} = e^{-at}.}$

Example 41: Find the inverse Laplace transform of the following:

(*i*) $\dfrac{5}{s^2} + \left(\dfrac{\sqrt{s}-1}{s}\right)^2 - \dfrac{7}{3s+2}$ (*ii*) $\dfrac{3}{s^2-3} + \dfrac{3s+2}{s^3} - \dfrac{3s-27}{s^2+9} + \dfrac{6-30\sqrt{s}}{s^4}$.

Solution:

(*i*) $$L^{-1}\left\{\frac{5}{s^2}+\left(\frac{\sqrt{s}-1}{s}\right)^2-\frac{7}{3s+2}\right\}=5L^{-1}\left\{\frac{1}{s^2}\right\}+L^{-1}\left\{\frac{s-2\sqrt{s}+1}{s^2}\right\}-7L^{-1}\left\{\frac{1}{3s+2}\right\}$$

$$=6L^{-1}\left\{\frac{1}{s^2}\right\}+L^{-1}\left\{\frac{1}{s}\right\}-2L^{-1}\left\{\frac{1}{s^{3/2}}\right\}-\frac{7}{3}L^{-1}\left\{\frac{1}{s+\frac{2}{3}}\right\}$$

$$=6\cdot\frac{t^{2-1}}{\lfloor 1}+1-2\frac{t^{3/2-1}}{\lfloor 3/2}-\frac{7}{3}e^{-2t/3}=6t+1-4\sqrt{\frac{t}{\pi}}-\frac{7}{3}e^{-2t/3}.$$ **Ans.**

(*ii*) $$L^{-1}\left\{\frac{3}{s^2-3}+\frac{3s+2}{s^3}-\frac{3s-27}{s^2+9}+\frac{6-30\sqrt{s}}{s^4}\right\}$$

$$=3L^{-1}\left\{\frac{1}{s^2-3}\right\}+3L^{-1}\left\{\frac{1}{s^2}\right\}+2L^{-1}\left\{\frac{1}{s^3}\right\}-3L^{-1}\left\{\frac{s}{s^2+9}\right\}$$

$$+27L^{-1}\left\{\frac{1}{s^2+9}\right\}+6L^{-1}\left\{\frac{1}{s^4}\right\}-30L^{-1}\left\{\frac{1}{s^{7/2}}\right\}$$

$$=3L^{-1}\left\{\frac{1}{s^2-(\sqrt{3})^2}\right\}+3L^{-1}\left\{\frac{1}{s^2}\right\}+2L^{-1}\left\{\frac{1}{s^3}\right\}-3L^{-1}\left\{\frac{s}{s^2+3^2}\right\}$$

$$+27L^{-1}\left\{\frac{1}{s^2+3^2}\right\}+6L^{-1}\left\{\frac{1}{s^4}\right\}-30L^{-1}\left\{\frac{1}{s^{7/2}}\right\}$$

$$=3\cdot\frac{1}{\sqrt{3}}\cdot\sin h(\sqrt{3})t+3\cdot\frac{t^{2-1}}{\lfloor 2-1}+2\frac{t^{3-1}}{\lfloor 3-1}-3\cos 3t+27\cdot\frac{1}{3}\sin 3t+6\cdot\frac{t^{4-1}}{\lfloor (4-1)}-30\cdot\frac{t^{7/2-1}}{\lfloor 7/2}$$

$$=\sqrt{3}\sin h\sqrt{3}t+3t+t^2-3\cos 3t+9\sin 3t+t^3-16t^2\sqrt{(t/\pi)}.$$ **Ans.**

Example 42: Evaluate the following:

$$L^{-1}\left\{\frac{6}{2s-3}-\frac{3+4s}{9s^2-16}+\frac{8-6s}{16s^2+9}\right\}.$$ (*U.P.T.U. 2001*)

Solution: We have $L^{-1}\left\{\frac{6}{2s-3}\right\}-L^{-1}\left\{\frac{3+4s}{9s^2-16}\right\}+L^{-1}\left\{\frac{8-6s}{16s^2+9}\right\}$

$$= \frac{6}{2}L^{-1}\left\{\frac{1}{s-\frac{3}{2}}\right\} - \frac{3}{9}L^{-1}\left\{\frac{1}{s^2-\left(\frac{4}{3}\right)^2}\right\} - \frac{4}{9}L^{-1}\left\{\frac{s}{s^2-\left(\frac{4}{3}\right)^2}\right\}$$

$$+\frac{8}{16}L^{-1}\left\{\frac{1}{s^2+\frac{9}{16}}\right\} - \frac{6}{16}L^{-1}\left\{\frac{s}{s^2+\frac{9}{16}}\right\}$$

$$= 3e^{3t/2} - \frac{1}{3}\cdot\frac{3}{4}\sin h\frac{4t}{3} - \frac{4}{9}\cos h\frac{4t}{3} + \frac{1}{2}\cdot\frac{4}{3}\cdot\sin\frac{3t}{4} - \frac{3}{8}\cos\frac{3t}{4}$$

$$= 3e^{3t/2} - \frac{1}{4}\sin h\frac{4t}{3} - \frac{4}{9}\cos h\frac{4t}{3} + \frac{2}{3}\sin\frac{3t}{4} - \frac{3}{8}\cos\frac{3t}{4}.$$ **Ans.**

2.25 FIRST TRANSLATION OR SHIFTING THEOREM

If $L^{-1}\{f(s)\} = F(t)$, then $L^{-1}\{f(s-a)\} = e^{at}\cdot F(t) = e^{at}\cdot L^{-1}\{f(s)\}$

Proof: We have $f(s) = \int_0^\infty e^{-st}\cdot F(t)\,dt$

or $$f(s-a) = \int_0^\infty e^{-(s-a)t}\cdot F(t)dt = \int_0^\infty e^{-st}\{e^{at}\cdot F(t)\}dt$$

$\Rightarrow$ $$f(s-a) = L\{e^{at}\cdot F(t)\} \quad \Rightarrow \quad \boxed{L^{-1}\{f(s-a)\} = e^{at}\cdot F(t) = e^{at}L^{-1}\{f(s)\}}$$

Alter: We have $L\{e^{at}\cdot F(t)\} = f(s-a)$ (By 1st shifting theorem of L.T.)

$\Rightarrow$ $$L^{-1}\{f(s-a)\} = e^{at}\cdot F(t) = e^{at}\cdot L^{-1}\{f(s)\}.$$ **Proved.**

2.26 SECOND TRANSLATION OR SHIFTING THEOREM

If $L^{-1}\{f(s)\} = F(t)$ and $G(t) = \begin{cases} F(t-a), & t > a \\ 0, & t < a \end{cases}$

then, $$L^{-1}\{e^{-as}\cdot f(s)\} = G(t)$$

Proof: By definition, we have

$$f(s) = \int_0^\infty e^{-st} \cdot F(t)\, dt$$

$$\therefore \quad e^{-as} f(s) = \int_0^\infty e^{-as} \cdot e^{-st} \cdot F(t)\, dt = \int_0^\infty e^{-(t+a)s} \cdot F(t)\, dt$$

Putting $t + a = x \quad \Rightarrow \quad dt = dx$

$$\Rightarrow \quad e^{-as} f(s) = \int_a^\infty e^{-sx} \cdot F(x-a)\, dx = \int_a^\infty e^{-st} \cdot F(t-a)\, dt \qquad \text{(By definite integral)}$$

$$= \int_0^a e^{-st} \cdot 0\, dt + \int_a^\infty e^{-st} \cdot F(t-a)\, dt$$

$$\Rightarrow \quad e^{-as} \cdot f(s) = \int_0^\infty e^{-st} \cdot G(t)\, dt = L\{G(t)\}$$

$$\Rightarrow \quad \boxed{L^{-1}\{e^{-as} \cdot f(s)\} = G(t).}$$

Alter: By second shifting theorem of Laplace transform, we have

$$L\{G(t)\} = e^{-as} \cdot f(s)$$

$$\Rightarrow \quad L^{-1}\{e^{-as} \cdot f(s)\} = G(t). \qquad \textbf{Proved.}$$

Note : We may write $G(t)$ interms of unit step as $F(t-a) \cdot U(t-a)$

$$\therefore \quad L^{-1}\{e^{-as} f(s)\} = F(t-a) \cdot U(t-a).$$

2.27 CHANGE OF SCALE PROPERTY

If $L^{-1}\{f(s)\} = F(t)$, then $L^{-1}\{f(as)\} = \dfrac{1}{a} F\left(\dfrac{t}{a}\right)$

Proof: We have $f(s) = \displaystyle\int_0^\infty e^{-st} \cdot F(t)\, dt$

$$f(as) = \int_0^\infty e^{-ast} \cdot F(t)\, dt$$

Putting $at = x$ such that $dt = \dfrac{dx}{a}$

$$= \int_0^\infty e^{-sx} \cdot F\left(\frac{x}{a}\right) \cdot \frac{dx}{a} = \frac{1}{a}\int_0^\infty e^{-st} . F\left(\frac{t}{a}\right) dt \qquad \text{(By definite integral)}$$

$$\therefore \qquad f(as) = \frac{1}{a} L\left\{F\left(\frac{t}{a}\right)\right\}$$

Hence $\boxed{L^{-1}\{f(as)\} = \frac{1}{a} F\left(\frac{t}{a}\right)}$

Alter: By change of scale of Laplace transform, we have

$$L\{F(at)\} = \frac{1}{a} f(s/a)$$

Take $a = \frac{1}{k}$ then

$$F\left(\frac{t}{k}\right) = L^{-1}\{kf(ks)\} \quad \Rightarrow \quad L^{-1}\{f(ks)\} = \frac{1}{k} F\left(\frac{t}{k}\right) \qquad \textbf{Proved.}$$

Example 43: Find the inverse Laplace transform of the following:

(*i*) $\frac{1}{s^2 - 6s + 10}$ (*ii*) $\frac{s}{(s+1)^{3/2}}$ (*iii*) $\frac{3s-2}{s^2 - 4s + 20}$.

Solution: (*i*) $L^{-1}\left\{\frac{1}{s^2 - 6s + 10}\right\} = L^{-1}\left\{\frac{1}{(s-3)^2 + 1}\right\} = e^{3t} L^{-1}\left\{\frac{1}{s^2 + 1}\right\}$ (By first shifting)

$$\Rightarrow \qquad L^{-1}\left\{\frac{1}{s^2 - 6s + 10}\right\} = e^{3t} \cdot \sin t. \quad \textbf{Ans.}$$

(*ii*) $L^{-1}\left\{\frac{s}{(s+1)^{3/2}}\right\} = L^{-1}\left\{\frac{s+1-1}{(s+1)^{3/2}}\right\} = e^{-t} L^{-1}\left\{\frac{s-1}{s^{3/2}}\right\}$

$$= e^{-t}\left[L^{-1}\left\{\frac{1}{s^{1/2}}\right\} - L^{-1}\left\{\frac{1}{s^{3/2}}\right\}\right]$$

$$L^{-1}\left\{\frac{s}{(s+1)^{3/2}}\right\} = e^{-t} \frac{t^{\frac{1}{2}-1}}{\left|\frac{1}{2}\right.} - e^{-t} \cdot \frac{t^{3/2-1}}{\left|3/2\right.} = e^{-t}\sqrt{\frac{1}{t\pi}} - 2e^{-t}\sqrt{\frac{t}{\pi}}. \quad \textbf{Ans.}$$

(*iii*) $L^{-1}\left\{\dfrac{3s-2}{s^2-4s+20}\right\} = L^{-1}\left\{\dfrac{3s-2}{(s-2)^2+16}\right\} = 3L^{-1}\left\{\dfrac{s-2}{(s-2)^2+4^2}\right\} + 4L^{-1}\left\{\dfrac{1}{(s-2)^2+4^2}\right\}$

$$= 3e^{2t}L^{-1}\left\{\frac{s}{s^2+4^2}\right\} + 4e^{2t}L^{-1}\left\{\frac{1}{s^2+4^2}\right\} = 3e^{2t}\cos 4t + e^{2t}.\sin 4t. \quad \textbf{Ans.}$$

Inverse by Partial Fraction

Example 44: Find the inverse Laplace transform of

$$\frac{5s+3}{(s-1)(s^2+2s+5)}. \qquad (U.P.T.U.\ 2005)$$

Solution: Let $\dfrac{5s+3}{(s-1)(s^2+2s+5)} = \dfrac{A}{s-1} + \dfrac{Bs+C}{s^2+2s+5}$

$$\Rightarrow \qquad 5s+3 = A(s^2+2s+5)+(Bs+C)(s-1)$$

$$5s+3 = s^2(A+B)+s(2A-B+C)+(5A-C)$$

Equating the coefficient of like powers of s on both sides, we get

$A + B = 0$.....(*i*), $2A - B + C = 5$.....(*ii*), $5A - C = 3$.....(*iii*)

On solving these equations, we get

$$\boxed{A=1},\ \boxed{B=-1} \text{ and } \boxed{C=2.}$$

$$\therefore \quad \frac{5s+3}{(s-1)(s^2+2s+5)} = \frac{1}{s-1} + \frac{-s+2}{s^2+2s+5} = \frac{1}{s-1} - \frac{s-2}{(s+1)^2+2^2}$$

$$= \frac{1}{s-1} - \frac{s+1}{(s+1)^2+2^2} + \frac{3}{(s+1)^2+2^2}$$

Now $L^{-1}\left\{\dfrac{5s+3}{(s-1)(s^2+2s+5)}\right\} = L^{-1}\left\{\dfrac{1}{s-1}\right\} - L^{-1}\left\{\dfrac{s+1}{(s+1)^2+2^2}\right\} + 3L^{-1}\left\{\dfrac{1}{(s+1)^2+2^2}\right\}$

$$= e^t - e^{-t}L^{-1}\left\{\frac{s}{s^2+2^2}\right\} + 3e^{-t}L^{-1}\left\{\frac{1}{s^2+2^2}\right\}$$

or $\quad L^{-1}\left\{\dfrac{5s+3}{(s-1)(s^2+2s+5)}\right\} = e^t - e^{-t}\cos 2t + \dfrac{3}{2}e^{-t}\cdot\sin 2t.$ **Ans.**

Example 45: Find the inverse Laplace transform of the following functions:

(*i*) $\dfrac{2s^2-1}{(s^2+1)(s^2+4)}$ [*U.P.T.U. (C.O.) 2004*] (*ii*) $\dfrac{2s^2-6s+5}{s^3-6s^2+11s-6}$. (*U.P.T.U. 2004*)

Solution: (*i*) $\dfrac{2s^2-1}{(s^2+1)(s^2+4)} = \dfrac{2x-1}{(x+1)(x+4)}$ | Let $s^2 = x$

$$\Rightarrow \quad \frac{2x-1}{(x+1)(x+4)} = \frac{A}{(x+1)} + \frac{B}{(x+4)} \Rightarrow (2x-1) = A(x+4) + B(x+1)$$

Putting $x + 4 = 0$ or $x = -4$, we get

$$(-8-1) = B(-4+1)$$

$$\Rightarrow \quad \boxed{B = 3.}$$

and putting $x + 1 = 0$ or $x = -1$, we get

$$(-2-1) = A(-1+4) \Rightarrow \boxed{A = -1.}$$

So $$\frac{2x-1}{(x+1)(x+4)} = -\frac{1}{x+1} + \frac{3}{x+4}$$

or $$\frac{2s^2-1}{(s^2+1)(s^2+4)} = -\frac{1}{s^2+1} + \frac{3}{s^2+4}$$

$$\therefore \quad L^{-1}\left\{\frac{2s^2-1}{(s^2+1)(s^2+4)}\right\} = -L^{-1}\left\{\frac{1}{s^2+1}\right\} + 3L^{-1}\left\{\frac{1}{s^2+2^2}\right\}$$

$$= -\sin t + \frac{3}{2}\sin 2t. \quad \textbf{Ans.}$$

(***ii***) $$\frac{2s^2-6s+5}{s^3-6s^2+11s-6} = \frac{2s^2-6s+5}{(s-1)(s-2)(s-3)} = \frac{A}{(s-1)} + \frac{B}{(s-2)} + \frac{C}{(s-3)}$$

$$\Rightarrow \quad 2s^2-6s+5 = A(s-2)(s-3) + B(s-1)(s-3) + C(s-2)(s-1)$$

Putting $s - 2 = 0 \quad \Rightarrow \quad s = 2$, we get

$$8-12+5 = B(2-1)(2-3) \Rightarrow 1 = -B \Rightarrow \boxed{B = -1.}$$

Again putting $s - 3 = 0 \Rightarrow s = 3$

$$18-18+5 = C(3-2)(3-1) \Rightarrow \boxed{C = \frac{5}{2}.}$$

and finally putting $s - 1 = 0 \Rightarrow s = 1$, we get

$$1 = A(-1)(-2) \Rightarrow \boxed{A = \frac{1}{2}.}$$

So $\dfrac{2s^2-6s+5}{s^3-6s^2+11s-6} = \dfrac{1}{2(s-1)}+(-1)\cdot\dfrac{1}{(s-2)}+\dfrac{5}{2(s-3)}$

Taking inverse Laplace transform, we get

$$L^{-1}\left\{\frac{2s^2-6s+5}{s^3-6s^2+11s-6}\right\} = \frac{1}{2}L^{-1}\left\{\frac{1}{(s-1)}\right\}-L^{-1}\left\{\frac{1}{s-2}\right\}+\frac{5}{2}L^{-1}\left\{\frac{1}{s-3}\right\}$$

$$= \frac{1}{2}e^{t}-e^{2t}+\frac{5}{2}e^{3t}. \quad \textbf{Ans.}$$

Example 46: Find the inverse Laplace transform of $\dfrac{e^{-cs}}{s^2(s+a)}$. *(U.P.T.U. 2002)*

Solution: $\dfrac{1}{s^2(s+a)} = \dfrac{A}{s}+\dfrac{B}{s^2}+\dfrac{C}{s+a}$

or $$1 = As\,(s+a)+B\,(s+a)+Cs^2$$

Let $s + a = 0$ s.t. $s = -a$

$$1 = 0+0+Ca^2 \Rightarrow \boxed{C=\frac{1}{a^2}.}$$

Again let $s = 0$

$\Rightarrow$ $$1 = B\,(0+a) \Rightarrow \boxed{B=\frac{1}{a}.}$$

Equating the coefficient of s^2, we get

$$A + C = 0 \Rightarrow A = -C \Rightarrow \boxed{A=-\frac{1}{a^2}.}$$

$\therefore$ $$\frac{1}{s^2(s+a)} = -\frac{1}{a^2 s}+\frac{1}{as^2}+\frac{1}{a^2(s+a)}$$

or $$L^{-1}\left\{\frac{1}{s^2(s+a)}\right\} = -\frac{1}{a^2}L^{-1}\left\{\frac{1}{s}\right\}+\frac{1}{a}L^{-1}\left\{\frac{1}{s^2}\right\}+\frac{1}{a^2}L^{-1}\left\{\frac{1}{s+a}\right\}$$

$$= -\frac{1}{a^2}+\frac{1}{a}\cdot t+\frac{1}{a^2}e^{-at} = F(t)$$

Therefore, $L^{-1}\left\{\dfrac{e^{-cs}}{s^2\cdot(s+a)}\right\} = \begin{cases} F(t-c), t<c \\ 0, \quad t>c \end{cases}$

As $L^{-1}\{e^{-as}\cdot f(s)\} = G(t)$

where $G(t) = \begin{cases} F(t-a), t<a \\ 0 \quad, t>a \end{cases}$

$= F(t-a)\cdot U(t-a)$

$$= \begin{cases} \dfrac{-1}{a^2} + \dfrac{1}{a}(t-c) + \dfrac{1}{a^2}e^{-a(t-c)}, t < c \\ 0, \qquad\qquad t > c \end{cases}$$

$$\Rightarrow \quad L^{-1}\left\{\frac{e^{-cs}}{s^2(s+a)}\right\} = \left[-\frac{1}{a^2} + \frac{1}{a}(t-c) + \frac{1}{a^2}e^{-a(t-c)}\right] \cdot U(t-c) + 0$$

$$= -\frac{1}{a^2}U(t-c) + \frac{1}{a}(t-c)U(t-c) + \frac{1}{a^2}e^{-a(t-c)} \cdot U(t-c). \qquad \textbf{Ans.}$$

Example 47: Prove that $L^{-1}\{f(as+b)\} = \dfrac{1}{a}e^{-\frac{bt}{a}} \cdot F(t/a)$.

Solution: $f(as+b) = \displaystyle\int_0^\infty e^{-(as+b)t} \cdot F(t)\,dt = \int_0^\infty e^{-ast} \cdot e^{-bt} \cdot F(t)\,dt$

Let $at = u \quad \Rightarrow \quad dt = \dfrac{du}{a}$

$$\therefore \qquad f(as+b) = \int_0^\infty e^{-su} \cdot e^{-\frac{b}{a}u} \cdot F\left(\frac{u}{a}\right) \cdot \frac{du}{a}$$

$$= \frac{1}{a}\int_0^\infty e^{-su}\left\{e^{-\frac{b}{a}u} \cdot F\left(\frac{u}{a}\right)\right\}du$$

$$= \frac{1}{a}\int_0^\infty e^{-st}\left\{e^{-\frac{b}{a}t} \cdot F\left(\frac{t}{a}\right)\right\}dt \qquad \text{(By definite integral)}$$

Hence $\qquad f(as+b) = \dfrac{1}{a}L\left\{e^{-\frac{b}{a}t} \cdot F\left(\dfrac{t}{a}\right)\right\}$

$$\Rightarrow \qquad L^{-1}\{f(as+b)\} = \frac{1}{a}e^{-\frac{b}{a}t} \cdot F\left(\frac{t}{a}\right). \qquad \textbf{Proved.}$$

Example 48: Find the inverse Laplace transform of $\dfrac{e^{-s}}{\sqrt{s+1}}$.

Solution: Let $f(s) = \dfrac{1}{\sqrt{s+1}}$

$$\therefore \qquad L^{-1}\{f(s)\} = L^{-1}\left\{\frac{1}{(s+1)^{1/2}}\right\} = e^{-t}L^{-1}\left\{\frac{1}{s^{1/2}}\right\} = e^{-t}\cdot\frac{t^{-1/2}}{\sqrt{\pi}}$$

$$\Rightarrow \qquad F(t) = \frac{e^{-t}}{\sqrt{t\pi}}$$

By second shifting theorem, we have

$$L^{-1}\left\{\frac{e^{-s}}{\sqrt{s+1}}\right\} = G(t) = \begin{cases} F(t-1), t>1 \\ 0, \quad t<1 \end{cases} = F(t-1)\cdot U(t-1)$$

$$= \begin{cases} \dfrac{e^{-(t-1)}}{\sqrt{\pi(t-1)}}, t>1 \\ 0, t<1 \end{cases} = \frac{e^{-(t-1)}}{\sqrt{\pi(t-1)}}U(t-1). \quad \textbf{Ans.}$$

Example 49: Find $L^{-1}\left\{\dfrac{1}{s^2-e^{-cs}}\right\}$.

Solution:
$$\frac{1}{s^2-e^{-cs}} = \frac{1}{s^2}\left[1-\frac{e^{-cs}}{s^2}\right]^{-1}$$

$$= \frac{1}{s^2}\left[1+\frac{e^{-cs}}{s^2}+\left(\frac{e^{-cs}}{s^2}\right)^2+\ldots+\left(\frac{e^{-cs}}{s^2}\right)^n+\ldots\right] = \frac{1}{s^2}\sum_{n=0}^{\infty}\left(\frac{e^{-cs}}{s^2}\right)^n$$

$$= \sum_{n=0}^{\infty}\frac{e^{-cns}}{s^{2n+2}} = \sum_{n=0}^{\infty}\frac{e^{-cns}}{s^{(2n+1)+1}}$$

$$\Rightarrow \qquad L^{-1}\left\{\frac{1}{s^2-e^{-cs}}\right\} = L^{-1}\left\{\sum_{n=0}^{\infty}\frac{e^{-cns}}{s^{(2n+1)+1}}\right\} = \left\{\sum_{n=0}^{\infty}L^{-1}\left\{\frac{e^{-cns}}{s^{(2n+1)+1}}\right\}\right.$$

$$= \sum_{n=0}^{\infty}\frac{(t-cn)^{2n+1}}{\lfloor 2n+1}U(t-cn). \quad \textbf{Ans.}$$

Example 50: Find inverse Laplace transform of $\dfrac{s^2+2s+3}{(s^2+2s+2)(s^2+2s+5)}$.

Solution:
$$\frac{s^2+2s+3}{(s^2+2s+2)(s^2+2s+5)}$$

$$= \frac{As+B}{s^2+2s+2} + \frac{Cs+D}{s^2+2s+5}$$

$$\Rightarrow \quad s^2+2s+3 = (As+B)(s^2+2s+5)+(Cs+D)(s^2+2s+2)$$

$$= (A+C)s^3 + (2A+B+2C+D)s^2$$

$$+(5A+2B+2C+2D)s+5B+2D$$

Comparing coefficients of s on either side, we get

$$A + C = 0,$$

$$2A + B + 2C + D = 1,$$

$$5A + 2B + 2C + 2D = 2,$$

$$5B + 2D = 3$$

By solving these equations, we get

$$A = 0, \; B = \frac{1}{3}, C = 0, D = \frac{2}{3}$$

Thus $\quad L^{-1}\left\{\frac{s^2+2s+3}{(s^2+2s+2)(s^2+2s+5)}\right\}$

$$= L^{-1}\left\{\frac{\frac{1}{3}}{s^2+2s+2}\right\} + L^{-1}\left\{\frac{\frac{2}{3}}{s^2+2s+5}\right\}$$

$$= \frac{1}{3}L^{-1}\left\{\frac{1}{(s+1)^2+1}\right\} + \frac{2}{3}L^{-1}\left\{\frac{1}{(s+1)^2+4}\right\}$$

$$= \frac{1}{3}e^{-t}\sin t + \frac{2}{3}\cdot\frac{1}{2}\cdot e^{-t}\cdot\sin 2t$$

$$= \frac{1}{3}e^{-t}(\sin t + \sin 2t). \quad \textbf{Ans.}$$

Alter: Reduction of non-repeated quadratic factors to non-repeated linear factors using complex numbers

$$\frac{s^2+2s+3}{[(s^2+2s+2)]\,[(s^2+2s+5)]}$$

$$= \frac{s^2+2s+3}{[(s+1-i)(s+1+i)]\,[(s+1-2i)(s+1+2i)]}$$

$$= \frac{A}{s+1-i} + \frac{B}{s+1+i} + \frac{C}{s+1-2i} + \frac{D}{s+1+2i}$$

Solving $\quad A = \frac{1}{6i}, B = -\frac{1}{6i}, C = \frac{1}{6i}, D = \frac{-1}{6i}$

$$L^{-1}\left\{\frac{s^2+2s+3}{(s^2+2s+2)\,(s^2+2s+5)}\right\}$$

$$= \frac{e^{-(1-i)t}}{6i} - \frac{e^{-(1+i)t}}{6i} + \frac{e^{-(1-2i)t}}{6i} - \frac{e^{-(1+2i)t}}{6i}$$

$$= \frac{1}{3}e^{-t}\left(\frac{e^{it}-e^{-it}}{2i}\right) + \frac{1}{3}e^{-t}\left(\frac{e^{2it}-e^{-2it}}{2i}\right)$$

$$= \frac{1}{3}e^{-t}\cdot\sin t + \frac{1}{3}e^{-t}\sin 2t. \quad \textbf{Ans.}$$

Example 51: If $L^{-1}\left\{\frac{e^{-1/s}}{s^{1/2}}\right\} = \frac{\cos 2\sqrt{t}}{\sqrt{\pi t}}$, find $L^{-1}\left\{\frac{e^{-a/s}}{s^{1/2}}\right\}, a > 0.$

Solution: We have $L^{-1}\left\{\frac{e^{-1/s}}{s^{1/2}}\right\} = \frac{\cos 2\sqrt{t}}{\sqrt{\pi t}}$

$$\therefore \quad L^{-1}\left\{\frac{e^{-\frac{1}{sk}}}{(sk)^{1/2}}\right\} = \frac{1}{k}\frac{\cos 2\sqrt{\frac{t}{k}}}{\sqrt{\pi t/k}}$$

By change of scale property $L^{-1}\{f(ks) = \frac{1}{k}F\left(\frac{t}{k}\right)$

$$= \frac{\sqrt{k}}{k}\frac{\cos 2\sqrt{\frac{t}{k}}}{\sqrt{\pi t}}$$

$$\Rightarrow \quad L^{-1}\left\{\frac{e^{-\frac{1}{sk}}}{(s)^{1/2}}\right\} = \cos 2\sqrt{\frac{t}{k}}\Big/\sqrt{\pi t}$$

Let $\quad \frac{1}{k} = a$

$$\Rightarrow \quad L^{-1}\left\{\frac{e^{-\frac{a}{s}}}{s^{1/2}}\right\} = \frac{\cos 2\sqrt{at}}{\sqrt{\pi t}}. \quad \textbf{Ans.}$$

Example 52: Evaluate $L^{-1}\left\{\dfrac{a(s^2-2a^2)}{s^4+4a^4}\right\}$.

Solution: $L^{-1}\left\{\dfrac{a(s^2-2a^2)}{s^4+4a^4}\right\} = L^{-1}\left\{\dfrac{a(s^2-2a^2)}{(s^2+2as+2a^2)(s^2-2as+2a^2)}\right\}$ | As $A^2+B^2=(A+B)^2-2AB$

$$= L^{-1}\left\{-\frac{1}{2}\frac{(s+a)}{(s^2+2as+2a^2)}+\frac{1}{2}\frac{(s-a)}{(s^2-2as+2a^2)}\right\}$$ | By partial fraction

$$= -\frac{1}{2}L^{-1}\left\{\frac{(s+a)}{(s+a)^2+a^2}\right\}+\frac{1}{2}L^{-1}\left\{\frac{s-a}{(s-a)^2+a^2}\right\}$$

$$= -\frac{1}{2}e^{-at}L^{-1}\left\{\frac{s}{s^2+a^2}\right\}+\frac{1}{2}e^{at}L^{-1}\left\{\frac{s}{s^2+a^2}\right\}$$

$$= -\frac{1}{2}e^{-at}\cos at+\frac{1}{2}e^{at}\cos at=\cos at\left(\frac{e^{at}-e^{-at}}{2}\right)$$

or $$L^{-1}\left\{\frac{a(s^2-2a^2)}{s^4+4a^4}\right\} = \cos at.\sin h\, at.$$ **Ans.**

Example 53: Prove that $L^{-1}\left\{\dfrac{s}{s^4+s^2+1}\right\}=\dfrac{2}{\sqrt{3}}\sin h\dfrac{t}{2}.\sin\dfrac{1}{2}\sqrt{3}t$

Solution: $L^{-1}\left\{\dfrac{s}{s^4+s^2+1}\right\} = L^{-1}\left\{\dfrac{s}{(s^2+1)^2-s^2}\right\}=L^{-1}\left\{\dfrac{s}{(s^2-s+1)(s^2+s+1)}\right\}$

$$= L^{-1}\left\{\frac{1}{2(s^2-s+1)}-\frac{1}{2(s^2+s+1)}\right\}=\frac{1}{2}L^{-1}\left\{\frac{1}{\left(s-\frac{1}{2}\right)^2+\frac{3}{4}}\right\}-\frac{1}{2}L^{-1}\left\{\frac{1}{\left(s+\frac{1}{2}\right)^2+\frac{3}{4}}\right\}$$

$$= \frac{1}{2}e^{t/2}L^{-1}\left\{\frac{1}{s^2+\left(\frac{\sqrt{3}}{2}\right)^2}\right\}-\frac{1}{2}e^{-t/2}L^{-1}\left\{\frac{1}{s^2+\left(\frac{\sqrt{3}}{2}\right)^2}\right\}$$

$$= \frac{1}{2}\left[e^{t/2}\cdot\frac{2}{\sqrt{3}}\sin\frac{1}{2}\sqrt{3}t - e^{-t/2}\frac{2}{\sqrt{3}}\sin\frac{1}{2}\sqrt{3}t\right] = \frac{2}{\sqrt{3}}\sin\frac{1}{2}\sqrt{3}t\left[\frac{e^{t/2}-e^{-t/2}}{2}\right]$$

$$\Rightarrow \quad L^{-1}\left\{\frac{s}{s^4+s^2+1}\right\} = \frac{2}{\sqrt{3}}\sinh\frac{t}{2}\cdot\sin\frac{1}{2}\sqrt{3}t. \quad \textbf{Proved.}$$

Example 54: Prove that $L^{-1}\left\{\dfrac{s^3-3s^2+6s-4}{(s^2-2s+2)^2}\right\} = e^t(\cos t + t\sin t)$.

Solution: $$\frac{s^3-3s^2+6s-4}{(s^2-2s+2)^2} = \frac{As+B}{(s^2-2s+2)} + \frac{Cs+D}{(s^2-2s+2)^2}$$

$$\Rightarrow \quad s^3-3s^2+6s-4 = (As+B)(s^2-2s+2)+(Cs+D)$$

Equating like power coefficient of s and solving, we get

$A = 1, \quad B = -1, C = 2$ and $D = -2$

$$\therefore \quad L^{-1}\left\{\frac{s^3-3s^2+6s-4}{(s^2-2s+2)^2}\right\} = L^{-1}\left\{\frac{s-1}{s^2-2s+2}\right\} + 2L^{-1}\left\{\frac{s-1}{(s^2-2s+2)^2}\right\}$$

$$= L^{-1}\left\{\frac{(s-1)}{(s-1)^2+1}\right\} + 2L^{-1}\left\{\frac{(s-1)}{((s-1)^2+1)^2}\right\}$$

$$= e^tL^{-1}\left\{\frac{s}{(s^2+1)}\right\} + 2e^tL^{-1}\left\{\frac{s}{(s^2+1)^2}\right\}$$

$$= e^t\cdot\cos t + 2e^t\cdot\frac{t}{2}\sin t.$$

$$\Rightarrow \quad L^{-1}\left\{\frac{s^3-3s^2+6s-4}{(s^2-2s+2)^2}\right\} = e^t(\cos t + t\sin t). \quad \textbf{Proved.}$$

Example 55: Find $L^{-1}\left\{\dfrac{5-3e^{-2s}-2e^{-7s}}{s}\right\}$.

Solution: $$L^{-1}\left\{\frac{5-3e^{-2s}-2e^{-7s}}{s}\right\} = 5L^{-1}\left\{\frac{1}{s}\right\} - 3L^{-1}\left\{\frac{e^{-2s}}{s}\right\} - 2L^{-1}\left\{\frac{e^{-7s}}{s}\right\}$$

Applying second shifting theorem

$$= 5\cdot 1 - 3\cdot U(t-2) - 2U(t-7). \quad \textbf{Ans.}$$

or $$L^{-1}\left\{\frac{5-3e^{-2s}-2e^{-7s}}{s}\right\} = 5-3\begin{cases}1, & t>2\\ 0, & 0<t<2\end{cases} -2\begin{cases}1, & t>7\\ 0, & 0<t<7\end{cases}$$

$$= \begin{cases}5-3\cdot 0-2\cdot 0, & 0<t<2\\ 5-3\cdot 1-2\cdot 0, & 2<t<7\\ 5-3\cdot 1-2\cdot 1, & t>7\end{cases}$$

$$= \begin{cases}5, & 0<t<2\\ 2, & 2<t<7\\ 0, & t>7\end{cases} \quad \textbf{Ans.}$$

Example 56: Evaluate $L^{-1}\left\{\frac{(s+1)e^{-s\pi}}{s^2+s+1}\right\}$.

Solution: First of all, we find

$$L^{-1}\left\{\frac{s+1}{s^2+s+1}\right\} = L^{-1}\left\{\frac{\left(s+\frac{1}{2}\right)+\frac{1}{2}}{\left(s+\frac{1}{2}\right)^2+\frac{3}{4}}\right\} = e^{-t/2}L^{-1}\left\{\frac{s+\frac{1}{2}}{s^2+\frac{3}{4}}\right\}$$

$$= e^{-t/2}L^{-1}\left\{\frac{s}{s^2+\left(\frac{\sqrt{3}}{2}\right)^2}\right\}+\frac{1}{2}e^{-t/2}L^{-1}\left\{\frac{1}{s^2+\left(\frac{\sqrt{3}}{2}\right)^2}\right\}$$

$$= e^{-t/2}\cos\frac{\sqrt{3}}{2}t+\frac{1}{2}e^{-t/2}\cdot\frac{2}{\sqrt{3}}\sin\frac{\sqrt{3}}{2}t$$

$$= \frac{e^{-t/2}}{\sqrt{3}}\left[\sqrt{3}\cos\frac{\sqrt{3}}{2}t+\sin\frac{\sqrt{3}}{2}t\right]$$

Therefore, $$L^{-1}\left\{\frac{(s+1)e^{-\pi s}}{s^2+s+1}\right\} = \begin{cases}\frac{1}{\sqrt{3}}e^{-(t-\pi)/2}\left[\sqrt{3}\cos\frac{\sqrt{3}}{2}(t-\pi)+\sin\frac{\sqrt{3}}{2}(t-\pi)\right], t>\pi\\ 0, t<\pi\end{cases}$$

$$= \frac{1}{\sqrt{3}}e^{-(t-\pi)/2}\left[\sqrt{3}\cos\frac{\sqrt{3}}{2}(t-\pi)+\sin\frac{\sqrt{3}}{2}(t-\pi)\right]\cdot U(t-\pi). \quad \textbf{Ans.}$$

Example 57: Find the inverse Laplace transform of

$$\frac{s^3+6s^2+14s}{(s+2)^4}.$$

Solution: We have

$$\frac{s^3+6s^2+14s}{(s+2)^4} = \frac{A}{(s+2)^4}+\frac{B}{(s+2)^3}+\frac{C}{(s+2)^2}+\frac{D}{(s+2)}$$

$$s^3+6s^2+14s = A+B(s+2)+C(s+2)^2+D(s+2)^3$$

$$= Ds^3+(6D+C)s^2+(12D+4C+B)s+(8D+4C+2B+A)$$

Equating the coefficients of s

$$A=-12,\ B=2,\ C=0,\ D=1$$

$$\therefore \quad L^{-1}\left\{\frac{s^3+6s^2+14s}{(s+2)^4}\right\} = -12L^{-1}\left\{\frac{1}{(s+2)^4}\right\}+2L^{-1}\left\{\frac{1}{(s+2)^3}\right\}+L^{-1}\left\{\frac{1}{s+2}\right\}$$

$$= -12e^{2t}L^{-1}\left\{\frac{1}{s^4}\right\}+2e^{-2t}L^{-1}\left\{\frac{1}{s^3}\right\}+e^{-2t}$$

$$= -2e^{-2t}t^3+e^{-2t}t^2+e^{-2t}$$

$$= e^{-2t}\{1+t^2-2t^3\}.$$ **Ans.**

EXERCISE 2.4

Find the inverse Laplace transforms of the following functions:

1. $\frac{1}{2s-5}$. $\left[\textbf{Ans. } \frac{1}{2}e^{5t/2}\right]$

2. $\left(\frac{\sqrt{s}-1}{s}\right)^2$. $\left[\textbf{Ans. } 1+t-4\sqrt{\frac{t}{\pi}}\right]$

3. $\frac{2s+9}{s^2+9}$. $[\textbf{Ans. } 2\cos 3t+3\sin 3t]$

4. $\left\{\frac{5}{s^2}+\left(\frac{\sqrt{s}-1}{s}\right)^2-\frac{7}{3s+2}\right\}$. $\left[\textbf{Ans. } 1+6t-4\sqrt{\frac{t}{\pi}}-\frac{7}{3}e^{-2t/3}\right]$

5. $\left\{\dfrac{3}{s^2-3}+\dfrac{3s+2}{s^3}-\dfrac{3s-27}{s^2+9}+\dfrac{6-30\sqrt{s}}{s^4}\right\}.$

$\left[\textbf{Ans. } \sqrt{3}\sin h\sqrt{3}t+3t+t^2-3\cos 3t+9\sin 3t+t^3-16t^2\sqrt{\dfrac{t}{\pi}}\right]$

6. $\left\{\dfrac{1}{s}\cos\dfrac{1}{s}\right\}.$ $\left[\textbf{Ans. } 1-\dfrac{t^2}{(\lfloor 2)^2}+\dfrac{t^4}{(\lfloor 4)^2}-\dfrac{t^6}{(\lfloor 6)^2}+........\right]$

7. $\dfrac{3s-2}{s^{5/2}}-\dfrac{7}{3s+2}.$ $\left[\textbf{Ans. } 6\sqrt{\dfrac{t}{\pi}}-\dfrac{8t^{3/2}}{3\sqrt{\pi}}-\dfrac{7}{3}e^{-2t/3}\right]$

8. $\dfrac{2s-5}{4s^2+25}+\dfrac{4s-18}{9-s^2}.$ $\left[\textbf{Ans. } \dfrac{1}{2}\left(\cos\dfrac{5t}{2}-\sin\dfrac{5t}{2}\right)-4\cos h3t+6\sin h3t\right]$

9. $\dfrac{3(s^2-2)^2}{2s^5}.$ $\left[\textbf{Ans. } \dfrac{3}{2}-3t^2+\dfrac{1}{2}t^4\right]$

10. $\dfrac{1}{s^{7/2}}.$ $\left[\textbf{Ans. } \dfrac{8}{15}t^2\sqrt{\dfrac{t}{\pi}}\right]$

11. $\dfrac{3s+5\sqrt{2}}{s^2+8}.$ $\left[\textbf{Ans. } 3\cos(2\sqrt{2}t)+\dfrac{5}{2}\sin(2\sqrt{2}t)\right]$

12. $\dfrac{4s+2}{s^2+8s+16}.$ $\left[\textbf{Ans. } 4e^{-4t}(1-t)\right]$

13. $\dfrac{s}{(s+1)^{5/2}}.$ $\left[\textbf{Ans. } \dfrac{2t^{1/2}(3-2t)}{3\sqrt{\pi}}\right]$

14. $\dfrac{1}{\sqrt[3]{8s-27}}.$ $\left[\textbf{Ans. } \dfrac{t^{-2/3}e^{27t/8}}{2\sqrt{1/3}}\right]$

15. $\dfrac{3s+2}{4s^2+12s+9}.$ $\left[\textbf{Ans. } \dfrac{3}{4}e^{-\frac{3t}{2}}-\dfrac{5}{8}te^{-\frac{3t}{2}}\right]$

16. $\dfrac{6s-4}{s^2-4s+20}.$ $\left[\textbf{Ans. } 2e^{2t}(3\cos 4t+\sin 4t)\right]$

17. $\dfrac{s^2}{(s+2)^3}.$ $\left[\textbf{Ans. } e^{-2t}(1-4t+2t^2)\right]$

18. $\dfrac{(s-4)}{4(s-3)^2+16}$. $\left[\textbf{Ans. } \dfrac{1}{4}e^{3t}\cos 2t - \dfrac{1}{8}e^{3t}\sin 2t\right]$

19. $\dfrac{(5s+2)}{(s^2e^{4s})}$. $\left[\textbf{Ans. } \begin{cases} 2t-3, & t>4 \\ 0, & t<4 \end{cases}\right]$

20. $\dfrac{se^{-s/2}+\pi e^{-s}}{s^2+\pi^2}$. $\left[\textbf{Ans. } \sin \pi t\left\{U\left(t-\dfrac{1}{2}\right)-U(t-1)\right\}\right]$

21. $\dfrac{e^{4-3s}}{(s+4)^{5/2}}$. $\left[\textbf{Ans. } \dfrac{4(t-3)^{3/2}e^{-4(t-4)}}{3\sqrt{\pi}}U(t-3)\right]$

22. $\dfrac{3}{s}-\dfrac{4e^{-s}}{s^2}+\dfrac{4e^{-3s}}{s^2}$. $[\textbf{Ans. } 3-4(t-1)U(t-1)+4(t-3)U(t-3)]$

23. $\dfrac{e^{-3s}}{(s+1)^3}$ Evaluate $F(2), F(5), F(7)$. $[\textbf{Ans. } 0, 2e^{-2}, 8e^{-4}]$

24. $\left\{\dfrac{3(1+e^{-s\pi})}{s^2+9}\right\}$. $\left[\textbf{Ans. } \cos 3\left(t-\dfrac{2\pi}{3}\right)U\left(t-\dfrac{2\pi}{3}\right)\right]$

25. If $L^{-1}\left\{\dfrac{e^{-\sqrt{s}}}{\sqrt{s}}\right\}=\dfrac{e^{-\frac{1}{4t}}}{\sqrt{\pi t}}$, show that

$$L^{-1}\left\{\frac{e^{-k\sqrt{s}}}{\sqrt{s}}\right\}=\frac{e^{-k^2/4t}}{\sqrt{\pi t}}.$$

26. $\dfrac{3s^3-3s^2-40s+36}{(s^2-4)^2}$. $[\textbf{Ans. } (5t+3)e^{-2t}-2te^{2t}]$

27. $\dfrac{s^2}{s^4+4a^4}$. $\left[\textbf{Ans. } \dfrac{1}{2a}[\sinh at\cos at+\cosh at\sin at\right]$

28. $\dfrac{s^2+8s+27}{(s+1)(s^2+4s+13)}$. $[\textbf{Ans. } 2e^{-t}+e^{-2t}(\sin 3t-\cos 3t)]$

29. $\dfrac{10}{s(s^2-2s+5)}$. $[\textbf{Ans. } 2-e^{t}(2\cos 2t-\sin 2t)]$

30. Prove that $L^{-1}\left\{\dfrac{se^{-2s\pi/3}}{s^2+9}\right\}=\cos 3\left(t-\dfrac{2\pi}{3}\right)U(t-2\pi/3)$.

2.28 INVERSE LAPLACE TRANSFORM OF DERIVATIVES

If $$L^{-1}\{f(s)\} = F(t),$$

then $$L^{-1}\{f^n(s)\} = L^{-1}\left\{\frac{d^n}{ds^n} f(s)\right\} = (-1)^n t^n F(t), n = 1, 2, 3, \ldots$$

Proof: Since

$$L\{t^n F(t)\} = (-1)^n \frac{d^n}{ds^n} f(s)$$

$\Rightarrow$ $$t^n F(t) = (-1)^n L^{-1}\left\{\frac{d^n}{ds^n} f(s)\right\}$$

$\Rightarrow$ $$\boxed{L^{-1}\left\{\frac{d^n}{ds^n} f(s)\right\} = (-1)^n t^n F(t).}$$

2.29 INVERSE LAPLACE TRANSFORM OF INTEGRALS

If $$L^{-1}\{f(s)\} = F(t),$$

then $$L^{-1}\left\{\int_s^\infty f(s)\, ds\right\} = \frac{F(t)}{t}$$

Proof: We have

$$L\left\{\frac{F(t)}{t}\right\} = \int_s^\infty f(s) ds \Rightarrow \frac{F(t)}{t} = L^{-1}\left\{\int_s^\infty f(s)\, ds\right\}$$

or $$\boxed{L^{-1}\left\{\int_s^\infty f(s)\, ds\right\} = \frac{F(t)}{t}.}$$

2.30 MULTIPLICATION BY POWERS OF *s*

If $$L^{-1}\{f(s)\} = F(t)$$

and $$F(0) = 0,$$

then $$L^{-1}\{s\, f(s)\} = \frac{d}{dt} F(t).$$

Proof: We have

$$L\{F'(t)\} = sL\{F(t)\} - F(0) = sL(F(t)\} \qquad |\text{As } F(0) = 0$$

$\Rightarrow$ $$L\{F'(t)\} = s\, f(s) \Rightarrow L^{-1}\{sf(s)\} = F'(t)$$

or $$L^{-1}\{s\,f(s)\} = \frac{d}{dt}F(t). \qquad \ldots(i)$$

Remark 1: If $F(0) \neq 0$, then

$$L^{-1}\{s\,f(s) - F(0)\} = F'(t)$$

$$\Rightarrow \quad L^{-1}\{s\,f(s)\} - F(0)L^{-1}\{1\} = F'(t)$$

$$\Rightarrow \quad L^{-1}\{s\,f(s)\} = F'(t) + F(0)\,\delta(t). \qquad \left| L^{-1}\{1\} = \delta(t) \right.$$

Remark 2: General form of (*i*) may be written as

$$L^{-1}\{s^n\,f(s)\} = \frac{d^n}{dt^n}F(t), \quad F(0) = 0.$$

2.31 DIVISION BY POWERS OF *s*

If $$L^{-1}\{f(s)\} = F(t),$$

then $$L^{-1}\left\{\frac{f(s)}{s}\right\} = \int_0^t F(x)\,dx$$

Proof: Suppose

$$G(t) = \int_0^t F(x)\,dx \qquad \ldots(i)$$

Then by Leibnitz's rule for differentiation under the integral sign, we have

$$G'(t) = \int_0^t \frac{\partial}{\partial x}F(x)\,dx = \int_0^t F'(x)\,dx = [F(x)]_0^t$$

$$\Rightarrow \quad G'(t) = F(t) - F(0) \quad \Rightarrow \quad G'(t) = F(t) \qquad \left| F(0) = 0 \right.$$

Now $$L\{G'(t)\} = L\{F(t)\}$$

$$\Rightarrow \quad s\,g(s) - G(0) = f(s)$$

$$\Rightarrow \quad s\,g(s) = f(s) \quad [\because G(0) = 0]$$

$$\Rightarrow \quad g(s) = \frac{f(s)}{s}$$

$$\Rightarrow \quad L\{G(t)\} = \frac{f(s)}{s} \quad \Rightarrow \quad G(t) = L^{-1}\left\{\frac{f(s)}{s}\right\}$$

$$\Rightarrow \quad L^{-1}\left\{\frac{f(s)}{s}\right\} = \int_0^t F(x)\,dx$$

Next, we want to find $L^{-1}\left\{\frac{f(s)}{s^2}\right\}$ *i.e.,* $L^{-1}\left\{\frac{f(s)}{s^2}\right\} = \int_0^t \int_0^y F(x)\,dx\,dy$

Let $$G(t) = \int_0^t \int_0^y F(x)\,dx\,dy \;\Rightarrow\; G'(t) = \int_0^t F(x)dx \;\Rightarrow\; G''(t) = F(t)$$

Since $$G(0) = G'(0) = 0$$

Now $$L\{G''(t)\} = s^2 L\{G(t)\} - sG(0) - G'(0) = s^2 L\{G(t)\}$$

or $$f(s) = s^2 L\{G(t)\} \qquad \left| \text{As } L\{G''(t)\} = L\{F(t)\} = f(s) \right.$$

$\Rightarrow$ $$L\{G(t)\} = \frac{f(s)}{s^2}$$

$$L\left\{\int_0^t \int_0^y F(x)\,dx\,dy\right\} = \frac{f(s)}{s^2}$$

or $$L^{-1}\left\{\frac{f(s)}{s^2}\right\} = \int_0^t \int_0^y F(x)\,dx\,dy$$

which may also be written as

$$L^{-1}\left\{\frac{f(s)}{s^2}\right\} = \int_0^t \int_0^t F(t)\,dt^2$$

In general $$L^{-1}\left\{\frac{f(s)}{s^n}\right\} = \int_0^t \int_0^t \cdots\cdots \int_0^t F(t)\,dt^n.$$

2.32 CONVOLUTION

If $F(t)$ and $G(t)$ are two functions of t, then the convolution of these two functions denoted by $F*G$ and is defined as

$$F*G = \int_0^t F(x)\cdot G(t-x)\,dx.$$

The relation $F*G$ is also called faltung[1] of $F(t)$ and $G(t)$ and can be regarded as a 'generalized product' of these functions.

2.33 CONVOLUTION THEOREM

(U.P.T.U. 2002, 2007)

If $$L\{F(t)\} = f(s),\; L\{G(t)\} = g(s)$$

then $$L^{-1}\{f(s)\cdot g(s)\} = F*G = \int_0^t F(x)G(t-x)\,dx$$

1. German for folding.

or $$L\{F * G\} = f(s) \cdot g(s)$$

Proof: We have,

$$L\{F(t)\} = \int_0^{\infty} e^{-st} F(t)\, dt$$

$$\Rightarrow \quad L\{F * G\} = \int_0^{\infty} e^{-st} (F * G)\, dt$$

$$= \int_0^{\infty} e^{-st} \left[\int_0^t F(x) \cdot G(t-x)\, dx \right] dt$$

$$= \int_0^{\infty}\int_0^t e^{-st} \cdot F(x) G(t-x)\, dx\, dt \quad \text{...(1)}$$

Here $t = 0$, $t = \infty$ and $x = 0, x = t$

By changing the order of integration from equation (1), we have

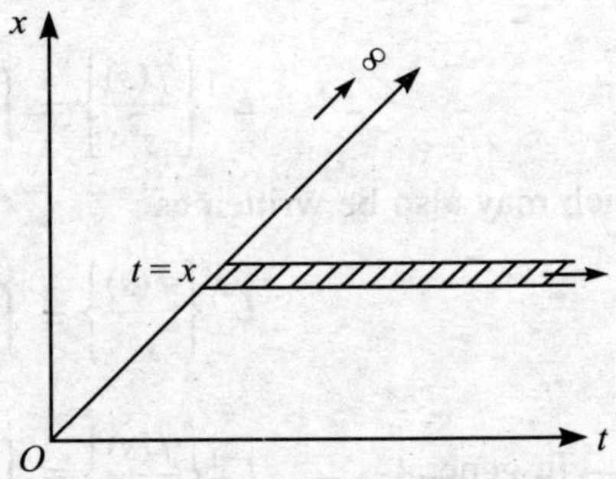

$$L\{F * G\} = \int_{x=0}^{\infty} \int_{t=x}^{\infty} e^{-st} \cdot F(x) \cdot G(t-x)\, dt\, dx$$

$$= \int_{x=0}^{\infty} \int_{t=x}^{\infty} e^{-s(t-x)} \cdot e^{-sx} F(x) \cdot G(t-x)\, dt\, dx$$

$$= \int_{x=0}^{\infty} e^{-sx} \cdot F(x) \left\{ \int_{t=x}^{\infty} e^{-s(t-x)} \cdot G(t-x)\, dt \right\} dx$$

$$= \int_{x=0}^{\infty} e^{-sx} \cdot F(x) \left\{ \int_0^{\infty} e^{-su} \cdot G(u)\, du \right\} dx \quad \left| \begin{array}{l} \text{Putting } t - x = u \\ dt = du \end{array} \right.$$

$$= \int_{x=0}^{\infty} e^{-sx} \cdot F(x) \cdot g(s)\, dx = g(s) \int_0^{\infty} e^{-sx} \cdot F(x)\, dx$$

$$L\{F * G\} = g(s) f(s) = f(s) g(s)$$

or $$\boxed{L^{-1}\{f(s) \cdot g(s)\} = F * G = \int_0^t F(x) G(t-x)\, dx.}$$

2.34 HEAVISIDE'S EXPANSION THEOREM (OR FORMULA)

Let $F(s)$ and $G(s)$ be two polynomials in s where $F(s)$ has degree less than that of $G(s)$. If $G(s)$ has n distinct zeroes $\alpha_1, \alpha_2, ..., \alpha_r$ *i.e.*, $G(s) = (s - \alpha_1)(s - \alpha_2) .... (s - \alpha_n)$, then

$$L^{-1}\left\{\frac{F(s)}{G(s)}\right\} = \sum_{r=1}^{n}\frac{F(\alpha_r)}{G'(\alpha_r)}e^{\alpha_r t}$$

Proof: $\dfrac{F(s)}{G(s)} = \dfrac{F(s)}{(s-\alpha_1)(s-\alpha_2)\cdots\cdots(s-\alpha_n)} = \dfrac{A_1}{(s-\alpha_1)} + \dfrac{A_2}{(s-\alpha_2)} + \cdots + \dfrac{A_r}{(s-\alpha_r)} + \cdots + \dfrac{A_n}{(s-\alpha_n)}$

Multiplying both sides by $(s-\alpha_r)$ and taking limit as $s \to \alpha_r$, we get

$$A_r = \lim_{s\to\alpha_r}\frac{F(s)(s-\alpha_r)}{G(s)} = \frac{0}{0}$$

$$\Rightarrow \quad A_r = F(\alpha_r)\cdot\lim_{s\to\alpha_r}\cdot\frac{1}{G'(s)} \qquad \text{[By } L\text{-Hospital's rule]}$$

$$= F(\alpha_r)\cdot\frac{1}{G'(\alpha_r)}$$

$$\therefore \quad \frac{F(s)}{G(s)} = \frac{F(\alpha_1)}{G'(\alpha_1)}\cdot\frac{1}{(s-\alpha_1)} + \frac{F(\alpha_2)}{G'(\alpha_2)}\cdot\frac{1}{(s-\alpha_2)} + \cdots + \frac{F(\alpha_n)}{G'(\alpha_n)}\cdot\frac{1}{(s-\alpha_n)}$$

Hence

$$L^{-1}\left\{\frac{F(s)}{G(s)}\right\} = \frac{F(\alpha_1)}{G'(\alpha_1)}L^{-1}\left\{\frac{1}{s-\alpha_1}\right\} + \frac{F(\alpha_2)}{G'(\alpha_2)}L^{-1}\left\{\frac{1}{s-\alpha_2}\right\} + \cdots + \frac{F(\alpha_n)}{G'(\alpha_n)}L^{-1}\left\{\frac{1}{s-\alpha_n}\right\}$$

$$= \frac{F(\alpha_1)}{G'(\alpha_1)}\cdot e^{\alpha_1 t} + \frac{F(\alpha_2)}{G'(\alpha_2)}e^{\alpha_2 t} + \cdots + \frac{F(\alpha_n)}{G'(\alpha_n)}\cdot e^{\alpha_n t}$$

$$\Rightarrow \quad \boxed{L^{-1}\left\{\frac{F(s)}{G(s)}\right\} = \sum_{r=1}^{n}\frac{F(\alpha_r)}{G'(\alpha_r)}e^{\alpha_r t}.}$$

Example 58: Find the function whose Laplace transform is $\log\left(1+\frac{1}{s}\right)$. *(U.P.T.U. 2007)*

Solution: We have

$$f(s) = \log\left(1+\frac{1}{s}\right) = \log\frac{(s+1)}{s}$$

$$= \log(s+1) - \log s$$

$$\Rightarrow \quad \frac{d}{ds}f(s) = \frac{1}{s+1} - \frac{1}{s}$$

$$L^{-1}\left\{\frac{d}{ds}f(s)\right\} = L^{-1}\left\{\frac{1}{s+1}\right\} - L^{-1}\left\{\frac{1}{s}\right\}$$

$$= e^{-t} - 1$$

Using inverse Laplace transform of derivatives

$$\therefore \quad L^{-1}\left\{\frac{d}{ds}f(s)\right\} = e^{-t} - 1 = -t\,F(t)$$

$$\Rightarrow \quad F(t) = \frac{1-e^{-t}}{t}. \quad \textbf{Ans.}$$

Example 59: A function $F(t)$ obeys the equation $F(t) + 2\int_0^t F(t)\,dt = \cosh 2t.$ Find the Laplace transform of $F(t)$. (*U.P.T.U. 2006*)

Solution: We have

$$F(t) + 2\int_0^t F(t)\,dt = \cosh 2t$$

Taking Laplace transform on both sides, we get

$$L\{F(t)\} + 2L\left\{\int_0^t F(t)\,dt\right\} = L\{\cosh 2t\}$$

$$\Rightarrow \quad L\{F(t)\} + 2\cdot\frac{f(s)}{s} = \frac{s}{s^2-4} \qquad \text{As } L^{-1}\left\{\frac{f(s)}{s}\right\} = \int_0^t F(x)\,dx$$

$$\Rightarrow \quad L\{F(t)\}\left[1+\frac{2}{s}\right] = \frac{s}{s^2-4} \qquad \{L\{F(t) = f(s)\}$$

$$\Rightarrow \quad L\{F(t)\} = \frac{s}{(s+2)(s-2)} \times \frac{s}{(s+2)} = \frac{s^2}{(s+2)^2(s-2)}. \quad \textbf{Ans.}$$

Example 60: Find $L^{-1}\left\{\cot^{-1}\frac{(s+a)}{b}\right\}.$

Solution: We have

$$f(s) = \cot^{-1}\frac{(s+a)}{b}$$

Differentiate w.r.t. 's', we get

$$\frac{d}{ds}f(s) = -\frac{1}{1+\frac{(s+a)^2}{b^2}} \cdot \frac{1}{b} = -\frac{b}{b^2+(s+a)^2}$$

$$\therefore \qquad L^{-1}\left\{\frac{d}{ds}f(s)\right\} = -bL^{-1}\left\{\frac{1}{(s+a)^2+b^2}\right\} = -be^{-at}L^{-1}\left\{\frac{1}{s^2+b^2}\right\}$$

$$= -\frac{b}{b}e^{-at}\sin bt = -e^{-at}\sin bt$$

But we have

$$L^{-1}\left\{\frac{d}{ds}f(s)\right\} = -t\,F(t)$$

$$\Rightarrow \qquad L^{-1}\left\{\frac{d}{ds}f(s)\right\} = -t\,F(t) = -e^{-at}\cdot\sin bt$$

$$\Rightarrow \qquad F(t) = \frac{e^{-at}}{t}\sin bt. \quad \textbf{Ans.}$$

Example 61: Evaluate $L^{-1}\left\{\log\frac{s+a}{s+b}\right\}$. *(U.P.T.U. 2003)*

Solution: Let $f(s) = \log\frac{s+a}{s+b} \Rightarrow f(s) = \log(s+a) - \log(s+b)$

Differentiate w.r.t. 's', we get

$$\frac{d}{ds}f(s) = \frac{1}{s+a} - \frac{1}{s+b} \Rightarrow L^{-1}\left\{\frac{d}{ds}f(s)\right\} = (e^{-at} - e^{-bt})$$

Hence $$L^{-1}\left\{\frac{d}{ds}f(s)\right\} = -t\,F(t) = e^{-at} - e^{-bt}$$

$$\Rightarrow \qquad F(t) = \frac{e^{-bt} - e^{-at}}{t}. \quad \textbf{Ans.}$$

Example 62: Find $L^{-1}\left\{\frac{1}{(s-a)^3}\right\}$.

Solution: Suppose $f(s) = \frac{1}{(s-a)} \Rightarrow f'(s) = -\frac{1}{(s-a)^2}$

and $$f''(s) = \frac{2}{(s-a)^3}$$

Now $$L^{-1}\left\{\frac{1}{s-a}\right\} = e^{at} = F(t)\text{(say)}$$

$\therefore$ $$L^{-1}\left\{\frac{d^2}{ds^2} f(s)\right\} = 2L^{-1}\left\{\frac{1}{(s-a)^3}\right\} = (-1)^2 t^2 \cdot e^{at}$$

$\Rightarrow$ $$L^{-1}\left\{\frac{1}{(s-a)^3}\right\} = \frac{t^2}{2} e^{at}. \quad \textbf{Ans.}$$

Example 63: Find the inverse Laplace transform of $\log\left\{\sqrt{\frac{s+1}{s-1}}\right\}$.

Solution: Let

$$f(s) = \log\left\{\sqrt{\frac{s+1}{s-1}}\right\} = \log\left(\frac{s+1}{s-1}\right)^{1/2} = \frac{1}{2}\log\left(\frac{s+1}{s-1}\right)$$

$\Rightarrow$ $$f(s) = \frac{1}{2}\left[\log(s+1) - \log(s-1)\right]$$

Differentiating with respect to s, we get

$$f'(s) = \frac{1}{2}\left[\frac{1}{s+1} - \frac{1}{s-1}\right]$$

or $$L^{-1}\{f'(s)\} = \frac{1}{2} L^{-1}\left\{\frac{1}{s+1} - \frac{1}{s-1}\right\}$$

$$= \frac{1}{2}(e^{-t} - e^{t})$$

or $$-tF(t) = \frac{1}{2}(e^{-t} - e^{t}) \qquad \Big| \text{ As } L^{-1}\{f'(s)\} = -tF(t)$$

$\Rightarrow$ $$F(t) = \frac{1}{t}\left(\frac{e^{t} - e^{-t}}{2}\right) = \frac{\sin ht}{t}$$

$\Rightarrow$ $$L^{-1}\{f(s)\} = \frac{\sin ht}{t}$$

Hence,

$$L^{-1}\left\{\log\sqrt{\frac{s+1}{s-1}}\right\} = \frac{\sin ht}{t}. \quad \textbf{Ans.}$$

Example 64: Evaluate $L^{-1}\left\{\int_s^\infty \log\left(\frac{s+2}{s+1}\right)ds\right\}$.

Solution: Let $$f(s) = \log\frac{(s+2)}{(s+1)}$$

$$\therefore \quad L^{-1}\{f(s)\} = L^{-1}\left\{\log\frac{s+2}{s+1}\right\} = \frac{e^{-t}-e^{-2t}}{t} = F(t)\,(\text{say})$$

Hence $$L^{-1}\left\{\int_s^\infty \log\left(\frac{s+2}{s+1}\right)ds\right\} = \frac{F(t)}{t} = \frac{e^{-t}-e^{-2t}}{t^2}.$$ **Ans.**

$$\left|\text{As } L^{-1}\left\{\int_s^\infty F(s)\,ds\right\} = \frac{F(t)}{t}\right.$$

Example 65: Evaluate $L^{-1}\left\{\frac{s^2}{s^2+1}\right\}$.

Solution: Let $$f(s) = \frac{1}{s^2+1}$$

$$\therefore \quad L^{-1}\left\{\frac{1}{s^2+1}\right\} = \sin t = F(t)\,(\text{say})$$

Hence $$L^{-1}\left\{\frac{s^2}{s^2+1}\right\} = \frac{d^2}{dt^2}\cdot(\sin t) = -\sin t.$$ **Ans.**

$$\left|\begin{array}{l} F(0) = 0 \\ L^{-1}\{s^n f(s)\} = \dfrac{d^n}{dt^n}F(t) \end{array}\right.$$

Example 66: Evaluate $L^{-1}\left\{\frac{s}{(s^2+1)^2}\right\}$, where $L^{-1}\left\{\frac{1}{(s^2+1)^2}\right\} = \frac{\sin t - t\cos t}{2}$.

Solution: By multiplication formula, we have

$$L^{-1}\{s\,f(s)\} = \frac{d}{dt}F(t)$$

$$\Rightarrow \quad L^{-1}\left\{\frac{s}{(s^2+1)^2}\right\} = \frac{1}{2}\frac{d}{dt}(\sin t - t\cos t)$$

$$= \frac{1}{2}(\cos t - \cos t + t\sin t)$$

$$= \frac{1}{2}\cdot t\sin t.$$ **Ans.**

Example 67: Find $L^{-1}\left\{\frac{1}{s^3(s^2+1)}\right\}$.

Solution: $L^{-1}\left\{\frac{1}{s^2+1}\right\} = \sin t$

By division formula, we have

$$L^{-1}\left\{\frac{1}{s(s^2+1)}\right\} = \int_0^t \sin x dx = [-\cos x]_0^t = (1-\cos t) \qquad \left| L^{-1}\left\{\frac{f(s)}{s}\right\} = \int_0^t F(x)dx\right.$$

Again, $$L^{-1}\left\{\frac{1}{s^2(s^2+1)}\right\} = L^{-1}\left\{\frac{\frac{1}{s(s^2+1)}}{s}\right\} = \int_0^t (1-\cos x)\,dx = t - \sin t$$

and $$L^{-1}\left\{\frac{1}{s^3(s^2+1)}\right\} = L^{-1}\left\{\frac{\frac{1}{s^2(s^2+1)}}{s}\right\} = \int_0^t (x-\sin x)\,dx = \frac{t^2}{2} + \cos t - 1. \quad \textbf{Ans.}$$

Example 68: Show that

$$1*1*1*........*1(n\,\text{times}) = \frac{t^{n-1}}{\lfloor n-1}, \text{ where } n = 1, 2,.....$$

Solution: Let $F(t) = 1$ and $G(t) = 1$

Apply the convolution of two function, we get

$$F*G = \int_0^t F(x)\cdot G(t-x)\,dx$$

$$1*1 = \int_0^t dx = t$$

Again let $F(t) = t$ and $G(t) = 1$

then $$F(t)*G(t) = \int_0^t x\,dx = \frac{t^2}{2}$$

$\Rightarrow$ $$1*1*1 = t^2/\lfloor 2$$

and $$1*1*1*1 = \frac{t^3}{\underline{|3}}$$

Proceeding n times, we get

$$1*1*1.....*1 \ (n \text{ times}) = \frac{t^{n-1}}{\underline{|n-1}}. \quad \textbf{Proved.}$$

Example 69: Use convolution theorem to find the inverse of the function $\frac{1}{(s^2+a^2)^2}$.

(U.P.T.U. 2009)

Solution: Let $$f(s) = \frac{1}{(s^2+a^2)}$$

$$g(s) = \frac{1}{(s^2+a^2)}$$

$\therefore$ $$L^{-1}\{f(s)\} = L^{-1}\left\{\frac{1}{s^2+a^2}\right\} = \frac{1}{a}\sin at = F(t)$$

and $$L^{-1}\{g(s)\} = \frac{1}{a}\sin at = G(t)$$

By convolution theorem

$$L^{-1}\{f(s)\cdot g(s)\} = \int_0^t F(x)\cdot G(t-x)dx$$

$$\Rightarrow \quad L^{-1}\left\{\frac{1}{(s^2+a^2)^2}\right\} = \int_0^t \frac{1}{a^2}\sin ax \sin(at-ax)dx$$

$$= \frac{1}{2a^2}\int_0^t \{\cos(at-2ax)-\cos(at)\}dx$$

| As $2\sin A\sin B = \cos(A-B)-\cos(A+B)$

$$= \frac{1}{2a^2}\left[\left\{-\frac{\sin(at-2ax)}{2a}\right\}_0^t - \cos at\cdot(x)_0^t\right]$$

$$= \frac{1}{2a^2}\left[\frac{\sin at}{2a}+\frac{\sin at}{2a}-t\cos at\right] \quad \Big| \text{ As } \sin(-\theta) = -\sin\theta$$

$$= \frac{1}{2a^2}\left[\frac{\sin at}{a}-t\cos at\right]$$

Hence

$$L^{-1}\left\{\frac{1}{(s^2+a^2)}\right\} = \frac{1}{2a^3}[\sin at - at\cos at] \quad \textbf{Ans.}$$

Example 70: Using convolution theorem, find $L^{-1}\left\{\frac{s^2}{(s^2+1)(s^2+4)}\right\}$. (*U.P.T.U. 2002*)

Solution : Let $f(s) = \frac{s}{s^2+4}$ and $g(s) = \frac{s}{s^2+1}$

$$\therefore \quad L^{-1}\{f(s)\} = L^{-1}\left\{\frac{s}{s^2+4}\right\} = \cos 2t \Rightarrow F(t) = \cos 2t$$

and

$$L^{-1}\{g(s)\} = L^{-1}\left\{\frac{s}{s^2+1}\right\} = \cos t \Rightarrow G(t) = \cos t$$

Hence, By convolution theorem, we have

$$L^{-1}\{f(s)\cdot g(s)\} = \int_0^t \cos 2x \cdot \cos(t-x)dx$$

$$\Rightarrow \quad L^{-1}\left\{\frac{s^2}{(s^2+1)(s^2+4)}\right\} = \frac{1}{2}\int_0^t \{\cos(x+t)+\cos(3x-t)\}dx$$

$$= \frac{1}{2}\left[\sin(x+t)+\frac{1}{3}\sin(3x-t)\right]_0^t$$

$$= \frac{1}{2}\left[\sin 2t+\frac{1}{3}\sin 2t-\sin t+\frac{1}{3}\sin t\right]$$

$$= \frac{1}{2}\left[\frac{4}{3}\sin 2t-\frac{2}{3}\sin t\right] = \frac{1}{3}[2\sin 2t-\sin t]. \quad \textbf{Ans.}$$

Example 71: Using the convolution theorem find $L^{-1}\left\{\frac{s^2}{(s^2+a^2)(s^2+b^2)}\right\}$, $a \neq b$.

(*U.P.T.U. 2006, 2004*)

Solution: Let $f(s) = \frac{s}{s^2+a^2}$ and $g(s) = \frac{s}{s^2+b^2}$

$$\therefore \quad L^{-1}\{f(s)\} = L^{-1}\left\{\frac{s}{s^2+a^2}\right\} = \cos at \Rightarrow F(t) = \cos at$$

and $$L^{-1}\{g(s)\} = L^{-1}\left\{\frac{s}{s^2+b^2}\right\} = \cos bt \quad \Rightarrow \quad G(t) = \cos bt$$

Now, applying convolution theorem, we have

$$L^{-1}\{f(g)\cdot g(s)\} = L^{-1}\left\{\frac{s^2}{(s^2+a^2)(s^2+b^2)}\right\} = \int_0^t \cos ax.\cos b(t-x)\,dx$$

$$= \frac{1}{2}\int_0^t \cos(ax+bt-bx)+\cos(ax-bt+bx)\}\,dx$$

$$= \frac{1}{2}\int_0^t \cos[(a-b)x+bt]\,dx + \frac{1}{2}\int_0^t \cos[(a+b)x-bt]\,dx$$

$$= \frac{1}{2}\left[\frac{\sin[(a-b)x+bt]}{(a-b)}\right]_0^t + \frac{1}{2}\left[\frac{\sin[(a+b)x-bt]}{(a+b)}\right]_0^t$$

$$= \frac{1}{2}\left[\frac{\sin at-\sin bt}{(a-b)}\right] + \frac{1}{2}\left[\frac{\sin at+\sin bt}{(a+b)}\right]$$

$$\Rightarrow L^{-1}\left\{\frac{s^2}{(s^2+a^2)(s^2+b^2)}\right\} = \frac{a\sin at-b\sin bt}{a^2-b^2}.$$ **Ans.**

Example 72: Using convolution theorem, prove that $L^{-1}\left\{\frac{1}{s^3(s^2+1)}\right\} = \frac{t^2}{2}+\cos t-1.$

(U.P.T.U. 2005)

Solution: Let $f(s) = \frac{1}{s^2+1}$ and $g(s) = \frac{1}{s^3}$

$$\therefore \quad L^{-1}\{f(s)\} = L^{-1}\left\{\frac{1}{s^2+1}\right\} = \sin t \quad \Rightarrow \quad F(t) = \sin t$$

and $$L^{-1}\left\{\frac{1}{s^3}\right\} = \frac{t^2}{\underline{|2}} \quad \Rightarrow \quad G(t) = \frac{t^2}{\underline{|2}}$$

By convolution theorem, we have

$$L^{-1}\left\{\frac{1}{s^3(s^2+1)}\right\} = \frac{1}{\underline{|2}}\int_0^t \sin x\cdot(t-x)^2\,dx$$

$$= \frac{1}{2}\int_0^t \sin x\cdot(t^2-2tx+x^2)\,dx$$

$$= \frac{t^2}{2}\int_0^t \sin x\,dx - t\int_0^t x\sin x\,dx + \frac{1}{2}\int_0^t x^2 \sin x\,dx$$

$$= \frac{t^2}{2}[-\cos x]_0^t + t[t\cos t - \sin t] + \frac{1}{2}[-x^2\cos x]_0^t + \int_0^t x\cos x\,dx$$

$$= \frac{t^2}{2}[-\cos t + 1] + t[t\cos t - \sin t] + \frac{1}{2}[-t^2\cos t] + [x\sin x]_0^t + [\cos x]_0^t$$

$$= -\frac{t^2}{2}\cos t + \frac{t^2}{2} + t^2\cos t - t\sin t - \frac{t^2}{2}\cos t + t\sin t + \cos t - 1$$

$$= \frac{t^2}{2} + \cos t - 1. \quad \textbf{Ans.}$$

Example 73: Evaluate $L^{-1}\left\{\dfrac{2s^2 - 6s + 5}{s^3 - 6s^2 + 11s - 6}\right\}$.

Solution: Let $F(s) = 2s^2 - 6s + 5$

$$G(s) = s^3 - 6s^2 + 11s - 6 \quad \Rightarrow \quad G'(s) = 3s^2 - 12s + 11$$

For the zeros of $G(s)$,

$$s^3 - 6s^2 + 11s - 6 = 0$$

$$\Rightarrow \quad (s-1)(s-2)(s-3) = 0$$

$$\Rightarrow \quad s = 1, 2, 3, \text{ i.e., } \alpha_1 = 1,\ \alpha_2 = 2,\ \alpha_3 = 3$$

Now, applying Heaviside's inverse formula, we have

$$L^{-1}\left\{\frac{F(s)}{G(s)}\right\} = \sum_{r=1}^{n}\frac{F(\alpha_r)}{G'(\alpha_r)}e^{\alpha_r t} = \frac{F(\alpha_1)}{G'(\alpha_1)}e^{\alpha_1 t} + \frac{F(\alpha_2)}{G'(\alpha_2)}e^{\alpha_2 t} + \frac{F(\alpha_3)}{G'(\alpha_3)}e^{\alpha_3 t}$$

$$\Rightarrow \quad L^{-1}\left\{\frac{2s - 6s + 5}{s^3 - 6s^2 + 11s - 6}\right\} = \frac{F(1)}{G'(1)}e^{t} + \frac{F(2)}{G'(2)}e^{2t} + \frac{F(3)}{G'(3)}e^{3t}$$

$$= \frac{1}{2}e^{t} - e^{2t} + \frac{5}{2}e^{3t}. \quad \textbf{Ans.}$$

Example 74: Using convolution theorem, find the inverse Laplace transform of the following $\frac{s}{(s^2+a^2)^3}$. *(U.P.T.U. 2008)*

Solution: Let $f(s) = \frac{s}{(s^2+a^2)^2}$ and $g(s) = \frac{1}{(s^2+a^2)}$

$$\therefore \quad L^{-1}\{f(s)\} = L^{-1}\left\{\frac{s}{s^2+a^2}\right\} = \cos at$$

Differentiate w.r. to a on both side, we get

$$L^{-1}\left\{\frac{d}{da}\left(\frac{s}{s^2+a^2}\right)\right\} = \frac{d}{da}\cos at$$

$$\text{or} \quad L^{-1}\left\{\frac{-2as}{(s^2+a^2)^2}\right\} = -t\sin at$$

$$\text{or} \quad L^{-1}\left\{\frac{s}{(s^2+a^2)^2}\right\} = \frac{t}{2a}\sin at = F(t) \text{ (say)}$$

$$\text{and} \quad L^{-1}\left\{\frac{1}{s^2+a^2}\right\} = \frac{1}{a}\sin at = G(t)$$

Apply convolution theorem

$$L^{-1}\left\{\frac{s}{(s^2+a^2)^3}\right\} = \int_0^t \frac{x\sin ax}{2a}\cdot\frac{1}{a}\sin a(t-x)dx = \frac{1}{2a^2}\int_0^t x\sin ax\sin(at-ax)dx$$

$$= \frac{1}{2a^2}\int_0^t x\cdot\frac{1}{2}[\cos(2ax-at)-\cos(at)]dx$$

$$= \frac{1}{4a^2}\left[\left\{\frac{x}{2a}\sin(2ax-at)\right\}_0^t - \frac{1}{2a}\int_0^t \sin(2ax-at)dx - \left\{\frac{x^2}{2}\cos at\right\}_0^t\right]$$

$$= \frac{1}{4a^2}\left[\frac{t}{2a}\sin at + \frac{1}{4a^2}\{\cos(2ax-at)\}_0^t - \frac{t^2}{2}\cos at\right]$$

$$= \frac{1}{4a^2}\left[\frac{t}{2a}\sin at + \frac{1}{4a^2}\cos at - \frac{1}{4a^2}\cos at - \frac{t^2}{2}\cos at\right]$$

$$= \frac{t}{8a^2}\left[\frac{1}{2}\sin at - t\cos at\right] \quad \textbf{Ans.}$$

Example 75: Use the convolution theorem to find

(*i*) $L^{-1}\left\{\dfrac{s^2}{(s^2+4)^2}\right\}$ (*ii*) $L^{-1}\left\{\dfrac{s}{(s^2+4)^3}\right\}$.

Solution: (*i*) Let $f(s) = \dfrac{s}{s^2+4}$ and $g(s) = \dfrac{s}{s^2+4}$.

$$\therefore \quad L^{-1}\{f(s)\} = L^{-1}\left\{\frac{s}{s^2+4}\right\} = \cos 2t$$

Here $F(t) = G(t) = \cos 2t$ | As $f(s) = g(s)$

By convolution theorem

$$L^{-1}\left\{\frac{s^2}{(s^2+4)^2}\right\} = \int_0^t \cos 2x \cos 2(t-x)\,dx$$

$$= \frac{1}{2}\int_0^t 2\cos 2x \cos 2(t-x)\,dx$$

$$= \frac{1}{2}\int_0^t [\cos 2t + \cos(4x-2t)]\,dx$$

$$= \frac{1}{2}\cos 2t \int_0^t dx + \frac{1}{2}\int_0^t \cos(4x-2t)\,dx$$

$$= \frac{t}{2}\cos 2t + \frac{1}{2\times 4}[\sin(4x-2t)]_0^t$$

$$= \frac{t}{2}\cos 2t + \frac{1}{8}[\sin 2t + \sin 2t]$$

$$= \frac{t\cos 2t}{2} + \frac{\sin 2t}{4}. \quad \textbf{Ans.}$$

(*ii*) $L^{-1}\left\{\dfrac{s}{(s^2+4)^3}\right\}$

Let $f(s) = \dfrac{s}{(s^2+4)^2}$ and $g(s) = \dfrac{1}{(s^2+4)}$

$$L^{-1}\left\{\frac{s}{(s^2+4)^2}\right\} = \frac{t\sin 2t}{4} = F(t) \text{ (say)}$$ | By convolution theorem

and $$L^{-1}\left\{\frac{1}{s^2+4}\right\} = \frac{1}{2}\sin 2t = G(t)$$

Apply convolution theorem

$$L^{-1}\left\{\frac{s}{(s^2+4)^3}\right\} = \int_0^t \frac{x\sin 2x}{4}\cdot\frac{1}{2}\sin 2(t-x)dx = \frac{1}{8}\int_0^t x\sin 2x\sin(2t-2x)\,dx$$

$$= \frac{1}{8}\int_0^t x\cdot\frac{1}{2}[\cos(4x-2t)-\cos(2t)]dx \qquad \left|\begin{array}{l}2\sin A\sin B = \cos(A-B)\\ \quad -\cos(A+B)\end{array}\right.$$

$$= \frac{1}{16}\left[\left\{\frac{x}{4}\sin(4x-2t)\right\}_0^t - \frac{1}{4}\int_0^t \sin(4x-2t)\,dx - \left\{\frac{x^2}{2}\cos 2t\right\}_0^t\right]$$

$$= \frac{1}{16}\left[\frac{t}{4}\sin 2t + \frac{1}{16}\{\cos(4x-2t)\}_0^t - \frac{t^2}{2}\cos 2t\right]$$

$$= \frac{1}{16}\left[\frac{t}{4}\sin 2t + \frac{1}{16}\cos 2t - \frac{1}{16}\cos 2t - \frac{t^2}{2}\cos 2t\right]$$

$$= \frac{t}{32}\left[\frac{1}{2}\sin 2t - t\cos 2t\right]. \quad \textbf{Ans.}$$

Example 76: Find $L^{-1}\left\{\frac{1}{\sqrt{s}(s-a)}\right\}$ by the convolution integral and deduce the value of $L^{-1}\left\{\frac{1}{s\sqrt{s+a}}\right\}$.

Solution: Let $f(s) = \frac{1}{\sqrt{s}}$ and $g(s) = \frac{1}{(s-a)}$

so that $$L^{-1}\{f(s)\} = \frac{1}{\sqrt{\pi t}} = F(t)\text{(say)}$$

and $$L^{-1}\{g(s)\} = L^{-1}\left\{\frac{1}{s-a}\right\} = e^{at} = G(t) \text{ (say)}$$

∴ By the convolution theorem, we have

$$L^{-1}\left\{\frac{1}{\sqrt{s}(s-a)}\right\} = \int_0^t F(x)\cdot G(t-x)dx = \int_0^t \frac{1}{\sqrt{\pi x}}\cdot e^{a(t-x)}dx$$

$$= \frac{e^{at}}{\sqrt{\pi}} \int_0^{\sqrt{ta}} \frac{\sqrt{a}}{u} e^{-u^2} \cdot 2u \frac{du}{a} \qquad \text{(Putting } ax = u^2\text{)}$$

$$= \frac{e^{at}}{\sqrt{a}} \cdot \frac{2}{\sqrt{\pi}} \int_0^{\sqrt{ta}} e^{-u^2} \cdot du = \frac{e^{at}}{\sqrt{a}} erf(\sqrt{at}). \quad \textbf{Ans.}$$

Deduction: $L^{-1}\left\{\frac{1}{s\sqrt{s+a}}\right\} = L^{-1}\left\{\frac{1}{(s+a-a)\sqrt{s+a}}\right\} = e^{-at} L^{-1}\left\{\frac{1}{(s-a)\sqrt{s}}\right\}$

$$\Rightarrow \quad L^{-1}\left\{\frac{1}{s\sqrt{s+a}}\right\} = e^{-at} \cdot \frac{e^{at}}{\sqrt{a}} erf(\sqrt{at}) = \frac{1}{\sqrt{a}} erf(\sqrt{at}). \quad \textbf{Ans.}$$

Example 77: Prove that $\int_0^\infty e^{-x^2} \cdot dx = \frac{1}{2}\sqrt{\pi}.$

Solution: Let $\quad F(t) = \int_0^\infty e^{-tx^2} dx$

$$\therefore \quad L\{F(t)\} = \int_0^\infty L\left\{e^{-tx^2}\right\} dx = \int_0^\infty \frac{dx}{s+x^2}$$

$$= \left(\frac{1}{\sqrt{s}} \tan^{-1} \frac{x}{\sqrt{s}}\right)_0^\infty = \frac{\pi}{2\sqrt{s}}$$

or $\quad F(t) = \frac{\pi}{2} L^{-1}\left\{\frac{1}{\sqrt{s}}\right\} = \frac{\pi}{2} \cdot \frac{1}{\sqrt{\pi t}}$

$$\Rightarrow \quad F(t) = \frac{1}{2}\sqrt{\frac{\pi}{t}}$$

$$\Rightarrow \quad \int_0^\infty e^{-tx^2} dx = \frac{1}{2}\sqrt{\frac{\pi}{t}}$$

Taking $t = 1$, we have

$$\int_0^\infty e^{-x^2} dx = \frac{\sqrt{\pi}}{2}. \quad \textbf{Proved.}$$

Example 78: Show that $\int_0^{\infty} \cos x^2 dx = \frac{1}{2}\left(\frac{\sqrt{\pi}}{2}\right)$.

Solution: Solve as example 77.

Example 79: Solve the integral equation

$$F(t) = 1 + \int_0^t F(u) \cdot \sin(t-u)\, du.$$

Solution: The given equation can be expressed as

$$F(t) = 1 + F(t) * \sin t$$

Taking L.T. on both sides, we have

$$L\{F(t)\} = L\{1\} + L\{F(t) * \sin t\}$$

$$L\{F(t)\} = \frac{1}{s} + L\{F(t)\} \cdot \frac{1}{s^2+1} \quad \Rightarrow \quad L\{F(t)\}\left(1 - \frac{1}{s^2+1}\right) = \frac{1}{s}$$

$$\Rightarrow \quad L\{F(t)\} = \frac{s^2+1}{s^3} \quad \Rightarrow \quad F(t) = L^{-1}\left\{\frac{s^2+1}{s^3}\right\}$$

$$F(t) = L^{-1}\left\{\frac{1}{s}\right\} + L^{-1}\left\{\frac{1}{s^3}\right\}$$

$$= 1 + \frac{t^2}{\lfloor 3} = 1 + \frac{t^2}{2}. \quad \textbf{Ans.}$$

Example 80: Solve for $y(t)$ the equation

$$y(t) = 1 + \int_0^t y(\tau) \cos(t-\tau)\, d\tau.$$ *(U.P.T.U. 2007)*

Solution: The given equation can be expressed as

$$y(t) = 1 + y(t) * \cos t$$

Taking Laplace transform on both side, we have

$$L\{y(t)\} = L\{1\} + L\{y(t) * \cos t\} = \frac{1}{s} + L\{y(t)\} \cdot \frac{s}{s^2+1}$$

$$\Rightarrow \quad L\{y(t)\}\left\{1 - \frac{s}{s^2+1}\right\} = \frac{1}{s}$$

$$\Rightarrow \quad L\{y(t)\}\left(\frac{s^2-s+1}{s^2+1}\right) = \frac{1}{s} \quad \Rightarrow \quad L\{y(t)\} = \frac{s^2+1}{s(s^2-s+1)}$$

$$\Rightarrow \quad y(t) = L^{-1}\left\{\frac{s^2+1}{s(s^2-s+1)}\right\} = L^{-1}\left\{\frac{(s^2-s+1)+s}{s(s^2-s+1)}\right\}$$

$$\Rightarrow \quad y(t) = L^{-1}\left\{\frac{1}{s}\right\} + L^{-1}\left\{\frac{1}{(s^2-s+1)}\right\} = 1 + L^{-1}\left\{\frac{1}{s^2-s+\left(\frac{1}{2}\right)^2+\frac{3}{4}}\right\}$$

$$= 1 + L^{-1}\left\{\frac{1}{\left(\left(s-\frac{1}{2}\right)^2+\left(\frac{\sqrt{3}}{2}\right)^2\right)}\right\} = 1 + e^{t/2}L^{-1}\left\{\frac{1}{\left(s^2+\left(\frac{\sqrt{3}}{2}\right)^2\right)}\right\}$$

$$\Rightarrow \quad y(t) = 1 + \frac{2}{\sqrt{3}}e^{t/2}\cdot\sin\frac{\sqrt{3}}{2}t. \quad \textbf{Ans.}$$

EXERCISE 2.5

Find the inverse Laplace transform of the following:

1. $\log\dfrac{s(s+1)}{(s^2+4)}$. $\left[\textbf{Ans.} \dfrac{2\cos 2t - e^{-t} - 1}{t}\right]$

2. $\cot^{-1}\left(\dfrac{s+3}{2}\right)$. $\left[\textbf{Ans.} \dfrac{2}{t}e^{-3t}\cdot\sin 2t\right]$

3. $\tan^{-1}\dfrac{2}{s^2}$. $\left[\textbf{Ans.} \dfrac{2}{t}\cdot\sinh t\cdot\sin t\right]$

4. $\dfrac{s}{(s^2+a^2)^2}$. $\left[\textbf{Ans.} \dfrac{t\sin at}{2a}\right]$

5. $\dfrac{1}{s(s^2+a^2)}$. $\left[\textbf{Ans.} \dfrac{1}{a^2}(1-\cos at)\right]$

6. $\dfrac{1}{s}\log\dfrac{s+3}{s+2}$. $\left[\textbf{Ans.} \displaystyle\int_0^t \frac{e^{-2x}-e^{-3x}}{x}\,dx\right]$

7. Evaluate $L^{-1}\left\{\int_s^{\infty}\left(\frac{1}{x}-\frac{1}{x+1}\right)dx\right\}$. $\left[\textbf{Ans. } \frac{1-e^{-t}}{t}\right]$

8. Evaluate $L^{-1}\left\{\int_s^{\infty}\tan^{-1}\frac{2}{x^2}\,dx\right\}$. $\left[\textbf{Ans. } \frac{2\sin t\sin h t}{t^3}\right]$

9. Find $L^{-1}\left\{s\log\frac{s}{\sqrt{s^2+1}}+\cot^{-1}s\right\}$. $\left[\textbf{Ans. } \frac{1-\cos t}{t^2}\right]$

10. Find $L^{-1}\left\{\frac{s+1}{(s^2+6s+13)^2}\right\}$. $\left[\textbf{Ans. } \frac{e^{-3t}}{8}[2t\sin 2t+2t\cos 2t-\sin 2t]\right]$

11. Evaluate $L^{-1}\left\{\frac{s}{(s^2+1)^2}\right\}$. $\left[\textbf{Ans. } \frac{t}{2}\sin t\right]$

12. Evaluate $L^{-1}\left\{\frac{s^2+4}{s^2+9}\right\}$. $\left[\textbf{Ans. } -\frac{5}{3}\sin 3t+1\right]$

13. Evaluate $L^{-1}\left\{\frac{s}{4s^2-25}\right\}$. $\left[\textbf{Ans. } \frac{1}{4}\cos h\frac{5}{2}t\right]$

Find the inverse Laplace transform of the following:

14. $\frac{s^2}{(s^2+a^2)^2}$. $\left[\textbf{Ans. } \frac{1}{2a}\{\sin at+at\cos at\}\right]$

15. $\frac{1}{s}\left(\frac{s-a}{s+a}\right)$ $\left[\textbf{Ans. } 2e^{-at}--1\right]$

16. $\frac{1}{s^2(s^2+a^2)}$. $\left[\textbf{Ans. } \frac{1}{a^2}\left(t-\frac{\sin at}{a}\right)\right]$

17. $\frac{1}{s(s^2+a^2)}$. $\left[\textbf{Ans. } \frac{1-\cos at}{a^2}\right]$

18. $\frac{s}{(s-2)^5(s+1)}$. $\left[\textbf{Ans. } e^{2t}\left(\frac{t^4}{36}+\frac{t^3}{54}-\frac{t^2}{54}+\frac{t}{81}-\frac{1}{243}\right)+\frac{e^{-t}}{243}\right]$

19. $\dfrac{s^2+2}{s(s^2+4)}$. $\left[\textbf{Ans.}\cos^2 t\right]$

20. $\dfrac{1}{s(s^2+a^2)}$. $\left[\textbf{Ans.}\dfrac{1-\cos at}{a^2}\right]$

Apply convolution theorem to find the inverse Laplace transform of the following:

21. $\dfrac{1}{(s^2+1)(s+1)}$. $\left[\textbf{Ans.}\dfrac{1}{2}(\sin t-\cos t+e^{-t})\right]$

22. $\dfrac{s^2}{(s^2+a^2)^2}$. $\left[\textbf{Ans.}\dfrac{1}{2a}(at\cos at+\sin at)\right]$

23. $\dfrac{1}{\sqrt{s}\,(s+1)}$. $\left[\textbf{Ans.}\, erf(\sqrt{t})\right]$

24. $\dfrac{s}{(s^2+a^2)^2}$. $\left[\textbf{Ans.}\dfrac{t\sin at}{2a}\right]$

25. $\dfrac{s+1}{(s^2+2s+2)^2}$. $\left[\textbf{Ans.}\dfrac{1}{2}te^{-t}\sin t\right]$

26. $\dfrac{1}{s^3(s^2+1)}$. $\left[\textbf{Ans.}\dfrac{1}{2}t^2-1+\cos t\right]$

27. $\dfrac{1}{(s+a)(s+b)}$. $\left[\textbf{Ans.}\dfrac{e^{-at}-e^{-bt}}{b-a}\right]$

28. $\dfrac{1}{s(s+1)^3}$. $\left[\textbf{Ans.}\,1-e^{-t}\left(1+t+\dfrac{1}{2}t^2\right)\right]$

Using Heaviside's expansion formula find the inverse Laplace transform of the following:

29. $\dfrac{193+37}{(s+1)(s-2)(s+3)}$. $\left[\textbf{Ans.}\,5e^{2t}-3e^{-t}-2e^{-3t}\right]$

30. $\dfrac{s-1}{(s+3)(s^2+2s+2)}$. $\left[\textbf{Ans.}\dfrac{1}{5}e^{-t}(4\cos t-3\sin t)-\dfrac{4}{5}e^{-3t}\right]$

31. $\dfrac{1}{s^3+1}$. $\left[\textbf{Ans.}\dfrac{1}{3}\left[e^{-t}-e^{t/2}\left(\cos\dfrac{\sqrt{3}}{2}t-\sqrt{3}\sin\dfrac{\sqrt{3}}{2}t\right)\right]\right]$

32. $\dfrac{11s^2 - 2s + 5}{2s^3 - 3s^2 - 3s + 2}$. $\left[\textbf{Ans. } 2e^{-2t} + 5e^{2t} - \dfrac{3}{2}e^{t/2}\right]$

33. $\dfrac{s^2 - 6}{s^3 + 4s^2 + 3s}$. $\left[\textbf{Ans. } \dfrac{1}{2}e^{-3t} + \dfrac{5}{2}e^{-t} - 2\right]$

34. Apply the convolution theorem to show that $\int_0^t \sin u . \cos(t - u)\, du = \dfrac{t}{2}\sin t.$

35. Evaluate $\int_0^\infty \sin 2x^2 dx.$ $\left[\textbf{Ans. } \dfrac{\sqrt{\pi}}{4}\right]$

36. Evaluate $\int_0^\infty e^{-5x^2} dx.$ $\left[\textbf{Ans. } \dfrac{1}{2}\sqrt{\dfrac{\pi}{5}}\right]$

37. Solve the integral equation for $F(t)$.

$$F(t) = e^{-t} - 2\int_0^t F(x) \cdot \cos(t - x) \cdot dx. \qquad \left[\textbf{Ans. } e^{-t}(1-t)^2\right]$$

38. Solve the integral equation for $Y(t)$

$$Y'(t) + 5\int_0^t \cos 2(t - x) Y(x) dx = 10,\ Y(0) = 2. \qquad \left[\textbf{Ans. } \frac{1}{9}(8 + 40t + \frac{50}{3}\sin 3t + 10\cos 3t)\right]$$

39. Evaluate $L^{-1}\left\{\dfrac{1}{\sqrt{p^2 - 4p + 20}}\right\}$. $\left[\textbf{Ans. } e^{2t} \cdot J_o(4t)\right]$

40. Evaluate $L^{-1}\left\{\dfrac{2s^2 - 6s + 5}{s^3 - 6s^2 + 11s - 6}\right\}$. $\left[\textbf{Ans. } \dfrac{1}{2}e^t - e^{2t} + \dfrac{5}{2}e^{3t}\right]$ *(U.P.T.U. 2004)*

41. Evaluate $L^{-1}\left\{\dfrac{s}{s+5}\right\}$. $\left[\textbf{Ans. } -5e^{-5t} + 1\right]$

42. Evaluate $L^{-1}\left\{\dfrac{2s}{3s+6}\right\}$. $\left[\textbf{Ans. } \dfrac{2}{3}[-2e^{-2t} + 1]\right]$

2.35 SOLUTION OF DIFFERENTIAL EQUATIONS BY LAPLACE TRANSFORMS

Method: First of all take Laplace transform on both sides of given problem and then use the formula of Laplace transform of derivatives *i.e.,*

$$L\{y'(t)\} = sL\{y(t)\} - y(0) \qquad \text{Here } F(t) = y(t)$$

$$L\{y''(t)\} = s^2 L\{y(t)\} - sy(0) - y'(0)$$

$$L\{y'''(t)\} = s^3 L\{y(t)\} - s^2 y(0) - sy'(0) - y''(0)$$

and so on.

Example 81: Solve by Laplace transform method

$$y''(t) + 4y'(t) + 4y(t) = 6e^{-t}, y(0) = -2, y'(0) = 8.$$ *(U.P.T.U. 2007)*

Solution: We have

$$y''(t) + 4y'(t) + 4y(t) = 6e^{-t}$$

Taking Laplace transform on both sides, we get

$$L\{y''(t)\} + 4L\{y'(t)\} + 4L\{y(t)\} = 6L\{e^{-t}\}$$

$$\Rightarrow \quad s^2 L\{y(t)\} - s\,y(0) - y'(0) + 4sL\{y(t)\} - 4y(0) + 4L\{y(t)\} = \frac{6}{s+1}$$

$$\Rightarrow \quad (s^2 + 4s + 4)\bar{y} + 2s = \frac{6}{s+1} \qquad |L\{y(t)\} = \bar{y}$$

$$\Rightarrow \quad \bar{y} = \frac{6}{(s+1)(s+2)^2} - \frac{2s}{(s+2)^2}$$

$$\therefore \quad y = 6L^{-1}\left\{\frac{1}{(s+1)(s+2)^2}\right\} - 2L^{-1}\left\{\frac{s}{(s+2)^2}\right\} \qquad \ldots(1)$$

Now we find $L^{-1}\left\{\dfrac{1}{(s+1)(s+2)^2}\right\}$

$$L^{-1}\left\{\frac{1}{(s+1)(s+2)}\right\} = \int_0^t e^{-2x} \cdot e^{-(t-x)} dx = e^{-t}\int_0^t e^{-x} dx \qquad \text{As } L^{-1}\left\{\frac{1}{s+2}\right\} = e^{-2t} = F(t), \quad L^{-1}\left\{\frac{1}{s+1}\right\} = e^{-t} = G(t)$$

$$= e^{-t}[-e^{-x}]_0^t = e^{-t}(1 - e^{-t})$$

Next suppose $\quad F(t) = e^{-t}(1 - e^{-t})$ and $G(t) = L^{-1}\left\{\dfrac{1}{s+2}\right\} = e^{-2t}$

$$\Rightarrow \quad L^{-1}\left\{\frac{1}{(s+1)(s+2)^2}\right\} = \int_0^t (e^{-x} - e^{-2x})e^{-2(t-x)} dx$$

$$= e^{-2t}\int_0^t (e^x - 1)\,dx = e^{-2t}[e^x - x]_0^t$$

$$\Rightarrow \quad L^{-1}\left\{\frac{1}{(s+1)(s+2)^2}\right\} = e^{-2t}(e^t - t - 1) \qquad ...(2)$$

and $$L^{-1}\left\{\frac{s}{(s+2)^2}\right\} = L^{-1}\left\{\frac{1}{s+2}\right\} - 2L^{-1}\left\{\frac{1}{(s+2)^2}\right\} = e^{-2t} - 2e^{-2t}L^{-1}\left\{\frac{1}{s^2}\right\}$$

$$\Rightarrow \quad L^{-1}\left\{\frac{s}{(s+2)^2}\right\} = e^{-2t} - 2te^{-2t} = e^{-2t}(1-2t) \qquad ...(3)$$

Hence using (2) and (3) in (1), we get the required solution of given equation

$$y = 6e^{-2t}(e^t - t - 1) - 2e^{-2t}(1-2t) = 6e^{-t} - 6te^{-2t} - 6e^{-2t} - 2e^{-2t} + 4te^{-2t}$$

$$\Rightarrow \quad y = 6e^{-t} - 8e^{-2t} - 2te^{-2t}.$$ **Ans.**

Example 82: Using Laplace transformation, solve the following differential equation:

$$\frac{d^2x}{dt^2} + 9x = \cos 2t \quad \text{if} \quad x(0) = 1,\ x\left(\frac{\pi}{2}\right) = -1. \qquad (U.P.T.U.\ 2002)$$

Solution: The given equation can be written as

$$x'' + 9x = \cos 2t$$

Taking Laplace transform on both sides, we get

$$L\{x''\} + 9L\{x\} = L\{\cos 2t\}$$

$$\Rightarrow \quad s^2L\{x\} - s\,x(0) - x'(0) + 9L\{x\} = \frac{s}{s^2+4}$$

Let $x'(0) = a$, $L\{x\} = \bar{x}$ and using $x(0) = 1$

$$\Rightarrow \quad s^2\bar{x} - s - a + 9\bar{x} = \frac{s}{s^2+4}$$

$$\Rightarrow \quad (s^2+9)\bar{x} = s + \frac{s}{s^2+4} + a$$

$$\Rightarrow \quad \bar{x} = \frac{s}{s^2+9} + \frac{s}{(s^2+9)(s^2+4)} + \frac{a}{(s^2+9)}$$

$$\bar{x} = \frac{4}{5}\frac{s}{(s^2+9)} + \frac{1}{5}\frac{s}{(s^2+4)} + \frac{a}{(s^2+9)} \qquad \text{(By partial fraction)}$$

$$\Rightarrow \quad x = \frac{4}{5}L^{-1}\left\{\frac{s}{s^2+9}\right\} + \frac{1}{5}L^{-1}\left\{\frac{s}{s^2+4}\right\} + aL^{-1}\left\{\frac{1}{s^2+9}\right\} \qquad (\text{As } L(x) = \bar{x})$$

$$x = \frac{4}{5}\cos 3t + \frac{1}{5}\cos 2t + \frac{a}{3}\sin 3t \qquad ...(1)$$

On putting $x(\pi/2) = -1$ in (1), we get

$$-1 = \frac{4}{5}\cos\frac{3\pi}{2} + \frac{1}{5}\cos\pi + \frac{a}{3}\sin\frac{3\pi}{2}$$

$$\Rightarrow \quad -1 = -\frac{1}{5} - \frac{a}{3} \quad \Rightarrow \quad a = \frac{12}{5}$$

Putting the value of a in equation (1), we get the required solution of given equation

$$x = \frac{4}{5}\cos 3t + \frac{1}{5}\cos 2t + \frac{4}{5}\sin 3t$$

$$= \frac{1}{5}(4\cos 3t + \cos 2t + 4\sin 3t). \qquad \textbf{Ans.}$$

Example 83: Using Laplace transform, solve the following differential equation:

$$y'' + 2ty' - y = t, \; y(0) = 0 \text{ and } y'(0) = 1. \qquad (U.P.T.U.\ 2003)$$

Solution: We have

$$y'' + 2ty' - y = t$$

Taking Laplace transform on both sides, we get

$$L\{y''\} + 2L\{ty'\} - L\{y\} = L\{t\}$$

$$\Rightarrow \quad s^2\bar{y} - sy(0) - y'(0) + 2\left\{-\frac{d}{ds}L(y')\right\} - \bar{y} = \frac{1}{s^2} \qquad (\bar{y} = L\{y\})$$

$$\Rightarrow \quad s^2\bar{y} - 1 - 2\frac{d}{ds}[s\bar{y} - y(0)] - \bar{y} = \frac{1}{s^2} \qquad (\text{using } y(0) = 0,\ y'(0) = 1)$$

$$\Rightarrow \quad s^2\bar{y} - 1 - 2\left[s\frac{d\bar{y}}{ds} + \bar{y}\right] - \bar{y} = \frac{1}{s^2} \qquad (y(0) = 0)$$

$$\Rightarrow \quad \frac{d\bar{y}}{ds} + \left(\frac{3 - s^2}{2s}\right)\bar{y} = -\frac{1}{2s^3} - \frac{1}{2s} \qquad ...(1)$$

which is Linear differential equation of first order in $\bar{y}$

$$\therefore \quad \text{I.F.} = e^{\int \frac{3-s^2}{2s} \cdot ds} = e^{\frac{1}{2}\int\left(\frac{3}{s} - s\right)ds} = e^{\frac{1}{2}\left(3\log s - \frac{s^2}{2}\right)} = e^{-(s^2/4)} \cdot s^{3/2}$$

The solution of equation (1) is

$$\bar{y} \cdot e^{-s^2/4} \cdot s^{3/2} = -\frac{1}{2}\int\left(\frac{1}{s^3} + \frac{1}{s}\right)s^{3/2} \cdot e^{-s^2/4} ds$$

$$= -\frac{1}{2}\int\left(\sqrt{s}+\frac{1}{s^{3/2}}\right)e^{-s^2/4}ds$$

Let $\quad s^2 = 4t \quad \Rightarrow \quad ds = \dfrac{dt}{\sqrt{t}}$

$$\Rightarrow \quad \bar{y}e^{-s^2/4}\cdot s^{3/2} = -\frac{1}{2}\left(\sqrt{2}t^{1/4}+\frac{1}{2\sqrt{2}}t^{-3/4}\right)e^{-t}.\frac{dt}{\sqrt{t}}$$

$$= -\frac{1}{\sqrt{2}}\int\left(t^{-1/4}+\frac{1}{4}t^{-5/4}\right)e^{-t}dt$$

$$= -\frac{1}{\sqrt{2}}\left[-t^{-1/4}e^{-t}-\frac{1}{4}\int t^{-5/4}.e^{-t}dt+\frac{1}{4}\int t^{-5/4}e^{-t}dt\right]$$

$$= \frac{1}{\sqrt{2}}t^{-\frac{1}{4}}e^{-t} = \frac{1}{\sqrt{2}}\left(\frac{s^2}{4}\right)^{-1/4}\cdot e^{-s^2/4} = \frac{1}{\sqrt{s}}e^{-s^2/4}$$

$\Rightarrow \quad \bar{y} = \dfrac{1}{s^2}$ | The value of constant of integration must be vanish if $\bar{y}$ is a transform since $\bar{y}$ must vanish as $s \to \infty$ Theorem 1 on page 262.

$$\Rightarrow \quad y = L^{-1}\left\{\frac{1}{s^2}\right\} = t. \quad \textbf{Ans.}$$

Example 84: Solve $(D^3 - D^2 - D + 1)y = 8te^{-t}$

If $\quad y = D^2y = 0, \quad Dy = 1 \quad$ when $t = 0$.

Solution: The given equation may be written as

$$y''' - y'' - y' + y = 8te^{-t}$$

Taking Laplace transform on both sides, we get

$$L\{y'''\} - L\{y''\} - L\{y'\} + L\{y\} = 8L\{te^{-t}\}$$

or $\quad s^3L\{y\} - s^2y(0) - sy'(0) - y''(0) - [s^2L\{y\} - sy(0) - y'(0)] - [sL\{y\} - y(0)] + L\{y\} = -8\dfrac{d}{ds}\left[L\{e^{-t}\}\right]$

or $\quad s^3\bar{y} - s - s^2\bar{y} + 1 - s\bar{y} + \bar{y} = -8\dfrac{d}{ds}\left(\dfrac{1}{s+1}\right)$ | $y(0) = y''(0) = 0$, $y'(0) = 1$ and $L\{y\} = \bar{y}$

$$\Rightarrow \quad (s^3 - s^2 - s + 1)\bar{y} = s - 1 + \frac{8}{(s+1)^2}$$

$$\Rightarrow \quad (s-1)^2(s+1)\bar{y} = (s-1) + \frac{8}{(s+1)^2}$$

$$\bar{y} = \frac{1}{(s-1)(s+1)} + \frac{8}{(s-1)^2(s+1)^3}$$

$$\Rightarrow \quad y = L^{-1}\left\{\frac{1}{(s-1)(s+1)}\right\} + 8L^{-1}\left\{\frac{1}{(s-1)^2(s+1)^3}\right\}$$

$$= \frac{1}{2}L^{-1}\left\{\frac{1}{s-1} - \frac{1}{s+1}\right\} + 8\cdot\frac{1}{8}L^{-1}\left\{-\frac{3}{2(s-1)} + \frac{1}{(s-1)^2} + \frac{3}{2(s+1)} + \frac{2}{(s+1)^2} + \frac{2}{(s+1)^3}\right\}$$

$$= -L^{-1}\left\{\frac{1}{s-1}\right\} + L^{-1}\left\{\frac{1}{s+1}\right\} + L^{-1}\left\{\frac{1}{(s-1)^2}\right\} + 2L^{-1}\left\{\frac{1}{(s+1)^2}\right\} + 2L^{-1}\left\{\frac{1}{(s+1)^3}\right\}$$

$$= -e^t + e^{-t} + e^t L^{-1}\left\{\frac{1}{s^2}\right\} + 2e^{-t}L^{-1}\left\{\frac{1}{s^2}\right\} + 2e^{-t}L^{-1}\left\{\frac{1}{s^3}\right\}$$

$$= -e^t + e^{-t} + e^t\cdot t + 2e^{-t}\cdot t + 2e^{-t}\cdot t^2/\lfloor 2$$

$$\Rightarrow \quad y = -e^t + e^{-t} + e^t\cdot t + 2e^{-t}\cdot t + e^{-t}\cdot t^2$$

$$= (1+2t+t^2)e^{-t} - (1-t)e^t. \quad \textbf{Ans.}$$

Example 85: Solve $ty'' + (2t+3)y' + (t+3)y = ae^{-t}$ if $y(t)$ and its derivatives have transforms.

Solution: Taking Laplace transform on both sides, we get

$$L\{ty''\} + 2L\{ty'\} + 3L\{y'\} + L\{ty\} + 3L\{y\} = aL\{e^{-t}\}$$

or $$-\frac{d}{ds}L\{y''\} + (-2)\frac{d}{ds}L\{y'\} + 3L\{y'\} - \frac{d}{ds}L\{y\} + 3L\{y\} = \frac{a}{s+1}$$

or $$-\frac{d}{ds}\left[s^2L\{y\} - sy(0) - y'(0)\right] - 2\frac{d}{ds}\left[sL\{y\} - y(0)\right] + 3[sL\{y\} - y(0)] - \frac{d}{ds}L\{y\} + 3L\{y\} = \frac{a}{s+1}$$

or $$-\frac{d}{ds}[s^2\bar{y} - sA - B] - 2\frac{d}{ds}[s\bar{y} - A] + 3[s\bar{y} - A] - \frac{d\bar{y}}{ds} + 3\bar{y} = \frac{a}{s+1}$$

Let $L\{y\} = \bar{y}$, $y(0) = A$ and $y'(0) = B$

or $$-s^2\frac{d\bar{y}}{ds} - 2s\bar{y} + A - 2s\frac{d\bar{y}}{ds} - 2\bar{y} + 3s\bar{y} - 3A - \frac{d\bar{y}}{ds} + 3\bar{y} = \frac{a}{s+1}$$

or $$-(s^2+2s+1)\frac{d\bar{y}}{ds} + (s+1)\bar{y} = 2A + \frac{a}{s+1}$$

or
$$\frac{d\bar{y}}{ds}-\frac{\bar{y}}{(s+1)}=-\frac{2A}{(s+1)^2}-\frac{a}{(s+1)^3} \qquad \text{...(1)}$$

which is a linear differential equation of first order

$$\therefore \qquad \text{I.F.}=e^{-\int\frac{1}{s+1}\cdot ds}=e^{-\log(s+1)}=e^{\log\frac{1}{s+1}}=\frac{1}{s+1}$$

Therefore, solution of equation (1) is

$$\bar{y}\cdot\frac{1}{(s+1)}=-2A\int\frac{ds}{(s+1)^3}-a\int\frac{ds}{(s+1)^4}+C$$

$$\Rightarrow \qquad \bar{y}\cdot\frac{1}{(s+1)}=\frac{A}{(s+1)^2}+\frac{a}{3(s+1)^3}+C$$

$$\Rightarrow \qquad \bar{y}=L\{y\}=\frac{A}{(s+1)}+\frac{a}{3(s+1)^2}+C(s+1)$$

$$y=AL^{-1}\left\{\frac{1}{s+1}\right\}+\frac{a}{3}L^{-1}\left\{\frac{1}{(s+1)^2}\right\}+CL^{-1}\{s+1\}$$

$$=Ae^{-t}+\frac{a}{3}e^{-t}L^{-1}\left\{\frac{1}{s^2}\right\}+C\left[L^{-1}\{s\}+L^{-1}\{1\}\right]$$

$$=Ae^{-t}+\frac{a}{3}e^{-t}t+0$$

As $L^{-1}\{s^n f(s)\}=F^n(t), n=0,1,2,....$
$\Rightarrow L^{-1}\{s^n\}=0, n=0,1,2,..........$
or Theorem 1 on page 262, C must be zero

$$=\left(A+\frac{at}{3}\right)e^{-t}. \quad \textbf{Ans.}$$

Example 86: Find $L^{-1}\{s^2\}$.

Solution: Let $\qquad f(s)=1$

$$\Rightarrow \qquad L^{-1}\{f(s)\}=L^{-1}\{1\}=\delta(t)$$

As $L\{\delta(t-a)\}=e^{-as}$
$\Rightarrow L\{\delta(t)\}=1$

$$\Rightarrow \qquad F(t)=\delta(t)=\begin{cases}\infty & \text{for } t=0\\ 0 & \text{for } t\neq 0\end{cases}$$

$$\therefore \qquad L^{-1}\{s^2 f(s)\}=F''(t)=\frac{d^2}{dt^2}\delta(t)=0, t\neq 0$$

Hence $\qquad L^{-1}\{s^2\}=0$

In general $\qquad L^{-1}\{s^n\}=0, \quad n=0,1,2,........$

Example 87: Solve $\frac{d^2y}{dx^2}+2\frac{dy}{dx}+5y=e^{-x}\sin x$ where $y(0) = 0$, $y'(0) = 1$.

[*U.P.T.U. 2004, 2008 (Special) 2009*]

Solution: The given equation can be written as

$$y''(x)+2y'(x)+5y(x) = e^{-x}\sin x$$

Taking Laplace transform on both sides

$$L\{y''(x)\}+2L\{y'(x)\}+5L\{y(x)\} = L\{e^{-x}\sin x\}$$

$$\Rightarrow \quad s^2\bar{y}-sy(0)-y'(0)+2s\bar{y}-2y(0)+5\bar{y} = \frac{1}{(s+1)^2+1}$$

$$\Rightarrow \quad (s^2+2s+5)\bar{y} = 1+\frac{1}{(s^2+2s+2)} \qquad \left|\begin{array}{l}\text{As } y(0)=0\\ y'(0)=1\end{array}\right.$$

$$\bar{y} = \frac{1}{(s^2+2s+5)}+\frac{1}{(s^2+2s+5)(s^2+2s+2)}$$

$$L\{y\} = \frac{2}{3}\cdot\frac{1}{s^2+2s+5}+\frac{1}{3}\cdot\frac{1}{s^2+2s+2} \qquad \text{(By partial fraction)}$$

$$\Rightarrow \quad y= \frac{2}{3}L^{-1}\left\{\frac{1}{s^2+2s+5}\right\}+\frac{1}{3}L^{-1}\left\{\frac{1}{s^2+2s+2}\right\}$$

$$= \frac{2}{3}L^{-1}\left\{\frac{1}{(s+1)^2+2^2}\right\}+\frac{1}{3}L^{-1}\left\{\frac{1}{(s+1)^2+1^2}\right\}$$

$$= \frac{2}{3}e^{-x}L^{-1}\left\{\frac{1}{s^2+2^2}\right\}+\frac{1}{3}e^{-x}L^{-1}\left\{\frac{1}{s^2+1^2}\right\}$$

$$= \frac{2}{3}e^{-x}\cdot\frac{1}{2}\sin 2x+\frac{1}{3}e^{-x}\sin x$$

$$\Rightarrow \quad y= \frac{1}{3}e^{-x}(\sin x+\sin 2x). \quad \textbf{Ans.}$$

Example 88: A particle moves in a line so that its displacement x from a fixed point O at any time t, is given by $\frac{d^2x}{dt^2}+4\frac{dx}{dt}+5x = 80 \sin 5t$. Using Laplace transform find its displacement at any time t if initially particle is at rest at $x = 0$.

(*U.P.T.U. 2009*)

Solution: The given equation can be written as

$$x''(t)+4x'(t)+5x(t) = 80\sin 5t$$

Taking Laplace transform on both sides

$$L\{x''(t)\}+4L\{x'(t)\}+5L\{x(t)\}=80L\{\sin 5t\}$$

$$\Rightarrow \quad [s^2\bar{x}-sx(0)-x'(0)]+4[s\bar{x}-x(0)]+5\bar{x}=80\cdot\frac{5}{s^2+25}$$

$$\Rightarrow \quad (s^2+4s+5)\bar{x}-(s+4)x(0)=\frac{400}{s^2+25} \qquad \Big|\ \text{As } x'(0)=0$$

$$\Rightarrow \quad \bar{x}=\frac{400}{(s^2+4s+5)(s^2+25)}+\frac{(s+4)}{(s^2+4s+5)}x(0)$$

$$\Rightarrow \quad x=400\,L^{-1}\left\{\frac{1}{(s^2+4s+5)(s^2+25)}\right\}+L^{-1}\left\{\frac{s+4}{s^2+4s+5}\right\}x(0) \qquad \ldots(i)$$

Now $\dfrac{1}{(s^2+4s+5)(s^2+25)}=\dfrac{As+B}{(s^2+4s+5)}+\dfrac{Cs+D}{s^2+25}$

$$\Rightarrow \quad 1=(As+B)(s^2+25)+(Cs+D)(s^2+4s+5)$$

Equating the coefficients like powers of s from both sides,

we get $\quad A+C=0,\quad B+4C+D=0$

$\quad 25A+5C+4D=0,\quad 25B+5D=1$

On solving these equations, we get

$A=\dfrac{1}{200},\quad B=\dfrac{9}{200},\quad C=-\dfrac{1}{200}$ and $D=-\dfrac{1}{40}$

From (i)

$$x=\frac{400}{200}L^{-1}\left[\frac{s+9}{s^2+4s+5}-\frac{s+5}{s^2+25}\right]+L^{-1}\left\{\frac{s+2+2}{(s+2)^2+1}\right\}x(0)$$

$$x=2L^{-1}\left\{\frac{s+2}{(s+2)^2+1}+\frac{7}{(s+2)^2+1}-\frac{s}{s^2+5^2}-\frac{5}{s^2+5^2}\right\}$$

$$+L^{-1}\left\{\frac{s+2}{(s+2)^2+1}+\frac{2}{(s+2)^2+1}\right\}x(0)$$

$$=2\left\{e^{-2t}L^{-1}\left(\frac{s}{s^2+1}\right)+7e^{-2t}L^{-1}\left(\frac{1}{s^2+1}\right)-\cos 5t-\sin 5t\right\}$$

$$+\left\{e^{-2t}L^{-1}\left(\frac{s}{s^2+1}\right)+2e^{-2t}L^{-1}\left(\frac{1}{s^2+1}\right)\right\}x(0)$$

$$\Rightarrow \quad x=e^{-2t}(2\cos t+14\sin t)-2\cos 5t-2\sin 5t+e^{-2t}(\cos t+2\sin t)x(0).$$ **Ans.**

Example 89: Solve the following simultaneous differential equations by Laplace transform

$$\frac{dx}{dt}+4\frac{dy}{dt}-y=0$$

$$\frac{dx}{dt}+2y=e^{-t}, \quad \text{with condition } x(0)=y(0)=0 \qquad (U.P.T.U.\ 2008)$$

Solution: Taking the Laplace transformation of given equations

we get $\quad s\bar{x}-x(0)+4s\bar{y}-y(0)-\bar{y}=0$

$$\Rightarrow \quad s\bar{x}+(4s-1)\bar{y}=0 \qquad |\text{As } x(0)=y(0)=0 \qquad \ldots(1)$$

and $\quad s\bar{x}-x(0)+2\bar{y}=\dfrac{1}{s+1}$

$$\Rightarrow \quad s\bar{x}+2\bar{y}=\frac{1}{s+1} \qquad \ldots(2)$$

Solving (1) and (2) for $\bar{x}$ and $\bar{y}$, we get

$$\bar{x}=\frac{4s-1}{s(s+1)(4s-3)}=\frac{1}{3s}-\frac{5}{7(s+1)}+\frac{32}{21(4s-3)} \qquad |\text{By P.F.}$$

$$\bar{y}=-\frac{1}{(s+1)(4s-3)}=\frac{1}{7(s+1)}-\frac{4}{7(4s-3)}$$

$$\therefore \quad x=\frac{1}{3}L^{-1}\left\{\frac{1}{s}\right\}-\frac{5}{7}L^{-1}\left\{\frac{1}{s+1}\right\}+\frac{32}{21\times 4}L^{-1}\left\{\frac{1}{s-\frac{3}{4}}\right\}$$

or $\quad x=\dfrac{1}{3}-\dfrac{5}{7}e^{-t}+\dfrac{8}{21}e^{\frac{3}{4}t}.$ **Ans.**

and $\quad y=\dfrac{1}{7}L^{-1}\left\{\dfrac{1}{s+1}\right\}-\dfrac{4}{7\times 4}L^{-1}\left\{\dfrac{1}{s-\frac{3}{4}}\right\}$

or $\quad y=\dfrac{1}{7}e^{-t}-\dfrac{1}{7}e^{\frac{3}{4}t}.$ **Ans.**

Example 90: Solve the simultaneous equations [*U.P.T.U. (C.O.) 2004*]

$$(D^2-3)x-4y=0$$

$$x+(D^2+1)y=0 \quad \text{for} \quad t>0$$

Given that $$x=y=\frac{dy}{dt}=0 \quad \text{and} \quad \frac{dx}{dt}=2 \quad \text{at} \quad t=0.$$

Solution: Taking the Laplace transformation of given equations, we get

$$s^2\bar{x}-sx(0)-x'(0)-3\bar{x}-4\bar{y}=0$$

$$\Rightarrow \quad (s^2-3)\bar{x}-4\bar{y}=2 \quad \ldots(1) \left[\begin{array}{l}\text{As } x(0)=0\\ x'(0)=2\end{array}\right.$$

and $$\bar{x}+s^2\bar{y}-sy(0)-y'(0)+\bar{y}=0$$

$$\Rightarrow \quad \bar{x}+(s^2+1)\bar{y}=0 \quad \ldots(2)\ [\text{As } y(0)=y'(0)=0]$$

Solving (1) and (2) for $\bar{x}$ and $\bar{y}$, we get

$$\bar{x}=\frac{2(s^2+1)}{(s^2-1)^2}=\frac{1}{(s-1)^2}+\frac{1}{(s+1)^2} \quad \text{(By Partial Fraction)}$$

$$\bar{y}=-\frac{2}{(s^2-1)^2}=-\frac{1}{2}\left[\frac{1}{s+1}-\frac{1}{s-1}-\frac{1}{(s+1)^2}+\frac{1}{(s-1)^2}\right]$$

$$\Rightarrow \quad x=L^{-1}\left\{\frac{1}{(s-1)^2}+\frac{1}{(s+1)^2}\right\}=e^tL^{-1}\left\{\frac{1}{s^2}\right\}+e^{-t}L^{-1}\left\{\frac{1}{s^2}\right\}$$

$$=te^t+te^{-t}=2t\left(\frac{e^t+e^{-t}}{2}\right)=2t\cos ht$$

and $$y=-\frac{1}{2}L^{-1}\left\{\frac{1}{s+1}-\frac{1}{s-1}-\frac{1}{(s+1)^2}+\frac{1}{(s-1)^2}\right\}$$

$$=-\frac{1}{2}\left[e^{-t}-e^t-te^{-t}+e^tt\right]=(1-t)\left(\frac{e^t-e^{-t}}{2}\right)$$

$$y=(1-t)\sin ht$$

Hence $$x=2t\cos ht, \quad y=(1-t)\sin ht. \quad \textbf{Ans.}$$

Example 91: The coordinates (x, y) of a particle moving along a plane curve at any time t are given by

$$\frac{dy}{dt}+2x=\sin 2t, \quad \frac{dx}{dt}-2y=\cos 2t; \ (t>0)$$

It is given that at $t = 0$, $x = 1$ and $y = 0$. Show using transforms that the particle moves along the curve $4x^2 + 4xy + 5y^2 = 4$. *(U.P.T.U. 2003)*

Solution: Taking the Laplace transform of given equations, we get

$$L\{y'\} + 2L\{x\} = L\{\sin 2t\}$$

$$\Rightarrow \quad sL\{y\} - y(0) + 2L\{x\} = \frac{2}{s^2+4}$$

$$\Rightarrow \quad s\bar{y} + 2\bar{x} = \frac{2}{(s^2+4)} \quad \text{...(1)} \quad \left| \begin{array}{r} \text{As } y(0)=0 \\ \text{and let } L\{y\} = \bar{y} \\ L\{x\} = \bar{x} \end{array} \right.$$

and

$$L\{x'\} - 2L\{y\} = L\{\cos 2t\}$$

$$\Rightarrow \quad sL\{x\} - x(0) - 2L\{y\} = \frac{s}{s^2+4}$$

$$\Rightarrow \quad s\bar{x} - 2\bar{y} = 1 + \frac{s}{s^2+4} \quad \text{...(2)} \quad | \text{ As } x(0) = 1$$

Solving equation (1) and (2) for $\bar{x}$ and $\bar{y}$, we get

$$\bar{x} = \frac{1+s}{4+s^2} = \frac{1}{s^2+4} + \frac{s}{s^2+4}$$

and

$$\bar{y} = -\frac{2}{s^2+4}$$

$$\Rightarrow \quad x = L^{-1}\left\{\frac{1}{s^2+4}\right\} + L^{-1}\left\{\frac{s}{s^2+4}\right\}$$

$$x = \frac{1}{2}\sin 2t + \cos 2t$$

and

$$y = -2L^{-1}\left\{\frac{1}{s^2+4}\right\} = -\frac{2}{2}\sin 2t = -\sin 2t$$

Hence

$$x = \frac{1}{2}\sin 2t + \cos 2t$$

$$y = -\sin 2t. \quad \textbf{Ans.}$$

Next,

$$4x^2 + 4xy + 5y^2 = 4\left(\frac{1}{2}\sin 2t + \cos 2t\right)^2 + 4\left(\frac{1}{2}\sin 2t + \cos 2t\right)(-\sin 2t) + 5\sin^2 2t$$

$$= 4(\sin^2 2t + \cos^2 2t)$$

$$\Rightarrow \quad 4x^2 + 4xy + 5y^2 = 4. \quad \textbf{Proved.}$$

Example 92: Solve

$$(D-2)x-(D+1)y = 6e^{3t}$$

$$(2D-3)x+(D-3)y = 6e^{3t} \qquad (U.P.T.U.\ 2001)$$

Given $\quad x = 3,\ y = 0$ when $t = 0$.

Solution: Taking the Laplace transform of both sides of the two equations, we have

$$L\{x'\}-2L\{x\}-L\{y'\}-L\{y\} = 6L\{e^{3t}\}$$

$$\Rightarrow \quad s\bar{x}-x(0)-2\bar{x}-s\bar{y}+y(0)-\bar{y} = 6\cdot\frac{1}{s-3} \qquad \Big|\ L(x)=\bar{x},\ L(y)=\bar{y}$$

$$\Rightarrow \quad (s-2)\bar{x}-(s+1)\bar{y} = 3+\frac{6}{s-3} \qquad ...(1)\ \Big|\ \text{As } x(0)=3$$

and

$$2L\{x'\}-3L\{x\}+L\{y'\}-3L\{y\} = 6L\{e^{3t}\}$$

$$\Rightarrow \quad 2[s\bar{x}-x(0)]-3\bar{x}+s\bar{y}-y(0)-3\bar{y} = \frac{6}{s-3}$$

$$\Rightarrow \quad (2s-3)\bar{x}+(s-3)\bar{y} = 6+\frac{6}{s-3} \qquad ...(2)\ \Big|\ \text{As } y(0)=0$$

Solving equation (1) and (2) for $\bar{x}$ and $\bar{y}$, we have

$$\bar{x} = \frac{3s^2(6s-1)}{(s-3)(s-1)^2} = \frac{1}{s-1}+\frac{2}{(s-1)^2}+\frac{2}{s-3}$$

and

$$\bar{y} = \frac{-3s+5}{(s-3)(s-1)^2} = \frac{1}{s-1}-\frac{1}{(s-1)^2}-\frac{1}{(s-3)}$$

$$\therefore \quad x = L^{-1}\left\{\frac{1}{s-1}\right\}+2L^{-1}\left\{\frac{1}{(s-1)^2}\right\}+2L^{-1}\left\{\frac{1}{s-3}\right\}$$

$$= e^t+2te^t+2e^{3t}$$

and

$$y = L^{-1}\left\{\frac{1}{s-1}\right\}-L^{-1}\left\{\frac{1}{(s-1)^2}\right\}-L^{-1}\left\{\frac{1}{s-3}\right\}$$

$$= e^t-te^t-e^{3t}$$

Hence $\quad x = e^t+2te^t+2e^{3t},\ y = e^t-te^t-e^{3t}$. **Ans.**

Example 93: Solve the following equation by Laplace transform:

$$(D^2+n^2)x = a\sin(nt+\alpha);\ x = Dx = 0 \text{ at } t=0. \qquad [U.P.T.U.\ (C.O.)\ 2001]$$

Solution: The given equation can be written as

$$x''(t)+n^2x(t) = a\sin(nt+\alpha)$$

Taking Laplace transform on both sides, we get

$$L\{x''(t)\}+n^2L\{x(t)\} = aL\{\sin(nt+\alpha)\}$$

$$\Rightarrow \quad s^2L\{x(t)\}-sx(0)-x'(0)+n^2L\{x(t)\} = a\cos\alpha\cdot\frac{n}{s^2+n^2}+a\sin\alpha\frac{s}{s^2+n^2}$$

$$\Rightarrow \quad (s^2+n^2)\bar{x} = \frac{an\cos\alpha}{s^2+n^2}+\frac{as\sin\alpha}{s^2+n^2} \qquad \Big| \text{ As } x(0)=x'(0)=0$$

$$\Rightarrow \quad \bar{x} = \frac{an\cos\alpha}{(s^2+n^2)^2}+\frac{as\sin\alpha}{(s^2+n^2)^2}$$

$$\Rightarrow \quad x = a\cos\alpha\, L^{-1}\left\{\frac{n}{(s^2+n^2)^2}\right\}+a\sin\alpha\, L^{-1}\left\{\frac{s}{(s^2+n^2)^2}\right\} \qquad \ldots(1)$$

Now $\qquad L^{-1}\left\{\dfrac{1}{s^2+n^2}\right\} = \dfrac{1}{n}\sin nt$

Differentiate w.r.t. n, we get

$$\Rightarrow \quad L^{-1}\left\{\frac{d}{dn}\left(\frac{1}{s^2+n^2}\right)\right\} = \frac{d}{dn}\left\{\frac{\sin nt}{n}\right\}$$

$$\Rightarrow \quad L^{-1}\left\{\frac{-2n}{(s^2+n^2)^2}\right\} = \frac{nt\cos nt-\sin nt}{n^2}$$

$$\Rightarrow \quad L^{-1}\left\{\frac{n}{(s^2+n^2)^2}\right\} = \frac{1}{2n^2}(\sin nt-nt\cos nt)$$

and $\qquad L^{-1}\left\{\dfrac{d}{ds}\left(\dfrac{1}{s^2+n^2}\right)\right\} = -\dfrac{t}{n}\sin nt \qquad \Big| \text{ As } L^{-1}\left\{\dfrac{d}{ds}f(s)\right\}=-t\,F(t)$

$$\Rightarrow \quad L^{-1}\left\{\frac{-2s}{(s^2+n^2)^2}\right\} = -\frac{t}{n}\sin nt$$

$$\Rightarrow \quad L^{-1}\left\{\frac{s}{(s^2+n^2)^2}\right\} = \frac{t}{2n}\sin nt$$

Hence, from equation (1), we get

$$x = a\cos\alpha\cdot\frac{1}{2n^2}(\sin nt-nt\cos nt)+a\sin\alpha\cdot\frac{t}{2n}\sin nt$$

$$x = \frac{a}{2n^2}[\cos\alpha \sin nt - nt\cos(\alpha + nt)]. \quad \textbf{Ans.}$$

Example 94: Solve $ty'' + y' + 4ty = 0$, given that $y(0) = 3, y'(0) = 0$.

Solution: Taking the Laplace transform of given equation on both sides, we get

$$L\{ty''\} + L\{y'\} + 4L\{ty\} = 0$$

$$\Rightarrow \quad -\frac{d}{ds}L\{y''\} + sL\{y\} - y(0) + 4\left(-\frac{d}{ds}L\{y\}\right) = 0$$

$$\Rightarrow \quad -\frac{d}{ds}\left[s^2 L\{y\} - sy(0) - y'(0)\right] + sL\{y\} - y(0) - 4\frac{d}{ds}L\{y\}$$

Using $y(0) = 3, y'(0) = 0$ and let $L\{y\} = \bar{y}$, we get

$$-\frac{d}{ds}(s^2\bar{y} - 3s) + (s\bar{y} - 3) - 4\frac{d\bar{y}}{ds} = 0$$

$$\Rightarrow \quad -(s^2 + 4)\frac{d\bar{y}}{ds} - s\bar{y} = 0$$

$$\Rightarrow \quad \frac{d\bar{y}}{\bar{y}} + \frac{s}{(s^2 + 4)}ds = 0$$

On integrating, we get

$$\log\bar{y} + \frac{1}{2}\log(s^2 + 4) = \log C_1$$

$$\Rightarrow \quad \bar{y} = \frac{C_1}{\sqrt{s^2 + 4}}$$

$$\Rightarrow \quad y = C_1 L^{-1}\left\{\frac{1}{\sqrt{s^2 + 4}}\right\}$$

$$\Rightarrow \quad y = C_1 J_0(2t) \qquad \left| L^{-1}\left\{\frac{1}{\sqrt{s^2 + a^2}}\right\} = J_0(at)\right.$$

Since $\quad y(0) = 3$

$\therefore \quad 3 = C_1 J_0(0) \Rightarrow \quad C_1 = 3 \quad (\because J_0(0) = 1)$

Hence, the required solution of given equation is

$$y = 3J_0(2t). \quad \textbf{Ans.}$$

Example 95: Solve the simultaneous equation

$$\frac{dx}{dt} - y = e^t, \quad \frac{dy}{dt} + x = \sin t, \text{ given } x(0) = 1, \ y(0) = 0. \qquad (U.P.T.U.\ 2006)$$

Solution: Taking the Laplace transform of given equations, we get

$$L\{x'\} - L\{y\} = L\{e^t\} \quad \Rightarrow \quad sL\{x\} - x(0) - L\{y\} = \frac{1}{s-1}$$

$$\Rightarrow \quad s\bar{x} - 1 - \bar{y} = \frac{1}{s-1}$$

$$\Rightarrow \quad s\bar{x} - \bar{y} = \frac{s}{s-1} \qquad \ldots(1) \quad \Big| \begin{array}{l} \text{As } x(0) = 1 \\ \text{and } L\{x\} = \bar{x}, L\{y\} = \bar{y} \end{array}$$

and

$$L\{y'\} + L\{x\} = L\{\sin t\} \quad \Rightarrow \quad sL\{y\} - y(0) + L\{x\} = \frac{1}{s^2+1}$$

$$\Rightarrow \quad \bar{x} + s\bar{y} = \frac{1}{s^2+1} \qquad \ldots(2) \quad \Big| \text{ As } y(0) = 0$$

Solving (1) and (2) for $\bar{x}$ and $\bar{y}$, we get

$$\bar{x} = \frac{s^2}{(s-1)(s^2+1)} + \frac{1}{(s^2+1)^2} = \frac{1}{2}\left[\frac{1}{(s-1)} + \frac{s}{(s^2+1)} + \frac{1}{s^2+1}\right] + \frac{1}{(s^2+1)^2} \quad \text{(By P.F.)}$$

and

$$\bar{y} = \frac{s}{(s^2+1)^2} - \frac{s}{(s-1)(s^2+1)} = \frac{s}{(s^2+1)^2} - \frac{1}{2}\left[\frac{1}{s-1} - \frac{s}{s^2+1} + \frac{1}{s^2+1}\right] \quad \text{(By P.F.)}$$

Now

$$x = \frac{1}{2}\left[L^{-1}\left\{\frac{1}{s-1}\right\} + L^{-1}\left\{\frac{s}{s^2+1}\right\} + L^{-1}\left\{\frac{1}{s^2+1}\right\}\right] + L^{-1}\left\{\frac{1}{(s^2+1)^2}\right\}$$

$$\Rightarrow \quad x = \frac{1}{2}[e^t + \cos t + \sin t] + \frac{1}{2}(\sin t - t\cos t)$$

$$\Rightarrow \quad x = \frac{1}{2}[e^t + \cos t + 2\sin t - t\cos t]$$

and

$$y = L^{-1}\left\{\frac{s}{(s^2+1)^2}\right\} - \frac{1}{2}\left[L^{-1}\left\{\frac{1}{s-1}\right\} - L^{-1}\left\{\frac{s}{s^2+1}\right\} + L^{-1}\left\{\frac{1}{s^2+1}\right\}\right]$$

$$\Rightarrow \quad y = \frac{1}{2}t\sin t - \frac{1}{2}\left[e^t - \cos t + \sin t\right]$$

$$y = \frac{1}{2}\left[t\sin t - e^t + \cos t - \sin t\right]$$

Hence, the required solutions of given equations are

$$x = \frac{1}{2}\left[e^t + \cos t + 2\sin t - t\cos t\right]$$

$$y = \frac{1}{2}\left[t\sin t - e^t + \cos t - \sin t\right]. \quad \textbf{Ans.}$$

Example 96: Solve the following system of differential equation:

$$x''(t) - y''(t) + y'(t) - x(t) = e^t - 2$$

$$2x''(t) - y''(t) - 2x'(t) + y(t) = -t$$

Given $\quad x(0) = x'(0) = y(0) = y'(0) = 0$

Solution: Taking the Laplace transform of given equations and using the initial conditions, we obtain

$$s^2\bar{x} - s^2\bar{y} + s\bar{y} - \bar{x} = \frac{1}{s-1} - \frac{2}{s}$$

$$\Rightarrow \quad (s+1)\bar{x} - s\bar{y} = -\frac{s-2}{s(s-1)^2} \qquad ...(1)$$

and $\quad 2s^2\bar{x} - s^2\bar{y} - 2s\bar{x} + \bar{y} = -\frac{1}{s^2}$

$$\Rightarrow \quad 2s\bar{x} - (s+1)\bar{y} = -\frac{1}{s^2(s-1)} \qquad ...(2)$$

Solving (1) and (2) for $\bar{x}$ and $\bar{y}$, we get

$$\bar{x} = \frac{1}{s(s-1)^2} = \frac{1}{s} - \frac{1}{s-1} + \frac{1}{(s-1)^2} \qquad \text{(By P.F.)}$$

and $\quad \bar{y} = \frac{2s-1}{s^2(s-1)^2} = -\frac{1}{s^2} + \frac{1}{(s-1)^2}$

so $\quad x = L^{-1}\left\{\frac{1}{s}\right\} - L^{-1}\left\{\frac{1}{s-1}\right\} + L^{-1}\left\{\frac{1}{(s-1)^2}\right\} = 1 - e^t + te^t$

and $\quad y = -L^{-1}\left\{\frac{1}{s^2}\right\} + L^{-1}\left\{\frac{1}{(s-1)^2}\right\} = -t + te^t$

Hence, the required solutions of given equations are

$$x = 1 - e^t + te^t$$

$$y = -t + te^t. \quad \textbf{Ans.}$$

Example 97: Solve the problem

$$ty''(t) + 2y'(t) + ty(t) = 0, \quad \text{given} \quad y(0) = 1,\ y(\pi) = 0.$$

Solution: Taking the Laplace transform of given equation, we get

$$L\{ty''(t)\} + 2L\{y'(t)\} + L\{ty(t)\} = 0$$

$$\Rightarrow -\frac{d}{ds}[s^2L(y) - sy(0) - y'(0)] + 2[sL(y) - y(0)] - \frac{d}{ds}L\{y\} = 0$$

$$\Rightarrow \quad -\frac{d}{ds}[s^2\bar{y} - s - a] + 2[s\bar{y} - 1] - \frac{d\bar{y}}{ds} = 0 \qquad \left|\begin{array}{l}\text{As } y(0) = 1 \\ \text{and let } y'(0) = a\end{array}\right.$$

$$\Rightarrow \quad -s^2\frac{d\bar{y}}{ds} - 2s\bar{y} + 1 + 2s\bar{y} - 2 - \frac{d\bar{y}}{ds} = 0$$

$$\Rightarrow \quad -(s^2+1)\frac{d\bar{y}}{ds} - 1 = 0 \Rightarrow d\bar{y} = -\frac{ds}{s^2+1}$$

On integrating, we get

$$\bar{y} = -\tan^{-1}s + C \qquad \text{...(1)}$$

Now $\lim_{s\to\infty}\bar{y}(s) = 0$ (refer Theorem 1, page 174). Therefore from (1), we get

$$\lim_{s\to\infty}\bar{y}(s) = -\tan^{-1}\infty + C$$

$$\Rightarrow \quad 0 = -\frac{\pi}{2} + C \Rightarrow \boxed{C = \frac{\pi}{2}}$$

From (1) $$\bar{y} = \frac{\pi}{2} - \tan^{-1}s = \cot^{-1}s = \tan^{-1}\frac{1}{s}$$

$$\Rightarrow \quad y = L^{-1}\left\{\tan^{-1}\frac{1}{s}\right\} = \frac{\sin t}{t}. \quad \textbf{Ans.}$$

Example 98: In an electrical circuit with e.m.f. $E(t)$, resistance R and inductance L, the current i builds up at the rate given by $L\frac{di}{dt} + Ri = E(t)$. If the switch is connected at $t = 0$ and disconnected at $t = a$, find the current i at any instant. *(U.P.T.U. 2001)*

Solution: Taking the Laplace transform of given equation, we get

$$L(s\bar{i} - i(0)) + R\bar{i} = L\{E(t)\}$$

$$\Rightarrow \quad (Ls + R)\bar{i} = \int_0^\infty e^{-st}.E(t)\,dt \qquad |\text{ As } i = 0 \text{ at } t = 0$$

Now

$$E(t) = \begin{cases} E \text{ for } 0 < t < a \\ 0 \text{ for } t > a \end{cases}$$

$$\Rightarrow \qquad (Ls+R)\bar{i} = \int_0^a e^{-st}\cdot E\,dt = E\left[\frac{e^{-st}}{-s}\right]_0^a = \frac{E}{s}(1-e^{-as})$$

or

$$\bar{i} = \frac{E}{s(Ls+R)} - \frac{E e^{-as}}{s(Ls+R)}$$

$$\Rightarrow \qquad i = L^{-1}\left\{\frac{E}{s(Ls+R)}\right\} - L^{-1}\left\{\frac{E e^{-as}}{s(Ls+R)}\right\}$$

$$i = \frac{E}{R}\left[L^{-1}\left\{\frac{1}{s}\right\} - L^{-1}\left\{\frac{1}{s+\frac{R}{L}}\right\} - \frac{E}{R}[1-e^{-R(t-a)/L}U(t-a)\right]$$

$$= \frac{E}{R}(1-e^{-\frac{Rt}{L}}) - \frac{E}{R}[1-e^{-R(t-a)/L}]U(t-a). \quad \textbf{Ans.}$$

EXERCISE 2.6

Solve the following differential equations by Laplace transform:

1. $y''+9y = 6\cos 3t,\ y(0)=2, y'(0)=0.$ $\left[\textbf{Ans. } y = 2\cos 3t + t\sin 3t\right]$ *(U.P.T.U. 2006)*

2. $\dfrac{d^2x}{dt^2} + x = F(t)$, if $x = x' = 0$ for $t = 0$. $\left[\textbf{Ans. } x = \int_0^t \sin(t-x)\,f(x)\,dx\right]$

3. $\dfrac{d^2y}{dt^2} + y = t\cos 2t, t>0$ given that $y = y' = 0$ for $t = 0$. *(U.P.T.U. 2005)*

   $\left[\textbf{Ans. } y = -\frac{5}{9}\sin t + \frac{4}{9}\sin 2t - \frac{t}{3}\cos 2t\right]$

4. $(D^2 - D - 6)y = 2, t>0$, if $y=1, y'=0$, when $t=0$. $\left[\textbf{Ans. } y = -\frac{1}{3} + \frac{8}{15}e^{3t} + \frac{4}{3}e^{-2t}\right]$

5. $\dfrac{d^2y}{dt^2} + y = U(t-\pi) - U(t-2\pi), y(0) = y'(0) = 0.$

   $\left[\textbf{Ans. } y = (1+\cos t)U(t-\pi) - (1-\cos t)U(t-2\pi)\right]$

6. $\dfrac{d^2y}{dt^2} + y = \sin t\,\sin 2t,\ t>0.$ $\left[\textbf{Ans. } y = \frac{15}{16}\cos t + \frac{t}{4}\sin t + \frac{1}{16}\cos 3t\right]$

7. $\frac{d^3y}{dt^3} - 3\frac{d^2y}{dt^2} + 3\frac{dy}{dt} - y = t^2 e^{2t}$, where $y(0) = 1, y'(0) = 0, y''(0) = -2$.

$$\left[\textbf{Ans. } y = e^{2t}(t^2 - 6t + 12) - e^t(15t^2 + 7t + 11)\right]$$

8. $\frac{d^2y}{dt^2} + 2y = F(t), y(0) = 0; y'(0) = 0$ where $F(t) = \begin{cases} 0 & t \geq 1 \\ 1 & 0 \leq t < 1 \end{cases}$ $\left[\textbf{Ans. } y = \frac{1}{2} - \frac{1}{2}\cos\sqrt{2}t\right]$

9. $(D^3 - D)y = 2\cos t; y(0) = 3, \quad Dy(0) = 2, \quad D^2y(0) = 1.$ $\left[\textbf{Ans. } y = 2 + \cos ht + 3\sin ht - \sin t\right]$

10. $(D^2 + 2D + 1)y = t$, if $y(0) = -3$ and $y(1) = -1$. $\left[\textbf{Ans. } y = t - 2 - e^{-t} + te^{-t}\right]$

11. $(D^2 + 4D + 4)x = \sin \omega t, t > 0$ with x_0 and x_1 for the values of x and Dx when $t = 0$.

$$\left[\textbf{Ans.} - \frac{4\omega}{(4 + \omega^2)^2}\cos\omega t + \frac{(4 - \omega^2)}{(4 + \omega^2)^2}\sin\omega t\right]$$

12. $(D^2 - D - 2)y = 20\sin 2t, y = -1, Dy = 2$ when $t = 0$. $\left[\textbf{Ans. } y = 2e^{2t} - 4e^{-t} + \cos 2t - 3\sin 2t\right]$

13. $(D^2 - 3D + 2)y = 1 - e^{2t}, \quad y(0) = 1, y'(0) = 0.$ $\left[\textbf{Ans. } y = \frac{1}{2} + \frac{1}{2}e^{2t} - te^{2t}\right]$

14. $(D^3 + 1)y = 1, t > 0, \ y(0) = y'(0) = y''(0) = 0.$ $\left[\textbf{Ans. } 1 + \frac{1}{3}e^{-t} - \frac{2}{3}e^{t/2}\cos\left(\frac{\sqrt{3}}{2}t\right)\right]$

15. $(D^3 - D^2 - D + 1)y = 8te^{-t}$ if $y(0) = y''(0) = 0, y'(0) = 1.$ $\left[\textbf{Ans. } (1 + 2t + t^2)e^{-t} - (1 - t)e^t\right]$

16. $\frac{d^2y}{dt^2} - 2k\frac{dy}{dt} + k^2y = F(t).$ $\left[\textbf{Ans. } y = Ae^{kt} + Cte^{kt} + e^{kt}\int_0^t (t - x)e^{-kx}F(x)\,dx\right]$

17. $t\frac{d^2y}{dt^2} + (t - 1)\frac{dy}{dt} - y = 0$, if $y(0) = 5, y(\infty) = 0.$ $\left[\textbf{Ans. } y = 5e^{-t}\right]$

18. $t\,y''(t) + (2t + 3)y'(t) + (t + 3)y(t) = 3e^{-t}.$ $\left[\textbf{Ans. } y = (C + t)e^{-t}\right]$

19. $y''(t) + at\,y'(t) - 2ay(t) = 1, y(0) = y'(0) = 0, a > 0.$ $\left[\textbf{Ans. } y = \frac{1}{2}t^2\right]$

20. $Dx + Dy = t, D^2x - y = e^{-t}$, if $x(0) = 3, x'(0) = -2, y(0) = 0.$

$$\left[\textbf{Ans. } x = 2 + \frac{1}{2}t^2 + \frac{1}{2}e^{-t} + \frac{1}{2}\cos t - \frac{3}{2}\sin t;\right.$$

$$\left. y = 1 - \frac{1}{2}e^{-t} - \frac{1}{2}\cos t + \frac{3}{2}\sin t\right]$$

21. $(D^2+3)x+Dy=15e^{-t}, -4Dx+(D^2+3)y=15\sin 2t,$ if $x(0)=35,\ x'(0)=-48,\ y(0)=27.$

$\left[\textbf{Ans. } x=30\cos t-15\sin 3t+3e^{-t}+2\cos 2t;\right.$
$\left. y=30\cos 3t-60\sin t-3e^{-t}+\sin 2t\right]$

22. $\dfrac{dx}{dt}+y=\sin t,\ \dfrac{dy}{dt}+x=\cos t.$ *(U.P.T.U. 2004)*

given that $x = 2$ and $y = 0$ when $t = 0$. $\left[\textbf{Ans. } x=e^t+e^{-t}, y=e^{-t}-e^t+\sin t\right]$

23. The small oscillations of a certain system with two degrees of freedom are given by the equations $D^2x+3x-2y=0$ and $D^2x+D^2y-3x+5y=0,$ if $x(0)=0, y(0)=0, x'(0)=3,$ $y'(0)=2.$

$\left[\textbf{Ans. } x=-\dfrac{11}{4}\sin t+\dfrac{1}{12}\sin 3t, y=\dfrac{11}{4}\sin t-\dfrac{1}{4}\sin 3t\right]$

24. $D^2x+y=-5\cos 2t,\ D^2y+x=5\cos 2t,$ if $x(0)=x'(0)=y'(0)=1$ and $y(0)=-1.$

$\left[\textbf{Ans. } x=\sin t+\cos 2t, y=\sin t-\cos 2t\right]$

25. $D^2x+2x-Dy=2t+5,\ Dx-x+Dy+y=-(2t+1).$ if $x(0)=3, x'(0)=0, y(0)=-3.$

$\left[\textbf{Ans. } x=t+2+e^{-2t}+\sin t, y=1-t-3e^{-2t}-\cos t\right]$

26. $(D^3-3D^2+3D-1)y=t^2e^t, y(0)=1, y'(0)=0, y''(0)=-2.$ $\left[\textbf{Ans. } y=e^t-te^t-\dfrac{t^2e^t}{2}+\dfrac{t^5e^t}{60}\right]$

27. $y''+9y=9U(t-3), y(0)=y'(0)=0.$ $\left[\textbf{Ans. } y=[1-\cos 3(t-3)]U(t-3)\right]$

28. $(D^2+3D+2)y=0,\ y(0)=y_0,\ y'(0)=y_1.$ $\left[\textbf{Ans. } y=(2y_0+y_1)e^{-t}-(y_0+y_1)e^{-2t}\right]$

29. $y_1''=y_1+3y_2, y_2''=4y_1-4e^t,$ if $y_1(0)=2, y_1'(0)=3, y_2(0)=1, y_2'(0)=2.$

$\left[\textbf{Ans. } y_1=e^t+e^{2t}, y_2=e^{2t}\right]$

30. $ty''+(t-1)y'+y=0, y(0)=0.$ $\left[\textbf{Ans. } y=Ct^2e^{-t}\right]$

OBJECTIVE TYPE QUESTIONS

A. Pick the correct answer of the given choices below:

1. The Laplace transform of the function *(U.P.T.U. 2009)*

$$f(t)=\begin{cases}1, & 0\le t<2\\ -1, & 2\le t<4, f(t+4)=f(t)\end{cases}$$

is given as

(*i*) $\dfrac{1-e^{-2s}}{s(1+e^{-2s})}$ (*ii*) $\dfrac{1+e^{-2s}}{s(1+e^{-2s})}$ (*iii*) 0 (*iv*) $\dfrac{s+1}{s-1}$

2. The inverse Laplace transform of $\log\left(\dfrac{s+1}{s-1}\right)$ is given by (*U.P.T.U. 2009*)

(*i*) $\dfrac{2}{t}\cosh t$ (*ii*) $\dfrac{2}{t}\sinh t$ (*iii*) $2t\cos t$ (*iv*) $2t\sin t$

3. If $L\{J_0(\sqrt{t})\} = \dfrac{e^{-1/4s}}{s}$ then $L\{J_0 2(\sqrt{t})\}$ is equal to

(*i*) $\dfrac{1}{s}$ (*ii*) $\dfrac{1}{2s}$ (*iii*) $\dfrac{1}{s^2}e^{-1/s}$ (*iv*) $\dfrac{1}{s}e^{-1/s}$

4. $L^{-1}\left\{\dfrac{1}{s}\right\}$ is equal to

(*i*) s (*ii*) 2 (*iii*) t^2 (*iv*) 1

5. $L^{-1}\left\{\dfrac{1}{s^2-6s+10}\right\}$ is equal to

(*i*) $\sin t$ (*ii*) $e^t \sin t$ (*iii*) $e^{-3t}\sin t$ (*iv*) $e^{3t}\sin t$

6. $L^{-1}\left\{\dfrac{s^2}{s^2+1}\right\}$ is equal to

(*i*) $\sin t$ (*ii*) $\cos t$ (*iii*) $-\sin t$ (*iv*) None

7. $L\left\{\dfrac{1}{\sqrt{\pi t}}\right\}$ is equal to

(*i*) $\dfrac{1}{\sqrt{s}}$ (*ii*) $\dfrac{1}{s^{3/2}}$ (*iii*) $s^{-5/2}$ (*iv*) s

8. $L\{t\,u(t-2)\}$ is equal to

(*i*) $\dfrac{1}{s}$ (*ii*) $\left(\dfrac{1}{s^2}+\dfrac{2}{s}\right)e^{-2s}$ (*iii*) $\dfrac{1}{s-2}$ (*iv*) None of these

9. $L^{-1}\left\{\dfrac{1}{(s-4)^4}\right\}$ is equal to

(*i*) $\dfrac{t^3}{6}e^{4t}$ (*ii*) $-\dfrac{t^3}{\lfloor 3}e^{-4t}$ (*iii*) $\dfrac{t^2}{\lfloor 5}e^{t}$ (*iv*) $\dfrac{1}{t}$

10. The value of $\int_0^\infty t\,e^{-2t}\sin t\,dt$ is equal to

(*i*) $\frac{2}{25}$ (*ii*) $\frac{4}{25}$ (*iii*) $-\frac{4}{25}$ (*iv*) $\frac{3}{25}$

11. $L^{-1}\left\{\frac{\boxed{n}}{(s-a)^n}\right\}, s>a$ is equal to

(*i*) $e^{at}\cdot t^{n-1}$ (*ii*) $e^{-at}t^{n+1}$ (*iii*) $e^{at}\cdot t^{n-1}$ (*iv*) t^{n-1}

B. Fill in the blanks:

1. The Laplace transform of the function $\int_0^t\int_0^t\int_0^t \cos au\,du\,du\,dv$ is given as ________ .

[*U.P.T.U. (Summer) 2009*]

2. The Inverse Laplace Transform of the function $f(s)=\frac{2s+1}{s^2-4}$ is ________ .

[*U.P.T.U. (Summer) 2009*]

3. If $F(t)=\int_t^\infty \frac{e^{-u}}{u}du$ then $L\{F(t)\}=$ ________ .

4. $L^{-1}\left\{\frac{e^{-3s}}{s^3}\right\}=$ ________ .

5. $L^{-1}\left\{\frac{s^2}{s^2+1}\right\}=$ ________ .

6. $L\left\{\frac{\sin t}{t}\right\}=$ ________ .

7. $L\{5\sin 3t+2\}=$ ________ .

8. $\int_0^\infty e^{-3t}\cdot t\,dt=$ ________ .

9. $1 * 1 * 1 * 1 =$ ________ .

10. $L^{-1}\left\{\int_s^\infty f(s)ds\right\}=$ ________ .

11. The value of $\int_{-\infty}^{\infty} e^{-3t} \cdot \delta'(t-2)dt$ is ________ .

C. State True or False for the following statements:

1. De Laplace was a French Mathematician.

2. The function e^{t^2} is not of exponential order as $t \to \infty$.

3. If $F(t)$ is of class A function and if $L\{F(t)\} = f(s)$ then $\lim_{s\to\infty} f(s) \neq 0$.

4. Dirac's delta function or unit impulse function denoted by $\delta(t - a)$ is defined as the limit of $f_\varepsilon(t-a)$ as $\varepsilon \to 0$.

5. The conditions in existence theorem of Laplace transform are both necessary and sufficient.

6. The unit step function is also called Heaviside's unit function.

7. $L\{U(t-a)\} = \frac{1}{s} e^{as}$.

8. $L\{\delta(t-a)\} = e^{-as}$.

9. $L\{y''(t)\} = s^2 L\{y(t)\} - sy(0) + y(0)$.

10. $L\{1\} = \frac{1}{s^2}$.

D. Match the following:

1. (*i*) $L\{1\}$ — (*a*) $\frac{1}{s-a}$

(*ii*) $L\{e^{at}\}$ — (*b*) $\frac{\lfloor n}{s^{n+1}}$

(*iii*) $L\{t^n\}$ — (*c*) $\frac{1}{s}$

(*iv*) $L\{e^{-at}\}$ — (*d*) $\frac{1}{s+a}$

2. (*i*) $L^{-1}\left\{\frac{1}{s+2}\right\}$ — (*a*) $\frac{1}{a} e^{-as}$

(*ii*) $L\{U(t-a)\}$ — (*b*) e^{-2t}

(*iii*) $L\{tF(t)\}$ — (*c*) $\lim_{s\to 0} sL\{F(t)\}$

(*iv*) $\lim_{s\to\infty} F(t)$ — (*d*) $-\frac{d}{ds} f(s)$

ANSWERS TO OBJECTIVE TYPE QUESTIONS

A. Pick the correct answers:

1. (*i*) **2.** (*ii*) **3.** (*iii*) **4.** (*iv*) **5.** (*iv*)
6. (*iii*) **7.** (*i*) **8.** (*ii*) **9.** (*i*) **10.** (*ii*)
11. (*i*)

B. Fill in the blanks:

1. $\dfrac{-2s}{(s^2+a^2)^2}$ **2.** $2\cosh 2t + \dfrac{1}{2}\sinh 2t$

3. $\dfrac{1}{s}\log(1+s)$ **4.** $\dfrac{(t-1)^2}{2}u(t-3)$

5. $-\sin t$ **6.** $\cot^{-1} s$ **7.** $\dfrac{15}{s^2+9}+\dfrac{2}{s}$

8. $\dfrac{1}{9}$ **9.** $\dfrac{t^3}{\lfloor 3}$ **10.** $\dfrac{F(t)}{t}$

11. $3e^{-6}$

C. True/False:

1. T **2.** T **3.** F **4.** T **5.** F
6. T **7.** F **8.** T **9.** F **10.** F

D. Match the following:

1. $i \to c$, $ii \to a$, $iii \to b$, $iv \to d$
2. $i \to b$, $ii \to a$, $iii \to d$, $iv \to c$

UNIT III

Infinite Series

3.1 INTRODUCTION

The finite series generally do not involve any difficulty in respect of the validity of application of algebraic operations there to, as compared to the infinite series. The application of algebraic operations to infinite series requires the additional concept of the convergence of series. Violation of condition of convergence of infinite series may cause serious complications. For example, successive division gives

$$\frac{1}{x-1} = \frac{1}{x} + \frac{1}{x^2} + \frac{1}{x^3} + \ldots$$

Hence, if we put $x = 1/2$, we get

$$-2 = 2 + 2^2 + 2^3 + \ldots$$

as absurd result. (The condition violated is $|x| > 1$).

The theory behind the validity of such expansions is covered under the study of convergence.

3.2 SEQUENCES

Let N be the set of natural numbers and R is any set of real numbers then the mapping f from N to R *i.e.*, $f : N \to R$ is called sequence. Since the sequence is uniquely and completely determined by the values $f(n) = u_n \forall n \in R$. A sequence is usually denoted by $\langle u_n \rangle$, $\{u_1, u_2, \ldots\}$ or $(u_1, u_2, \ldots)$. The value u_n is called the nth term of the sequence $\langle u_n \rangle$. For example, the sequence $\langle 1, 2, 1, 1, 2, 1, 2, \ldots \rangle$ in R (set of real number) is the same as the same as the function $f : N \to R$ defined by

$$f(n) = \begin{cases} 1 & \text{if } n \text{ is odd} \\ 2 & \text{if } n \text{ is even} \end{cases}$$

Infinite Series

If $\langle u_n \rangle$ be a sequence of real numbers, then the sum of the infinite numbers of this sequence *i.e.*,

$u_1 + u_2 + u_3 + \ldots + u_n + \ldots \infty$ is called an infinite series. This infinite series is generally denoted by $\sum_{n=1}^{\infty} u_n$ or $\sum u_n$. Also the 'series' means an infinite series.

Positive term series

The series $\sum u_n$ is called a positive term series if each term of this series is positive.

Alternating series

The series $\sum u_n$ is called an alternating series if it has alternately positive and negative terms. *e.g.*, $u_1 - u_2 + u_3 - u_4 + \ldots$ is alternating series.

Convergent series

An infinite series $\sum u_n$ is said to be convergent if the sum of its first n terms *i.e.*, S_n tends to unique and finite limit as n tends to infinity. Thus

$$\boxed{\sum u_n \text{ is convergent if } \lim_{n\to\infty} S_n = \text{finite and unique}}$$

Divergent series

A series $\sum u_n$ is said to be divergent if the sum of its first n terms tends to plus or minus infinity as n tends to infinity *i.e.*,

$$\boxed{\sum u_n \text{ is divergent if } \lim_{n\to\infty} S_n = +\infty \text{ or } -\infty}$$

Oscillating series

An infinite series $\sum u_n$ is said to be oscillating series if $\lim_{n\to\infty} S_n$ is not unique whether finite or infinite. The series which oscillates between finite limits is called finitely oscillatory series while if one or both limits are infinite numbers the series is called infinitely oscillatory series.

3.3 SOME USEFUL RESULTS

1. $\lim_{n\to\infty}\left(1+\frac{r}{n}\right)^n = e^r$
2. $\lim_{n\to\infty}\left(1+\frac{1}{n}\right)^n = e$
3. $\lim_{n\to\infty}\frac{\log n}{n} = 0$
4. $\lim_{n\to\infty}\frac{x^n}{n!} = 0, \lim_{n\to\infty} n^{1/n} = 1, \lim_{n\to\infty}\left(\frac{1}{n}\right)^{1/n} = 1$

Convergence of Geometric series

If $a + ar + ar^2 + \ldots + ar^{n-1} + \ldots \infty$

then

(*i*) If $r > 1$

In this case $S_n = \dfrac{a(r^n - 1)}{r - 1}$ $\left[\because \lim_{n\to\infty} S_n = \infty\right]$

Hence the series is divergent

(*ii*) If $r < 1$, in this case

$$S_n = \frac{a(1 - r^n)}{1 - r}$$

$$\therefore \quad \lim_{n\to\infty} S_n = \frac{a}{1 - r} \qquad \text{(finite and unique)}$$

Hence $\sum u_n$ is convergent

(*iii*) If $r = 1$, (divergent)

convergence of auxiliary series $\sum \dfrac{\mathbf{1}}{\boldsymbol{n^p}}$

(*i*) If $p > 1$, then A.S. is convergent

(*ii*) If $p \le 1$, then A.S. is divergent

3.4 COMPARISON TEST

If $\sum u_n$ and $\sum v_n$ are two series of positive terms such that $\lim_{n\to\infty}\left(\dfrac{u_n}{v_n}\right)$ is finite and non-zero then $\sum u_n$ and $\sum v_n$ are either both convergent or both divergent.

Proof: Let

$$\sum u_n = u_1 + u_2 + u_3 + u_4 + \ldots + u_n + \ldots$$

and

$$\sum v_n = v_1 + v_2 + v_3 + v_4 + \ldots + v_n + \ldots$$

As $\dfrac{u_1}{v_1}, \dfrac{u_2}{v_2}, \dfrac{u_3}{v_3}, \ldots$ are all finite, let $\dfrac{u_3}{v_3}$ has the greatest value u amongst the fractions $\dfrac{u_1}{v_1}, \dfrac{u_2}{v_2}, \dfrac{u_3}{v_3}$, etc.

Then $\dfrac{u_1}{v_1} < u$ or $u_1 < v_1 u$

$\dfrac{u_2}{v_2} < u$ or $u_2 < v_2 u$

$$\frac{u_3}{v_3} = u \quad \text{or} \quad u_3 = v_3 u$$

$$\frac{u_4}{v_4} < u \quad \text{or} \quad u_4 < v_4 u \text{ and so on.}$$

Adding these results, we have

$$u_1 + u_2 + u_3 + u_4 + \ldots < u\,(v_1 + v_2 + v_3 + \ldots)$$

or $$\frac{u_1 + u_2 + u_3 + u_4 + \ldots}{v_1 + v_2 + v_3 + v_4 \ldots} < u \quad \text{or} \quad \frac{\sum u_n}{\sum v_n} < u \qquad \ldots(i)$$

in a similar way if u_2 / v_2 has the least value u' amongst the fractions $\frac{u_1}{v_1}, \frac{u_2}{v_2}, \frac{u_3}{v_3}, \ldots$ etc.

then $$\frac{u_1}{v_1} > u' \quad \text{or} \quad u_1 > v_1 u'$$

$$\frac{u_2}{v_2} = u' \quad \text{or} \quad u_2 = v_2 u'$$

$$\frac{u_3}{v_3} > u' \quad \text{or} \quad u_3 > v_3 u' \text{ and so on.}$$

Adding these results, we have

$$\frac{u_1 + u_2 + u_3 + \ldots}{v_1 + v_2 + v_3 + \ldots} > u'$$

or $$\frac{\sum u_n}{\sum v_n} > u' \qquad \ldots(ii)$$

From equations (*i*) and (*ii*), we get

$$u' < \frac{\sum u_n}{\sum v_n} < u$$

i.e., the value of $\frac{\sum u_n}{\sum v_n}$ lies between the least value u' and the greatest value u of the functions $\frac{u_1}{v_1}, \frac{u_2}{v_2}, \frac{u_3}{v_3}, \ldots$ and each of these fractions is finite.

$$\therefore \quad \frac{\sum u_n}{\sum v_n} = \lambda \qquad \text{(a finite quantity)}$$

$$\Rightarrow \quad \sum u_n = \lambda \sum v_n \qquad \ldots(iii)$$

From equation (*iii*) it is clear that if the series $\sum v_n$ is convergent then the series $\sum u_n$ is also convergent.

Similarly, if the series $\sum v_n$ is divergent then its sum $\to +\infty$(or $-\infty$) and λ being finite from (*iii*), we find that, the sum of the series $\sum u_n$ also tends to $+\infty$ (or $-\infty$) or if the series $\sum v_n$ is divergent then the series $\sum u_n$ is also divergent.

Working Rule

(*i*) Write down the *n*th term u_n of the series $\sum u_n$

(*ii*) Find out the *n*th term v_n of auxiliary series $\sum v_n$

where
$$v_n = \frac{\text{Term containing highest power of } n \text{ in the numerator of } u_n}{\text{Term containing highest power of } n \text{ in the denominator of } u_n}$$

(*iii*) Find $\lim\limits_{n\to\infty} \dfrac{u_n}{v_n}$

(*iv*) If the above limit is finite and non-zero then the nature of the given series $\sum u_n$ will be the same as that of auxiliary series $\sum v_n$. [The series $\sum v_n$ *i.e.*, $\sum \dfrac{1}{n^p}$ will be convergent if $p > 1$ and divergent if $p \le 1$.]

WORKED OUT EXAMPLES

Example 1: Test the series $\dfrac{1}{5}+\dfrac{\sqrt{2}}{7}+\dfrac{\sqrt{3}}{9}+\dfrac{\sqrt{4}}{11}+\dots$

Solution: If the given series be $\sum u_n$, then $u_n = \sqrt{n}/(2n+3)$

then
$$v_n = \frac{\text{Term containing highest power of } n \text{ in the numerator of } u_n}{\text{Term containing highest power of } n \text{ in the denominator of } u_n}$$

$$\therefore \qquad v_n = \frac{\sqrt{n}}{n} = \frac{1}{\sqrt{n}} = \frac{1}{n^{1/2}}$$

Then
$$\frac{u_n}{v_n} = \frac{\sqrt{n}}{(2n+3)} \times \frac{\sqrt{n}}{1} = \frac{n}{2n+3} = \frac{1}{2+(3/n)},$$

$$\therefore \qquad \lim_{n\to\infty} \frac{u_n}{v_n} = \lim_{n\to\infty}\left[\frac{1}{2+(3/n)}\right] = \frac{1}{2}, \text{ which being finite}$$

the series $\sum u_n$ and $\sum v_n$ are either both convergent or both divergent.

But the auxiliary series $\sum v_n = \sum_n \left(\frac{1}{n^{1/2}}\right) = \sum \left(\frac{1}{n^p}\right) \Rightarrow p = \frac{1}{2} < 1.$

Therefore $\sum v_n$ is divergent when $p < 1$. Hence the given series is also divergent. **Ans.**

Example 2: Test the series $\sum_{n=1}^{\infty} \frac{(n+1)}{(n+2)^2}$

Solution: If the given series be $\sum u_n$, then

$$u_n = (n+1)/(n+2)^2$$

$$\therefore \quad v_n = \frac{n}{n^2} = \frac{1}{n}$$

Then
$$\frac{u_n}{v_n} = \frac{(n+1)}{(n+2)^2} \times \frac{n}{1} = \frac{n(n+1)}{(n+2)^2}$$

$$= \frac{1+(1/n)}{[1+(2/n)]^2}, \text{ dividing numerator, and denominator by } n^2$$

$\therefore \quad \lim_{n\to\infty} \frac{u_n}{v_n} = \lim_{n\to\infty} \left[\frac{1+(1/n)}{\{1+(2/n)\}^2}\right] = 1,$ which being finite the series $\sum u_n$ and $\sum v_n$ converge or diverge together.

But the auxiliary series $\sum v_n = \sum (1/n) = \sum \frac{1}{n^p} \Rightarrow p = 1$ is a divergent series.

Hence the given series is also divergent. **Ans.**

Example 3: Test the series $\frac{1}{a \cdot 1^2 + b} + \frac{2}{a \cdot 2^2 + b} + \frac{3}{a \cdot 3^2 + b} + \ldots$

Solution: Let the given series be denoted by $\sum u_n$. Then we have $u_n = n/(an^2 + b)$, which contains powers of n only.

$$\therefore \quad v_n = \frac{\text{Highest power of } n \text{ in the numerator of } u_n}{\text{Highest power of } n \text{ in the denominator of } u_n} = \frac{n}{n^2} = \frac{1}{n}$$

Now
$$\frac{u_n}{v_n} = \frac{n/(an^2+b)}{1/n} = \frac{n^2}{an^2+b} = \frac{1}{a+(b/n^2)}$$

$$\therefore \quad \lim_{n\to\infty} \frac{u_n}{v_n} = \lim_{n\to\infty} \frac{1}{a+(b/n^2)} = \frac{1}{a}, \text{ which is finite,} \qquad [\because a \neq 0]$$

But the auxiliary series $\sum v_n = \frac{1}{n} = \sum \frac{1}{n^p} \Rightarrow p = 1$

Therefore $\sum v_n$ is divergent

Hence the given series is also divergent. **Ans.**

Example 4: Test the series

$$\frac{1}{2} + \frac{\sqrt{2}}{3} + \frac{\sqrt{3}}{8} + \ldots + \frac{\sqrt{n}}{n^2 - 1} + \ldots$$

Solution: Here $u_n = \dfrac{\sqrt{n}}{n^2 - 1}$

$\therefore \quad v_n = \dfrac{\sqrt{n}}{n^2} = \dfrac{1}{n^{3/2}},$ $\quad$ As $v_n = \dfrac{\text{Highest power of } n \text{ in the numerator of } u_n}{\text{Highest power of } n \text{ in the denominator of } u_n}$

$$\therefore \quad \frac{u_n}{v_n} = \frac{\sqrt{n}}{n^2 - 1} \times \frac{n^{3/2}}{1} = \frac{n^2}{n^2 - 1} = \frac{1}{1 - (1/n^2)},$$

Then $\quad \lim\limits_{n \to \infty} \dfrac{u_n}{v_n} = \lim\limits_{n \to \infty} \dfrac{1}{1 - (1/n^2)} = 1,$ which being a finite quantity,

But the auxiliary series $\sum v_n = \sum \dfrac{1}{n^{3/2}} = \sum \dfrac{1}{n^p} \Rightarrow p = \dfrac{3}{2}$

i.e., $\quad p > 1$

Therefore $\sum v_n$ is convergent

Hence the given series is also convergent. **Ans.**

Example 5: Test the series $\dfrac{\sqrt{1}}{1 + \sqrt{1}} + \dfrac{\sqrt{2}}{2 + \sqrt{2}} + \dfrac{\sqrt{3}}{3 + \sqrt{3}} + \ldots$

Solution: Here the nth term of given series is

$$u_n = \frac{\sqrt{n}}{n + \sqrt{n}} = \frac{1}{\sqrt{n} + 1}$$

$$\therefore \quad v_n = \frac{1}{\sqrt{n}} = \frac{1}{n^{1/2}},$$

Now $\quad \dfrac{u_n}{v_n} = \dfrac{1/(\sqrt{n} + 1)}{1/\sqrt{n}} = \dfrac{\sqrt{n}}{\sqrt{n} + 1} = \dfrac{\sqrt{n}}{\sqrt{n}\left(1 + \dfrac{1}{\sqrt{n}}\right)} = \dfrac{1}{1 + \dfrac{1}{\sqrt{n}}}$

$\Rightarrow \quad \lim\limits_{n \to \infty} \dfrac{u_n}{v_n} = \lim\limits_{n \to \infty} \dfrac{1}{1 + (1/\sqrt{n})} = 1,$ which is a finite.

But the auxiliary series $\sum v_n = \sum \frac{1}{n^{1/2}} = \sum \frac{1}{n^p} \Rightarrow p = \frac{1}{2} < 1$

Therefore $\sum v_n$ is divergent when $p < 1$.

Hence the given series is also divergent. **Ans.**

Example 6: Test the series $1 + \frac{1+2}{1+2^2} + \frac{1+3}{1+3^2} + \ldots$

Solution: Here $u_n = \frac{1+n}{1+n^2}$.

$$\therefore \qquad v_n = \frac{n}{n^2} = \frac{1}{n}.$$

Then $\frac{u_n}{v_n} = \frac{(1+n)}{(1+n^2)} \times \frac{n}{1} = \frac{1+(1/n)}{1+(1/n^2)}$, dividing num. and denom, by n^2.

$$\therefore \qquad \lim_{n\to\infty} \frac{u_n}{v_n} = \lim_{n\to\infty} \frac{1+(1/n)}{1+(1/n^2)} = 1, \text{ which is finite,}$$

But $\sum v_n = \sum \frac{1}{n} = \sum \frac{1}{n^p} \Rightarrow p = 1$ $\left[\because \sum v_n \text{ is divergent}\right]$

Hence $\sum u_n$ is also divergent. **Ans.**

Example 7: Test the series $\frac{1}{1\cdot 2\cdot 3} + \frac{3}{2\cdot 3\cdot 4} + \frac{5}{3\cdot 4\cdot 5} + \ldots$

Solution: If the given series be denoted by $\sum u_{n,}$ then

$$u_n = \frac{(2n-1)}{n(n+1)(n+2)} \qquad [n\text{th term of } 1, 3, 5 = 1 + (n-1)2 = 2n-1]$$

Compare the given series $\sum u_n$ with the auxiliary series $\sum v_n$,

where $v_n = \frac{\text{Highest power of } n \text{ in numerator of } u_n}{\text{Highest power of } n \text{ in denominator of } u_n} = \frac{n}{n^3} = \frac{1}{n^2}$.

Now $\frac{u_n}{v_n} = \frac{(2n-1)}{n(n+1)(n+2)} \times \frac{n^2}{1} = \frac{(2n-1)n}{(n+1)(n+2)}$

$$\therefore \qquad \lim_{n\to\infty} \frac{u_n}{v_n} = \lim_{n\to\infty} \frac{[2-(1/n)]}{[1+(1/n)][1+(2/n)]}, \text{ dividing each term by } n$$

$$= 2, \text{ which being a finite quantity,}$$

But $$\sum v_n = \sum \frac{1}{n^2} = \sum \frac{1}{n^p} \Rightarrow p = 2 \text{ i.e., } p > 1$$

$\Rightarrow \sum v_n$ is convergent.

Hence the given series is convergent. **Ans.**

Example 8: Test for convergence the series $\frac{2^p}{1^q} + \frac{3^p}{2^q} + \frac{4^p}{3^q} + \ldots$ p and q being +ve numbers.

Solution: Let the given series be denoted by $\sum u_n$ where $u_n = \frac{(n+1)^p}{n^q}$.

$$\therefore \quad v_n = \frac{n^p}{n^q} = \frac{1}{n^{q-p}}.$$

Then $$\frac{u_n}{v_n} = \frac{(n+1)^p}{n^q} \times \frac{n^{q-p}}{1} = (n+1)^p \times n^{-p} = \left(\frac{n+1}{n}\right)^p = \left(1 + \frac{1}{n}\right)^p$$

$$\therefore \quad \lim_{n\to\infty} \frac{u_n}{v_n} = \lim_{n\to\infty} \left(1 + \frac{1}{n}\right)^p = 1, \text{ a finite quantity.}$$

Hence $\sum u_n$ and $\sum v_n$ are either both convergent or both divergent. But the auxiliary series $\sum v_n$ is $\sum \frac{1}{n^{q-p}}$, which is convergent if $q - p > 1$ and divergent if $q - p \leqslant 1$. Hence the given series $\sum u_n$ is convergent if $q - p > 1$ and divergent if $q - p \leqslant 1$ *i.e.*, **convergent** if $q > p + 1$ and **divergent** if $q \leqslant (p + 1)$. **Ans.**

Example 9: Test the series whose nth term is $\frac{\sqrt[3]{(2n^2 - 1)}}{\sqrt[4]{(3n^3 + 2n + 5)}}$

Solution: Let $u_n = \frac{\sqrt[3]{[(2n^2 - 1)]}}{\sqrt[4]{[(3n^3 + 2n + 5)]}} = \frac{(2n^2 - 1)^{1/3}}{(3n^3 + 2n + 5)^{1/4}}$.

$$\therefore \quad v_n = \frac{(n^2)^{1/3}}{(n^3)^{1/4}} = \frac{n^{2/3}}{n^{3/4}} = \frac{1}{n^{3/4-2/3}} = \frac{1}{n^{1/12}}.$$

But $$\sum v_n = \sum \frac{1}{n^{1/12}} = \sum \frac{1}{n^p}$$

$$\Rightarrow \quad p = \frac{1}{12} < 1$$

$\Rightarrow \quad \sum v_n$ is divergent.

Then $\quad \dfrac{u_n}{v_n} = \dfrac{(2n^2-1)^{1/3}}{(3n^3+2n+5)^{1/4}} \times \dfrac{n^{3/4}}{n^{2/3}}$

$$= \left(\frac{2n^2-1}{n^2}\right)^{1/3}\left(\frac{n^3}{3n^3+2n+5}\right)^{1/4}$$

$$= \left(2-\frac{1}{n^2}\right)^{1/3}\left\{\frac{1}{3+(2/n^2)+(5/n^3)}\right\}^{1/4}$$

dividing each term in the first bracket by n^2 and that in second by n^3

$\therefore \quad \lim\limits_{n\to\infty}\dfrac{u_n}{v_n} = \lim\limits_{n\to\infty}\left(2-\dfrac{1}{n^2}\right)^{1/3}\left(\dfrac{1}{3+2/n^2+5/n^3}\right)^{1/4} = \dfrac{2^{1/3}}{3^{1/4}},$

which is a finite quantity.

Hence the given series is divergent.

Example 10: Test the convergence of $\sum[\sqrt{(n+1)}-\sqrt{n}]$.

Solution: If $\sum u_n$ be the given series, then

$$u_n = \sqrt{(n+1)} - \sqrt{n}$$

$$= \sqrt{\left[n\left(1+\frac{1}{n}\right)\right]} - \sqrt{n} = \sqrt{n}\left[\sqrt{1+\frac{1}{n}}-1\right]$$

$$= \sqrt{n}\left[\left(1+\frac{1}{n}\right)^{1/2}-1\right]$$

$$= \sqrt{n}\left[\left\{1+\frac{1}{2}\cdot\frac{1}{n}+\frac{\frac{1}{2}\left(-\frac{1}{2}\right)}{2!}\left(\frac{1}{n}\right)^2+\ldots\right\}-1\right]$$

| As $(1+x)^n = 1+nx+\dfrac{n(n-1)}{2!}x^2+\ldots$

or $\quad u_n = n^{1/2}\left[\dfrac{1}{2}\cdot\dfrac{1}{n}-\dfrac{1}{8}\cdot\dfrac{1}{n^2}+\ldots\right] = \dfrac{1}{2}\cdot\dfrac{1}{n^{1/2}}-\dfrac{1}{8}\cdot\dfrac{1}{n^{3/2}}+\ldots$

Retaining only the highest power of n, we have the auxilary series

$$\sum v_n = \sum \frac{1}{n^{1/2}}$$

where $$v_n = \frac{1}{n^{1/2}}$$

Now $$\frac{u_n}{v_n} = \frac{\frac{1}{2}(1/n^{1/2}) - \frac{1}{8}(1/n^{3/2}) + \ldots}{(1/n^{1/2})} = \frac{1}{2} - \frac{1}{8}\left(\frac{1}{n}\right) + \ldots$$

$\therefore$ $$\lim_{n\to\infty} \frac{u_n}{v_n} = \lim_{n\to\infty}\left[\frac{1}{2} - \frac{1}{8}\left(\frac{1}{n}\right) + \ldots\right]$$

$$= \frac{1}{2}, \text{ a finite quantity.}$$

But $$\sum v_n = \sum \frac{1}{n^{1/2}} = \frac{1}{n^p} \Rightarrow p = \frac{1}{2} < 1 \quad \left[\because \sum v_n \text{ is divergent}\right]$$

Thus, the given series is also divergent.

Example 11: Test the series whose general term is $\sqrt{(n^4+1)} - \sqrt{(n^4-1)}$.

Solution: If $\sum u_n$ be the given series, then we have

$$u_n = \sqrt{(n^4+1)} - \sqrt{n^4-1)}$$

$$= \sqrt{\left[n^4\left(1+\frac{1}{n^4}\right)\right]} - \sqrt{\left[n^4\left(1-\frac{1}{n^4}\right)\right]}$$

$$= n^2\left(1+\frac{1}{n^4}\right)^{1/2} - n^2\left(1-\frac{1}{n^4}\right)^{1/2}$$

$$= n^2\left[\left\{1 + \frac{1}{2}\cdot\frac{1}{n^4} + \frac{\frac{1}{2}\left(-\frac{1}{2}\right)}{2!}\cdot\frac{1}{n^8} + \ldots\right\}\right.$$

$$\left. - \left\{1 - \frac{1}{2}\cdot\frac{1}{n^4} + \frac{\frac{1}{2}\left(-\frac{1}{2}\right)}{2!}\cdot\frac{1}{n^8} - \ldots\right\}\right]$$

$$= n^2\left[\left\{1+\frac{1}{2n^4}-\frac{1}{8n^8}+\ldots\right\}-\left\{1-\frac{1}{2n^4}-\frac{1}{8n^8}-\ldots\right\}\right]$$

$$= n^2\left[2\left\{\frac{1}{2n^4}-\frac{1}{16n^{12}}+\ldots\right\}\right]=\frac{1}{n^2}+\frac{1}{8n^{10}}+\ldots$$

Retaining only the highest power of n in u_n, we have the auxiliary series $\sum v_n = \sum(1/n^2)$, where $v_n = 1/n^2$

$$\therefore \qquad \lim_{n\to\infty}\frac{u_n}{v_n} = \lim_{n\to\infty}\frac{(1/n^2)+(1/8n^{10})+\ldots}{(1/n^2)} = \lim_{n\to\infty}\left[1+\frac{1}{8n^8}+\ldots\right]$$

$= 1$, which is a finite quantity.

But $$\sum v_n = \sum\frac{1}{n^2} = \sum\frac{1}{n^p} \Rightarrow p = 2 > 1. \qquad \left[\because \sum v_n \text{ is convergent}\right]$$

Thus the given series is convergent. **Ans.**

Example 12: Show that the series $1+\frac{1}{2^2}+\frac{2^2}{3^3}+\frac{3^3}{4^4}+\frac{4^4}{5^5}+\ldots$ is divergent.

Solution: Neglecting the first term, let the given series be denoted by $\sum u_n$ (neglecting of a finite number of terms does not affect the convergency or divergency of a series).

Then $$u_n = \frac{n^n}{(n+1)^{n+1}}$$

$$v_n = \frac{\text{Highest power of } n \text{ in the numerator of } u_n}{\text{Highest power of } n \text{ in the denominator of } u_n}$$

i.e., $$v_n = \frac{n^n}{n^{n+1}} = \frac{1}{n}$$

Then $$\lim_{n\to\infty}\frac{u_n}{v_n} = \lim_{n\to\infty}\frac{n^n}{(n+1)^{n+1}}\times\frac{n}{1},$$

$$= \lim_{n\to\infty}\frac{n^{n+1}}{(n+1)^{n+1}} = \lim_{n\to\infty}\left(\frac{n}{n+1}\right)^n\left(\frac{n}{n+1}\right)$$

$$= \lim_{n\to\infty}\frac{1}{\{1+(1/n)\}^n}\cdot\frac{1}{\{1+(1/n)\}} = \frac{1}{e}\cdot\frac{1}{1}. \quad \left|\ \text{As } \lim_{n\to\infty}\left(1+\frac{1}{x}\right)^x = e\right.$$

But $$\sum v_n = \sum \frac{1}{n} = \sum \frac{1}{n^p} \Rightarrow p = 1$$

i.e., $\sum v_n$ is divergent.

Hence the given series is also divergent. **Hence proved.**

Example 13: Test the series $\sum [\sqrt{n} + \sqrt{(n+1)}]^{-1}$.

Solution: Let the given series be denoted by $\sum u_n$.

Then $$u_n = \frac{1}{\sqrt{n} + \sqrt{(n+1)}} = \frac{\sqrt{n} - \sqrt{(n+1)}}{[\sqrt{n} + \sqrt{(n+1)}]\,[\sqrt{n} - \sqrt{n+1})]},$$

| Rationalising the denominator.

$$= \frac{\sqrt{n} - \sqrt{(n+1)}}{n-(n+1)} = \sqrt{(n+1)} - \sqrt{n} = \sqrt{n}\left[\left(1+\frac{1}{n}\right)^{1/2} - 1\right]$$

$$= \sqrt{n}\left[\left(1 + \frac{1}{2n} - \frac{1}{8n^2} + \ldots\right) - 1\right] = \frac{1}{2\sqrt{n}} - \frac{1}{8n\sqrt{n}} + \ldots$$

Retaining the highest power of n, the auxiliary series is

$$\sum v_n = \sum (1/\sqrt{n}) \quad \text{or} \quad \sum (1/n^{1/2})$$

$$\therefore \quad \frac{u_n}{v_n} = \frac{\frac{1}{2\sqrt{n}} - \frac{1}{8n\sqrt{n}} + \ldots}{1/\sqrt{n}} = \sqrt{n}\left(\frac{1}{2\sqrt{n}} - \frac{1}{8n\sqrt{n}} + \ldots\right) = \frac{1}{2} - \frac{1}{8n} + \ldots$$

$$\therefore \quad \lim_{n\to\infty} \frac{u_n}{v_n} = \lim_{n\to\infty}\left(\frac{1}{2} - \frac{1}{8n} + \ldots\right) = \frac{1}{2}, \text{ a finite quantity.}$$

Hence $\sum u_n$ and $\sum v_n$ are either both convergent or both divergent.

But $\sum v_n = \sum (1/n^{1/2})$ is a divergent series as $\sum (1/n^p)$ is divergent $\left(\text{As } p = \frac{1}{2} < 1\right)$

Hence the given series is divergent. **Ans.**

Example 14: Test the series $\frac{1}{x} + \frac{1}{x-1} + \frac{1}{x+1} + \frac{1}{x-2} + \frac{1}{x+2} + \ldots x$ being a positive function.

Solution: Neglecting the first term, denote the series by $\sum u_n$.

Then $$\sum u_n = \left(\frac{1}{x-1}+\frac{1}{x+1}\right)+\left(\frac{1}{x-2}+\frac{1}{x+2}\right)+\ldots$$

$$= \left(\frac{2x}{x^2-1^2}\right)+\left(\frac{2x}{x^2-2^2}\right)+\left(\frac{2x}{x^2-3^2}\right)+\ldots$$

Its general term $u_n = 2x/(x^2 - n^2)$

$\therefore$ $$v_n = \frac{1}{n^2}$$

Then $$\frac{u_n}{v_n} = \frac{2x}{x^2-n^2}\times\frac{n^2}{1}=\frac{2x}{[(x^2/n^2)-1]}$$

$\therefore$ $$\lim_{n\to\infty}\frac{u_n}{v_n} = \lim_{n\to\infty}\left\{\frac{2x}{(x^2/n^2)-1}\right\}=\frac{2x}{-1}=-2x,$$ (a finite quantity)

Here $$\sum v_n = \sum\frac{1}{n^2}=\sum\frac{1}{n^p} \Rightarrow p=2>1,$$ *i.e.*, the $\sum v_n$ is convergent.

Hence the given series is convergent. **Ans.**

Example 15: Test the series whose nth term is $\left(\frac{1}{n}-\log\frac{n+1}{n}\right)$.

Solution: If $\sum u_n$ be the given series, then we have

$$u_n = \frac{1}{n}-\log\left(\frac{n+1}{n}\right)=\frac{1}{n}-\log\left(1+\frac{1}{n}\right)$$

$$= \frac{1}{n}-\left\{\frac{1}{n}-\frac{1}{2n^2}+\frac{1}{3n^3}-\ldots\right\},$$ | As $\log(1+x) = x-\frac{1}{2}x^2+\frac{1}{3}x^3-\ldots$

$$= \frac{1}{2n^2}-\frac{1}{3n^3}+\ldots$$

$\therefore$ $$v_n = \frac{1}{n^2}$$

$\therefore$ $$\lim_{n\to\infty}\frac{u_n}{v_n} = \lim_{n\to\infty}\frac{(1/2n^2)-(1/3n^3)+\ldots}{(1/n^2)}$$

$$= \lim_{n\to\infty}\left[\frac{1}{2}-\frac{1}{3n}+\ldots\right]$$

$$= \frac{1}{2},$$ which is a finite quantity.

Here $\sum v_n = \sum \frac{1}{n^2} = \sum \frac{1}{n^p} \Rightarrow p = 2 > 1$, *i.e.*, the $\sum v_n$ is convergent.

Hence the given series is convergent. **Ans.**

Example 16: Test the series $\sum \sin(1/n)$.

Solution: Let the series be $\sum u_n$, where

$$u_n = \sin\left(\frac{1}{n}\right) = \frac{1}{n} - \frac{1}{3!}\cdot\frac{1}{n^3} + \frac{1}{5!}\cdot\frac{1}{n^5} - \ldots,$$

$$\left| \text{As } \sin x = x - (x^3/3!) + (x^5/5!) - \ldots \right.$$

Retaining only the highest power of n in u_n, we have the auxiliary series $\sum v_n = \sum (1/n)$, where $v_n = 1/n$.

Then
$$\lim_{n\to\infty} \frac{u_n}{v_n} = \lim_{n\to\infty} \frac{\frac{1}{n} - \frac{1}{3!}\cdot\frac{1}{n^3} + \frac{1}{5!}\cdot\frac{1}{n^5} - \ldots}{(1/n)}$$

$$= \lim_{n\to\infty} \left[1 - \frac{1}{6n^2} + \frac{1}{120n^4} - \ldots\right]$$

$= 1$, which is a finite quantity.

Hence $\sum u_n$ and $\sum v_n$ converge or diverge together. But the auxiliary series $\sum v_n = \sum (1/n)$ is divergent as $\sum (1/n^p)$ is divergent if $p = 1$.

Hence the given series $\sum \sin(1/n)$ is divergent. **Ans.**

Example 17: Test the series whose nth term is given by $u_n = \frac{1}{n}\sin\left(\frac{1}{n}\right)$.

Solution: $u_n = \frac{1}{n}\sin\left(\frac{1}{n}\right) = \frac{1}{n}\left[\left(\frac{1}{n}\right) - \frac{1}{3!}\left(\frac{1}{n}\right)^3 + \frac{1}{5!}\left(\frac{1}{n}\right)^5 - \ldots\right]$

$$= \frac{1}{n^2} - \frac{1}{6n^4} + \frac{1}{120n^6} - \ldots$$

Compare the series $\sum u_n$ with the auxiliary series $\sum v_n = \sum (1/n^2)$.

Then
$$\lim_{n\to\infty} \frac{u_n}{v_n} = \lim_{n\to\infty} \left[\frac{(1/n^2) - \frac{1}{6}(1/n^4) + \ldots}{(1/n^2)}\right]$$

$$= \lim_{n\to\infty}\left[1-\frac{1}{6n^2}+\ldots\right]=1, \text{ a finite quanity.}$$

Hence $\sum u_n$ and $\sum v_n$ are either both convergent or both divergent. But the auxiliary series $\sum v_n = \sum(1/n^2)$ is a convergent series as $\sum(1/n^p)$ is convergent if $p > 1$.

Hence the given series is convergent. **Ans.**

Example 18: Test the series whose nth term is cos $(1/n)$.

Solution: Let the given series be $\sum u_n$, then $u_n = \cos(1/n)$.

or $$u_n = 1-\frac{1}{n^2\cdot 2!}+\frac{1}{n^4\cdot 4!}-\ldots, \quad \left| \text{ As } \cos x = 1-\frac{x^2}{2!}+\frac{x^4}{4!}-\ldots \right.$$

or $$u_n = 1-\frac{1}{2n^2}+\frac{1}{24n^4}-\ldots=\frac{1}{n^0}-\frac{1}{2n^2}+\frac{1}{24n^4}-\ldots$$

Compare the given series $\sum u_n$ with the auxiliary series

$$\sum v_n = \sum(1/n^0)$$

Then $$\lim_{n\to\infty}\frac{u_n}{v_n} = \lim_{n\to\infty}\left\{\frac{\frac{1}{n^0}-\frac{1}{2n^2}+\frac{1}{24n^4}\ldots}{1/n^0}\right\}$$

$$= \lim_{n\to\infty}\left[1-\frac{1}{2n^2}+\frac{1}{24n^4}-\ldots\right]=1, \text{ a finite quantity.}$$

Hence $\sum u_n$ and $\sum v_n$ are either both convergent or both divergent. But the auxiliary series $\sum v_n = \sum(1/n^0)$ is divergent as $\sum(1/n^p)$ is divergent if $p < 1$ and here $p = 0$.

Hence the given series $\sum \cos(1/n)$ is divergent. **Ans.**

Example 19: Test the series whose nth term is $\tan^{-1}(1/n)$.

Solution: Let the given series be $\sum u_n$, where

$$u_n = \tan^{-1}\left(\frac{1}{n}\right)=\frac{1}{n}-\frac{1}{3}\left(\frac{1}{n}\right)^3+\frac{1}{5}\left(\frac{1}{n}\right)^5-\ldots,$$

$$\left| \because \tan^{-1}x = x-(x^3/3)+(x^5/5)-\ldots \right.$$

Retaining only the highest power of n in u_n, we have the auxiliary series $\sum v_n = \sum (1/n)$, where $v_n = 1/n$.

Then
$$\lim_{n\to\infty} \frac{u_n}{v_n} = \lim_{n\to\infty} \frac{(1/n)-(1/3n^2)+(1/5n^5)-\ldots}{(1/n)}$$
$$= \lim_{n\to\infty}\left[1-\frac{1}{3n^2}+\frac{1}{5n^4}-\ldots\right]$$
= 1, which is a finite quantity.

Hence $\sum u_n$ and $\sum v_n$ converge or diverge together. But the auxiliary series $\sum v_n = \sum (1/n)$ is a divergent series as the series $\sum (1/n^p)$ is divergent if $p = 1$.

Hence the given series $\sum \tan^{-1}(1/n)$ is divergent. **Ans.**

Example 20: Test the series whose general term is $\frac{1}{\sqrt{n}}\tan\left(\frac{1}{n}\right)$.

Solution: Let $\sum u_n$ be the given series, then $u_n = \frac{1}{\sqrt{n}}\tan\left(\frac{1}{n}\right)$

or
$$u_n = \frac{1}{\sqrt{n}}\left[\frac{1}{n}+\frac{1}{3n^3}+\ldots\right]\ldots \qquad \left|\ \text{As } \tan x = x+\frac{x^3}{3}+\ldots\right.$$
$$= \frac{1}{n^{3/2}}+\frac{1}{3n^{7/2}}+\ldots$$

Compare the series $\sum u_n$ with the auxiliary series
$$\sum v_n = \sum (1/n^{3/2}).$$

Then
$$\lim_{n\to\infty}\frac{u_n}{v_n} = \lim_{n\to\infty}\left\{\frac{\frac{1}{n^{3/2}}+\frac{1}{3n^{7/2}}+\ldots}{1/n^{3/2}}\right\} = 1, \text{ a finite quantity.}$$

Hence $\sum u_n$ and $\sum v_n$ converge or diverge together. But $\sum v_n = \sum 1/n^{3/2}$ is a convergent series as $\sum 1/n^p$ is convergent if $p > 1$ and here $p = 3/2$. Hence the given series is convergent. **Ans.**

EXERCISE 3.1

Test the following series for convergent and divergent:

1. $\frac{14}{1^3}+\frac{24}{2^3}+\frac{34}{3^3}+\ldots+\frac{10n+4}{n^3}+\ldots$ [**Ans.** Convergent]

2. $\frac{1}{(x-d)^2}+\frac{1}{(x-2d)^2}+\frac{1}{(x-3d)^2}+\ldots$ [**Ans.** Convergent]

Test the convergence divergence of series where *n*th term is:

3. $\sqrt{n^2+1}-n$ [**Ans.** Divergent]

4. $\sqrt{n^2+1}-\sqrt{n^2-1}$ [**Ans.** Divergent]

5. $\sqrt{\frac{1}{2^3}}-\sqrt{\frac{2}{3^3}}+\sqrt{\frac{3}{4^3}}+\ldots$ [**Ans.** Divergent]

6. $\frac{2}{1^p}+\frac{3}{2^p}+\frac{4}{3^p}+\ldots$ [**Ans.** Divergent]

7. $[(n^3+1)^{1/3}-n]$ [**Ans.** Convergent]

8. $\sqrt{n^3+1}-\sqrt{n^3}$ [**Ans.** Convergent]

9. Test the series $\frac{1}{1+\sqrt{2}}+\frac{2}{1+2\sqrt{3}}+\frac{3}{1+3\sqrt{4}}+\ldots$ [**Ans.** Divergent]

10. Test the series whose *n*th term is $\left[1+\frac{1}{n}\right]^{-1}$. [**Ans.** Divergent]

Test the convergence of the following series:

11. $\frac{4}{2}+\frac{7}{9}+\frac{10}{28}+\frac{13}{65}+\ldots+\frac{3n+1}{n^3+1}+\ldots$ [**Ans.** Convergent]

12. $1+\frac{1}{2^2}+\frac{2^2}{3^3}+\frac{3^3}{4^4}+\frac{4^4}{5^5}+\ldots+\frac{n^n}{(n+1)^{n+1}}+\ldots$ [**Ans.** Divergent]

13. $\frac{1}{1\cdot2\cdot3}+\frac{3}{2\cdot3\cdot4}+\frac{5}{3\cdot4\cdot5}+\ldots+\frac{2n-1}{n(n+1)(n+2)}+\ldots$ [**Ans.** Convergent]

14. Test the convergence of the series whose *n*th term is

$$\frac{2n(2n-1)}{(2n+1)^2(2n+2)^2}$$ [**Ans.** Convergent]

3.5 D'ALEMBERT'S RATIO TEST

If $\sum u_n$ is a series of positive terms such that

(*a*) $\lim_{n\to\infty} \frac{u_n}{u_{n+1}} = r$, then

(*i*) $\sum u_n$ is convergent if $r > 1$

(*ii*) $\sum u_n$ is divergent if $r < 1$

(*iii*) $\sum u_n$ may converge or diverge if $r = 1$ (Test fail)

(*b*) $\lim_{n\to\infty} \frac{u_n}{u_{n+1}} = \infty$, then $\sum u_n$ is convergent.

Proof: Let the series beginning from fixed term be $u_1 + u_2 + u_3 + \ldots + u_n + \ldots \infty$

and let $\frac{u_1}{u_2} > r$ or $\frac{u_2}{u_1} < \frac{1}{r} = k$ (say), $k < 1$

so that $\frac{u_2}{u_1} < k, \frac{u_3}{u_2} < k, \frac{u_4}{u_3} < k, \ldots$

Now
$$\sum u_n = u_1 + u_2 + u_3 + \ldots + \infty$$
$$= u_1\left(1 + \frac{u_2}{u_1} + \frac{u_3}{u_1} + \frac{u_4}{u_1} + \ldots \infty\right)$$
$$= u_1\left(1 + \frac{u_2}{u_1} + \frac{u_3}{u_2} \cdot \frac{u_2}{u_1} + \frac{u_4}{u_2} \cdot \frac{u_2}{u_1} + \ldots \infty\right)$$
$$< u_1\left(1 + k + k^2 + k^3 + \ldots \infty\right)$$
$$< \frac{u_1}{1-k} \text{ which is fixed number.}$$

| sum of an infinite in G.P. $= \frac{a}{1-r}$

Hence the given series $\sum u_n$ is convergent, in general $\lim_{n\to\infty} \frac{u_n}{u_{n+1}} > 1$, then $\sum u_n$ is convergent.

Next if $\frac{u_1}{u_2} \le 1, \frac{u_2}{u_3} \le 1, \frac{u_3}{u_4} \le 1, \ldots$

i.e., if $\frac{u_2}{u_1} \geq 1, \frac{u_3}{u_2} \geq 1, \frac{u_4}{u_3} \geq 1, \ldots$

i.e., if $u_2 \geq u_1, u_3 \geq u_2, u_4 \geq u_3, \ldots$

then sum of n terms *i.e.*, S_n is given by

$$S_n = u_1 + u_2 + \ldots + u_n$$

$$S_n > u_1 + u_1 + \ldots + u_1 \text{ (} n \text{ terms)}$$

$$S_n > u_1(1 + 1 + 1 + \ldots n \text{ terms)}$$

$$S_n > nu_1$$

$$\therefore \quad \lim_{n\to\infty} S_n = \infty$$

Hence the given series $\sum u_n$ is divergent.

In general if $\lim_{n\to\infty} \frac{u_n}{u_{n+1}} < 1$, then $\sum u_n$ is divergent.

If $\lim_{n\to\infty} \frac{u_n}{u_{n+1}} = 1$, the series may be either convergent or divergent and hence case fail.

WORKED OUT EXAMPLES

Example 1: Test the convergence of the series $\frac{x}{1\cdot 2} + \frac{x^2}{2\cdot 3} + \frac{x^3}{3\cdot 4} + \frac{x^4}{4\cdot 5} + \ldots, x > 0$

Solution: Here nth term of given series $u_n = \frac{x^n}{n(n+1)}$

$$\therefore \quad u_{n+1} = \frac{x^{n+1}}{(n+1)(n+2)}$$

and

$$\frac{u_n}{u_{n+1}} = \frac{x^n}{n(n+1)} \times \frac{(n+1)(n+2)}{x^n \cdot x} = \frac{(n+2)}{n} \cdot \frac{1}{x}$$

$$= \left(1 + \frac{2}{n}\right) \cdot \frac{1}{x}$$

$$\therefore \quad \lim_{n\to\infty} \frac{u_n}{u_{n+1}} = \frac{1}{x}$$

Hence if $\frac{1}{x} > 1$ *i.e.,* $x < 1$, the series is convergent and if $\frac{1}{x} < 1$ *i.e.,* $x > 1$, the series is divergent.

But if $x = 1$ the test fails. In this case

$$u_n = \frac{1}{n(n+1)}$$

$$\therefore \quad v_n = \frac{1}{n^2}$$

Now
$$\frac{u_n}{v_n} = \frac{n^2}{n(n+1)} = \frac{n^2}{n^2\left(1+\frac{1}{n}\right)} = \frac{1}{1+\frac{1}{n}}$$

$$\therefore \quad \lim_{n\to\infty} \frac{u_n}{v_n} = 1, \text{ which is finite and unique.}$$

Now
$$\sum v_n = \sum \frac{1}{n^2} = \sum \frac{1}{n^p} \Rightarrow p = 2 > 1 \left[\because \sum v_n \text{ is convergent}\right]$$

Therefore the given series is also convergent for $x = 1$

Thus the given series is convergent for $x \le 1$ and divergent for $x > 1$. **Ans.**

Example 2: Test the convergence of the series $2 + \frac{3x}{2} + \frac{4x^2}{3} + \frac{5x^3}{4} + \ldots$

Solution: Here $u_n = \left(\frac{n+1}{n}\right)x^{n-1}$ and $u_{n+1} = \left(\frac{n+2}{n+1}\right)x^n$.

Now
$$\frac{u_n}{u_{n+1}} = \left(\frac{n+1}{n}\right)x^{n-1} \times \frac{(n+1)}{(n+2)}\frac{1}{x^n} = \frac{(n+1)^2}{n(n+2)} \cdot \frac{1}{x}$$

$$= \frac{n^2\left(1+\frac{1}{n}\right)^2}{n^2\left(1+\frac{2}{n}\right)} \cdot \frac{1}{x} = \frac{\left(1+\frac{1}{n}\right)^2}{\left(1+\frac{2}{n}\right)} \cdot \frac{1}{x}$$

$$\therefore \quad \lim_{n\to\infty} \frac{u_n}{u_{n+1}} = \frac{1}{x}.$$

If $\frac{1}{x} > 1$ *i.e.,* $x < 1$, the series is convergent and if $\frac{1}{x} < 1$ *i.e.,* $x > 1$, the series is divergent.

However, if $x = 1$, the above test fails and in this case $u_n = \frac{n+1}{n}$ and we shall apply comparison test. If v_n is the nth term of the auxiliary series $\sum_n v_n$ then $v_n = \frac{1}{n^0}$.

Now $$\frac{u_n}{v_n} = \frac{n+1}{n} = 1 + \frac{1}{n}.$$

$\therefore$ $$\lim_{n\to\infty} \frac{u_n}{v_n} = 1, \text{ which is finite and non-zero.}$$

Hence the given series $\sum_n u_n$ is divergent because the auxiliary series $\sum_n \frac{1}{n^0}$ is divergent.

Thus the given series is convergent for $x < 1$ and divergent for $x \geq 1$. **Ans.**

Example 3: Test the convergence of the series $1 + \frac{2^p}{2!} + \frac{3^p}{3!} + \frac{4^p}{4!} + \ldots$

Solution: Here $u_n = \frac{n^p}{n!}$ and $u_{n+1} = \frac{(n+1)^p}{(n+1)!}$

Now $$\frac{u_n}{u_{n+1}} = \frac{n^p}{n!} \times \frac{(n+1)!}{(n+1)^p} = \frac{(n+1)n^p}{(n+1)^p} = (n+1)\left(\frac{n}{n+1}\right)^p$$

$$= (n+1)\left[\frac{1}{1+\frac{1}{n}}\right]^p.$$

$\therefore$ $$\lim_{n\to\infty} \frac{u_n}{u_{n+1}} = \infty$$

which is greater than 1.

Hence the given series is convergent. **Ans.**

Example 4: Test for convergence the series

$$2x + \frac{3x^2}{8} + \frac{4x^3}{27} + \ldots + \frac{(n+1)}{n^3}x^n + \ldots$$

Solution: If $\sum u_n$ be the given series, then we have

$$u_n = \frac{(n+1)x^n}{n^3} \text{ and } u_{n+1} = \frac{(n+1+1)x^{n+1}}{(n+1)^3} = \frac{(n+2)x^{n+1}}{(n+1)^3}$$

$\therefore$ $$\frac{u_n}{u_{n+1}} = \frac{(n+1)x^n}{n^3} \times \frac{(n+1)^3}{(n+2)x^{n+1}} = \frac{(n+1)^4}{n^3(n+2)} \cdot \frac{1}{x} = \frac{[1+(1/n)]^4}{[1+(2/n)]x}$$

$$\therefore \qquad \lim_{n\to\infty} \frac{u_n}{u_{n+1}} = \lim_{n\to\infty} \frac{[1+(1/n)]^4}{[1+(2/n)]x} = \frac{1}{x}$$

Hence if $\frac{1}{x} > 1$ *i.e.,* $x < 1$ the series is convergent and if $\frac{1}{x} < 1$ *i.e.,* $x > 1$, the series is divergent.

But if $x = 1$, this test fails.
In this case

$$u_n = \frac{n+1}{n^3} \qquad \text{(Put } x = 1\text{)}$$

$$\therefore \qquad v_n = \frac{n}{n^3} = \frac{1}{n^2}$$

Then

$$\lim_{n\to\infty} \frac{u_n}{v_n} = \lim_{n\to\infty}\left[\frac{n+1}{n^3} \times \frac{n^2}{1}\right] = \lim_{n\to\infty} \frac{n+1}{n}$$

$$= \lim_{n\to\infty}\left(1+\frac{1}{n}\right)$$

= 1, which is a finite quantity.

Since

$$\sum v_n = \sum \frac{1}{n^2} = \sum \frac{1}{n^p} \quad \Rightarrow \quad p = 2 > 1$$

$\therefore$ $\sum v_n$ is convergent

Therefore the given series is convergent at $x = 1$

Hence, the given series $\sum u_n$ is convergent if $x \le 1$ and divergent if $x > 1$. **Ans.**

Example 5: Test the convergence of the series whose *n*th term is $\frac{n!}{n^n}$.

Solution: We have $u_n = \frac{n!}{n^n}$

So

$$u_{n+1} = \frac{(n+1)!}{(n+1)^{n+1}}$$

$$\therefore \qquad \lim_{n\to\infty} \frac{u_n}{u_{n+1}} = \lim_{n\to\infty}\left[\frac{n!}{n^n} \times \frac{(n+1)(n+1)^n}{(n+1)n!}\right] = \lim_{n\to\infty}\left(\frac{n+1}{n}\right)^n$$

$$= \lim_{n\to\infty} \frac{n^n\left(1+\frac{1}{n}\right)^n}{n^n} = \lim_{n\to\infty}\left(1+\frac{1}{n}\right)^n = e$$

$$\Rightarrow \qquad \lim_{n\to\infty} \frac{u_n}{u_{n+1}} = e = 2.718 > 1$$

Hence from ratio test the given series $\sum u_n$ is convergent. **Ans.**

Example 6: Test the series $\dfrac{1}{2\sqrt{1}} + \dfrac{x^2}{3\sqrt{2}} + \dfrac{x^4}{4\sqrt{3}} + \dfrac{x^6}{5\sqrt{4}} + \ldots$

Solution: If $\sum u_n$ be the given series, then we have

$$u_n = \frac{x^{2n-2}}{(n+1)\sqrt{n}} \text{ and so } u_{n+1} = \frac{x^{2n}}{(n+2)\sqrt{(n+1)}}$$

$$\therefore \qquad \frac{u_n}{u_{n+1}} = \frac{x^{2n-2}}{(n+1)\sqrt{n}} \times \frac{(n+2)\sqrt{(n+1)}}{x^{2n}} = \left(\frac{n+2}{n+1}\right)\sqrt{\left(\frac{n+1}{n}\right)} \cdot \frac{1}{x^2}$$

$$\therefore \qquad \lim_{n\to\infty} \frac{u_n}{u_{n+1}} = \lim_{n\to\infty}\left[\frac{1+(2/n)}{1+(1/n)}\right] \cdot \sqrt{\left(1+\frac{1}{n}\right)} \cdot \frac{1}{x^2} = \frac{1}{x^2}$$

If $\dfrac{1}{x^2} > 1$ *i.e.*, $x^2 < 1$, the series is convergent and if $\dfrac{1}{x^2} < 1$ *i.e.*, $x^2 > 1$, the series is divergent.

If $x^2 = 1$, then test fails

$$\therefore \qquad u_n = \frac{1}{(n+1)\sqrt{n}} \quad \text{so} \quad v_n = \frac{1}{n\sqrt{n}} = \frac{1}{n^{3/2}}$$

Then $$\lim_{n\to\infty} \frac{u_n}{v_n} = \lim_{n\to\infty}\left[\frac{1}{(n+1)\sqrt{n}} \cdot \frac{n\sqrt{n}}{1}\right] = \lim_{n\to\infty}\left[\frac{1}{1+(1/n)}\right]$$

$= 1$, which is a finite quantity.

But $$\sum v_n = \sum \frac{1}{n^{3/2}} = \sum \frac{1}{n^p} \quad \Rightarrow \quad p = 3/2 > 1 \quad \left[\because \sum v_n \text{ is convergent}\right]$$

And hence the given series $\sum u_n$ will be convergent at $x^2 = 1$

Thus, we conclude that the given series is convergent if $x^2 \le 1$ and divergent $x^2 > 1$. **Ans.**

Example 7: Test the series $\dfrac{2}{1!} + \dfrac{3}{2!} + \dfrac{4}{3!} + \dfrac{5}{4!} + \ldots$

Solution: Let the given series be $\sum u_n$.

Then $$u_n = \frac{n+1}{n!}, \quad \therefore u_{n+1} = \frac{n+2}{(n+1)!}$$

$$\therefore \quad \frac{u_n}{u_{n+1}} = \frac{(n+1)}{n!} \times \frac{(n+1)!}{(n+2)} = \frac{(n+1)(n+1)}{(n+2)}$$

$$= (n+1)\left[\frac{1+(1/n)}{1+(2/n)}\right]$$

$$\therefore \quad \lim_{n\to\infty} \frac{u_n}{u_{n+1}} = \lim_{n\to\infty} (n+1)\left[\frac{1+(1/n)}{1+(2/n)}\right] = \infty$$

Hence from ratio test, the given series is convergent. **Ans.**

Example 8: Show that the exponential series $1 + x + \frac{x^2}{2!} + \frac{x^3}{3!} + \ldots$ is convergent for all values of x.

Solution: Here neglecting the first term $u_n = \frac{x^{n-1}}{(n-1)}$ and $u_{n+1} = \frac{x^n}{n!}$

$$\therefore \quad \lim_{n\to\infty} \frac{u_n}{u_{n+1}} = \lim_{n\to\infty} \left[\frac{x^{n-1}}{(n-1)!} \times \frac{n!}{x^n}\right]$$

$$= \lim_{n\to\infty} \left[\frac{n}{x}\right] = \infty, \text{ whatever be the finite value of } x.$$

$\therefore$ From ratio test, the given series $\sum u_n$ is convergent as $\infty > 1$. **Hence proved.**

Example 9: Test the series $1 + \frac{3}{2}x + \frac{5}{9}x^2 + \frac{7}{28}x^3 + \ldots + \frac{2n+1}{n^3+1}x^n + \ldots$

Solution: Neglecting the first term, if $\sum u_n$ be the given series, then

$$u_n = \frac{2n+1}{n^3+1}x^n \text{ and so } u_{n+1} = \frac{2n+3}{(n+1)^3+1}x^{n+1}$$

$$\therefore \quad \lim_{n\to\infty} \frac{u_n}{u_{n+1}} = \lim_{n\to\infty} \left[\frac{(2n+1)x^n}{(n^3+1)} \times \frac{\{(n+1)^3+1\}}{(2n+3)x^{n+1}}\right]$$

$$= \lim_{n\to\infty} \left[\left(\frac{2n+1}{2n+3}\right)\left\{\frac{(n+1)^3+1}{n^3+1}\right\} \cdot \frac{1}{x}\right]$$

$$= \lim_{n\to\infty} \left[\left\{\frac{2+(1/n)}{2+(3/n)}\right\}\left\{\frac{(1+1/n)^3+(1/n^3)}{1+(1/n^3)}\right\} \cdot \frac{1}{x}\right] = \frac{1}{x}$$

If $\frac{1}{x} > 1$ *i.e.*, $x < 1$ the series is convergent and if $\frac{1}{x} < 1$ *i.e.*, $x > 1$, the series is divergent.

If $x = 1$, then this test fails, in this case

$$u_n = \frac{2n+1}{n^3+1} \text{ so } v_n = \frac{1}{n^2}$$

$$\therefore \quad \lim_{n\to\infty} \frac{u_n}{v_n} = \lim_{n\to\infty}\left[\frac{2n+1}{n^3+1}\times\frac{n^2}{1}\right] = \lim_{n\to\infty}\left[\frac{2+\frac{1}{n}}{1+\frac{1}{n^3}}\right]$$

$$= 2, \text{ which is a finite quantity}$$

Since $\sum v_n = \sum \frac{1}{n^2} = \sum \frac{1}{n^p} \Rightarrow p = 2 > 1 \quad \left[\because \sum v_n \text{ is convergent}\right]$

Thus, the series $\sum u_n$ is convergent when $x = 2$.

Hence the given series is convergent or divergent according as $x \leq 1$ or $x > 1$. **Ans.**

Example 10: Show that the series $1 + \frac{x}{1!}(\log a) + \frac{x^2}{2!}(\log a)^2 + \ldots$ is convergent for all values of x.

Solution: If $\sum u_n$ be the given series, then we have

$$u_n = \frac{x^{n-1}}{(n-1)!}(\log a)^{n-1} \text{ and so } u_{n+1} = \frac{x^n}{n!}(\log a)^n$$

$$\therefore \quad \lim_{n\to\infty}\frac{u_n}{u_{n+1}} = \lim_{n\to\infty}\left[\frac{x^{n-1}(\log a)^{n-1}}{(n-1)!}\times\frac{n!}{x^n(\log a)^n}\right]$$

$$= \lim_{n\to\infty}\left[\frac{n}{x(\log a)}\right] = \infty > 1 \text{ for all } x.$$

Hence from ratio test the given series $\sum u_n$ is convergent. **Hence proved.**

Example 11: Test the series whose nth term is $(-1)^n x^n/n$.

Solution: If $\sum u_n$ be the given series, then we have

$$u_n = \frac{(-1)^n x^n}{n} \text{ and so } u_{n+1} = \frac{(-1)^{n+1} x^{n+1}}{(n+1)}$$

$$\therefore \quad \lim_{n\to\infty}\frac{u_n}{u_{n+1}} = \lim_{n\to\infty}\frac{(-1)^n x^n}{n}\times\frac{(n+1)}{(-1)^{n+1}x^{n+1}}$$

$$= \lim_{n\to\infty}\left[-\left(\frac{n+1}{n}\right)\cdot\frac{1}{x}\right] = \lim_{n\to\infty}\left[-\left(1+\frac{1}{n}\right)\cdot\frac{1}{x}\right]$$

$$= -\frac{1}{x} = \frac{1}{x}, \text{ numerically.}$$

If $\frac{1}{x} > 1$ *i.e.*, $x < 1$ the series is convergent and if $\frac{1}{x} < 1$ *i.e.*, $x > 1$ the series is divergent.

If $x = 1$, this test fails and we have $u_n = (-1)^n/n$

$$\therefore \quad \sum u_n = \sum \frac{(-1)^n}{n} = -\frac{1}{1} + \frac{(-1)^2}{2} + \frac{(-1)^3}{3} + \frac{(-1)^4}{4} + \ldots \text{ to } \infty$$

$$= -1 + \frac{1}{2} - \frac{1}{3} + \frac{1}{4} - \ldots \text{ to } \infty = -\left[1 - \frac{1}{2} + \frac{1}{3} - \frac{1}{4} + \ldots \text{ to } \infty\right]$$

$$= -\log 2, \text{ which is a finite quantity.}$$

Hence we find that the sum of the given series is finite if $x = 1$ therefore this series is convergent.

Also if $(1/x) = -1$ or $x = -1$, then $u_n = \frac{(-1)^n(-1)^n}{n} = \frac{1}{n}$ and so the series in this case is $\sum u_n = \sum (1/n)$, which is a divergent series as $\sum (1/n^p)$ is divergent if $p = 1$. Thus the given series is divergent when $x = -1$.

Hence, we find that the given series is convergent if $|x| < 1$ or $x = 1$ and divergent if $|x| > 1$ or $x = -1$. **Ans.**

EXERCISE 3.2

1. Test the series whose nth term is $\frac{(3n-1)}{2^n}$. **[Ans.** Convergent]

2. Test the series $x + \frac{3}{5}x^2 + \frac{8}{10}x^3 + \ldots + \frac{n^2-1}{n^2+1}x^n + \ldots$

[Ans. Convergent if $x < 1$ and divergent if $x \geqslant 1$]

3. Check the convergence of the series $2 + \frac{3x}{2} + \frac{4x^2}{3} + \frac{5x^3}{4} + \ldots$

[Ans. Convergent if $x < 1$, divergent if $x \geq 1$]

4. Examine the convergence or divergence of the series $\frac{x^2}{2\sqrt{1}} + \frac{x^3}{3\sqrt{2}} + \frac{x^4}{4\sqrt{3}} + \ldots (x > 0)$

[Ans. Convergent if $x \leqslant 1$ and divergent if $x > 1$]

5. $\frac{1}{2}+\frac{2!}{2^3}+\frac{3!}{2^5}+\ldots$ **[Ans.** Divergent]

6. Test the series whose nth term is $\frac{5^n}{n^2+7}$ **[Ans.** Divergent]

7. Test the convergence of the series $\sum\frac{1}{3^n+x}, x>0,$ **[Ans.** Divergent]

8. $\frac{1}{4}+\frac{1\cdot 3}{4\cdot 7}+\frac{1\cdot 3\cdot 5}{4\cdot 7\cdot 10}+\frac{1\cdot 3\cdot 5\ldots(2n-1)}{4\cdot 7\cdot 10\ldots(3n+1)}+\ldots$ **[Ans.** Convergent]

9. $\frac{2}{2}+\frac{4x}{5}+\frac{6x^2}{10}+\frac{8x^3}{17}+\ldots+\frac{2n}{n^2+1}x^{n-1}+\ldots$ **[Ans.** Convergent for $x<1$, divergent $\geqslant 1$]

10. $1+\frac{x}{2}+\frac{x^2}{5}+\frac{x^3}{10}+\ldots+\frac{x^n}{n^2+1}+\ldots$ **[Ans.** Convergent if $x\leqslant 1$ and divergent if $x>1$]

11. $\sum\frac{\sqrt{n}}{\sqrt{n^2+1}}x^n (x>0)$ **[Ans.** Convergent if $x<1$ and divergent if $x\geqslant 1$]

3.6 COMPARISON OF RATIOS

Statement: If $\sum u_n$ and $\sum v_n$ be two infinite series of positive terms and if from and after some fixed term.

(*i*) $\frac{u_{n+1}}{u_n}<\frac{v_{n+1}}{v_n}$ for all n then $\sum_n u_n$ will be convergent provided that $\sum_n v_n$ is convergent.

(*ii*) $\frac{u_{n+1}}{u_n}>\frac{v_{n+1}}{v_n}$ for all n then $\sum_n u_n$ will be divergent provided that $\sum_n v_n$ is divergent.

Proof: (*i*) Let the series beginning from some particular term be

$$u_1+u_2+u_3+\ldots+u_n+\ldots$$

and

$$v_1+v_2+v_3+\ldots+v_n+\ldots$$

We are given that $\frac{u_{n+1}}{u_n}<\frac{v_{n+1}}{v_n}$ for all n, *i.e.*,

$$\frac{u_2}{u_1}<\frac{v_2}{v_1},\frac{u_3}{u_2}<\frac{v_3}{v_2},\frac{u_4}{u_3}<\frac{v_4}{v_3}, \text{etc.}$$

Now $$\sum_n u_n = u_1 + u_2 + u_3 + u_4 + \ldots$$

$$= u_1\left[1 + \frac{u_2}{u_1} + \frac{u_3}{u_1} + \frac{u_4}{u_1} + \ldots\right]$$

$$= u_1\left[1 + \frac{u_2}{u_1} + \frac{u_3}{u_2}\cdot\frac{u_2}{u_1} + \frac{u_4}{u_3}\cdot\frac{u_3}{u_2}\cdot\frac{u_2}{u_1} + \ldots\right]$$

$$< u_1\left[1 + \frac{v_2}{v_1} + \frac{v_3}{v_2}\cdot\frac{v_2}{v_1} + \frac{v_4}{v_3}\cdot\frac{v_3}{v_2}\cdot\frac{v_2}{v_1} + \ldots\right]$$

$$< u_1\left[1 + \frac{v_2}{v_1} + \frac{v_3}{v_1} + \frac{v_4}{v_1} + \ldots\right]$$

$$< \frac{u_1}{v_1}[v_1 + v_2 + v_3 + v_4 + \ldots]$$

$$< \frac{u_1}{v_1}\sum_n v_n.$$

If $\frac{u_1}{v_1} = k$, a finite quantity, then $\sum_n u_n < k\sum_n v_n$.

Hence $\sum u_n$ is convegent provided that $\sum v_n$ is convergent.

(*ii*) As shown in case (*i*), it can be deduced that $\sum u_n > k\sum v_n$.

This shows that $\sum u_n$ is divergent provided that $\sum v_n$ is divergent.

3.7 RAABE'S TEST

Statement: An infinite series $\sum u_n$ of positive terms is convergent or divergent according as

$$\lim_{n\to\infty}\left[n\left(\frac{u_n}{u_{n+1}} - 1\right)\right] > 1 \text{ or } < 1.$$

Proof. Let us compare the given series $\sum u_n$ with the auxiliary series $\sum v_n$, where

$$\sum v_n = \sum\left(\frac{1}{n^p}\right) = \frac{1}{1^p} + \frac{1}{2^p} + \frac{1}{3^p} + \ldots\ldots + \frac{1}{n^p} + \ldots$$

It is known that $\sum v_n$ is convergent if $p > 1$.

Now the given series $\sum u_n$ will also be convergent if

$$\frac{u_n}{u_{n+1}} > \frac{v_n}{v_{n+1}}$$

or
$$\frac{u_n}{u_{n+1}} > \frac{1/n^p}{1/(n+1)^p}$$

or
$$\frac{u_n}{u_{n+1}} > \left(1+\frac{1}{n}\right)^p$$

or
$$\frac{u_n}{u_{n+1}} > 1+p\cdot\frac{1}{n}+\frac{p(p-1)}{2!}\cdot\frac{1}{n^2}+\ldots$$

or
$$\frac{u_n}{u_{n+1}}-1 > \frac{1}{n}\left[p+\frac{p(p-1)}{2!}\cdot\frac{1}{n^2}+\ldots\right]$$

or
$$n\left[\frac{u_n}{u_{n+1}}-1\right] > p+\frac{p(p-1)}{2!}\cdot\frac{1}{n}+\ldots$$

or
$$\lim_{n\to\infty}\left[n\left(\frac{u_n}{u_{n+1}}-1\right)\right] > p$$

But $p > 1$, hence $\sum u_n$ will be convergent if

$$\lim_{n\to\infty}\left[n\left(\frac{u_n}{u_{n+1}}-1\right)\right] > 1.$$

WORKED OUT EXAMPLES

Example 1: Test for the convergence the series

$$\frac{1^2}{4^2}+\frac{1^2\cdot 5^2}{4^2\cdot 8^2}+\frac{1^2\cdot 5^2\cdot 9^2}{4^2\cdot 8^2\cdot 12^2}+\frac{1^2\cdot 5^2\cdot 9^2\cdot 13^2}{4^2\cdot 8^2\cdot 12^2\cdot 16^2}+\ldots$$

Solution: Here the nth term of 1, 5, 9, 13, ... = 1 + $(n - 1)$ 4 = $4n - 3$
and nth term of 4, 8, 12, 16, ... = 4 + $(n - 1)$ 4 = $4n$

$$\therefore \qquad u_n = \frac{1^2\cdot 5^2\cdot 9^2\ldots(4n-3)^2}{4^2\cdot 8^2\cdot 12^2\ldots(4n)^2}$$

and
$$u_{n+1} = \frac{1^2\cdot 5^2\cdot 9^2\ldots(4n+1)^2}{4^2\cdot 8^2\cdot 12^2\ldots(4n+4)^2}$$

$$\therefore \quad \frac{u_n}{u_{n+1}} = \frac{1^2 \cdot 5^2 \cdot 9^2 \ldots (4n-3)^2}{4^2 \cdot 8^2 \cdot 12^2 \ldots (4n)^2} \times \frac{4^2 \cdot 8^2 \cdot 12^2 \ldots (4n)^2 \cdot (4n+4)^2}{1^2 \cdot 5^2 \cdot 9^2 \ldots (4n-3)^2 \cdot (4n+1)^2}$$

$$\frac{u_n}{u_{n+1}} = \frac{(4n+4)^2}{(4n+1)^2} = \frac{16n^2+32n+16}{16n^2+8n+1} \qquad \ldots(i)$$

or
$$\lim_{n\to\infty}\left(\frac{u_n}{u_{n+1}}\right) = \lim_{n\to\infty}\left(\frac{16n^2+32n+16}{16n^2+8n+1}\right) = \lim_{n\to\infty}\left(\frac{16+\frac{32}{n}+\frac{16}{n^2}}{16+\frac{8}{n}+\frac{1}{n^2}}\right) = 1$$

$\therefore$ Ratio test fails, we have from (*i*)

$$\frac{u_n}{u_{n+1}} - 1 = \frac{16n^2+32n+16}{16n^2+8n+1} - 1 = \frac{16n^2+32n+16-16n^2-8n-1}{16n^2+8n+1}$$

or
$$n\left(\frac{u_n}{u_{n+1}} - 1\right) = n\left(\frac{24n+15}{16n^2+8n+1}\right) = \frac{n^2}{n^2}\left(\frac{24+\frac{15}{n}}{16+\frac{8}{n}+\frac{1}{n^2}}\right)$$

$$\therefore \quad \lim_{n\to\infty} n\left(\frac{u_n}{u_{n+1}} - 1\right) = \lim_{n\to\infty}\left(\frac{24+\frac{15}{n}}{16+\frac{8}{n}+\frac{1}{n^2}}\right) = \frac{24}{16} = \frac{3}{2} > 1$$

Hence from Raabe's test the given series is convergent. **Ans.**

Example 2: Test the convergence of the series

$$\frac{x}{1} + \frac{1}{2}\cdot\frac{x^2}{3} + \frac{1\cdot 3}{2\cdot 4}\cdot\frac{x^3}{5} + \frac{1\cdot 3\cdot 5}{2\cdot 4\cdot 6}\cdot\frac{x^4}{7} + \ldots$$

Solution: Here $u_n = \dfrac{1\cdot 3\cdot 5\ldots(2n-3)}{2\cdot 4\cdot 6\ldots(2n-2)}\cdot\dfrac{x^n}{(2n-1)}$

As $(n-1)$th term of $1\cdot 3\cdot 5\ldots$ $= 1+(n-1-1)2 = 2n-3$

$(n-1)$th term of $2\cdot 4\cdot 6\ldots$ $= 2+(n-1-1)2 = 2n-2$

and
$$u_{n+1} = \frac{1\cdot 3\cdot 5\ldots(2n-1)}{2\cdot 4\cdot 6\ldots(2n)}\cdot\frac{x^{n+1}}{2n+1}$$

$$\therefore \quad \frac{u_n}{u_{n+1}} = \frac{1\cdot 3\cdot 5\ldots(2n-3)}{2\cdot 4\cdot 6\ldots(2n-2)}\cdot\frac{x^n}{(2n-1)} \times \frac{2\cdot 4\cdot 6\ldots(2n-2)\cdot 2n}{1\cdot 3\cdot 5\ldots(2n-3)(2n-1)}\cdot\frac{(2n+1)}{x^n\cdot x}$$

$$\frac{u_n}{u_{n+1}} = \frac{2n(2n+1)}{(2n-1)(2n-1)} \cdot \frac{1}{x} = \frac{2n^2\left(2+\frac{1}{n}\right)}{n^2\left(2-\frac{1}{n}\right)^2} = \frac{2\left(2+\frac{1}{n}\right)}{\left(2-\frac{1}{n}\right)^2} \cdot \frac{1}{x}$$

$$\therefore \quad \lim_{n\to\infty} \frac{u_n}{u_{n+1}} = \lim_{n\to\infty} \frac{2\left(2+\frac{1}{n}\right)}{\left(2-\frac{1}{n}\right)^2} \cdot \frac{1}{x} = \frac{2}{x} = 2\left(\frac{1}{x}\right)$$

Hence if $\frac{1}{x} > 1$ *i.e.*, $x < 1$, then the given series is convergent and if $\frac{1}{x} < 1$ *i.e.*, $x > 1$, then the given series is divergent if $\frac{1}{x} = 1$ *i.e.*, $x = 1$, then the test fails.

Now for $x = 1$, we get

$$\frac{u_n}{u_{n+1}} = \frac{2n(2n+1)}{(2n-1)^2} = \frac{4n^2+2n}{4n^2-4n+1}$$

$$\therefore \quad n\left[\frac{u_n}{u_{n+1}} - 1\right] = n\left[\frac{4n^2+2n}{4n^2-4n+1} - 1\right] = n\left[\frac{6n-1}{4n^2-4n+1}\right]$$

$$= \frac{n^2\left(6-\frac{1}{n}\right)}{n^2\left(4-\frac{4}{n}+\frac{1}{n^2}\right)} = \frac{\left(6-\frac{1}{n}\right)}{\left(4-\frac{4}{n}+\frac{1}{n^2}\right)}$$

By Raabe's test

$$\lim_{n\to\infty} n\left[\frac{u_n}{u_{n+1}} - 1\right] = \lim_{n\to\infty} \frac{6-\frac{1}{n}}{4-\frac{4}{n}+\frac{1}{n^2}} = \frac{6}{4} = \frac{3}{2} > 1$$

Thus the given series is convergent for $x = 1$.
Hence the given series is convergent when $x \le 1$ and divergent when $x > 1$. **Ans.**

Example 3: Test for convergence the series

$$1 + a + \frac{a(a+1)}{1 \cdot 2} + \frac{a(a+1)(a+2)}{1 \cdot 2 \cdot 3} + \ldots$$

Solution: Leaving the first term, we get the nth term

$$u_n = \frac{a(a+1)(a+2)\ldots(a+n-1)}{1 \cdot 2 \cdot 3 \ldots n}$$

and $$u_{n+1} = \frac{a(a+1)(a+2)\ldots(a+n-1)(a+n)}{1\cdot 2\cdot 3\ldots n(n+1)}$$

$$\therefore \quad \frac{u_n}{u_{n+1}} = \frac{a(a+1)(a+2)\ldots(a+n-1)}{1\cdot 2\cdot 3\ldots n} \times \frac{1\cdot 2\cdot 3\ldots n(n+1)}{a(a+1)(a+2)\ldots(a+n-1)(a+n)}$$

$$\frac{u_n}{u_{n+1}} = \frac{n+1}{a+n} = \frac{n\left(1+\frac{1}{n}\right)}{n\left(1+\frac{a}{n}\right)} = \frac{1+\frac{1}{n}}{1+\frac{a}{n}}$$

$$\therefore \quad \lim_{n\to\infty} \frac{u_n}{u_{n+1}} = \lim_{n\to\infty} \frac{1+\frac{1}{n}}{1+\frac{a}{n}} = 1$$

Hence ratio test fails.

Now, we apply Raabe's test for this

$$n\left[\frac{u_n}{u_{n+1}} - 1\right] = n\left[\frac{n+1}{a+n} - 1\right] = \frac{n(1-a)}{(a+n)} = \frac{n(1-a)}{n\left(1+\frac{a}{n}\right)} = \frac{1-a}{1+\frac{a}{n}}$$

$$\therefore \quad \lim_{n\to\infty} n\left[\frac{u_n}{u_{n+1}} - 1\right] = \lim_{n\to\infty} \frac{1-a}{1+\frac{a}{n}} = 1-a$$

if $1 - a > 1$ *i.e.*, $a < 0$, the series is convergent

if $1 - a < 1$ *i.e.*, $a > 0$, the series is divergent

if $1 - a = 1$ *i.e.*, $a = 0$, then the given series is

$$\sum u_n = 1 + 0 + 0 + 0 + \ldots \infty$$

i.e., the sum of the series $\sum u_n = 1 \quad \Rightarrow \quad S_n = 1$

$$\therefore \quad \lim_{n\to\infty} S_n = 1. \text{ Hence } \sum u_n \text{ is convergent for } a = 0$$

Thus $\sum u_n$ is convergent for $a \leqslant 0$ and divergent for $a > 0$. **Ans.**

Example 4: Test for the convergence the series

$$x\log x + x^2 \log 2x + x^3 \log 3x + \ldots + x^n \log nx + \ldots$$

Solution: Here $u_n = x^n \log nx$ and $u_{n+1} = x^{n+1} \cdot \log(n+1)x$

$$\therefore \quad \frac{u_n}{u_{n+1}} = \frac{x^n \log nx}{x^{n+1} \cdot \log(n+1)x} = \frac{\log nx}{\log(n+1)x} \cdot \frac{1}{x}$$

$$= \left[\frac{\log(n+1)x}{\log nx}\right]^{-1} \cdot \frac{1}{x}$$

$$= \left[\frac{\log\left\{nx\left(1+\frac{1}{n}\right)\right\}}{\log nx}\right]^{-1} \cdot \frac{1}{x}$$

$$= \left[\frac{\log nx + \log\left(1+\frac{1}{n}\right)}{\log nx}\right]^{-1} \cdot \frac{1}{x}$$

$$= \left[1+\frac{\log\left(1+\frac{1}{n}\right)}{\log nx}\right]^{-1} \cdot \frac{1}{x}$$

$$= \left[1+\frac{\frac{1}{n}-\frac{1}{2n^2}+\frac{1}{3n^3}-\ldots}{\log nx}\right]^{-1} \cdot \frac{1}{x} \qquad \left| \text{ As } \log(1+x) = x - \frac{x^2}{2} + \frac{x^3}{3} \ldots \right.$$

$$= \left[1+\left\{\frac{1}{n\log nx} - \frac{1}{2n^2 \log nx} + \ldots\right\}\right]^{-1} \cdot \frac{1}{x}$$

$$\frac{u_n}{u_{n+1}} = \left[1-\left\{\frac{1}{n\log nx} - \frac{1}{2n^2 \log nx} + \ldots\right\} + \ldots\right] \cdot \frac{1}{x}$$

$$\left| \text{ As } (1+x)^{-1} = 1 - x + x^2 \ldots \right.$$

$$\therefore \qquad \lim_{n\to\infty} \frac{u_n}{u_{n+1}} = \frac{1}{x}$$

Therefore from the ratio test

if $\frac{1}{x} > 1$ *i.e.,* $x < 1$, then the series is convergent

if $\frac{1}{x} < 1$ *i.e.,* $x > 1$, then the series is divergent

if $\frac{1}{x} = 1$ *i.e.,* $x = 1$, the test fails

Now we apply the Raabe's test for this

$$\frac{u_n}{u_{n+1}} = 1 - \left\{\frac{1}{n\log n} - \frac{1}{2n^2 \log n} + \ldots\right\} + \ldots$$

$$\therefore \quad \frac{u_n}{u_{n+1}} - 1 = -\left\{\frac{1}{n\log n} - \frac{1}{2n^2 \log n} + \ldots\right\} + \ldots$$

or $$n\left(\frac{u_n}{u_{n+1}} - 1\right) = -\left\{\frac{1}{\log n} - \frac{1}{2n\log n} + \ldots\right\} + \ldots$$

$$\therefore \quad \lim_{n\to\infty} n\left(\frac{u_n}{u_{n+1}} - 1\right) = 0 < 1$$

$\therefore$ From Raabe's test the given series is divergent for $x = 1$

Hence the given series is convergent for $x < 1$ and divergent for $x \geqslant 1$. **Ans.**

Example 5: Test the convergence the series

$$\frac{1}{(\log 2)^p} + \frac{1}{(\log 3)^p} + \frac{1}{(\log 4)^p} + \ldots$$

Solution: Here $u_n = \dfrac{1}{\{\log(n+1)\}^p}$ and $u_{n+1} = \dfrac{1}{\{\log(n+2)\}^p}$

$$\therefore \quad \frac{u_n}{u_{n+1}} = \frac{\{\log(n+2)\}^p}{\{\log(n+1)\}^p} = \left[\frac{\log n\left(1+\frac{2}{n}\right)}{\log n\left(1+\frac{1}{n}\right)}\right]^p$$

$$= \left[\frac{\log n + \log\left(1+\frac{2}{n}\right)}{\log n + \log\left(1+\frac{1}{n}\right)}\right]^p = \left[\frac{\log n + \frac{2}{n} - \frac{4}{2n^2} + \ldots}{\log n + \frac{1}{n} - \frac{1}{2n^2} + \ldots}\right]^p$$

$$= \left[\frac{\log n}{\log n}\left\{\frac{1 + \frac{2}{n\log n} - \ldots}{1 + \frac{1}{n\log n} - \ldots}\right\}\right]^p = \left[\frac{1 + (2/n\log n) - \ldots}{1 + \frac{1}{n\log n} - \ldots}\right]^p$$

$$= \left[1 + \frac{2}{n\log n} - \ldots\right]^p \left[1 + \frac{1}{n\log n} - \ldots\right]^{-p}$$

$$\frac{u_n}{u_{n+1}} = \left[1 + \frac{2p}{n\log n} - \ldots\right]\left[1 - \frac{p}{n\log n} + \ldots\right] \quad \text{Expanding by Binomial Theorem}$$

or $$\frac{u_n}{u_{n+1}} = 1 + \frac{2p}{n\log n} - \ldots - \frac{p}{n\log n} + \ldots = 1 + \frac{p}{n\log n} - \ldots$$

$$\therefore \quad \lim_{n\to\infty} \frac{u_n}{u_{n+1}} = \lim_{n\to\infty}\left(1 + \frac{p}{n\log n} - \ldots\right) = 1$$

Hence ratio test fails
Now for Raabe's test

$$n\left(\frac{u_n}{u_{n+1}} - 1\right) = n\left(\frac{p}{n\log n} - \ldots\right) = \frac{p}{\log n} - \ldots$$

$$\therefore \quad \lim_{n\to\infty}\left[n\left(\frac{u_n}{u_{n+1}} - 1\right)\right] = \lim_{n\to\infty}\left(\frac{p}{\log n} - \ldots\right) = 0 < 1$$

Thus by Raabe's test the given series is divergent. **Ans.**

Example 6: Test for convergence the series

$$1 + \frac{1}{2}\cdot\frac{x^2}{4} + \frac{1\cdot 3\cdot 5}{2\cdot 4\cdot 6}\cdot\frac{x^4}{8} + \frac{1\cdot 3\cdot 5\cdot 7\cdot 9}{2\cdot 4\cdot 6\cdot 8\cdot 10}\cdot\frac{x^6}{12} + \ldots$$

Solution: For nth term, we neglect the first *i.e.,* 1 of given series

$$\therefore \quad u_n = \frac{1\cdot 3\cdot 5\ldots(2n-3)}{2\cdot 4\cdot 6\ldots(2n-2)}\cdot\frac{x^{2n}}{4n} \qquad \left|\begin{array}{l}(n-1)\text{th term of } 1\cdot 3\cdot 5\cdot 7 = (2n-3)\\ (n-1)\text{th term of } 2\cdot 4\cdot 6 = (2n-2)\end{array}\right.$$

$$\therefore \quad u_{n+1} = \frac{1\cdot 3\cdot 5\ldots(2n-3)(2n-1)}{2\cdot 4\cdot 6\ldots(2n-2)(2n)}\cdot\frac{x^{2n+2}}{4(n+1)}$$

$$\therefore \quad \frac{u_n}{u_{n+1}} = \frac{1\cdot 3\cdot 5\ldots(2n-3)}{2\cdot 4\cdot 6\ldots(2n-2)}\cdot\frac{x^{2n}}{4n}\times\frac{2\cdot 4\cdot 6\ldots(2n-2)(2n)}{1\cdot 3\cdot 5\ldots(2n-3)(2n-1)}\cdot\frac{(4n+4)}{x^{2n+2}}$$

$$= \frac{2n(4n+4)}{4n(2n-1)\cdot x^2} = \frac{(4n+4)}{2(2n-1)}\cdot\frac{1}{x^2} = \frac{\left(4 + \frac{4}{n}\right)}{\left(4 - \frac{2}{n}\right)}\cdot\frac{1}{x^2}$$

By ratio test

$$\lim_{n\to\infty}\frac{u_n}{u_{n+1}} = \lim_{n\to\infty}\frac{\left(4 + \frac{4}{n}\right)}{\left(4 - \frac{2}{n}\right)}\cdot\frac{1}{x^2} = \frac{1}{x^2}$$

If $\frac{1}{x^2} > 1$ *i.e.,* $x^2 < 1$, then series is convergent

If $\frac{1}{x^2} < 1$ *i.e.*, $x^2 > 1$, then series is divergent

If $\frac{1}{x^2} = 1$ *i.e.*, $x^2 = 1$, then the test fails.

Now, apply the Raabe's test for this, put $\frac{1}{x^2} = 1$

$$\frac{u_n}{u_{n+1}} = \frac{4+\frac{4}{n}}{4-\frac{2}{n}} \quad \text{or} \quad \frac{u_n}{u_{n+1}} - 1 = \frac{4+\frac{4}{n}-4+\frac{2}{n}}{4-\frac{2}{n}} = \frac{\frac{6}{n}}{4-\frac{2}{n}}$$

$$\therefore \qquad n\left(\frac{u_n}{u_{n+1}} - 1\right) = \frac{6}{\left(4-\frac{2}{n}\right)}$$

$$\Rightarrow \qquad \lim_{n\to\infty}\left[n\left(\frac{u_n}{u_{n+1}} - 1\right)\right] = \lim_{n\to\infty}\frac{6}{\left(4-\frac{2}{n}\right)} = \frac{6}{4} = \frac{3}{2} > 1$$

$\therefore$ From Raabe's test the given series is convergent for $x^2 = 1$

Hence the given series is convergent for $x^2 \le 1$ and divergent for $x^2 > 1$. **Ans.**

3.8 CAUCHY'S ROOT TEST

Statement: An infinite series $u_1 + u_2 + u_3 + \ldots + u_n + \ldots$ is convergent or divergent as according to $\lim_{n\to\infty}(u_n)^{1/n} < 1$ or > 1.

Proof: Let us consider an auxiliary series $\sum v_n$ which is given by

$$\sum v_n = r + r^2 + r^3 + \ldots + r^n + \ldots \qquad \ldots(i)$$

the series (i) is in G.P. which will be convergent if $r < 1$ and divergent if $r > 1$

Then, we know that if $\lim_{n\to\infty}\frac{u_n}{v_n}$ is finite quantity then $\sum u_n$ and $\sum v_n$ are either both convergent or both divergent.

Now let $\frac{u_n}{v_n} = k$ then $\frac{u_n}{r^n} = k$ or $u_n = kr^n$ or $(u_n)^{1/n} = (k)^{1/n}.\ r$

$$\therefore \qquad \lim_{n\to\infty}(u_n)^{1/n} = \lim_{n\to\infty}(k)^{1/n} \cdot r = k^0 \cdot r = r \qquad \ldots(ii)$$

But as discussed above if $r < 1$, the series $\sum v_n$ is convergent, hence the given series is also convergent.

Similarly, if $r > 1$, the given series is divergent.

Hence from (*ii*), we conclude that the series $\sum u_n$ is convergent or divergent as according to $\lim\limits_{n\to\infty} (u_n)^{1/n} < 1$ or > 1.

Example 7: Test the series whose *n*th term is $\dfrac{n^{n^2}}{(n+1)^{n^2}}$.

Solution: If u_n represents the *n*th term then

$$u_n = \frac{n^{n^2}}{(n+1)^{n^2}} = \left(\frac{n}{n+1}\right)^{n^2}$$

$$\therefore \qquad (u_n)^{1/n} = \left(\frac{n}{n+1}\right)^{n^2 \times \frac{1}{n}} = \left(\frac{n}{n+1}\right)^n = \left(\frac{1}{1+\frac{1}{n}}\right)^n$$

$$\Rightarrow \qquad \lim_{n\to\infty} (u_n)^{1/n} = \lim_{n\to\infty} \frac{1}{\left(1+\frac{1}{n}\right)^n} = \frac{1}{e} \qquad \left|\ \text{As} \lim_{n\to\infty}\left(1+\frac{1}{x}\right)^x = e\right.$$

$$= \frac{1}{2.718} < 1.$$

Hence the given series is convergent

Example 8: Test the convergence of the series whose *n*th term is $\dfrac{(n+1)^n}{n^{n+1}} x^n$.

Solution: Here $u_n = \dfrac{(n+1)^n \cdot x^n}{n^{n+1}} = \left(\dfrac{(n+1)x}{n}\right)^n \cdot \dfrac{1}{n}$

$$\therefore \qquad (u_n)^{1/n} = \frac{(n+1)x}{n} \cdot \frac{1}{n^{1/n}} = \frac{n\left(1+\frac{1}{n}\right)x}{n} \cdot \frac{1}{n^{1/n}}$$

or

$$(u_n)^{1/n} = \left(1+\frac{1}{n}\right) \cdot \frac{x}{n^{1/n}}$$

$$\therefore \qquad \lim_{n\to\infty} (u_n)^{1/n} = \lim_{n\to\infty} \left(1+\frac{1}{n}\right) \frac{x}{n^{1/n}} \ldots \qquad \ldots(i)$$

To evaluate this limit let $A = n^{1/n}$

$$\therefore \quad \log A = \frac{1}{n} \log n$$

$$\therefore \quad \lim_{n\to\infty} (\log A) = \lim_{n\to\infty} \left(\frac{\log n}{n} \right)$$

$$= \lim_{n\to\infty} \left(\frac{1/n}{1} \right) \qquad \text{| By L. Hospital's rules}$$

$$\lim_{n\to\infty} (\log A) = 0$$

or
$$\lim_{n\to\infty} A = e^0 = 1 \quad \Rightarrow \quad \lim_{n\to\infty} n^{1/n} = 1 \qquad \ldots(ii)$$

Using the result (*ii*) in equation (*i*), we get

$$\lim_{n\to\infty} (u_n)^{1/n} = (1+0) \cdot \frac{x}{1} = x$$

By Cauchy's *n*th root test, we have

if $x < 1$, then the given series is convergent

if $x > 1$, then the given series is divergent

if $x = 1$, then the test fails

$\therefore$ for $x = 1$ the *n*th term is

$$u_n = \frac{(n+1)^n}{n^{n+1}} \qquad \therefore \; v_n = \frac{n^n}{n^{n+1}} = \frac{1}{n}$$

Now
$$\frac{u_n}{v_n} = \frac{(n+1)^n}{n^n \cdot n} \times n = \frac{(n+1)^n}{n^n} = \frac{n^n \left(1+\frac{1}{n}\right)^n}{n^n}$$

$$\Rightarrow \quad \frac{u_n}{v_n} = \left(1+\frac{1}{n}\right)^n$$

$$\therefore \quad \lim_{n\to\infty} \frac{u_n}{v_n} = e \text{ which is finite.}$$

But
$$\sum v_n = \sum \frac{1}{n} = \sum \frac{1}{n^p} \quad \Rightarrow \quad p = 1 \qquad \left[\because \sum v_n \text{ is divergent}\right]$$

So $\sum u_n$ will also divergent.

Hence, the given series is convergent if $x < 1$ and divergent if $x \geqslant 1$. **Ans.**

Example 9: Test the convergence of the series

$$\left[\frac{2^2}{1^2}-\frac{2}{1}\right]^{-1}+\left[\frac{3^3}{2^3}-\frac{3}{2}\right]^{-2}+\left[\frac{4^4}{3^4}-\frac{4}{3}\right]^{-3}+\ldots$$

Solution: If the given series be denoted by $\sum u_n$, then

$$u_n = \left[\frac{(n+1)^{n+1}}{n^{n+1}}-\frac{(n+1)}{n}\right]^{-n}$$

$$\therefore \quad (u_n)^{1/n} = \left[\frac{(n+1)^n}{n^n}\cdot\frac{(n+1)}{n}-\frac{(n+1)}{n}\right]^{-n\times\frac{1}{n}} = \left[\frac{(n+1)^n}{n^n}\cdot\frac{(n+1)}{n}-\frac{(n+1)}{n}\right]^{-1}$$

$$= \left[\left(1+\frac{1}{n}\right)^n\cdot\left(1+\frac{1}{n}\right)-\left(1+\frac{1}{n}\right)\right]^{-1}$$

$$\therefore \quad \lim_{n\to\infty}(u_n)^{1/n} = \lim_{n\to\infty}\left[\left(1+\frac{1}{n}\right)^n\cdot\left(1+\frac{1}{n}\right)-\left(1+\frac{1}{n}\right)\right]^{-1}$$

$$=(e-1)^{-1}=\frac{1}{e-1}=\frac{1}{2.718-1}=\frac{1}{1.718}<1$$

Thus, by Cauchy's root test the given series is convergent. **Ans.**

Example 10: Discuss the convergence of the series.

$$a+b+a^2+b^2+a^3+b^3+\ldots$$

Solution: the nth term of given series is

$$u_n = a^n + b^n \qquad \ldots(i)$$

Let $\quad v_n = a^n \quad$ and $\quad w_n = b^n$

$\therefore \quad (v_n)^{1/n} = a \quad$ and $\quad (w_n)^{1/n} = b$

or $\quad \lim_{n\to\infty}(v_n)^{1/n} = a \quad$ and $\quad \lim_{n\to\infty}(w_n)^{1/n} = b$

Therefore from (*i*), we get

$$\lim_{n\to\infty}(u_n)^{1/n} = a+b$$

if (a and b) < 1, then the given series is convergent.
if (a and b) > 1, then the given series is divergent.
if $a = b = 1$ then the given series becomes

$$1+1+1^2+1^2+1^3+1^3+\ldots$$

i.e., $$1 + 1 + 1 + \ldots$$

$\therefore$ The sum $S_n = n \Rightarrow \lim\limits_{n\to\infty} S_n = \infty$

This shows that the series is divergent for $a = b = 1$

Hence the given series is convergent for (a and b) < 1 and divergent for (a and b) $\geqslant$ 1. **Ans.**

Example 11: Test the series $1+\dfrac{x}{2}+\dfrac{x^2}{3^2}+\dfrac{x^3}{4^3}+\ldots$

Solution: Neglecting the first term for nth term of given series, we have

$$u_n = \frac{x^n}{(n+1)^n} = \left[\frac{x}{(n+1)}\right]^n \Rightarrow (u_n)^{1/n} = \left(\frac{x}{n+1}\right)$$

$$\therefore \quad \lim_{n\to\infty} (u_n)^{1/n} = \lim_{n\to\infty}\left(\frac{x}{n+1}\right) = \lim_{n\to\infty}\frac{x}{\infty} = 0 < 1$$

Hence by Cauchy's root test the given series is convergent. **Ans.**

Example 12: Test for convergence the series $\sum\limits_{n=2}^{\infty}\dfrac{1}{(\log n)^n}$

Solution: Here $u_n = \dfrac{1}{(\log n)^n} \quad \therefore (u_n)^{1/n} = \dfrac{1}{\log n}$

$$\lim_{n\to\infty} (u_n)^{1/n} = \lim_{n\to\infty}\frac{1}{\log n} = 0 < 1$$

$\therefore$ From Cauchy's root test the given series is convergent. **Ans.**

EXERCISE 3.3

Raabe's Test

Test the convergence of the series:

1. $\dfrac{x}{1}+\dfrac{1}{2}\cdot\dfrac{x^3}{3}+\dfrac{1\cdot 3}{2\cdot 4}\cdot\dfrac{x^5}{5}+\dfrac{1\cdot 3\cdot 5}{2\cdot 4\cdot 6}\cdot\dfrac{x^7}{7}+\ldots$ [**Ans.** Convergent if $x^2 \le 1$, Divergent if $x^2 > 1$]

2. $1+\dfrac{3}{7}x+\dfrac{3\cdot 6}{7\cdot 10}x^2+\dfrac{3\cdot 6\cdot 9}{7\cdot 10\cdot 13}x^3+\dfrac{3\cdot 6\cdot 9\cdot 12}{7\cdot 10\cdot 13\cdot 16}x^4+\ldots$

   [**Ans.** Convergent if $x \le 1$, Divergent if $x > 1$]

3. $1+\dfrac{1}{2}x+\dfrac{1\cdot 3}{2\cdot 4}x^2+\dfrac{1\cdot 3\cdot 5}{2\cdot 4\cdot 6}x^3+\ldots$ [**Ans.** Convergent if $x < 1$, Divergent if $x \ge 1$]

4. $\frac{1^2}{4^2}+\frac{1^2\cdot 5^2}{4^2\cdot 8^2}+\frac{1^2\cdot 5^2\cdot 9^2}{4^2\cdot 8^2\cdot 12^2}+\frac{1^2\cdot 5^2\cdot 9^2\cdot 13^2}{4^2\cdot 8^2\cdot 12^2\cdot 16^2}+\ldots$ [**Ans.** Convergent]

5. $u_n=\frac{x^n}{a+\sqrt{n}}$ [**Ans.** Convergent if $x<1$, Divergent if $x\geq 1$]

6. $1+\frac{2}{3\cdot 4}+\frac{2\cdot 4}{3\cdot 5\cdot 6}+\frac{2\cdot 4\cdot 6}{3\cdot 5\cdot 7\cdot 8}+\ldots+\frac{2\cdot 4\cdot 6\ldots 2n}{3\cdot 5\cdot 7\ldots(2n+1)(2n+2)}$ [**Ans.** Convergent]

7. $\sum u_n=\sum\left[\frac{1\cdot 3\cdot 5\ldots(2n-1)}{2\cdot 4\cdot 6\ldots(2n)}\cdot\frac{x^{2n}}{2^n}\right].$ [**Ans.** Convergent if $x^2\leq 1$, Divergent if $x^2>1$]

8. $\frac{1}{2}+\frac{1\cdot 3}{2\cdot 4}+\frac{1\cdot 3\cdot 5}{2\cdot 4\cdot 6}+\ldots$ [**Ans.** Divergent]

Cauchy's Test

9. $\sum u_n=\sum\left(\frac{n}{n+1}\right)^{n^2}$ [**Ans.** Convergent]

10. $\sum u_n=\sum\left(\frac{1+nx}{n}\right)^n.$ [**Ans.** Convergent if $x>1$, Divergent if $x\geq 1$]

11. $1+\frac{x}{2}+\frac{x^2}{3^2}+\frac{x^3}{4^3}+\ldots$ [**Ans.** Convergent]

12. $\frac{1}{2}+\frac{2}{3}x+\left(\frac{3}{4}\right)^2x^2+\left(\frac{4}{5}\right)^3x^3+\ldots$ [**Ans.** Convergent if $x<1$, Divergent if $x>1$]

13. $\sum u_n=\sum\frac{(n+1)^n x^n}{n^{n+1}}.$ [**Ans.** Convergent if $x<1$, Divergent if $x\geq 1$]

14. $\sum 5^{-n-(-1)^n}.$ [**Ans.** Convergent]

15. $\sum_{1}^{\infty}3^{-2n-5(-1)^n}.$ [**Ans.** Convergent] $\left[\textbf{Hint: }\lim_{n\to\infty}u_n^{1/n}=\lim_{n\to\infty}3^{-2}\cdot 3^{-\frac{5(-1)^n}{n}}=\frac{1}{9}\right]$

16. $\sum\frac{1}{[\log(\log n)]^n}.$ [**Ans.** Convergent]

3.9 TEST FOR ALTERNATING SERIES

Definition: A series in which the term are alternately positive and negative is called an alternating series. In other words, the series $u_1 - u_2 + u_3 - u_4 + \ldots$ where $u_n > 0$ for each n is an alternating series and is briefly written as $\sum_{n=1}^{\infty} (-1)^{n-1} u_n$.

3.10 LEIBNITZ'S RULE

Statement: The alternating series $\sum_{n=1}^{\infty} (-1)^{n-1} \cdot u_n$ is convergent if:

(*i*) each term is numerically less than the preceding term *i.e.*, $u_{n+1} < u_n$ for all n and

(*ii*) $\lim_{n\to\infty} u_n = 0$

Proof: Consider the sum of an even number of terms, say $2n$, of given series.

$$S_{2n} = (u_1 - u_2) + (u_3 - u_4) + \ldots (u_{2n-1} - u_{2n}) \ldots \quad \ldots(i)$$

Since $u_1 > u_2 > u_3 \ldots$, therefore $u_1 - u_2, u_3 - u_4, \ldots$ all are positive.

Hence $S_{2n} > 0$.

We may express S_{2n} also as

$$S_{2n} = u_1 - (u_2 - u_3) - (u_4 - u_5) \ldots - u_{2n} \quad \ldots(ii)$$

The above representation of S_{2n} shows that its value is always less than u_1. Thus $0 < S_{2n} < u_1$.
Hence S_{2n} tends to a finite limit.
If we consider the sum of an odd number of terms, say $2n + 1$, then

$$S_{2n+1} = S_{2n} + u_{2n+1}$$

$$\therefore \quad \lim_{n\to\infty} S_{2n+1} = \lim_{n\to\infty} S_{2n} + \lim_{n\to\infty} u_{2n+1} \quad \ldots(iii)$$

Also, we are given that $\lim_{n\to\infty} u_n = 0$

In particular $\lim_{n\to\infty} u_{2n+1} = 0$

$\therefore$ From (*iii*), we get

$$\lim_{n\to\infty} S_{2n+1} = \lim_{n\to\infty} S_{2n}, \text{ which is finite.}$$

Hence the series is convergent.

WORKED OUT EXAMPLES

Example 1: Test the convergence of the series

$$1 - \frac{1}{2\sqrt{2}} + \frac{1}{3\sqrt{3}} - \frac{1}{4\sqrt{4}} + \ldots$$

Solution: In the given series, we find that

(*i*) the terms are alternately +ve and –ve

(*ii*) the terms are continuously decreasing and

(*iii*) $\lim_{n\to\infty} u_n = \lim_{n\to\infty} \cdot \frac{1}{n\sqrt{n}} = 0$

Hence all the conditions of alternating series test being satisfied the given series is convergent. **Ans.**

Example 2: Test the convergence of given series

$$1-\frac{1}{\sqrt{2}}+\frac{1}{\sqrt{3}}-\frac{1}{\sqrt{4}}+\ldots\infty$$

Solution: Here the given series is alternating series and each term is numerically less than the preceding term *i.e.,* $\frac{1}{\sqrt{3}}<\frac{1}{\sqrt{2}}, \frac{1}{\sqrt{4}}<\frac{1}{\sqrt{3}}$, etc.

and $$u_n = \frac{1}{\sqrt{n}}$$

$$\therefore \quad \lim_{n\to\infty} u_n = \lim_{n\to\infty} \frac{1}{\sqrt{n}} = 0$$

Hence by Leibnitz's test the given series is convergent. **Ans.**

Example 3: Test the convergence of given series

$$\frac{1}{1\cdot 3}-\frac{1}{2\cdot 4}+\frac{1}{3\cdot 5}-\frac{1}{4\cdot 6}+\ldots$$

Solution: Here the given series is alternating series and each term is numerically less than the preceding term *i.e.,* $\frac{1}{2\cdot 4}<\frac{1}{1\cdot 3}, \frac{1}{3\cdot 5}<\frac{1}{2\cdot 4}$, etc.

and $$u_n = \frac{1}{n(n+2)}$$

$$\therefore \quad \lim_{n\to\infty} \frac{1}{n(n+2)} = 0$$

Hence by Leibnitz's rule the given series is convergent. **Ans.**

Example 4: Test the convergence of given series

$$\frac{1}{\log 2}-\frac{1}{\log 3}+\frac{1}{\log 4}-\frac{1}{\log 5}+\ldots$$

Solution: Here the given series is an alternating series and each term is numerically less than the preceding term *i.e.,* $\frac{1}{\log 3}<\frac{1}{\log 2}, \frac{1}{\log 5}<\frac{1}{\log 4}$, etc.

as $\log 3 > \log 2, \log 5 > \log 4, \ldots$ etc.

and $$u_n = \frac{1}{\log(n+1)}$$

$$\therefore \quad \lim_{n\to\infty} u_n = \lim_{n\to\infty} \frac{1}{\log(n+1)} = \frac{1}{\infty} = 0$$

Hence by Leibnitz's rule the given series is convergent. **Ans.**

Example 5: Test the convergence of the series

$$\sum_n \frac{(-1)^{n+1}}{\sqrt{n^2+1}}$$

Solution: We have $\sum u_n = \sum_{n=1}^{\infty} \frac{(-1)^{n+1}}{\sqrt{n^2+1}}$

$$\therefore \quad \sum u_n = \frac{1}{\sqrt{2}} - \frac{1}{\sqrt{5}} + \frac{1}{\sqrt{10}} - \frac{1}{\sqrt{17}} + \ldots$$

Here the given series is an alternating series in which each term is numerically less than the preceding term *i.e.*, $\frac{1}{\sqrt{5}} < \frac{1}{\sqrt{2}}, \frac{1}{\sqrt{10}} < \frac{1}{\sqrt{5}}, \ldots,$ etc.

and $$\lim_{n\to\infty} u_n = \lim_{n\to\infty} \cdot \frac{1}{\sqrt{n^2+1}} = \frac{1}{\infty} = 0$$

Hence by Leibnitz's test the given series is convergent. **Ans.**

Example 6: Test the convergence of the series

$$\frac{1}{2^3} - \frac{1}{3^3}(1+2) + \frac{1}{4^3}(1+2+3)\ldots$$

Solution: The given series is

$$\sum_{n=1}^{\infty} u_n = \sum_{n=1}^{\infty} (-1)^{n-1} \cdot \frac{1}{(n+1)^3}(1+2+3+\ldots+n)$$

$$= \sum_{n=1}^{\infty} (-1)^{n-1} \cdot \frac{1}{(n+1)^3} \cdot \frac{n(n+1)}{2}$$

| Sum of natural numbers $= \frac{1}{2}n(n+1)$

$$\therefore \quad u_n = \frac{n}{2(n+1)^2} = \frac{n}{2n^2\left(1+\frac{1}{n}\right)^2} = \frac{1}{2n\left(1+\frac{1}{n}\right)^2}$$

and $$\lim_{n\to\infty} u_n = \lim_{n\to\infty} \frac{1}{2n\left(1+\frac{1}{n}\right)^2} = 0$$

Also the given series is alternating series in which each term is numerically less than the preceding term *i.e.*,

$$\frac{1}{3^3} < \frac{1}{2^3}\left(\frac{1}{27} < \frac{1}{8}\right), \frac{1+2+3}{4^3} < \frac{(1+2)}{3^3}, \ldots \text{ etc.}$$

Hence by Leibnitz's test the given series is convergent. **Ans.**

Example 7: Test the convergence of given series

$$\sum (-1)^{n-1} \cdot \sin\frac{1}{n}.$$

Solution: We have $\sum_{n=1}^{\infty} u_n = \sum_{n=1}^{\infty} (-1)^{n-1} \cdot \sin\frac{1}{n} = \sin 1 - \sin\frac{1}{2} + \sin\frac{1}{3} - \sin\frac{1}{4} + \ldots$

Here the given series is alternating series in which each term is numerically less than the preceding term *i.e.*,

$$\sin\frac{1}{2} < \sin 1, \sin\frac{1}{3} < \sin\frac{1}{2}, \ldots \text{etc.}$$

and numerically, $$u_n = \sin\frac{1}{n}$$

$$\therefore \quad \lim_{n\to\infty} u_n = \lim_{n\to\infty} \sin\frac{1}{n} = \sin\frac{1}{\infty} = \sin 0 = 0$$

Hence by Leibnitz's rule, the given series is convergent. **Ans.**

Example 8: Test the convergence of the series

$$2 - \frac{3}{2} + \frac{4}{3} - \frac{5}{4} + \ldots$$

Solution: Here we have

$$\sum_{n=1}^{\infty} u_n = \sum_{n=1}^{\infty} (-1)^{n-1}\left(\frac{n+1}{n}\right)$$

and the series is alternating series in which each term is numerically less than the preceding term *i.e.*,

$$\frac{3}{2} < 2, \frac{4}{3} < \frac{3}{2}, \frac{5}{4} < \frac{4}{3}, \ldots \text{ etc.}$$

$\therefore$ Numerically $$u_n = \frac{(n+1)}{n} = 1 + \frac{1}{n}$$

$$\Rightarrow \qquad \lim_{n\to\infty} u_n = \lim_{n\to\infty} 1+\frac{1}{n} = 1+\frac{1}{\infty} = 1 \neq 0$$

Since the second condition of Leibnitz's test is not satisfied the series is not convergent. **Ans.**

EXERCISE 3.4

Test the convergence of the following series:

1. $1-\frac{1}{2^2}+\frac{1}{3^2}-\frac{1}{4^2}+\ldots$ [**Ans.** Convergent]

2. $1-\frac{1}{2^3}(1+2)+\frac{1}{3^3}(1+2+3)\ldots$ [**Ans.** Convergent]

3. $1-\frac{1}{3}+\frac{1}{5}-\frac{1}{7}+\ldots$ [**Ans.** Convergent]

4. $1-\frac{1}{3!}+\frac{1}{5!}-\frac{1}{7!}+\ldots$ [**Ans.** Convergent]

5. $\sum_{n=1}^{\infty}(-1)^{n-1}\frac{n}{5n-7}$ [**Ans.** Not convergent]

6. $\sum_{n=1}^{\infty}\frac{(-1)^{n-1}n}{n+2}$ [**Ans.** Not convergent]

7. $\sum_{n=1}^{\infty}\frac{(-1)^{n-1}n}{10n-1}$ [**Ans.** Not convergent]

8. $\frac{1}{\sqrt{2}+1}-\frac{1}{\sqrt{3}+1}+\frac{1}{\sqrt{4}+1}-\frac{1}{\sqrt{5}+1}+\ldots$ [**Ans.** Convergent]

9. $\sum_{n=1}^{\infty}(-1)^n\cdot\frac{n+2}{2^n+5}$ [**Ans.** Convergent]

10. $\sum_{n=1}^{\infty}\frac{(-1)^{n+1}\cdot n}{n^2+1}$ [**Ans.** Convergent]

3.11 INTEGRAL TEST

In mathematics, the integral test for convergence is a method used to test infinite series of non-negative terms for convergence.

Statement: Let $f(x)$ be a non-negative monotonic decreasing function and $x \geqslant 1$. Then the series $\sum_{n=1}^{\infty} f(n)$ converges if and only if the integral $\int_1^{\infty} f(x)\,dx$ is finite.

In particular, if the integral diverges, then the series diverges as well.

Proof: Here we use the comparison test, comparing the term $f(n)$ with the integral of $f(x)$ over the intervals $[n-1, n]$ and $[n, n+1]$, respectively.

Since f is monotonic decreasing function, we know that

$$f(x) \le f(n) \text{ for } x \in [n, \infty]$$

and

$$f(n) \le f(x) \text{ for } x \in [1, n]$$

Hence for every n larger than 1

$$\int_n^{n+1} f(x)dx \le \int_n^{n+1} f(n)dx = f(n) = \int_{n-1}^{n} f(n)dx \le \int_{n-1}^{n} f(x)dx$$

Since the lower estimate (approximate) value is also valid for $f(1)$, we get by summation over all n from 1 to some larger integer N *i.e.*,

$$\int_1^{N+1} f(x)dx \le \sum_{n=1}^{N} f(n) \le f(1) + \int_1^{N} f(x)dx$$

Taking $N \to \infty$

then $\int_1^{\infty} f(x)dx$ is finite. Hence the result follows.

Monotonic Function

If a function f defined on a subset of the real numbers with real values is called monotonic (monotonically increasing), if for all x and y such that $x \leqslant y$ one has $f(x) \leqslant f(y)$. In similar way, a function is called monotonically decreasing if, $x \leqslant y$ and $f(x) \geqslant f(y)$.

WORKED OUT EXAMPLES

Example 1: Test the convergence of the series $\sum_{n=1}^{\infty} \frac{1}{n(n+1)}$.

Solution: Here $u_n = \frac{1}{n(n+1)}$

Let $$f(n) = u_n = \frac{1}{n(n+1)}$$

$$\therefore \quad \int_1^n f(x)dx = \int_1^n \frac{dx}{x(x+1)} = \int_1^n \left(\frac{1}{x} - \frac{1}{x+1}\right)dx$$

$$= [\log x - \log(x+1)]_1^n = \left[\log \frac{x}{x+1}\right]_1^n$$

$$= \log \frac{n}{n+1} - \log \frac{1}{2}$$

$$\Rightarrow \quad \int_1^\infty f(x)dx = \lim_{n\to\infty} \left[\log \frac{1}{1+\frac{1}{n}} - \log \frac{1}{2}\right]$$

$$= \log 1 - \log \frac{1}{2} = \log 2 \text{ (Finite)}$$

$\therefore \quad \int_1^\infty f(x)\,dx$ converges and hence by Integral test,

the series $\sum_{n=1}^{\infty} u_n$ is convergent. **Ans.**

Example 2: Test the convergence of the series $\sum_{n=1}^{\infty} \frac{1}{2n+3}$.

Solution: Here $u_n = f(n) = \frac{1}{2n+3} \quad \therefore \quad f(x) = \frac{1}{2x+3}$

$$\Rightarrow \quad \int_1^n f(x)dx = \int_1^n \frac{1}{2x+3}dx = \frac{1}{2}[\log(2x+3)]_1^n = \frac{1}{2}[\log(2n+3) - \log 5]$$

Taking $\lim_{n\to\infty}$, we get

$$\int_1^\infty f(x)\,dx = \frac{1}{2}\lim_{n\to\infty}[\log(2n+3) - \log 5] = \frac{1}{2}[\log\infty - \log 5] = \infty \text{ (Not finite)}$$

$\Rightarrow \quad \int_1^\infty f(x)\,dx$ diverges

Hence by integral test, the given series is divergent. **Ans.**

Example 3: Test for convergence the series

$$\sum_{n=1}^{\infty} \frac{2n^3}{n^4+3}$$

Solution: Here $u_n = \dfrac{2n^3}{n^4+3} = f(n)$

$$f(x) = \frac{2x^3}{x^4+3}$$

$$\therefore \quad \int_1^n f(x)\,dx = \int_1^n \frac{2x^3}{x^4+3}\,dx = \frac{1}{2}\int_1^n \frac{4x^3}{x^4+3}\,dx$$

$$= \frac{1}{2}[\log(x^4+3)]_1^n = \frac{1}{2}[\log(n^4+3) - \log 4]$$

Taking $\lim\limits_{n\to\infty}$

$$\therefore \quad \int_1^{\infty} f(x)dx = \lim_{n\to\infty} \frac{1}{2}[\log(n^4+3) - \log 4] = \frac{1}{2}[\infty - \log 4] = \infty$$

$$\Rightarrow \quad \int_1^{\infty} f(x)dx \text{ diverges} \Rightarrow \sum_{n=1}^{\infty} u_n \text{ diverges.}$$ **Ans.**

Example 4: Test the convergence of series $\displaystyle\sum_{n=1}^{\infty} \frac{n}{(n^2+1)^2}$.

Solution: Here $u_n = \dfrac{n}{(n^2+1)^2} = f(n) \quad \therefore f(x) = \dfrac{x}{(x^2+1)^2}$

$$\therefore \quad \int_1^n f(x)dx = \int_1^n \frac{x}{(x^2+1)^2}\,dx = \frac{1}{2}\int_1^n (x^2+1)^{-2}\cdot 2x\,dx$$

$$= \frac{1}{2}\left[\frac{(x^2+1)^{-1}}{-1}\right]_1^n = -\frac{1}{2}\left[\frac{1}{n^2+1} - \frac{1}{2}\right] = \frac{1}{2}\left[\frac{1}{2} - \frac{1}{n^2+1}\right]$$

Taking $\lim\limits_{n\to\infty}$

$$\int_1^{\infty} f(x)dx = \lim_{n\to\infty} \frac{1}{2}\left[\frac{1}{2} - \frac{1}{n^2+1}\right] = \frac{1}{2}\left(\frac{1}{2} - 0\right) = \frac{1}{4} \text{ (Finite)}$$

$$\Rightarrow \quad \int_1^{\infty} f(x)dx \text{ converges} \Rightarrow \sum_{n=1}^{\infty} u_n \text{ converges.}$$ **Ans.**

Example 5: Test for convergence the series $\sum_{n=2}^{\infty} \frac{1}{n\sqrt{n^2-1}}$.

Solution: Here $u_n = \frac{1}{n\sqrt{n^2-1}} = f(n),\ n \geqslant 2$

$$\therefore \quad f(x) = \frac{1}{x\sqrt{x^2-1}},\ x \geqslant 2$$

$$\int_2^n f(x)dx = \int_2^n \frac{dx}{x\sqrt{x^2-1}} = [\sec^{-1} x]_2^n = \sec^{-1} n - \sec^{-1} 2$$

$\Rightarrow$ Taking $\lim_{n\to\infty}$

$$\therefore \quad \int_2^{\infty} f(x)dx = \lim_{n\to\infty} I_n = \lim_{n\to\infty} (\sec^{-1} n - \sec^{-1} 2)$$

$$= \sec^{-1} \infty - \sec^{-1} 2 = \frac{\pi}{2} - \frac{\pi}{3} = \frac{\pi}{6} \text{ (Finite)}$$

$\Rightarrow \quad \int_2^{\infty} f(x)dx$ converges $\Rightarrow \sum_{n=2}^{\infty} u_n$ converges. **Ans.**

Example 6: Test the convergence of the series $\sum_{n=1}^{\infty} ne^{-n^2}$.

Solution: We have $f(n) = ne^{-n^2}$

$$\therefore \quad f(x) = xe^{-x^2}$$

$$\int_1^n f(x)dx = \int_1^n xe^{-x^2} dx$$

Let $\quad -x^2 = t \Rightarrow xdx = -\frac{dt}{2}$

$$\therefore \quad \int_1^n xe^{-x^2} dx = -\frac{1}{2}\int_{-1}^{-n^2} e^t \cdot dt$$

$$= -\frac{1}{2}\left[e^t\right]_{-1}^{-n^2} = -\frac{1}{2}(e^{-n^2} - e^{-1}) = \frac{1}{2}\left(\frac{1}{e} - e^{-n^2}\right)$$

Taking $\lim_{n\to\infty}$

$$\therefore \quad \int_1^{\infty} f(x)dx = \lim_{n\to\infty} \frac{1}{2}\left(\frac{1}{e} - e^{-n^2}\right) = \frac{1}{2}\left(\frac{1}{e} - 0\right) = \frac{1}{2e} \text{ (Finite)}$$

Hence the integral $\int_1^{\infty} f(x)dx$ is convergent so the given series is also convergent. **Ans.**

Example 7: Test the convergence of the series $\sum_{n=1}^{\infty} \frac{1}{(n+1)\log(n+1)}$.

Solution: Here $f(n) = \frac{1}{(n+1)\log(n+1)}$

$$\therefore \quad f(x) = \frac{1}{(x+1)\log(x+1)}$$

Now $$\int_1^n f(x)dx = \int_1^n \frac{1}{(x+1)\log(x+1)}dx$$

Let $$\log(x+1) = t \Rightarrow \left(\frac{1}{x+1}\right)dx = dt$$

$$\int_1^n f(x)dx = \int_{\log 2}^{\log(n+1)} \frac{1}{t}dt = [\log t]_{\log 2}^{\log(n+1)} = \log\log(n+1) - \log\log 2$$

Taking $\lim\limits_{n\to\infty}$, we get

$$\int_1^{\infty} f(x)dx = \lim_{n\to\infty}[\log\log(n+1) - \log\log 2]$$

$$= \log\log\infty - \log\log 2 = \infty \text{ (Not finite)}$$

i.e., $\int_1^{\infty} f(x)dx$ is divergent

Hence, the given series is divergent. **Ans.**

Example 8: Show that $\sum_{n=1}^{\infty} \frac{1}{n^p}$ converges is $p > 1$ and diverges if $0 < p < 1$.

Solution: Here $u_n = \frac{1}{n^p} = f(n) \quad \therefore f(x) = \frac{1}{x^p}$

$$\therefore \quad \int_1^n f(x)dx = \int_1^n \frac{1}{x^p}dx = \int_1^n x^{-p}\cdot dx = \left[\frac{x^{1-p}}{1-p}\right]_1^n$$

Case I: When $p > 1$ *i.e.,* $p - 1$ is positive then

$$\int_1^n f(x)dx = -\frac{1}{(p-1)}\left[\frac{1}{x^{p-1}}\right]_1^n = -\frac{1}{(p-1)}\left[\frac{1}{n^{p-1}} - 1\right] = \frac{1}{(p-1)}\left[1 - \frac{1}{n^{p-1}}\right]$$

Taking $\lim_{n\to\infty}$

$$\therefore \quad \int_1^\infty f(x)dx = \frac{1}{(p-1)}\lim_{n\to\infty}\left(1-\frac{1}{n^{p-1}}\right) = \frac{1}{p-1}(1-0) = \frac{1}{p-1} \text{ (Finite)}$$

Hence the given series is convergent when $p > 1$.

Case II: When $0 < p < 1$, *i.e.,* $1 - p$ is positive then

$$\int_1^n f(x)dx = \frac{1}{(1-p)}[n^{1-p} - 1]$$

Taking $\lim_{n\to\infty}$

$$\therefore \quad \int_1^\infty f(x)dx = \frac{1}{(1-p)}\lim_{n\to\infty}(\infty - 1) = \infty$$

Here the series is divergent.

Case III: When $p = 1$, then

$$\int_1^n f(x)dx = \int_1^n \frac{1}{x}dx = [\log x]_1^n = \log n - \log 1$$

$$= \log n \qquad \qquad | \text{ As } \log 1 = 0$$

Taking $\lim_{n\to\infty}$

$$\therefore \quad \int_1^\infty f(x)dx = \lim_{n\to\infty}(\log n) = \infty$$

So the series is divergent

Thus the given series converges if $p > 1$ and diverges if $0 < p \leq 1$. **Hence proved.**

Example 9: Using integral test, discuss the convergence of the series:

$$\sum_{n=2}^{\infty}\frac{1}{n(\log n)^p}, p > 0.$$

Solution: $u_n = \dfrac{1}{n(\log n)^p} = f(n) \quad \therefore f(x) = \dfrac{1}{x(\log x)^p}$

Now $$\int_2^n f(x)dx = \int_2^n \frac{dx}{x(\log x)^p} = \int_2^n (\log x)^{-p}\cdot\frac{1}{x}dx = \left[\frac{(\log x)^{1-p}}{1-p}\right]_2^n$$

Case I: When $p > 1$, $p - 1$ is positive so that

then $$\int_2^n f(x)dx = -\frac{1}{p-1}\left[\frac{1}{(\log x)^{p-1}}\right]_2^n = \frac{1}{p-1}\left[\frac{1}{(\log 2)^{p-1}} - \frac{1}{(\log n)^{p-1}}\right]$$

$$\therefore \quad \int_2^\infty f(x)dx = \lim_{n\to\infty}\frac{1}{p-1}\left[\frac{1}{(\log 2)^{p-1}} - \frac{1}{(\log n)^{p-1}}\right]$$

$$= \frac{1}{p-1}\left[\frac{1}{(\log 2)^{p-1}} - 0\right] = \frac{1}{(p-1)(\log 2)^{p-1}} \text{ (Finite)}$$

$\Rightarrow \int_2^\infty f(x)dx$ converges $\Rightarrow \sum_{n=2}^\infty u_n$ converges.

Case II: When $p < 1$, $1 - p$ is positive so that $\int_2^n f(x)dx = \frac{1}{1-p}[(\log n)^{1-p} - (\log 2)^{1-p}]$

Taking $\lim_{n\to\infty}$, we get

$$\int_2^\infty f(x)dx = \lim_{n\to\infty}\frac{1}{1-p}[(\log n)^{1-p} - (\log 2)^{1-p}] = \frac{1}{1-p}[\infty - (\log 2)^{1-p}] = \infty$$

$\Rightarrow \quad \int_2^\infty f(x)dx$ diverges $= \sum_{n=2}^\infty u_n$ diverges.

Case III: When $p = 1$, $f(x) = \frac{1}{x\log x}$

$$\therefore \quad \int_2^n f(x)dx = \int_2^n \frac{1}{x\log x}dx$$

Let $\log x = t \quad \therefore \frac{1}{x}dx = dt$

$$\Rightarrow \quad \int_2^n f(x)dx = \int_{\log 2}^{\log n}\frac{1}{t}dt = [\log t]_{\log 2}^{\log n} = \log(\log n) - \log(\log 2)$$

Taking $\lim_{n\to\infty}$, we get

$$\int_2^\infty f(x)dx = \lim_{n\to\infty}[\log(\log n) - \log(\log 2)]$$

$$= \infty - \log 2 = \infty$$

$\therefore \quad \sum u_n$ diverges.

Hence the series $\sum_{n=2}^{\infty} u_n$ convergence if $p > 1$ and diverges if $0 < p \leq 1$. **Ans.**

EXERCISE 3.5

Test for the convergence of the following series:

1. $\sum_{n=1}^{\infty} \frac{1}{\sqrt{n}}$. [**Ans.** Divergent]

2. $\sum_{n=1}^{\infty} \frac{1}{(n+1)^2}$. [**Ans.** Convergent]

3. $\sum_{n=1}^{\infty} \frac{1}{n^{2/3}}$. [**Ans.** Divergent]

4. $\sum_{n=2}^{\infty} \frac{1}{n(\log n)^3}$. [**Ans.** Convergent]

5. Discuss the convergence of the series:

$\sum_{n=3}^{\infty} \frac{1}{n \log n (\log \log n)^p}, p > 0.$ [**Ans.** Convergent ($p > 1$) : Divergent ($p \leqslant 1$)]

3.12 POWER SERIES

A power series is an infinite series of the form

$$f(x) = \sum_{n=0}^{\infty} a_n (x-a)^n = a_0 + a_1(x-a) + a_2)(x-a)^2 + \ldots + \ldots \qquad \ldots(i)$$

where a_n represents the coefficients of the nth term a is any constant and x varies around a

If $a = 0$ then the power series (i) is

$$f(x) = \sum_{n=0}^{\infty} a_n x^n = a_0 + a_1 x + a_2 x^2 + \ldots$$

Uniform Convergence

To define uniform convergence, we consider a series whose terms are functions of $f_0(x), f_1(x), \ldots$:

$$\therefore \qquad \sum_{n=0}^{\infty} f_n(x) = f_0(x) + f_1(x) + f_2(x) + \ldots \qquad \ldots(1)$$

Let $S_n(x)$ is the nth sum of series (1), then

$$S_n(x) = f_0(x) + f_1(x) + \ldots + f_n(x)$$

If the sequence $<S_n(x)>$ converges uniformly in some region R then the series (1) is said to be uniformly convergent in the region R.

Thus a series of functions $\sum f_n(x)$ converges uniformly to a function $f(x)$ in a region R if for every $\varepsilon > 0$ we can find an $N = N(\varepsilon)$ not depending on x, such that

$$|S(x) - S_n(x)| < \varepsilon \forall n > N(\varepsilon) \text{ and all } x \text{ in } R.$$

Weierstrass's* M-test

A series of functions $\sum_{n=1}^{\infty} f_n(x)$ converges uniformly in a region R if there exists a convergent series $\sum_{n=1}^{\infty} M_n$ of non-negative terms (*i.e.*, $M_n \geqslant 0 \forall n \in N$) such that

$$|f_n(x)| \leqslant M_n \forall n \in N.$$

Example 1: Show that the following series is uniformly convergent for all real x.

$$\sum_{n=1}^{\infty} \frac{\sin(x^2 + n^2 x)}{n(n+2)}.$$

Solution: Here $f_n(x) = \dfrac{\sin(x^2 + n^2 x)}{n(n+2)}$

$$\therefore \quad |f_n(x)| = \left|\frac{\sin(x^2 + n^2 x)}{n(n+2)}\right| = \frac{|\sin(x^2 + n^2 x)|}{n(n+2)}$$

But the maximum value of $\sin \theta$ is 1

$$\therefore \quad |f_n(x)| \leq \frac{1}{n(n+1)}$$

$$\Rightarrow \quad M_n = \frac{1}{n(n+1)}$$

Now we check the convergence of M_n as follows:

$$\therefore \quad M_{n+1} = \frac{1}{(n+1)(n+2)}$$

---

*Karl Weierstrass (1815 – 1897) great German mathematician, whose lifework was the development of complex analysis based on the concept of power series.

$$\frac{M_n}{M_{n+1}} = \frac{\dfrac{1}{n(n+1)}}{\dfrac{1}{(n+1)(n+2)}} = \frac{(n+1)(n+2)}{n(n+1)} = 1+\frac{2}{n}$$

$\therefore$ $$\lim_{n\to\infty}\frac{M_n}{M_{n+1}} = \lim_{n\to\infty}\left(1+\frac{2}{n}\right) = 1 \text{ (Test fails)}$$

Next $$n\left(\frac{M_n}{M_{n+1}}-1\right) = n\left(1+\frac{2}{n}-1\right) = 2$$

$\therefore$ $$\lim_{n\to\infty}\left\{n\left(\frac{M_n}{M_{n+1}}-1\right)\right\} = 2 > 1$$

Here the series $\sum M_n$ is converge by Raabe's test.

Hence the given series is converges uniformly on the region R by Weierstrass's M-test.

Example 2: Test for uniform convergence the series $\displaystyle\sum_{n=1}^{\infty}\frac{x}{(n+x^2)^2}$

Solution: Here $f_n(x) = \dfrac{x}{(n+x^2)^2}$

First of all we find the point where $f_n(x)$ is maximum for this, we get

$$\frac{df_n}{dx} = 0 \quad \text{and} \quad \frac{d^2 f_n}{dx} = -\text{ve}$$

$\therefore$ $$\frac{df_n(x)}{dx} = \frac{(n+x^2)^2\cdot 1 - x\cdot 2(n+x^2)\cdot 2x}{(n+x^2)^4} = \frac{(n+x^2)-4x^2}{(n+x^2)^3}$$

$$= \frac{n-3x^2}{(n+x^2)^3}$$

$\Rightarrow$ $$\frac{df_n(x)}{dx} = 0 \Rightarrow \frac{n-3x^2}{(n+x^2)^3} = 0 \;\Rightarrow\; n-3x^2 = 0 \;\Rightarrow\; x = \sqrt{\frac{n}{3}}$$

Also $$\frac{d^2 f_n(x)}{dx^2} = \frac{(n+x^2)^3(-6x)-(n-3x^2)\cdot 3(n+x^2)^2\cdot 2x}{(n+x^2)^6}$$

$$= \frac{-6x[(n+x^2)+(n-3x^2)]}{(n+x^2)^4}$$

$$\left[\frac{d^2 f_n(x)}{dx^2}\right]_{x=\sqrt{\frac{n}{3}}} = \frac{-6\sqrt{\frac{n}{3}}\left(n+\frac{n}{3}\right)}{\left(n+\frac{n}{3}\right)^4} = -\frac{27\sqrt{3}}{32n^{5/2}}$$

$\Rightarrow$ $f_n(x)$ is maximum at $x = \sqrt{n/3}$

$\therefore$ The maximum value of $f_n(x)$ is

$$= \frac{\sqrt{n/3}}{\left(n+\frac{n}{3}\right)^2} = \frac{3\sqrt{3}}{16n^{3/2}}$$

$$\Rightarrow \qquad |f_n(x)| \le \frac{3\sqrt{3}}{16n^{3/2}} < \frac{1}{n^{3/2}} = M_n$$

Here $$\sum_n M_n = \sum \frac{1}{n^{3/2}} = \sum \frac{1}{n^p} \Rightarrow p = \frac{3}{2} > 1$$

i.e., M_n is convergent.

Hence the given series is uniformly convergent for all real value of x. **Ans.**

Example 3: Show that the series $\sum\limits_{n=1}^{\infty} \frac{1}{1+n^2 x}$ converges in $1 \le x < \infty$.

Solution: Here $f_n(x) = \frac{1}{1+n^2 x}$

$$\therefore \qquad |f_n(x)| = \left|\frac{1}{1+n^2 x}\right| \le \frac{1}{1+n^2} < \frac{1}{n^2} = M_n \forall x \text{ in given range}$$

Since $\sum\limits_{n=1}^{\infty} M_n = \sum\limits_{n=1}^{\infty} \frac{1}{n^2}$ is convergent, by Weierstrass's M-test, the given series is uniformly convergent in $1 \le x < \infty$. **Hence proved.**

Example 4: Show that the series $\sum\limits_{n=1}^{\infty} \frac{a_n x^n}{1+x^{2n}}$ is uniformly convergent for all real x if $\sum\limits_{n=1}^{\infty} a_n$ is absolutely convergent.

Solution: Here $f_n(x) = \frac{a_n x^n}{1+x^{2n}}$

Then $$\frac{df_n}{dx} = \frac{(1+x^{2n}) \cdot nx^{n-1} - x^n \cdot 2nx^{2n-1}}{(1+x^{2n})^2}$$

$$= \frac{nx^{n-1}(1 + x^{2n} - 2x^{2n})}{(1 + x^{2n})^2} = \frac{nx^{n-1}(1 - x^{2n})}{(1 + x^{2n})^2}$$

For maximum or minimum $\frac{df_n}{dx} = 0 \Rightarrow x = 1$

Also

$$\frac{d^2 f_n}{dx^2} = \frac{(1+x^{2n})^2 \cdot [n(n-1)x^{n-2}(1-x^{2n}) + nx^{n-1}(-2nx^{2n-1})] - nx^{n-1}(1-x^{2n}) \cdot 2(1+x^{2n}) \cdot 2nx^{2n-1}}{(1+x^{2n})^4}$$

$$= \frac{(1+x^{2n})[n(n-1)x^{n-2}(1-x^{2n}) - 2n^2 x^{3n-2}] - 4n^2 x^{3n-2}(1-x^{2n})}{(1+x^{2n})^3}$$

$$\therefore \quad \left[\frac{d^2 f_n}{dx^2}\right]_{x=1} = \frac{2[0 - 2n^2] - 0}{(2)^3} = -\frac{n^2}{2} < 0$$

$\Rightarrow \quad f_n(x) = \frac{x^n}{1+x^{2n}}$ is maximum at $x = 1$ and the maximum value of y is $\frac{1}{2}$.

$$\therefore \quad |f_n(x)| = \left|\frac{a_n x^n}{1+x^{2n}}\right| = \left|\frac{x^n}{1+x^{2n}}\right| |a_n|$$

$$\leq \frac{1}{2}|a_n| < |a_n| = M_n \text{ for all real values of } x.$$

Since $\sum_{n=1}^{\infty} a_n$ is absolutely convergent, therefore, $\sum_{n=1}^{\infty} M_n = \sum_{n=1}^{\infty} |a_n|$ is convergent. Hence by Weierstrass's M-test, the given series is uniformly convergent for all real x. **Hence proved.**

Example 5: If the series $\sum a_n$ converges absolutely then prove that $\sum a_n \cos nx$ and $\sum a_n \sin nx$ are uniformly convergent on R.

Solution: Let us test $\sum a_n \cos nx$

Here $\quad f_n(x) = a_n \cos nx$

$$|f_n(x)| = |a_n \cos nx| = |a_n| |\cos nx| \leq |a_n| = M_n \text{ for all real values of } x.$$

Since $\sum a_n$ is absolutely convergent, therefore, $\sum M_n = \sum |a_n|$ is convergent. Hence by Weierstrass's M-test, the series $\sum a_n \cos nx$ is uniformly convergent on R.

Similarly, $\sum a_n \sin nx$ is uniformaly convergent on R. **Ans.**

Example 6: Show that the series $\sum_{n=1}^{\infty} \frac{(-1)^n}{n^p} \cdot \frac{x^{2n}}{1+x^{2n}}$ is uniformly convergent for all real x if $p > 1$.

Solution: Here $f_n(x) = \frac{(-1)^n}{n^p} \cdot \frac{x^{2n}}{1+x^{2n}}$

Since $\frac{x^{2n}}{1+x^{2n}} < 1 \forall x \in R$

$$\therefore \quad |f_n(x)| = \left| \frac{(-1)^n}{n^p} \cdot \frac{x^{2n}}{1+x^{2n}} \right| < \frac{1}{n^p} = M_n \text{ for all } x \in R.$$

Since $\sum_{n=1}^{\infty} M_n = \sum_{n=1}^{\infty} \frac{1}{n^p}$ is convergent if $p > 1$, therefore, by Weierstrass's M-test, the given series uniformly convergent for all real x if $p > 1$. **Hence proved.**

Example 7: Show that the series $1 + \frac{e^{-2x}}{2^2 - 1} + \frac{e^{-4x}}{4^2 - 1} + \ldots$ is uniformly convergent for $x \geqslant 0$.

Solution: Neglecting the first term, we have $f_n(x) = \frac{e^{-2nx}}{(2n)^2 - 1} = \frac{e^{-2nx}}{4n^2 - 1}$

For all $x \geq 0$, we have $e^{2nx} \geq 1 \Rightarrow e^{-2nx} \leq 1$

Also $\quad 3n^2 > 1 \forall n \quad \Rightarrow 4n^2 > n^2 + 1 \Rightarrow 4n^2 - 1 > n^2$

$$\therefore \quad |f_n(x)| = \left| \frac{e^{-2nx}}{4n^2 - 1} \right| < \frac{1}{n^2} = M_n.$$

Since $\sum_{n=1}^{\infty} M_n = \sum_{n=1}^{\infty} \frac{1}{n^2}$ is convergent, therefore, by Weierstrass's M-test, the given series is uniformly convergent for $x \geq 0$. **Hence proved.**

Example 8: Test for uniform convergence the series

$$\frac{1}{(1+x)^3} + \frac{2}{(2+x)^3} + \frac{3}{(3+x)^3} + \ldots, x \geq 0.$$

Solution: The given series is $\sum_{n=1}^{\infty} \frac{n}{(n+x)^3}$

Here $$f_n(x) = \frac{n}{(n+x)^3}$$

$\forall x \geq 0,$ $$|f_n(x)| = \left|\frac{n}{(n+x)^3}\right| = \frac{n}{(n+x)^3} < \frac{n}{n^3} = \frac{1}{n^2} = M_n$$

Since $\sum_{n=1}^{\infty} M_n = \sum_{n=1}^{\infty} \frac{1}{n^2}$ is convergent, therefore, by Weierstrass's M-test, the given series is uniformly convergent for all $x \geq 0$. **Ans.**

Example 9: Show that the series $\frac{2x}{1+x^2} + \frac{4x^3}{1+x^4} + \frac{8x^7}{1+x^8} + \ldots$ is uniformly convergent in (– 1, 1).

Solution: The given series is $\sum_{n=1}^{\infty} \frac{2^n x^{2^n-1}}{1+x^{2^n}} = \sum_{n=1}^{\infty} f_n(x)$

$$|f_n(x)| = \left|\frac{2^n \cdot x^{2^n-1}}{1+x^{2^n}}\right| \leq 2^n \cdot k^{2^n-1} = M_n \quad \text{for } |x| \leq k < 1 \qquad \ldots(i)$$

Now $$M_n = 2^n k^{2^n-1} \Rightarrow M_{n+1} = 2^{n+1} \cdot k^{2^{n+1}-1}$$

$$\therefore \quad \lim_{n\to\infty} \frac{M_n}{M_{n+1}} = \lim_{n\to\infty} \frac{k^{2^n-1}}{2 \cdot k^{2^{n+1}-1}} = \lim_{n\to\infty} \frac{1}{2k^{2^{n+1}-2^n}} = \lim_{n\to\infty} \frac{1}{2k^{2n}} = \infty \text{ since } k < 1$$

By ratio test, the series $\sum_{n=1}^{\infty} M_n$ is convergent. Hence, by Weierstrass's M-test, the given series is uniformly convergent in (–1, 1). **Hence proved.**

Properties of Uniformly Convergent Series

There are following properties of uniformly convergent series:

1. If a series of continuous terms is uniformly convergent, its sum is also continuous.
2. Under the same assumptions, termwise integration is permissible.

EXERCISE 3.7

1. Show that the following series is uniformly convergent $\sum_{n=1}^{\infty} \frac{\cos nx}{n^p}$

2. Test for uniform convergence of the series $\sum_{n=1}^{\infty} \frac{x}{n(1+nx^2)}$ [**Ans.** Uniformly convergent]

3. Test for uniform convergence the series $1+x+\frac{x^2}{2!}+\frac{x^3}{3!}+\ldots+\frac{x^n}{n!}+\ldots$ in [–1, 1]

   [**Ans.** Uniformly convergent]

4. Show that the series $\sum_{n=1}^{\infty} \frac{a_n x^{2n}}{1+x^{2n}}$ is uniformly convergent for all real x if $\sum a_n$ is absolutely convergent.

UNIT IV

Fourier Series and Partial Differential Equations

4.1 INTRODUCTION

The Fourier series introduced by 'Jean-Baptise Joseph Fourier' in 1807. If the periodic functions occur frequently in many physical investigations and engineering problems then their representation in terms of simple periodic functions such as sine and cosine, is a matter of great practical importance, which leads to Fourier series.

Thus Fourier series is an infinite series representation of periodic function in terms of trigonometric sine and cosine functions. It is a very powerful method to solve ordinary and partial differential equations particularly with periodic functions appearing as non-homogeneous terms. The periodic functions and discontinuous functions expand only by Fourier series but the continuous and differentiable functions may be expand by Taylor's series of expansion as well as Fourier series. Further, because of the periodic nature, Fourier series construct for one period is valid for all values.

4.2 PERIODIC FUNCTION

A function $f(x)$ is said to be periodic if $f(x + T) = f(x)$, for all real x and some positive number T. T is called the period of $f(x)$.

$$\Rightarrow \qquad f(x) = f(x+T) = f(x+2T) = \ldots\ldots\ldots\ldots = f(x+nT)$$

Example: $\sin x = \sin(x + 2\pi) = \sin(x + 4\pi) = \ldots\ldots\ldots\ldots\ldots = \sin(x + n\pi)$, is a periodic function with period 2π.

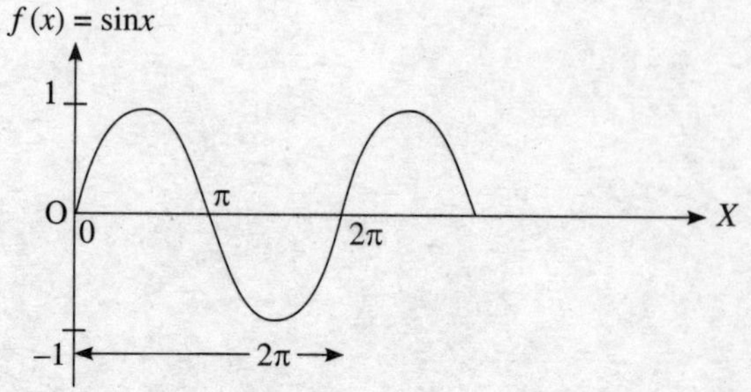

Period of tan θ and cot θ:

$$f(x) = \tan x = \frac{\sin x}{\cos x}$$

Here $f(x)$ will be undefined at the points where $\cos x = 0$

$$\Rightarrow \qquad \cos x = \cos(2n+1)\frac{\pi}{2}$$

$$\Rightarrow \qquad x = (2n+1)\frac{\pi}{2}, n = 0, 1, ...$$

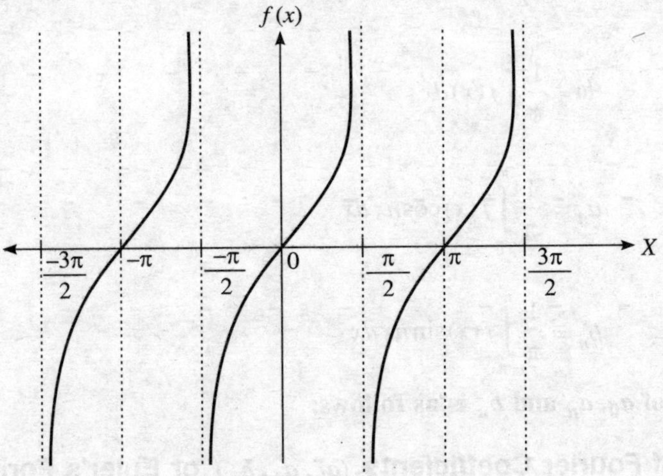

This shows that tan x with period π.
Similarly, we can show that cot x also has a period of π.

4.3 SOME CONCLUSIONS FOR PERIODIC FUNCTION

(*i*) The functions $h(x) = a\,f(x) + bg(x)$ has period T if $f(x)$ and $g(x)$ have period T.

(*ii*) If $f(x)$ is a periodic function of period T, then $f(ax)$ with $a \neq 0$, is a periodic function of period $\frac{T}{a}$.

Example: tan $2x$ has period $\frac{\pi}{2}$.

(*iii*) The period of a sum of a number of periodic functions is the least common multiple of the periods.

(*iv*) A constant function is periodic for any positive T.

Trigonometric Series:

It is a functional series of the form

$$\frac{a_0}{2} + a_1 \cos x + b_1 \sin x + a_2 \cos 2x + b_2 \sin 2x + ...$$

or $\frac{a_0}{2}+\sum_{n=1}^{\infty}(a_n \cos nx + b_n \sin nx)$, where the constants $a_0, a_1, b_1, a_2, b_2, \ldots\ldots\ldots a_n, b_n$ are called the coefficients.

4.4 FOURIER SERIES

The Fourier series of a periodic function $f(x)$ with period 2π is the trigonometric series in the interval $(-\pi, \pi)$, *i.e.*,

$$f(x) = \frac{a_0}{2}+\sum_{n=1}^{\infty}(a_n \cos nx + b_n \sin nx) \qquad \ldots(i)$$

where

$$a_0 = \frac{1}{\pi}\int_{-\pi}^{\pi} f(x)\,dx$$

$$a_n = \frac{1}{\pi}\int_{-\pi}^{\pi} f(x)\cos nx\,dx$$

$$b_n = \frac{1}{\pi}\int_{-\pi}^{\pi} f(x)\sin nx\,dx$$

The determination of a_0, a_n and b_n is as follows:

4.4.1 Determination of Fourier Coefficients (a_0, a_n, b_n) or Euler's Formulae

The coefficients a_0, a_n, b_n are also called Euler's formulae.

(*a*) Determination of a_0

Integrate both side equation (*i*) from $-\pi$ to π, we have

$$\int_{-\pi}^{\pi} f(x)\,dx = \frac{a_0}{2}\int_{-\pi}^{\pi} dx + \sum_{n=1}^{\infty}\left\{a_n \int_{-\pi}^{\pi}\cos nx\,dx + b_n\int_{-\pi}^{\pi}\sin nx\,dx\right\}$$

$$= \frac{a_0}{2}\times 2\pi + \sum_{n=1}^{\infty}\left[a_n\left\{\frac{\sin nx}{n}\right\}_{-\pi}^{\pi} + b_n\left\{-\frac{\cos nx}{n}\right\}_{-\pi}^{\pi}\right]$$

$$= a_0\pi + 0 + 0$$

$$\Rightarrow \qquad \boxed{a_0 = \frac{1}{\pi}\int_{-\pi}^{\pi} f(x)\,dx.} \qquad \ldots(ii)$$

(b) Determination of a_n

Multiply equation (*i*) by cos *mx*, where *m* is any fixed positive integer and then integrate from $-\pi$ to π, we get

$$\int_{-\pi}^{\pi} f(x)\cos mx\,dx = \frac{a_0}{2}\int_{-\pi}^{\pi}\cos mx\,dx + \sum_{n=1}^{\infty}\left\{a_n\int_{-\pi}^{\pi}\cos nx\cos mx\,dx + b_n\int_{-\pi}^{\pi}\sin nx\cos mx\,dx\right\} \quad ...(A)$$

As before, $\int_{-\pi}^{\pi}\cos mx\,dx = 0$

Next $$\int_{-\pi}^{\pi}\cos nx\cos mx\,dx = \frac{1}{2}\int_{-\pi}^{\pi}\{\cos(n+m)x + \cos(n-m)x\}dx$$

$$= \frac{1}{2}\left[\frac{\sin(n+m)x}{(n+m)} + \frac{\sin(n-m)x}{(n-m)}\right]_{-\pi}^{\pi} = 0, \text{ when } n \neq m$$

$$= \frac{1}{2}\int_{-\pi}^{\pi}(\cos 2mx + 1)\,dx, \quad \text{when} \quad n = m$$

$$= \frac{1}{2}\left[\frac{\sin 2mx}{2m} + x\right]_{-\pi}^{\pi} = \frac{1}{2}(0 + 2\pi) = \pi, \quad \text{when} \quad n = m$$

and $$\int_{-\pi}^{\pi}\sin nx\cos mx\,dx = \frac{1}{2}\int_{-\pi}^{\pi}\{\sin(n+m)x + \sin(n-m)x\}dx$$

$$= \frac{1}{2}\left[-\frac{\cos(n+m)x}{(n+m)} - \frac{\cos(n-m)x}{(n-m)}\right]_{-\pi}^{\pi} = 0, \text{ when } n \neq m$$

$$= \frac{1}{2}\int_{-\pi}^{\pi}\sin 2mx\,dx, \quad \text{when} \quad n = m$$

$$= \frac{1}{2}\left[-\frac{\cos 2mx}{2m}\right]_{-\pi}^{\pi} = 0, \quad \text{when} \quad (n = m)$$

Hence, making use of these, when $m = n$, we get

$$\int_{-\pi}^{\pi} f(x)\cos nx\,dx = 0 + \pi\cdot a_n + 0$$

$$\Rightarrow \qquad \boxed{a_n = \frac{1}{\pi}\int_{-\pi}^{\pi} f(x)\cos nx\,dx.} \qquad ...(iii)$$

(c) Determination of b_n

We multiply equation (*i*) by sin *mx* and then integrate from –π to π, we get

$$\int_{-\pi}^{\pi} f(x)\sin mx\,dx = \frac{a_0}{2}\int_{-\pi}^{\pi}\sin mx\,dx+\sum_{n=1}^{\infty}\left\{a_n\int_{-\pi}^{\pi}\cos nx\sin mx\,dx+b_n\int_{-\pi}^{\pi}\sin nx\sin mx\,dx\right\} \quad ...(B)$$

The first term of (*B*) will be zero, *i.e.*,

$$\int_{-\pi}^{\pi}\sin mx\,dx = 0$$

$$\int_{-\pi}^{\pi}\cos nx\sin mx\,dx = 0 \qquad \text{(shown in section } b\text{)}$$

and

$$\int_{-\pi}^{\pi}\sin nx\sin mx\,dx = \frac{1}{2}\int_{-\pi}^{\pi}\cos(n-m)x\,dx-\frac{1}{2}\int_{-\pi}^{\pi}\cos(n+m)\,x\,dx$$

$$= 0, \quad \text{when} \quad n \neq m$$

$$= \pi, \quad \text{when} \quad n = m$$

making use of these values, in (*B*), we get

$$\int_{-\pi}^{\pi} f(x)\sin nx\,dx = 0 + 0 + b_n\,\pi$$

$$\Rightarrow \qquad \boxed{b_n = \frac{1}{\pi}\int_{-\pi}^{\pi} f(x)\cdot\sin nx\,dx} \qquad ...(iv)$$

4.5 DIRICHLET'S CONDITIONS

A function $f(x)$ is said to be satisfy Dirichlet's conditions in the interval (*a*, *b*), if

(*i*) $f(x)$ is defined and is single-valued except possibly at a finite number of points in the interval (*a*, *b*) and

(*ii*) $f(x)$ and $f'(x)$ are piecewise continuous in the interval (*a*, *b*).

Remark: At point of discontinuity x_0, the Fourier series of $f(x)$ converges to the arithmetic mean of the left and right hand limits of $f(x)$ at x_0 *i.e.*, $f(x_0) = \frac{1}{2}\left[f(x_0-0)+f(x_0+0)\right]$.

4.6 FOURIER SERIES FOR THE INTERVAL $0 \le x \le 2\pi$

If $f(x)$ be a periodic with period 2π and defined in the interval (0, 2π) then the Fourier series defined like article (4.4) with coefficients

$$\left.\begin{aligned} a_0 &= \frac{1}{\pi}\int_0^{2\pi} f(x)\,dx \\ a_n &= \frac{1}{\pi}\int_0^{2\pi} f(x)\cos nx\,dx \\ b_n &= \frac{1}{\pi}\int_0^{2\pi} f(x)\sin nx\,dx \end{aligned}\right\} \quad \ldots(v)$$

Likewise, if $f(x)$ be defined in the interval $c < x < c + 2\pi$ then

$$a_0 = \frac{1}{\pi}\int_c^{c+2\pi} f(x)\,dx,$$

$$a_n = \frac{1}{\pi}\int_c^{c+2\pi} f(x)\cos nx\,dx,$$

$$b_n = \frac{1}{\pi}\int_c^{c+2\pi} f(x)\sin nx\,dx.$$

Example 1: Find a Fourier series of the periodic function $f(x)$ with period 2π in the interval $(-\pi, \pi)$ which is defined as follows:

$$f(x) = x, \ -\pi < x < \pi.$$

Solution: The graphical representation of given function may be as follows:

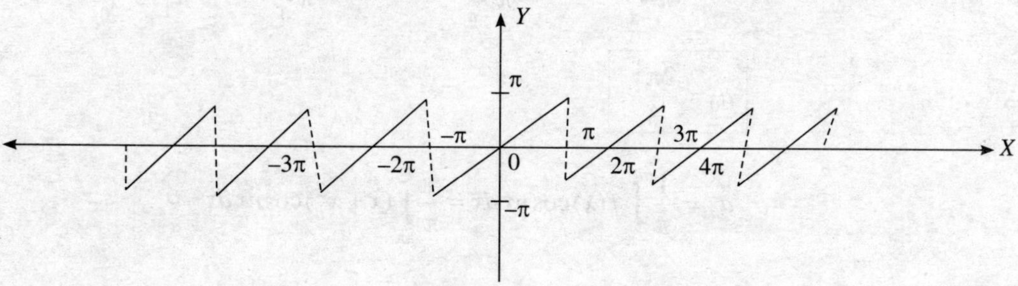

By Fourier series for the interval $(-\pi, \pi)$, we have

$$f(x) = \frac{a_0}{2} + \sum_{n=1}^{\infty}(a_n \cos nx + b_n \sin nx) \quad \ldots(1)$$

$$\therefore \quad a_0 = \frac{1}{\pi}\int_{-\pi}^{\pi} f(x)\,dx = \frac{1}{\pi}\int_{-\pi}^{\pi} x\,dx = \frac{1}{\pi}\left[\frac{x^2}{2}\right]_{-\pi}^{\pi} = 0$$

and

$$a_n = \frac{1}{\pi}\int_{-\pi}^{\pi} f(x)\cos nx\,dx$$

$$\Rightarrow \qquad a_n = \frac{1}{\pi}\int_{-\pi}^{\pi} x\cos nx\,dx = \frac{1}{\pi}\left[x\cdot\left(\frac{\sin nx}{n}\right)-1\left(-\frac{\cos nx}{n^2}\right)\right]_{-\pi}^{\pi} = 0$$

Finally
$$b_n = \frac{1}{\pi}\int_{-\pi}^{\pi} x\sin nx\,dx = \frac{1}{\pi}\left[x\cdot\left(-\frac{\cos nx}{n}\right)-\left(-\frac{\sin nx}{n^2}\right)\right]_{-\pi}^{\pi}$$

$$\Rightarrow \qquad b_n = (-1)^{n+1}\cdot\frac{2}{n},\ n = 1, 2, 3, \ldots\ldots$$

Hence, from equation (1), we get

$$f(x) = 2\left\{\frac{\sin x}{1}-\frac{\sin 2x}{2}+\frac{\sin 3x}{3}\ldots\ldots\ldots\ldots(-1)^{n+1}\frac{\sin nx}{n}\right\}. \quad \textbf{Ans.}$$

Example 2: Given that $f(x) = x + x^2$ for $-\pi < x < \pi$, find the Fourier expression of $f(x)$. Deduce that $\frac{\pi^2}{6} = 1 + \frac{1}{2^2} + \frac{1}{3^2} + \frac{1}{4^2} + \cdots\cdots$. *(U.P.T.U. 2003)*

Solution: By Fourier series

$$f(x) = \frac{a_0}{2} + \sum_{n=1}^{\infty}(a_n\cos nx + b_n\sin nx) \qquad \ldots(1)$$

$$a_0 = \frac{1}{\pi}\int_{-\pi}^{\pi} f(x)\,dx = \frac{1}{\pi}\int_{-\pi}^{\pi}(x+x^2)\,dx = \frac{2}{\pi}\int_0^{\pi} x^2\,dx = \frac{2}{\pi}\left[\frac{x^3}{3}\right]_0^{\pi} = \frac{2\pi^2}{3}$$

$$\Rightarrow \qquad \boxed{a_0 = \frac{2\pi^2}{3}.}$$

$$a_n = \frac{1}{\pi}\int_{-\pi}^{\pi} f(x)\cos nx\,dx = \frac{1}{\pi}\int_{-\pi}^{\pi}(x+x^2)\cos nx\,dx$$

$$= \frac{2}{\pi}\int_0^{\pi} x^2\cos nx\,dx \qquad \text{(As, } x\cos nx \text{ is odd)}$$

$$= \frac{2}{\pi}\left\{\left[\frac{x^2\sin nx}{n}\right]_0^{\pi} - \frac{2}{n}\int_0^{\pi} x\sin nx\,dx\right\}$$

$$= \frac{2}{n}\times\frac{2}{\pi}\left\{0+\left[\frac{x}{n}\cos nx\right]_0^{\pi} - \frac{1}{n^2}\{\sin nx\}_0^{\pi}\right\} = \frac{4}{n\pi}\left[\frac{\pi}{n}\cos n\pi\right]$$

$$\Rightarrow \qquad \boxed{a_n = \frac{4}{n^2}(-1)^n,\ n = 1, 2, ...} \qquad (\text{As } \cos n\pi = (-1)^n)$$

and
$$b_n = \frac{1}{\pi}\int_{-\pi}^{\pi} f(x)\sin nx\,dx = \frac{1}{\pi}\int_{-\pi}^{\pi}(x+x^2)\sin nx\,dx$$

$$\Rightarrow \qquad b_n = \frac{2}{\pi}\int_0^{\pi} x\sin nx\,dx \qquad (\text{As } x^2 \sin nx \text{ is odd})$$

$$= \frac{2}{\pi}\left\{\left[-\frac{x\cos nx}{n}\right]_0^{\pi} + \frac{1}{n}\int_0^{\pi}\cos nx\,dx\right\}$$

$$= \frac{2}{\pi}\left[-\frac{\pi}{n}\cos n\pi + \frac{1}{n^2}[\sin nx]_0^{\pi}\right] = -\frac{2}{n}\cos n\pi$$

$$\Rightarrow \qquad \boxed{b_n = -\frac{2}{n}(-1)^n,\ n = 1, 2, ...}$$

Putting the values of a_0, a_n and b_n in equation (1), we get

$$f(x) = x + x^2 = \frac{\pi^2}{3} + 4\left(-\cos x + \frac{1}{2^2}\cos 2x - \frac{1}{3^2}\cos 3x + ...\right)$$

$$-2\left(-\sin x + \frac{1}{2}\sin 2x - \frac{1}{3}\sin 3x + ...\right) \qquad ...(2)$$

Second part: At the point of discontinuity *i.e.*, $x = \pm\pi$, the sum of the series *i.e.*, the value of $f(x)$ at $x = \pi$ or $-\pi$,

$$= \frac{1}{2}\left[f(-\pi+0) + f(\pi-0)\right]$$

$$f(\pi \text{ or } -\pi) = \frac{1}{2}\left[-\pi + \pi^2 + \pi + \pi^2\right] = \pi^2$$

Thus putting $x = \pi$ or $x = -\pi$ on the right hand side of (2), we get

$$\pi^2 = \frac{\pi^2}{3} - 4\left(\frac{\cos\pi}{1^2} - \frac{\cos 2\pi}{2^2} + \frac{\cos 3\pi}{3^2} - ......\right) + 2.(0)$$

$$\Rightarrow \qquad \frac{\pi^2}{6} = \frac{1}{1^2} + \frac{1}{2^2} + \frac{1}{3^2} + \cdots\cdots = \sum_{n=1}^{\infty}\frac{1}{n^2}. \qquad \textbf{Proved.}$$

Example 3: Find the Fourier series for the function defined by

$$f(x) = \begin{cases} -1 & \text{for} \quad -\pi < x < 0 \\ 0 & \text{for} \quad x = 0 \\ 1 & \text{for} \quad 0 < x < \pi \end{cases}$$

Hence prove that

$$\frac{\pi}{4} = 1 - \frac{1}{3} + \frac{1}{5} - \frac{1}{7} + \frac{1}{9} - \cdots\cdots \qquad \textit{(U.P.T.U. 2005)}$$

Solution: By Fourier series, we have

$$f(x) = \frac{a_0}{2} + \sum_{n=1}^{\infty} (a_n \cos nx + b_n \sin nx) \qquad \text{...(1)}$$

$$a_0 = \frac{1}{\pi}\int_{-\pi}^{\pi} f(x)\,dx = \frac{1}{\pi}\left[\int_{-\pi}^{0} (-1)\,dx + \int_{0}^{\pi} dx\right] = \frac{1}{\pi}\left[[-x]_{-\pi}^{0} + [x]_0^{\pi}\right]$$

$$\Rightarrow \quad a_0 = \frac{1}{\pi}[-\pi + \pi] \quad \Rightarrow \quad \boxed{a_0 = 0.}$$

$$a_n = \frac{1}{\pi}\int_{-\pi}^{\pi} f(x)\cos nx\,dx = \frac{1}{\pi}\int_{-\pi}^{0} -\cos nx\,dx + \frac{1}{\pi}\int_{0}^{\pi} \cos nx\,dx$$

$$= \frac{1}{\pi}\left\{-\frac{1}{n}[\sin nx]_{-\pi}^{0} + \frac{1}{n}[\sin nx]_0^{\pi}\right\}$$

$$= \frac{1}{\pi}[-0+0] = 0 \quad \Rightarrow \quad \boxed{a_n = 0.}$$

and
$$b_n = \frac{1}{\pi}\int_{-\pi}^{\pi} f(x)\sin nx\,dx = \frac{1}{\pi}\int_{-\pi}^{0} -\sin nx\,dx + \frac{1}{\pi}\int_{0}^{\pi} \sin nx\,dx$$

$$= \frac{1}{\pi}\left\{\frac{1}{n}[\cos nx]_{-\pi}^{0} - \frac{1}{n}[\cos nx]_0^{\pi}\right\}$$

$$= \frac{1}{n\pi}[1 - \cos n\pi - \cos n\pi + 1]$$

$$= \frac{1}{n\pi}[2 - 2\cos n\pi] = \frac{2}{n\pi}[1 - (-1)^n]$$

$$\Rightarrow \quad \boxed{b_n = \frac{2}{n\pi}[1 - (-1)^n],\ n = 1, 2, \ldots\ldots}$$

Hence from (1)

$$f(x) = \sum_{n=1}^{\infty} b_n \sin nx = b_1 \sin x + b_2 \sin 2x + \cdots\cdots\cdots$$

$$\Rightarrow \quad f(x) = \frac{4}{\pi}\sin x + 0 + \frac{4}{3\pi}\sin 3x + 0 + \frac{4}{5\pi}\sin 5x + \cdots\cdots\cdots$$

$$\Rightarrow \quad f(x) = \frac{4}{\pi}\sin x + \frac{4}{3\pi}\sin 3x + \frac{4}{5\pi}\sin 5x + \cdots \qquad \text{...(2)}$$

The expansion (2) is required Fourier expansion.

Second part: The given function is continuous at $x = \frac{\pi}{2}$, therefore

$$f(\pi/2) = 1$$

Hence putting $x = \frac{\pi}{2}$, in equation (2), we get

$$1 = \frac{4}{\pi}\left(\sin\frac{\pi}{2} + \frac{1}{3}\sin\frac{3\pi}{2} + \frac{1}{5}\sin\frac{5\pi}{2} + \cdots\cdots\right)$$

$$\Rightarrow \quad \frac{\pi}{4} = 1 - \frac{1}{3} + \frac{1}{5} - \frac{1}{7} + \frac{1}{9} \cdots\cdots \qquad \textbf{Proved.}$$

Example 4: Find the Fourier series of the function defined as

$$f(x) = \begin{cases} x+\pi, & \text{for} \quad 0 \le x \le \pi \\ -x-\pi, & \text{for} \quad -\pi \le x < 0 \end{cases} \text{ and } f(x+2\pi) = f(x). \quad (\textit{U.P.T.U. 2006})$$

Solution: the Fourier series for the interval $(-\pi, \pi)$, is

$$f(x) = \frac{a_0}{2} + \sum_{n=1}^{\infty}(a_n \cos nx + b_n \sin nx) \qquad \text{...(1)}$$

$$a_0 = \frac{1}{\pi}\int_{-\pi}^{\pi} f(x)\,dx = \frac{1}{\pi}\int_{-\pi}^{0} -(x+\pi)\,dx + \frac{1}{\pi}\int_{0}^{\pi}(x+\pi)\,dx$$

$$= \frac{1}{\pi}\left[-\left\{\frac{x^2}{2} + \pi x\right\}_{-\pi}^{0} + \left\{\frac{x^2}{2} + \pi x\right\}_{0}^{\pi}\right] = \pi$$

$$\Rightarrow \quad \boxed{a_0 = \pi.}$$

$$a_n = \frac{1}{\pi}\int_{-\pi}^{\pi} f(x)\cos nx\,dx = \frac{1}{\pi}\int_{-\pi}^{0} -(x+\pi)\cos nx\,dx + \frac{1}{\pi}\int_{0}^{\pi}(x+\pi)\cos nx\,dx$$

$$= \frac{1}{\pi}\left[-(x+\pi)\frac{\sin nx}{n} - (-1)\left\{-\frac{\cos nx}{n^2}\right\}\right]_{-\pi}^{0} + \frac{1}{\pi}\left[(x+\pi)\frac{\sin nx}{n} + \frac{\cos nx}{n^2}\right]_{0}^{\pi}$$

$$= \frac{1}{\pi}\left[-\frac{1}{n^2} + \frac{(-1)^n}{n^2}\right] + \frac{1}{\pi}\left[\frac{(-1)^n}{n^2} - \frac{1}{n^2}\right] = \frac{2}{n^2\pi}\left[(-1)^n - 1\right]$$

$$\Rightarrow \quad \boxed{a_n = \begin{cases} -4/n^2\pi, & n \text{ is odd} \\ 0, & n \text{ is even} \end{cases}}$$

and

$$b_n = \frac{1}{\pi}\int_{-\pi}^{\pi} f(x)\sin nx\,dx = \frac{1}{\pi}\int_{-\pi}^{0} -(x+\pi)\sin nx\,dx + \frac{1}{\pi}\int_{0}^{\pi}(x+\pi)\sin nx\,dx$$

$$= \frac{1}{\pi}\left[-(x+\pi)\left(-\frac{\cos nx}{n}\right)-(-1)\left(-\frac{\sin nx}{n^2}\right)\right]_{-\pi}^{0}$$

$$+\frac{1}{\pi}\left[(x+\pi)\left(-\frac{\cos nx}{n}\right)-(1)\left(-\frac{\sin nx}{n^2}\right)\right]_{0}^{\pi}$$

$$= \frac{1}{\pi}\left[\frac{\pi}{n}\right]+\frac{1}{\pi}\left[-\frac{2\pi}{n}(-1)^n+\frac{\pi}{n}\right]=\frac{1}{n}+\left(-\frac{2}{n}\right)(-1)^n+\frac{1}{n}=\frac{2}{n}[1-(-1)^n]$$

$$\Rightarrow \quad \boxed{b_n = \begin{cases}\dfrac{4}{n}, & \text{if } n \text{ is odd}\\ 0, & \text{if } n \text{ is even.}\end{cases}}$$

Hence the required Fourier series is, from (1)

$$f(x) = \frac{\pi}{2}+a_1\cos x+a_2\cos 2x+\cdots\cdots+b_1\sin x+b_2\sin 2x+\cdots\cdots$$

$$= \frac{\pi}{2}-\frac{4}{\pi}\left(\frac{\cos x}{1^2}+\frac{\cos 3x}{3^2}+\cdots\cdots\right)+4\left(\frac{\sin x}{1}+\frac{\sin 3x}{3}+\cdots\cdots\right) \quad \textbf{Ans.}$$

Example 5: Obtain the Fourier series of $f(x) = (\pi - x)/2$ in the interval $(0, 2\pi)$. Deduce that

$$\frac{\pi}{4} = 1-\frac{1}{3}+\frac{1}{5}-\frac{1}{7}+\cdots\cdots.$$ *(U.P.T.U. 2009)*

Solution: The Fourier series for the interval $(0, 2\pi)$ is as follows:

$$f(x) = \frac{a_0}{2}+\sum_{n=1}^{\infty}(a_n\cos nx+b_n\sin nx) \quad \ldots(1)$$

$$a_0 = \frac{1}{\pi}\int_0^{2\pi} f(x)\,dx = \frac{1}{\pi}\int_0^{2\pi}\left(\frac{\pi-x}{2}\right)dx = \frac{1}{2\pi}\left[\pi x-\frac{x^2}{2}\right]_0^{2\pi} = 0$$

$$\Rightarrow \quad \boxed{a_0 = 0.}$$

$$a_n = \frac{1}{\pi}\int_0^{2\pi} f(x)\cos nx\,dx = \frac{1}{\pi}\int_0^{2\pi}\left(\frac{\pi-x}{2}\right)\cdot\cos nx\,dx$$

$$= \frac{1}{2\pi}\left[(\pi-x)\frac{\sin nx}{n}-(-1)\frac{(-\cos nx)}{n^2}\right]_0^{2\pi} = 0$$

$$\Rightarrow \quad \boxed{a_n = 0.}$$

and $$b_n = \frac{1}{\pi}\int_0^{2\pi} f(x)\sin nx\,dx = \frac{1}{\pi}\int_0^{2\pi}\left(\frac{\pi-x}{2}\right)\sin nx\,dx$$

$$= \frac{1}{2\pi}\left[(\pi - x)\left(-\frac{\cos nx}{n}\right) - (-1)\left(\frac{-1}{n}\right)\frac{\sin nx}{n}\right]_0^{2\pi}$$

$$= \frac{1}{2\pi}\left[\frac{\pi}{n} + \frac{\pi}{n}\right] = \frac{1}{n} \quad \Rightarrow \quad \boxed{b_n = \frac{1}{n}.}$$

Making use of the values a_0, a_n and b_n in (1), we get

$$f(x) = \frac{\pi - x}{2} = 0 + 0 + \sum_{n=1}^{\infty} \frac{1}{n} \sin nx$$

$$\Rightarrow \quad \frac{\pi - x}{2} = \sin x + \frac{1}{2}\sin 2x + \frac{1}{3}\sin 3x + \cdots\cdots \qquad ...(2)$$

which is the required Fourier expansion in the interval (0, 2π).

Now putting $x = \frac{\pi}{2}$ in (2), we get

$$\frac{\pi - \frac{\pi}{2}}{2} = \sin\frac{\pi}{2} + \frac{1}{2}\sin\pi + \frac{1}{3}\sin\frac{3\pi}{2} + \cdots\cdots$$

$$\Rightarrow \quad \frac{\pi}{4} = 1 - \frac{1}{3} + \frac{1}{5} - \frac{1}{7} + \cdots\cdots \qquad \textbf{Proved.}$$

Example 6: Find the Fourier series expansion for $f(x) = x + \frac{x^2}{4}, -\pi \le x \le \pi$ *(U.P.T.U. 2009)*

Solution: We have

$$f(x) = \frac{a_0}{2} + \sum_{n=1}^{\infty}(a_n \cos nx + b_n \sin nx) \qquad ...(i)$$

$$a_0 = \frac{1}{\pi}\int_{-\pi}^{\pi} f(x)dx = \frac{1}{\pi}\int_{-\pi}^{\pi}\left(x + \frac{x^2}{4}\right)dx$$

$$= \frac{1}{2\pi}\int_0^{\pi} x^2 dx \qquad \left| \text{As } \frac{1}{\pi}\int_{-\pi}^{\pi} x\,dx = 0 \right. \quad \text{(odd function)}$$

$$= \frac{1}{\pi}\left[\frac{x^3}{6}\right]_0^{\pi} = \frac{1}{\pi}\left[\frac{\pi^3}{6} - 0\right]$$

$$\Rightarrow \quad \boxed{a_0 = \frac{\pi^2}{6}}$$

$$a_n = \frac{1}{\pi}\int_{-\pi}^{\pi} f(x)\cos nx\,dx = \frac{1}{\pi}\int_{-\pi}^{\pi}\left(x + \frac{x^2}{4}\right)\cdot \cos nx\,dx$$

$$= \frac{1}{2\pi}\int_0^{\pi} x^2 \cos nx\, dx \qquad \Big|\ \text{As } \frac{1}{\pi}\int_{-\pi}^{\pi} x\cos nx\, dx = 0 \text{ (odd function)}$$

$$= \frac{1}{2\pi}\left[\left\{x^2 \sin nx \cdot \frac{1}{n}\right\}_0^{\pi} - \frac{2}{n}\int_0^{\pi} x \sin nx\, dx\right]$$

$$= -\frac{1}{\pi n}\left[\left(-x\cos nx \cdot \frac{1}{n}\right)_0^{\pi} + \frac{1}{n}\int_0^{\pi} \cos nx\, dx\right]$$

$$a_n = -\frac{1}{\pi n}\left[-\frac{\pi \cos n\pi}{n} + \frac{1}{n^2}(\sin nx)_0^{\pi}\right]$$

$$\boxed{a_n = \frac{1}{n^2}(-1)^n} \qquad \Big|\ \text{As } \cos n\pi = (-1)^n$$

$$b_n = \frac{1}{\pi}\int_{-\pi}^{\pi} f(x)\sin nx\, dx = \frac{1}{\pi}\int_{-\pi}^{\pi}\left(x + \frac{x^2}{4}\right)\sin nx\, dx$$

$$= \frac{2}{\pi}\int_0^{\pi} x \sin nx\, dx \qquad \Big|\ \text{As } \frac{1}{\pi}\int_{-\pi}^{\pi} \frac{x^2 \sin nx}{4}\, dx = 0 \text{ (odd function)}$$

$$= \frac{2}{\pi}\left[\left(-\frac{x}{n}\cos nx\right)_0^{\pi} + \left(\frac{1}{n^2}\sin nx\right)_0^{\pi}\right]$$

$$= \frac{2}{\pi}\left(-\frac{\pi}{n}\cos n\pi\right)$$

$$\Rightarrow \qquad \boxed{b_n = -\frac{2}{n}(-1)^n}$$

From (*i*)

$$f(x) = x + \frac{x^2}{4} = \frac{\pi^2}{12} + \sum_{n=1}^{\infty}\frac{(-1)^n}{n^2}\cos nx - 2\sum_{n=1}^{\infty}\frac{(-1)^n}{n}\sin nx. \quad \textbf{Ans.}$$

Example 7: If $f(x) = x$ in $\left(-\frac{\pi}{2}, \frac{\pi}{2}\right)$ and $f(x) = 0$ in $\left(\frac{\pi}{2}, \frac{3\pi}{2}\right)$ find the Fourier series of $f(x)$.

Deduce that $\frac{\pi^2}{8} = \sum_{n=1}^{\infty}\frac{1}{(2n-1)^2}$.

Solution: Here

$$f(x) = \begin{cases} x, & \frac{-\pi}{2} < x < \frac{\pi}{2} \\ 0, & \frac{\pi}{2} < x < \frac{3\pi}{2} \end{cases}$$

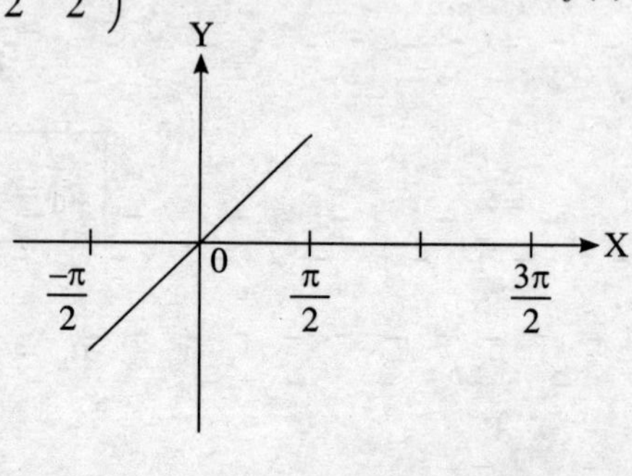

We have from Fourier series

$$f(x) = \frac{a_0}{2} + \sum_{n=1}^{\infty} (a_n \cos nx + b_n \sin nx) \quad \ldots(1)$$

$$a_0 = \frac{1}{\pi} \int_{-\pi/2}^{3\pi/2} f(x)\,dx = \frac{1}{\pi} \int_{-\pi/2}^{\pi/2} x\,dx + \frac{1}{\pi} \int_{\pi/2}^{3\pi/2} 0\,dx$$

$$a_0 = \frac{1}{\pi}\left[\frac{x^2}{2}\right]_{-\pi/2}^{\pi/2} = \frac{1}{2\pi}\left[\frac{\pi^2}{4} - \frac{\pi^2}{4}\right] = 0 \quad \Rightarrow \quad \boxed{a_0 = 0}$$

$$a_n = \frac{1}{\pi} \int_{-\pi/2}^{3\pi/2} f(x) \cos nx\,dx = \frac{1}{\pi} \int_{-\pi/2}^{\pi/2} x \cos nx\,dx + \frac{1}{\pi} \int_{\pi/2}^{3\pi/2} 0\,dx = \frac{1}{\pi} \int_{-\pi/2}^{\pi/2} x \cos nx\,dx$$

$$\Rightarrow \quad a_n = \frac{1}{\pi}\left[x \frac{\sin nx}{n} + \frac{1}{n} \cdot \frac{\cos nx}{n}\right]_{-\pi/2}^{\pi/2} = 0 \quad \Rightarrow \quad \boxed{a_n = 0.}$$

and

$$b_n = \frac{1}{\pi} \int_{-\pi/2}^{3\pi/2} f(x) \sin nx\,dx = \frac{1}{\pi} \int_{-\pi/2}^{\pi/2} x \sin nx\,dx + \frac{1}{\pi} \int_{\pi/2}^{3\pi/2} 0\,dx = \frac{1}{\pi} \int_{-\pi/2}^{\pi/2} x \sin nx\,dx$$

$$\Rightarrow \quad b_n = \frac{1}{\pi}\left[x \cdot \frac{(-\cos nx)}{n} + \frac{1}{n} \frac{\sin nx}{n}\right]_{-\pi/2}^{\pi/2}$$

$$\Rightarrow \quad \boxed{b_n = -\frac{1}{n}\left[\cos \frac{n\pi}{2}\right] + \frac{2}{n^2 \pi} \cdot \sin \frac{n\pi}{2}}$$

∴ From (1), we get

$$f(x) = \sum_{n=1}^{\infty} \left(\frac{-\cos n\pi/2}{n} + \frac{2}{n^2 \pi} \cdot \sin \frac{n\pi}{2} \right) \sin nx$$

$$= \frac{2}{\pi} \sin x + \frac{1}{2} \sin 2x - \frac{2}{3^2 \cdot \pi} \sin 3x - \frac{1}{4} \sin 4x + \frac{2}{5^2 \cdot \pi} \sin 5x + \cdots\cdots$$

$$\Rightarrow \quad f(x) = \frac{2}{\pi}\left\{\sin x - \frac{1}{3^2} \sin 3x + \frac{1}{5^2} \sin 5x - \cdots\cdots\right\} + \frac{1}{2}\left\{\sin 2x - \frac{1}{2} \sin 4x + \cdots\cdots\right\} \quad \ldots(2)$$

which is the required Fourier expansion

At $\quad x = \frac{\pi}{2}$ which is point of discontinuity

$$f(\pi/2) = \frac{1}{2}\left[f(\pi/2 - 0) + f(\pi/2 + 0)\right]$$

$$= \frac{1}{2}\left[\frac{\pi}{2}+0\right] = \pi/4.$$

Putting $x = \pi/2$ in (2), we get

$$\frac{\pi}{4} = \frac{2}{\pi}\left\{\sin\frac{\pi}{2} - \frac{1}{3^2}\sin\frac{3\pi}{2} + \frac{1}{5^2}\sin\frac{5\pi}{2} - \cdots\cdots\right\} + \frac{1}{2}\{\sin\pi - \frac{1}{2}\sin 2\pi + \cdots\cdots\}$$

$$\frac{\pi}{4} = \frac{2}{\pi}\left\{1 + \frac{1}{3^2} + \frac{1}{5^2} + \cdots\cdots\right\} + 0$$

$$\Rightarrow \quad \frac{\pi^2}{8} = \left\{1 + \frac{1}{3^2} + \frac{1}{5^2} + \cdots\cdots + \frac{1}{(2n-1)^2} + \cdots\cdots\right\}$$

$$\Rightarrow \quad \frac{\pi^2}{8} = \sum_{n=1}^{\infty}\frac{1}{(2n-1)^2}.$$ **Proved.**

Example 8: Show that, if p is neither zero nor an integer

$$\cos px = \frac{2p\sin p\pi}{\pi}\left\{\frac{1}{2p^2} - \frac{\cos x}{(p^2-1^2)} + \frac{\cos 2x}{(p^2-2^2)} - \cdots\cdots\right\} \text{ in } (-\pi, \pi)$$

Deduce that $$\frac{\pi}{\sin p\pi} = \sum_{-\infty}^{\infty}\frac{(-1)^n}{(p+n)}.$$

Solution: We know that, the Fourier expansion for function $f(x)$ in $(-\pi, \pi)$ is

$$f(x) = \frac{a_0}{2} + \sum_{n=1}^{\infty}(a_n\cos nx + b_n\sin nx) \qquad \ldots(1)$$

$$a_0 = \frac{1}{\pi}\int_{-\pi}^{\pi} f(x)\,dx = \frac{1}{\pi}\int_{-\pi}^{\pi}\cos px\,dx = \frac{1}{\pi}\left[\frac{\sin px}{p}\right]_{-\pi}^{\pi} = \frac{1}{\pi p}[2\sin p\pi]$$

$$\Rightarrow \quad \boxed{a_0 = \frac{2\sin p\pi}{p\pi}}, \ p \text{ is neither zero nor integer.}$$

$$a_n = \frac{1}{\pi}\int_{-\pi}^{\pi} f(x)\cos nx\,dx = \frac{1}{\pi}\int_{-\pi}^{\pi}\cos px\cos nx\,dx$$

$$= \frac{1}{2\pi}\int_{-\pi}^{\pi}\{\cos(p+n)x + \cos(p-n)x\}dx$$

$$\left|\text{As } 2\cos A\cos B = \cos(A+B) + \cos(A-B)\right.$$

$$= \frac{1}{\pi}\int_0^{\pi}\{\cos(p+n)x + \cos(p-n)x\}\,dx \quad \text{(By definite integral)}$$

$$= \frac{1}{\pi}\left[\frac{\sin(p+n)x}{(p+n)} + \frac{\sin(p-n)x}{(p-n)}\right]_0^{\pi} = \frac{1}{\pi}\left[\frac{\sin(p+n)\pi}{(p+n)} + \frac{\sin(p-n)\pi}{(p-n)}\right]$$

$$\Rightarrow \quad \boxed{a_n = \frac{(-1)^n}{\pi}\left[\frac{2p\sin p\pi}{(p^2-n^2)}\right]}$$

and
$$b_n = \frac{1}{\pi}\int_{-\pi}^{\pi} f(x)\sin nx\,dx = \frac{1}{\pi}\int_{-\pi}^{\pi}\cos px \sin nx\,dx$$

$$= \frac{1}{2\pi}\int_{-\pi}^{\pi}\{\sin(p+n)x - \sin(p-n)x\}dx \quad \left|\begin{array}{l}\text{As } 2\cos A\sin B \\ \quad = \sin(A+B) - \sin(A-B)\end{array}\right.$$

$= 0$ (By definite integral, since sin $(p \pm n)x$ is an odd function)

$\therefore$ From relation (1), we get

$$f(x) = \cos px = \frac{\sin p\pi}{p\pi} - \frac{1}{\pi}\frac{2p\sin p\pi}{(p^2-1^2)}\cos x + \frac{1}{\pi}\frac{2p\sin p\pi}{(p^2-2^2)}\cos 2x - \cdots\cdots$$

$$\Rightarrow \quad \cos px = \frac{2p\sin p\pi}{\pi}\left\{\frac{1}{2p^2} - \frac{\cos x}{p^2-1^2} + \frac{\cos 2x}{p^2-2^2} - \cdots\cdots\right\}. \quad \textbf{Proved.}$$

Deduction: Putting $x = 0$ in above result, we get

$$1 = \frac{2p\sin p\pi}{\pi}\left\{\frac{1}{2p^2} - \frac{1}{p^2-1^2} + \frac{1}{p^2-2^2} - \cdots\cdots\right\}$$

$$\Rightarrow \quad \frac{\pi}{\sin p\pi} = \left\{\frac{1}{p} - \frac{2p}{(p-1)(p+1)} + \frac{2p}{(p-2)(p+2)} - \cdots\cdots\right\}$$

$$= \left\{\frac{1}{p} - \left(\frac{1}{p+1} + \frac{1}{p-1}\right) + \left(\frac{1}{p+2} + \frac{1}{p-2}\right) - \cdots\cdots\right\}$$

$$= \left\{\cdots\cdots + \frac{1}{p-2} - \frac{1}{p-1} + \frac{1}{p} - \frac{1}{p+1} + \frac{1}{p+2} - \cdots\cdots\right\}$$

$$\Rightarrow \quad \frac{\pi}{\sin p\pi} = \sum_{n=-\infty}^{\infty}\frac{(-1)^n}{(p+n)}. \quad \textbf{Proved.}$$

4.7 FOURIER SERIES FOR FUNCTIONS HAVING ARBITRARY PERIOD

Fourier series of periodic function with period 2π has already discussed. Now we consider the Fourier series expansion of a function having period $2l$.

Let $f(x)$ be a periodic function with arbitrary period $2l$ defined in the interval $-l < x < l$.
Here we introduce a new variable 't' for which the function of t defined in the interval $-\pi < t < \pi$.

Let $$t = \frac{\pi x}{l} \quad i.e., \quad x = \frac{lt}{\pi} \qquad \text{...(i)}$$

then the function $f\left(\frac{l}{\pi}t\right)$ will be periodic with period 2π in the interval $-\pi < t < \pi$.

∴ By Fourier series, we have

$$f\left(\frac{l}{\pi}t\right) = \frac{a_0}{2} + \sum_{n=1}^{\infty}(a_n \cos nt + b_n \sin nt) \qquad \text{...(ii)}$$

where $$a_0 = \frac{1}{\pi}\int_{-\pi}^{\pi} f\left(\frac{l}{\pi}t\right)dt,$$

$$a_n = \frac{1}{\pi}\int_{-\pi}^{\pi} f\left(\frac{l}{\pi}t\right)\cos nt\, dt,$$

and $$b_n = \frac{1}{\pi}\int_{-\pi}^{\pi} f\left(\frac{l}{\pi}t\right)\sin nt\, dt.$$

Now let us return to the original variable x. From (*i*)

$$t = \frac{\pi}{l}x \quad \Rightarrow \quad dt = \frac{\pi}{l}dx$$

Therefore the Fourier series expansion (*ii*) takes the form

$$f(x) = \frac{a_0}{2} + \sum_{n=1}^{\infty}\left(a_n \frac{\cos n\pi x}{l} + b_n \sin\frac{n\pi x}{l}\right) \qquad \text{...(iii)}$$

where $$a_0 = \frac{1}{l}\int_{-l}^{l} f(x)\,dx,$$

$$a_n = \frac{1}{l}\int_{-l}^{l} f(x)\cos\frac{n\pi x}{l}\,dx,$$

and $$b_n = \frac{1}{l}\int_{-l}^{l} f(x)\sin\frac{n\pi x}{l}\,dx.$$

Remark: The interval of integration in (*iii*) may be replaced by any interval of $2l$.

4.8 HALF RANGE EXPANSIONS

4.8.1 Half Range Cosine Series in $0 < x < l$

If the function $f(x)$ is even in $(-l, l)$ and is periodic with period $2l$. Here we redefine this function in an arbitrary manner in the interval $-l < x < 0$.

Such that

$$f(-x) = f(x).$$

Therefore, the Fourier cosine series is defined as follows:

$$f(x) = \frac{a_0}{2} + \sum_{n=1}^{\infty} a_n \cos\frac{n\pi x}{l}, \quad (0 < x < l) \qquad |\text{As } b_n = 0$$

where $$a_0 = \frac{1}{l}\int_{-l}^{l} f(x)\,dx \quad \Rightarrow \quad a_0 = \frac{2}{l}\int_{0}^{l} f(x)\,dx$$

and $$a_n = \frac{1}{l}\int_{-l}^{l} f(x)\cos\frac{n\pi x}{l}\,dx \quad \Rightarrow \quad a_n = \frac{2}{l}\int_{0}^{l} f(x)\cos\frac{n\pi x}{l}\,dx$$

Particular case: If $l = \pi$, then the half range cosine series in the interval $0 < x < \pi$ is given by

$$f(x) = \frac{a_0}{2} + \sum_{n=1}^{\infty} a_n \cos nx$$

where $$a_0 = \frac{2}{\pi}\int_{0}^{\pi} f(x)\,dx$$

$$a_n = \frac{2}{\pi}\int_{0}^{\pi} f(x)\cos nx\,dx.$$

4.8.2 Half Range Sine Series in $0 < x < l$

If the function $f(x)$ is odd in the interval $(-l, l)$ and is periodic with period $2l$. Here we redefine this function in an arbitrary manner on the interval $-l < x < 0$ such that $f(-x) = -f(x)$.

Therefore, the Fourier sine series is defined as follows:

$$f(x) = \sum_{n=1}^{\infty} b_n \sin\frac{n\pi x}{l}, (0 < x < l) \qquad |\text{As } a_0 = a_n = 0$$

where $$b_n = \frac{1}{l}\int_{-l}^{l} f(x)\sin\frac{n\pi x}{l}\,dx \quad \Rightarrow \quad b_n = \frac{2}{l}\int_{0}^{l} f(x)\sin\frac{n\pi x}{l}\,dx$$

Particular case: If $l = \pi$, then the half range sine series in the interval $0 < x < \pi$ is given by

$$f(x) = \sum_{n=1}^{\infty} b_n \sin nx, \quad (0 < x < \pi)$$

where $$b_n = \frac{2}{\pi}\int_0^{\pi} f(x)\sin nx\,dx$$

Example 9: Find the Fourier series of

$$f(x) = \begin{cases} 0, & \text{if } -2 \le x \le -1 \\ 1+x, & \text{if } -1 \le x \le 0 \\ 1-x, & \text{if } 0 \le x \le 1 \\ 0, & \text{if } 1 \le x \le 2. \end{cases}$$

Solution: Here the function $f(x)$ is defined in the interval $(-2, 2)$ *i.e.*, $0 < x < 4$

$$\Rightarrow \qquad 2l = 4 \;\Rightarrow\; l = 2$$

By Fourier series, we have

$$f(x) = \frac{a_0}{2} + \sum_{n=1}^{\infty}\left(a_n\cos\frac{n\pi x}{2} + b_n\sin\frac{n\pi x}{2}\right) \qquad \ldots(1)$$

$$a_0 = \frac{1}{2}\int_{-2}^{2} f(x)\,dx = \frac{1}{2}\int_{-2}^{-1} 0\,dx + \frac{1}{2}\int_{-1}^{0}(1+x)\,dx + \int_0^1 (1-x)\,dx + \int_1^2 0\,dx$$

$$= \frac{1}{2}\left\{\left[x + \frac{x^2}{2}\right]_{-1}^{0} + \left[x - \frac{x^2}{2}\right]_0^1\right\} = \frac{1}{2}\left[\left(1-\frac{1}{2}\right)+\left(1-\frac{1}{2}\right)\right] = \frac{1}{2}$$

$$\Rightarrow \qquad \boxed{a_0 = \frac{1}{2}.}$$

$$a_n = \frac{1}{2}\int_{-2}^{-1} 0\,dx + \frac{1}{2}\int_{-1}^{0}(1+x)\cos\frac{n\pi x}{2}\,dx + \frac{1}{2}\int_0^1 (1-x)\cos\frac{n\pi x}{2}\,dx + \frac{1}{2}\int_1^2 0\,dx$$

$$= \frac{1}{2}\times 2\int_0^1 (1-x)\cos\frac{n\pi x}{2}\,dx$$ As the given function $f(x)$ is even in the interval $(-1, 1)$ *i.e.*, $f(-x) = f(x)$

$$= \int_0^1 (1-x)\cos\frac{n\pi x}{2}\,dx = \left[(1-x)\cdot\frac{2}{n\pi}\cdot\sin\left(\frac{n\pi x}{2}\right) - (-1)\frac{2}{n\pi}\cdot\frac{2}{n\pi}\cdot(-1)\cos\frac{n\pi x}{2}\right]_0^1$$

$$\boxed{a_n = \frac{-4}{n^2\pi^2}\left[\cos\frac{n\pi}{2} - 1\right]}$$

and $$b_n = 0$$ As $f(x)\cdot\sin\frac{n\pi x}{2}$ is odd in $-1 < x < 1$

Hence from expansion (1), we get

$$f(x) = \frac{1}{4} + \sum_{n=1}^{\infty} -\frac{4}{n^2\pi^2}\left[\cos\frac{n\pi}{2} - 1\right]\cos\frac{n\pi x}{2}$$

$$\Rightarrow \quad f(x) = \frac{1}{4} - \frac{4}{\pi^2}\left[-\frac{1}{1^2}\cos\frac{\pi x}{2} - \frac{2}{2^2}\cos\pi x - \frac{1}{3^2}\cos\frac{3\pi x}{2} - \frac{1}{5^2}\cos\frac{5\pi x}{2} - \frac{2}{6^2}\cos 3\pi x - \ldots\ldots\right]$$

$$f(x) = \frac{1}{4} + \frac{4}{\pi^2}\left[\left(\cos\frac{\pi x}{2} + \frac{1}{3^2}\cos\frac{3\pi x}{2} + \frac{1}{5^2}\cos\frac{5\pi x}{2} + \ldots\ldots\right)\right.$$

$$\left. + 2\left(\frac{1}{2^2}\cos\pi x + \frac{1}{6^2}\cos 3\pi x + \ldots\ldots\right)\right]. \quad \textbf{Ans.}$$

Example 10: Obtain the half range sine series for the function $f(x) = x^2$ in the interval $0 < x < 3$.

(U.P.T.U. 2002)

Solution: In order to find a half range Fourier sine series for the function $f(x) = x^2$ in the interval $0 < x < 3$, we must redefine it in the interval $-3 < x < 0$, so that $f(-x) = -f(x)$. Therefore

$$f(x) = \begin{cases} x^2 & \text{for} \quad 0 < x < 3 \\ -x^2 & \text{for} \quad -3 < x < 0 \end{cases}$$

$\therefore$ By the Fourier sine series, we have

$$f(x) = \sum_{n=1}^{\infty} b_n \sin\frac{n\pi x}{3} \qquad \ldots(1)$$

$$b_n = \frac{2}{3}\int_0^3 f(x)\sin\frac{n\pi x}{3}\,dx \qquad |\text{As } l = 3$$

$$= \frac{2}{3}\int_0^3 x^2 \sin\frac{n\pi x}{3}\,dx$$

$$= \frac{2}{3}\left[x^2\left(-\frac{3}{n\pi}\right)\left(\cos\frac{n\pi x}{3}\right)\right.$$

$$\left. + 2x \times \left(\frac{3}{n\pi}\right)\left(\frac{3}{n\pi}\right)\sin\frac{n\pi x}{3} - 2\left(\frac{3}{n\pi}\right)^3\cos\frac{n\pi x}{3}\right]_0^3$$

$$= \frac{2}{3}\left[\left\{-\frac{27}{n\pi}(-1)^n - \frac{54}{n^3\pi^3}(-1)^n\right\} + \frac{54}{n^3\pi^3}\right]$$

$$\Rightarrow \quad b_n = \frac{2}{3}\left[\frac{54}{n^3\pi^3}\left\{1 - (-1)^n\right\} - \frac{27}{n\pi}(-1)^n\right]$$

$$\Rightarrow \quad \boxed{\begin{aligned} b_n &= 2/3\left[\frac{108}{n^3\pi^3}+\frac{27}{n\pi}\right], \text{ when } n \text{ is odd} \\ &= -18/n\pi, \text{ when } n \text{ is even.}\end{aligned}}$$

Hence the half range sine series from (1) is

$$f(x) = \frac{2}{3}\left[\frac{108}{\pi^3}\left(\frac{\sin x}{1^3}+\frac{\sin 3x}{3^3}+\ldots\right)+\frac{27}{\pi}\left(\frac{\sin x}{1}+\frac{\sin 3x}{3}+\ldots\right)-\frac{18}{\pi}\left(\frac{\sin 2x}{2}+\frac{\sin 4x}{4}+\ldots\right)\right]. \quad \textbf{Ans.}$$

Example 11: Find a series of cosines of multiples of x which will represent $x \sin x$ in the interval $(0, \pi)$ and show that $\frac{1}{1\cdot 3}-\frac{1}{3\cdot 5}+\frac{1}{5\cdot 7}\ldots\ldots = \frac{\pi-2}{4}$. *(U.P.T.U. 2001)*

OR

Examine whether the function $f(x) = x \sin x$ is even or odd. Hence expand it in the form of Fourier series in the interval $(-\pi, \pi)$. *(U.P.T.U. 2008)*

Solution: $f(x)$ is even since $f(-x) = f(x)$. The Fourier series redues to cosine series for given function is as follows:

$$f(x) = x\sin x = \frac{a_0}{2}+\sum_{n=1}^{\infty} a_n \cos nx \qquad |\text{As } l = \pi$$

$$a_0 = \frac{2}{\pi}\int_0^{\pi} x\sin x\,dx = \frac{2}{\pi}[x(-\cos x)-1\cdot(-\sin x)]_0^{\pi} = \frac{2}{\pi}[-\pi\cos\pi] = 2$$

$$\Rightarrow \quad \boxed{a_0 = 2.}$$

$$a_n = \frac{2}{\pi}\int_0^{\pi} x\sin x\cos nx\,dx = \frac{1}{\pi}\int_0^{\pi} x[\sin(n+1)x-\sin(n-1)x]dx$$

$$= \frac{1}{\pi}\left[x\left\{-\frac{\cos(n+1)x}{(n+1)}+\frac{\cos(n-1)x}{(n-1)}\right\}-\left\{-\frac{\sin(n+1)x}{(n+1)^2}-\frac{\sin(n-1)x}{(n-1)^2}\right\}\right]_0^{\pi}$$

$$a_n = \frac{1}{\pi}\left[-\frac{\pi\cos(n+1)\pi}{(n+1)}+\frac{\pi\cos(n-1)\pi}{n-1}\right], \text{ when } n \neq 1$$

$$= -\frac{(-1)^{n+1}}{(n+1)}+\frac{(-1)^{n-1}}{(n-1)} = (-1)^{n-1}\left[\frac{1}{n-1}-\frac{1}{n+1}\right] = \frac{2(-1)^{n-1}}{(n-1)(n+1)}$$

$$\Rightarrow \quad \boxed{a_n = \frac{2(-1)^{n-1}}{(n-1)(n+1)}, \text{ when } n \neq 1}$$

When $n = 1$, we have

$$a_1 = \frac{2}{\pi}\int_0^{\pi} x\sin x\cos x\,dx = \frac{1}{\pi}\int_0^{\pi} x\sin 2x\,dx$$

$$= \frac{1}{\pi}\left[x\left(-\frac{\cos 2x}{2}\right) - 1\left(-\frac{\sin 2x}{2^2}\right)\right]_0^{\pi} = \frac{1}{\pi}\left[-\frac{\pi\cos 2\pi}{2}\right]$$

$$\Rightarrow \quad \boxed{a_1 = -\frac{1}{2}.}$$

Therefore, the half range cosine series is

$$x\sin x = 1 - \frac{1}{2}\cos x - 2\left(\frac{\cos 2x}{1\cdot 3} - \frac{\cos 3x}{2\cdot 4} + \frac{\cos 4x}{3\cdot 5} - \ldots\ldots\right)$$

Putting $x = \frac{\pi}{2}$, we get

$$\frac{\pi}{2} = 1 - 2\left(-\frac{1}{1\cdot 3} + \frac{1}{3\cdot 5} - \frac{1}{5\cdot 7} + \ldots\ldots\right)$$

$$\Rightarrow \quad 2\left(\frac{1}{1\cdot 3} - \frac{1}{3\cdot 5} + \frac{1}{5\cdot 7} - \ldots\ldots\right) = \frac{\pi}{2} - 1$$

$$\Rightarrow \quad \frac{1}{1\cdot 3} - \frac{1}{3\cdot 5} + \frac{1}{5\cdot 7} - \ldots\ldots = \frac{\pi - 2}{4}.$$ **Proved.**

Example 12: Let $$f(x) = \begin{cases} wx, & \text{when } 0 \le x \le \frac{l}{2} \\ w(l-x), & \text{when } \frac{l}{2} \le x \le l \end{cases}$$

Show that $$f(x) = \frac{4wl}{\pi^2}\sum_{n=0}^{\infty}\frac{(-1)^n}{(2n+1)^2}\sin\frac{(2n+1)\pi x}{l}.$$

Hence obtain the sum of the series

$$1 + \frac{1}{3^2} + \frac{1}{5^2} + \ldots\ldots$$ *(U.P.T.U. 2003)*

Solution: From the half range sine series, we have

$$f(x) = \sum_{n=1}^{\infty} b_n \sin\frac{n\pi x}{l} \quad \ldots(1)$$

$$b_n = \frac{2}{l}\int_0^{l} f(x)\sin\frac{n\pi x}{l}\,dx = \frac{2}{l}\left[\int_0^{l/2} wx\sin\frac{n\pi x}{l}\,dx + \int_{l/2}^{l} w(l-x)\sin\frac{n\pi x}{l}\,dx\right]$$

$$= \frac{2}{l}\left[w\left\{-\frac{x\cos\frac{n\pi x}{l}}{\frac{n\pi}{l}}\right\}_0^{l/2} + \frac{wl}{n\pi}\int_0^{l/2}\cos\frac{n\pi x}{l}\,dx + w\left\{(l-x)\cdot\left(\frac{-\cos\frac{n\pi x}{l}}{\frac{n\pi}{l}}\right)\right\}_{l/2}^{l} + \frac{lw}{n\pi}\int_{l/2}^{l} -\cos\frac{n\pi x}{l}\,dx\right]$$

$$= \frac{2}{l}\left[-\frac{l^2 w}{2n\pi}\cos\frac{n\pi}{2} + \frac{l^2 w}{n^2\pi^2}\left(\sin\frac{n\pi x}{l}\right)_0^{l/2} + \frac{l^2 w}{2n\pi}\cos\frac{n\pi}{2} - \frac{l^2 w}{n^2\pi^2}\left(\sin\frac{n\pi x}{l}\right)_{l/2}^{l}\right]$$

$$= \frac{2}{l}\left[\frac{wl^2}{n^2\pi^2}\sin\frac{n\pi}{2} - \frac{wl^2}{n^2\pi^2}\left(0-\sin\frac{n\pi}{2}\right)\right] = \frac{4wl}{n^2\pi^2}\sin\frac{n\pi}{2}$$

$$\Rightarrow \quad \boxed{b_n = \frac{4wl}{n^2\pi^2}\sin\frac{n\pi}{2}}$$

$\therefore$ From (1),

$$f(x) = \frac{4wl}{\pi^2}\sum_{n=1}^{\infty}\frac{1}{n^2}\sin\frac{n\pi}{2}\sin\frac{n\pi x}{l}$$

$$f(x) = \frac{4wl}{\pi^2}\sum_{n=0}^{\infty}\frac{1}{(2n+1)^2}\cdot\sin(2n+1)\frac{\pi}{2}\sin\frac{(2n+1)\pi x}{l}$$

$$= \frac{4wl}{\pi^2}\sum_{n=0}^{\infty}\frac{(-1)^n}{(2n+1)^2}\sin\frac{(2n+1)\pi x}{l} \quad \text{...(2)}$$

Putting $x = \frac{l}{2}$ in equation (2), we get

$$\frac{wl}{2} = \frac{4wl}{\pi^2}\sum_{n=0}^{\infty}\frac{(-1)^n}{(2n+1)^2}\sin(2n+1)\frac{\pi}{2} = \frac{4wl}{\pi^2}\sum_{n=0}^{\infty}\frac{(-1)^n}{(2n+1)^2}(-1)^n$$

$$\Rightarrow \quad \frac{\pi^2}{8} = \sum_{n=0}^{\infty}\frac{1}{(2n+1)^2}$$

or $$\frac{\pi^2}{8} = \frac{1}{1^2}+\frac{1}{3^2}+\frac{1}{5^2}+\cdots \quad \textbf{Ans.}$$

Example 13: Expand for $f(x) = k$ for $0 < x < 2$ in a half range

(1) Sine series

(2) Cosine series. *(U.P.T.U. 2007)*

Solution:

(1) Here $f(x) = k$ and $l = 2$

By half range sine series, we get

$$f(x) = k = \sum_{n=1}^{\infty} b_n \sin \frac{n\pi x}{2} \qquad ...(1)$$

$$b_n = \frac{2}{2}\int_0^2 k \cdot \sin \frac{n\pi x}{2}\, dx \qquad |\text{As } l = 2$$

$$= \frac{2k}{n\pi}\left[-\cos \frac{n\pi x}{2}\right]_0^2 = \frac{2k}{n\pi}[1 - \cos n\pi]$$

$$\Rightarrow \qquad \boxed{b_n = \frac{2k}{n\pi}[1-(-1)^n].}$$

$\therefore$ From (1), we get

$$k = \frac{2k}{\pi}\sum_{n=1}^{\infty}[1-(-1)^n] \sin \frac{n\pi x}{2}$$

$$\Rightarrow \qquad k = \frac{2k}{\pi}\left[2 \sin \frac{\pi x}{2} + 2 \sin \frac{3\pi x}{2} + 2 \sin \frac{5\pi x}{2} + \cdots\right]$$

$$\Rightarrow \qquad k = \frac{4k}{\pi}\left[\sin \frac{\pi x}{2} + \sin \frac{3\pi x}{2} + \sin \frac{5\pi x}{2} + \cdots\right]. \quad \textbf{Ans.}$$

(2) From half range cosine series, we get

$$f(x) = k = \frac{a_0}{2} + \sum_{n=1}^{\infty} a_n \cos \frac{n\pi x}{2} \qquad ...(2)$$

$$a_0 = \frac{2}{2}\int_0^2 k\, dx = k[x]_0^2 = 2k \qquad |\text{As } l = 2$$

and
$$a_n = \frac{2}{2}\int_0^2 k \cos \frac{n\pi x}{2}\, dx = \frac{2k}{n\pi}\left[\sin \frac{n\pi x}{2}\right]_0^2$$

$$\Rightarrow \qquad \boxed{a_n = 0}$$

$\therefore$ From (2), we get

$$f(x) = k = \frac{1}{2}(2k) + \sum_{n=1}^{\infty} 0 \cdot \cos \frac{n\pi x}{2}$$

$$\Rightarrow \qquad f(x) = k. \quad \textbf{Ans.}$$

Example 14: Find the Fourier half range cosine series of the function

$$f(t) = \begin{cases} 2t, & 0 < t < 1 \\ 2(2-t), & 1 < t < 2. \end{cases}$$

(U.P.T.U. 2001, 2006, 2007)

Solution: We know that

$$f(t) = \frac{a_0}{2} + \sum_{n=1}^{\infty} a_n \cos\frac{n\pi t}{l}$$

$$\Rightarrow \quad f(t) = \frac{a_0}{2} + \sum_{n=1}^{\infty} a_n \cos\frac{n\pi t}{2} \qquad ...(1) \;|\text{As } l = 2$$

$$a_0 = \frac{2}{l}\int_0^l f(t)\,dt = \int_0^2 f(t)\,dt = \int_0^1 2t\,dt + \int_1^2 2(2-t)\,dt$$

$$\Rightarrow \quad a_0 = [t^2]_0^1 + 2\left[2t - \frac{t^2}{2}\right]_1^2 = 1 + [(4t - t^2)]_1^2$$

$$\Rightarrow \quad a_0 = 1 + (8 - 4 - 4 + 1) = 2 \Rightarrow \boxed{a_0 = 2.}$$

and

$$a_n = \frac{2}{l}\int_0^l f(t)\cos\frac{n\pi t}{l}\,dt = \int_0^2 f(t)\cos\frac{n\pi t}{2}\,dt \qquad |\text{As } l = 2$$

$$= \int_0^1 2t\cos\frac{n\pi t}{2}\cdot dt + \int_1^2 2(2-t)\cos\frac{n\pi t}{2}\,dt$$

$$= \left[2t\left(\frac{2}{n\pi}\sin\frac{n\pi t}{2}\right) - 2\left(-\frac{4}{n^2\pi^2}\cos\frac{n\pi t}{2}\right)\right]_0^1$$

$$+ \left[(4-2t)\left(\frac{2}{n\pi}\sin\frac{n\pi t}{2}\right) - (-2)\left(-\frac{4}{n^2\pi^2}\cos\frac{n\pi t}{2}\right)\right]_1^2$$

$$= \left[\frac{4}{n\pi}\sin\frac{n\pi}{2} + \frac{8}{n^2\pi^2}\cos\frac{n\pi}{2} - \frac{8}{n^2\pi^2}\right] + \left[0 - \frac{8}{n^2\pi^2}\cos n\pi - \frac{4}{n\pi}\sin\frac{n\pi}{2} + \frac{8}{n^2\pi^2}\cos\frac{n\pi}{2}\right]$$

$$= \frac{16}{n^2\pi^2}\cos\frac{n\pi}{2} - \frac{8}{n^2\pi^2}(1 + \cos n\pi)$$

$$\Rightarrow \quad \boxed{a_n = \begin{cases} 0, & n \text{ is odd} \\ \dfrac{16}{n^2\pi^2}(-1)^{n/2} - \dfrac{16}{n^2\pi^2}, & n \text{ is even} \end{cases}} \qquad \left|\cos\frac{n\pi}{2} = (-1)^{n/2},\right. \text{ when } n \text{ is even}$$

Thus $a_2 = -\dfrac{32}{2^2\cdot\pi^2}$, $a_4 = 0$, $a_6 = -\dfrac{32}{6^2\pi^2}$, $a_8 = 0$, $a_{10} = -\dfrac{32}{10^2\cdot\pi^2}$ and so on.

Hence from (1) the half range Fourier cosine series is

$$f(t) = 1 - \frac{32}{\pi^2}\left(\frac{1}{2^2}\cos\pi t + \frac{1}{6^2}\cos 3\pi t + \frac{1}{10^2}\cos 5\pi t + \cdots\right)$$

or

$$f(t) = 1 - \frac{8}{\pi^2}\left(\frac{1}{1^2}\cos\pi t + \frac{1}{3^2}\cos 3\pi t + \frac{1}{5^2}\cos 5\pi t + \cdots\right) \quad \textbf{Ans.}$$

Example 15: Find the two half range expansions of

$$f(x) = \begin{cases} \dfrac{2kx}{L} & \text{if } 0 < x < \dfrac{L}{2} \\ \dfrac{2k(L-x)}{L} & \text{if } \dfrac{L}{2} < x < L \end{cases}$$

[*U.P.T.U. (Summer) 2009*]

Solution:

(*i*) Fourier half range cosine series:

$$f(x) = \frac{a_0}{2} + \sum_{n=1}^{\infty} a_n \cos\frac{n\pi x}{L} \qquad \ldots(i)$$

$$a_0 = \frac{2}{L}\int_0^L f(x)\,dx$$

$$= \frac{2}{L}\left[\int_0^{L/2} \frac{2kx}{L}\,dx + \int_{L/2}^{L} \frac{2k(L-x)}{L}\,dx\right]$$

$$= \frac{4k}{L^2}\left[\int_0^{L/2} x\,dx + \int_{L/2}^{L} (L-x)\,dx\right]$$

$$= \frac{4k}{L^2}\left[\left(\frac{x^2}{2}\right)_0^{L/2} + \left(Lx - \frac{x^2}{2}\right)_{L/2}^{L}\right]$$

$$= \frac{4k}{L^2}\left[\frac{L^2}{8} + L^2 - \frac{L^2}{2} - \frac{L^2}{2} + \frac{L^2}{8}\right]$$

$$a_0 = \frac{4k}{L^2}\left[\frac{L^2}{4}\right] \Rightarrow \boxed{a_0 = k}$$

$$a_n = \frac{2}{L}\int_0^L f(x)\cdot\cos\frac{n\pi x}{L}\,dx$$

$$= \frac{2}{L}\left[\frac{2k}{L}\int_0^{L/2} \cdot x\cos\frac{n\pi x}{L}\,dx + \frac{2k}{L}\int_{L/2}^{L}(L-x)\cos\frac{n\pi x}{L}\,dx\right]$$

$$a_n = \frac{4k}{L^2}\left[\left\{x\cdot\frac{L}{n\pi}\sin\frac{n\pi x}{L}\right\}_0^{L/2} + \frac{L^2}{n^2\pi^2}\left\{\cos\frac{n\pi x}{L}\right\}_0^{L/2}\right.$$

$$\left. + \left\{(L-x)\cdot\frac{L}{n\pi}\sin\frac{n\pi x}{L}\right\}_{L/2}^{L} - \frac{L^2}{n^2\pi^2}\left\{\cos\frac{n\pi x}{L}\right\}_{L/2}^{L}\right]$$

$$= \frac{4k}{L^2}\left[\frac{L^2}{2n\pi}\cdot\sin\frac{n\pi}{2} + \frac{L^2}{n^2\pi^2}\left(\cos\frac{n\pi}{2}-1\right)\right.$$

$$\left. - \frac{L^2}{2n\pi}\sin\frac{n\pi}{2} - \frac{L^2}{n^2\pi^2}\left(\cos n\pi - \cos\frac{n\pi}{2}\right)\right]$$

$$a_n = \frac{4k}{L^2}\left[\frac{L^2}{n^2\pi^2}\left(2\cos\frac{n\pi}{2} - \cos n\pi - 1\right)\right]$$

$$a_n = \frac{4k}{n^2\pi^2}\left(2\cos\frac{n\pi}{2} - \cos n\pi - 1\right)$$

Hence the required half range cosine series is

$$f(x) = \frac{k}{2} + \frac{4k}{\pi^2}\sum_{n=1}^{\infty}\frac{1}{n^2}\left(2\cos\frac{n\pi}{2} - \cos n\pi - 1\right)\cos\frac{n\pi x}{L}\cdot \quad \textbf{Ans.}$$

(*ii*) Fourier half range sine series:

$$f(x) = \sum_{n=1}^{\infty} b_n \sin\frac{n\pi x}{L} \qquad \text{...(ii)}$$

$$b_n = \frac{2}{L}\int_0^L f(x)\sin\frac{n\pi x}{L}\,dx$$

$$= \frac{2}{L}\left[\int_0^{L/2}\frac{2kx}{L}\cdot\sin\frac{n\pi x}{L}\,dx + \int_{L/2}^{L}\frac{2k(L-x)}{L}\cdot\sin\frac{n\pi x}{L}\,dx\right]$$

$$= \frac{4k}{L^2}\left[\left\{-x\cdot\frac{L}{n\pi}\cos\frac{n\pi x}{L}\right\}_0^{L/2} + \frac{L^2}{n^2\pi^2}\left\{\sin\frac{n\pi x}{L}\right\}_0^{L/2}\right.$$

$$\left. + \left\{-(L-x)\cdot\frac{L}{n\pi}\cos\frac{n\pi x}{L}\right\}_{L/2}^{L} - \frac{L^2}{n^2\pi^2}\left\{\sin\frac{n\pi x}{L}\right\}_{L/2}^{L}\right]$$

$$b_n = \frac{4k}{L^2}\left[-\frac{L^2}{2n\pi}\cos\frac{n\pi}{2} + \frac{L^2}{n^2\pi^2}\left(\sin\frac{n\pi}{2}\right) + \frac{L^2}{2n\pi}\cos\frac{n\pi}{2} + \frac{L^2}{n^2\pi^2}\left(\sin\frac{n\pi}{2}\right)\right]$$

$$= \frac{4k}{L^2}\left[\frac{2L^2}{n^2\pi^2}\sin\frac{n\pi}{2}\right]$$

$$\Rightarrow \quad b_n = \frac{8k}{n^2\pi^2}\sin\frac{n\pi}{2}$$

Hence $f(x) = \frac{8k}{\pi^2}\sum_{n=1}^{\infty}\frac{1}{n^2}\sin\frac{n\pi}{2}\sin\frac{n\pi x}{L}$. **Ans.**

Example 16: Obtain the Fourier series expansion of $f(x) = \frac{\pi - x}{2}$ in $0 < x < 2$.

Solution: Here the length of interval is $2l = 2$ *i.e.*, $l = 1$

$\therefore$ The Fourier series for $f(x)$ is

$$f(x) = \frac{\pi - x}{2} = \frac{a_0}{2} + \sum_{n=1}^{\infty}\left(a_n \cos\frac{n\pi x}{1} + b_n \sin\frac{n\pi x}{1}\right)$$

$$\Rightarrow \quad \frac{\pi - x}{2} = \frac{a_0}{2} + \sum_{n=1}^{\infty}(a_n \cos n\pi x + b_n \sin n\pi x) \quad \ldots(1)$$

$$a_0 = \frac{1}{1}\int_0^2\left(\frac{\pi - x}{2}\right)dx = \frac{1}{2}\left[\pi x - \frac{x^2}{2}\right]_0^2 = (\pi - 1) \quad \Rightarrow \quad \boxed{a_0 = (\pi - 1)}$$

$$a_n = \frac{1}{1}\int_0^2 f(x)\cos n\pi x\, dx = \int_0^2\left(\frac{\pi - x}{2}\right)\cos n\pi x dx$$

$$= \frac{1}{2}\left[(\pi - x)\left(\frac{1}{n\pi}\right)\sin n\pi x - (-1)\times\frac{1}{n\pi}(-1)\frac{\cos n\pi x}{n\pi}\right]_0^2$$

$$\Rightarrow \quad a_n = -\frac{1}{2n^2\pi^2}[\cos 2n\pi - \cos 0] = -\frac{1}{2n^2\pi^2}[1 - 1] = 0 \quad \Rightarrow \quad \boxed{a_n = 0.}$$

and

$$b_n = \frac{1}{1}\int_0^2 f(x)\cdot\sin n\pi x\, dx = \int_0^2\left(\frac{\pi - x}{2}\right)\sin n\pi x\, dx$$

$$= \frac{1}{2}\left[(\pi - x)\left(\frac{-\cos n\pi x}{n\pi}\right) - (-1)\cdot(-1)\frac{\sin n\pi x}{n^2\pi^2}\right]_0^2$$

$$= -\frac{1}{2n\pi}\left[(\pi-2)\cos 2n\pi - \pi\cdot\cos 0\right] = \frac{1}{n\pi}$$

$$\Rightarrow \qquad \boxed{b_n = \frac{1}{n\pi}}$$

$\therefore$ From (1), we get

$$\frac{\pi - x}{2} = \frac{\pi-1}{2} + 0 + \sum_{n=1}^{\infty} \frac{1}{n\pi} \sin n\pi x$$

$$\Rightarrow \qquad \frac{\pi - x}{2} = \frac{\pi-1}{2} + \frac{1}{\pi}\left(\sin \pi x + \frac{1}{2}\sin 2\pi x + \frac{1}{3}\sin 3\pi x + \cdots\cdots\right) \quad \textbf{Ans.}$$

Example 17: Find the half period sine series for $f(x)$ given in the range $(0, l)$ by the graph OPQ as shown in figure. *(U.P.T.U. 2009)*

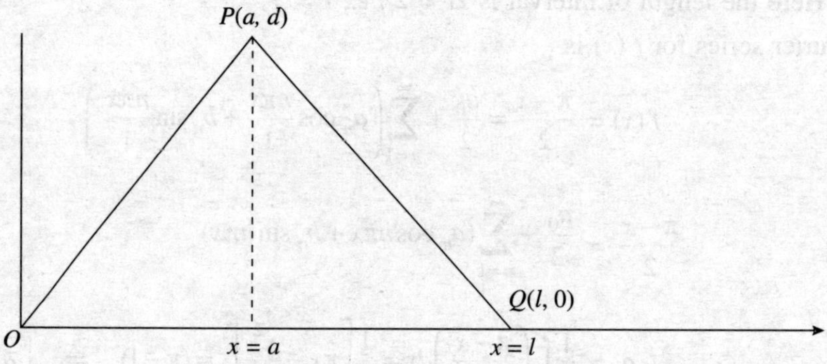

Solution: Equation of the line OP is

$$y = \frac{xd}{a}; \; 0 < x < a \qquad \left| \text{ As } y - y_1 = \frac{y_2 - y_1}{x_2 - x_1}(x - x_1) \right.$$

and equation of the line PQ is

$$y = \frac{d(l-x)}{l-a}, \, a < x < l$$

$$\therefore \qquad f(x) = \begin{cases} \dfrac{xd}{a}, & o < x < a \\ \dfrac{d(l-x)}{l-a}, & a < x < l \end{cases}$$

we know that

$$f(x) = \sum_{n=1}^{\infty} b_n \sin\frac{n\pi x}{l} \qquad ...(i)$$

$$b_n = \frac{2}{l}\int_0^l f(x)\sin\frac{n\pi x}{l}\cdot dx$$

$$= \frac{2}{l}\left[\int_0^a \frac{xd}{a}\sin\frac{n\pi x}{l}dx + \int_a^l \frac{d(l-x)}{(l-a)}\sin\frac{n\pi x}{l}dx\right]$$

$$= \frac{2}{l}\left[\frac{d}{a}\left\{\left(-x\cos\frac{n\pi x}{l}\times\frac{l}{n\pi}\right)_0^a + \frac{l^2}{n^2\pi^2}\left(\sin\frac{n\pi x}{l}\right)_0^a\right\}\right.$$

$$\left. + \frac{d}{(l-a)}\left\{\left(-(l-x)\cos\frac{n\pi x}{l}\times\frac{l}{n\pi}\right)_a^l - \frac{l^2}{n^2\pi^2}\left(\sin\frac{n\pi x}{l}\right)_a^l\right]$$

$$b_n = \frac{2}{l}\left[-d\cos\frac{n\pi a}{l}\cdot\frac{l}{n\pi} + \frac{dl^2}{an^2\pi^2}\sin\frac{n\pi a}{l} + \frac{dl}{n\pi}\cos\frac{n\pi a}{l}\right.$$

$$\left. + \frac{l^2}{n^2\pi^2}\cdot\frac{d}{l-a}\cdot\sin\frac{n\pi a}{l}\right]$$

$$b_n = \frac{2}{l}\left[\frac{dl^2}{n^2\pi^2}\left(\frac{1}{a}+\frac{1}{l-a}\right)\sin\frac{n\pi a}{l}\right]$$

$$= \frac{2dl}{n^2\pi^2}\left(\frac{l-a+a}{a(l-a)}\right)\sin\frac{n\pi a}{l}$$

$$\Rightarrow \qquad b_n = \frac{2dl^2}{a(l-a)n^2\pi^2}\sin\frac{n\pi a}{l}$$

Hence from (*i*)

$$f(x) = \frac{2dl^2}{a(l-a)\pi^2}\left[\sin\frac{\pi x}{l}\cdot\sin\frac{\pi a}{l} + \frac{1}{2^2}\cdot\sin\frac{2\pi x}{l}\cdot\sin\frac{2\pi a}{l} + \ldots\right]$$

Example 18: Find the Fourier series expansion for the function

$f(x) = x\cos x,\ -\pi < x < \pi.$ (*U.P.T.U. 2002*)

Solution: By Fourier series, we have

$$f(x) = \frac{a_0}{2} + \sum_{n=1}^{\infty}(a_n\cos nx + b_n\sin nx) \qquad ...(1)$$

$$a_0 = \frac{1}{\pi}\int_{-\pi}^{\pi} f(x)\,dx = \frac{1}{\pi}\int_{-\pi}^{\pi} x\cos x\,dx = 0 \qquad \left|\begin{array}{l}\text{As } x\cos x \\ \text{is odd function}\end{array}\right.$$

$\Rightarrow$ $$\boxed{a_0 = 0}$$

$$a_n = \frac{1}{\pi}\int_{-\pi}^{\pi} f(x)\cos nx\,dx = \frac{1}{\pi}\int_{-\pi}^{\pi} x\cos x\cos nx\,dx = 0 \qquad \left|\begin{array}{l}x\cos x\cos nx \\ \text{is odd function}\end{array}\right.$$

$$\boxed{a_n = 0}$$

and $$b_n = \frac{1}{\pi}\int_{-\pi}^{\pi} f(x)\sin nx\,dx = \frac{1}{\pi}\int_{-\pi}^{\pi} x\cos x\cdot\sin nx\,dx$$

$$= \frac{2}{\pi}\int_0^{\pi} x\cos x\sin nx\,dx \qquad \left|\begin{array}{l}\text{As } x\cos x\sin nx \\ \text{is even function}\end{array}\right.$$

$$= \frac{1}{\pi}\int_0^{\pi} x\{2\sin nx\cos x\}\,dx$$

$$= \frac{1}{\pi}\int_0^{\pi} x\{\sin(n+1)x + \sin(n-1)x\}\,dx \qquad \left|\begin{array}{l}2\sin A\cos B \\ = \sin(A+B) + \sin(A-B)\end{array}\right.$$

$$= \frac{1}{\pi}\left[x\left(\frac{-\cos(n+1)x}{n+1}\right) + \frac{\sin(n+1)x}{(n+1)^2}\right]_0^{\pi} + \frac{1}{\pi}\left[x\left(\frac{-\cos(n-1)x}{(n-1)}\right) + \frac{\sin(n-1)x}{(n-1)^2}\right]_0^{\pi}$$

$$= \frac{1}{\pi}\left[\pi\left\{-\frac{\cos(n+1)\pi}{n+1} - \frac{\cos(n-1)\pi}{n-1}\right\}\right] \qquad \left|\begin{array}{l}\text{As } \sin(n+1)\pi = 0 \\ \text{and } \sin(n-1)\pi = 0\end{array}\right.$$

$\Rightarrow$ $$b_n = \left\{-\frac{(-1)^{n+1}}{n+1} - \frac{(-1)^{n-1}}{n-1}\right\},\ n \neq 1$$

$\Rightarrow$ $$b_n = -(-1)^{n+1}\left\{\frac{1}{n+1} + \frac{1}{n-1}\right\}$$

$\Rightarrow$ $$b_n = -\left\{\frac{1}{(n+1)} + \frac{1}{(n-1)}\right\} = \frac{-2n}{n^2-1},\ n \text{ is odd}; n \neq 1$$

and $$b_n = \left\{\frac{1}{n+1} + \frac{1}{n-1}\right\} = \frac{2n}{n^2-1},\ n \text{ is even}$$

Thus $$b_n = \begin{cases} \dfrac{-2n}{n^2-1}, & n \text{ is odd}; n \neq 1 \\ \dfrac{2n}{n^2-1}, & n \text{ is even} \end{cases}$$

Now for $n = 1$

$$b_1 = \frac{2}{\pi}\int_0^{\pi} x\cos x \cdot \sin x\, dx = \frac{1}{\pi}\int_0^{\pi} x\sin 2x\, dx$$

$$= \frac{1}{\pi}\left[-\frac{x\cos 2x}{2} + \frac{\sin 2x}{4}\right]_0^{\pi} = \frac{1}{\pi}\left[\pi\left(\frac{-1}{2}\right)\right] = \frac{-1}{2}$$

$\Rightarrow$ $$b_1 = -\frac{1}{2}$$

Hence, from (1), we get

$$f(x) = -\frac{1}{2}\sin x + \frac{4\sin 2x}{2^2-1} - \frac{6\sin 3x}{3^2-1} + \cdots\cdots \quad \textbf{Ans.}$$

Example 19: Find the Fourier series expansion of the periodic function of period 2π

$$f(x) = x^2, -\pi \le x \le \pi.$$

Hence find the sum of the series $\frac{1}{1^2} - \frac{1}{2^2} + \frac{1}{3^2} - \frac{1}{4^2} + \cdots\cdots$ *(U.P.T.U. 2004, 2008)*

Solution: From Fourier series expansion, we have

$$f(x) = x^2 = \frac{a_0}{2} + \sum_{n=1}^{\infty}(a_n \cos nx + b_n \sin nx) \quad \ldots(1)$$

$$a_0 = \frac{1}{\pi}\int_{-\pi}^{\pi} x^2 dx = \frac{2}{\pi}\int_0^{\pi} x^2 dx = \frac{2}{\pi}\left[\frac{x^3}{3}\right]_0^{\pi} = \frac{2\pi^2}{3}$$

$\Rightarrow$ $$a_0 = \frac{2\pi^2}{3}.$$

$$a_n = \frac{1}{\pi}\int_{-\pi}^{\pi} x^2 \cos nx\, dx = \frac{2}{\pi}\int_0^{\pi} x^2 \cos nx\, dx \qquad \Big| \; x^2 \cos nx \text{ is even function}$$

$$= \frac{2}{\pi}\left[x^2\left(\frac{\sin nx}{n}\right) - (2x)\left(-\frac{\cos nx}{n^2}\right) + (2)\left(-\frac{\sin nx}{n^3}\right)\right]_0^{\pi}$$

$$= \frac{2}{\pi}\left[\frac{\pi^2 \sin n\pi}{n} + \frac{2\pi}{n^2}\cos n\pi - \frac{2\sin n\pi}{n^3} - 0\right]$$

$$\Rightarrow \quad \boxed{a_n = \frac{4(-1)^n}{n^2}}$$

and $$b_n = \frac{1}{\pi}\int_{-\pi}^{\pi} x^2 \sin nx\, dx = 0 \qquad \text{As } x^2 \sin nx \text{ is odd function}$$

$\therefore$ From equation (1), we get

$$f(x) = x^2 = \frac{\pi^2}{3} - 4\left[\frac{\cos x}{1^2} - \frac{\cos 2x}{2^2} + \frac{\cos 3x}{3^2} - \frac{\cos 4x}{4^2} + \ldots\ldots\right] \qquad \ldots(2)$$

which is the required Fourier series expansion.

Putting $x = 0$ in (2), we get

$$0 = \frac{\pi^2}{3} - 4\left[\frac{1}{1^2} - \frac{1}{2^2} + \frac{1}{3^2} - \frac{1}{4^2} + \ldots\ldots\right]$$

$$\Rightarrow \quad \frac{1}{1^2} - \frac{1}{2^2} + \frac{1}{3^2} - \frac{1}{4^2} + \ldots = \frac{\pi^2}{12}. \qquad \textbf{Ans.}$$

Example 20: Develop $\sin\left(\frac{\pi x}{l}\right)$ in half range cosine series in the range $0 < x < l$.

(*U.P.T.U. 2001*)

Solution: The half range cosine series for given function is as follows:

$$\sin\frac{\pi x}{l} = \frac{a_0}{2} + \sum_{n=1}^{\infty} a_n \cdot \cos\frac{n\pi x}{l} \qquad \ldots(1)$$

$$a_0 = \frac{2}{l}\int_0^l f(x)\, dx = \frac{2}{l}\int_0^l \sin\frac{\pi x}{l}\, dx = \frac{2}{l}\left[-\frac{\cos\frac{\pi x}{l}}{\frac{\pi}{l}}\right]_0^l$$

$$= -\frac{2}{\pi}[\cos\pi - \cos 0] = \frac{4}{\pi}$$

$$\Rightarrow \quad \boxed{a_0 = \frac{4}{n}}$$

and $$a_n = \frac{2}{l}\int_0^l f(x)\cos\frac{n\pi x}{l}\, dx = \frac{2}{l}\int_0^l \sin\frac{\pi x}{l}\cos\frac{n\pi x}{l}\, dx$$

$$= \frac{1}{l}\int_0^l \left[\sin(n+1)\frac{\pi x}{l} - \sin(n-1)\frac{\pi x}{l}\right] dx$$

$$= \frac{1}{l}\left[-\frac{l}{(n+1)\pi}\cos(n+1)\frac{\pi x}{l} + \frac{l}{(n-1)\pi}\cos(n-1)\frac{\pi x}{l}\right]_0^l$$

$$\Rightarrow \qquad a_n = \frac{1}{\pi}\left[\left\{-\frac{\cos(n+1)\pi}{n+1} + \frac{\cos(n-1)\pi}{n-1}\right\} - \left\{-\frac{1}{n+1} + \frac{1}{n-1}\right\}\right]$$

$$= \frac{1}{\pi}\left[-\frac{(-1)^{n+1}}{n+1} + \frac{(-1)^{n-1}}{n-1} + \frac{1}{n+1} - \frac{1}{n-1}\right] = 0, \text{ when } n \text{ is odd}$$

and
$$a_n = \frac{1}{\pi}\left[\frac{1}{n+1} - \frac{1}{n-1} + \frac{1}{n+1} - \frac{1}{n-1}\right] = \frac{2}{\pi}\left[\frac{1}{n+1} - \frac{1}{n-1}\right]$$

$$= -\frac{4}{\pi(n+1)(n-1)}, \text{ when } n \text{ is even}$$

$$\Rightarrow \qquad \boxed{\begin{aligned} a_n &= 0, \text{ when } n \text{ is odd} \\ &= -\frac{4}{\pi(n+1)(n-1)}, \text{ when } n \text{ is even} \end{aligned}}$$

∴ From (1), we get

$$\sin\left(\frac{\pi x}{l}\right) = \frac{2}{\pi} - \frac{4}{\pi}\left[\frac{\cos\frac{2\pi x}{l}}{1\cdot 3} + \frac{\cos\frac{4\pi x}{l}}{3\cdot 5} + \frac{\cos\frac{6\pi x}{l}}{5\cdot 7} + \cdots\right]. \quad \textbf{Ans.}$$

Example 21: Show that in the interval (0, 1)

$$\cos \pi x = \frac{8}{\pi}\sum_{n=1}^{\infty} \frac{n}{4n^2 - 1}\sin 2n\pi x.$$

Solution: This is a Fourier half range sine series representation of $\cos \pi x$ in the interval $0 < x < 1$.

Let $f(x) = \cos \pi x$

Here $l = 1$

∴ Fourier half range sine series is

$$f(x) = \sum_{n=1}^{\infty} b_n \sin n\pi x \qquad ...(1)$$

$$b_n = 2\int_0^1 f(x)\cdot \sin n\pi x \, dx$$

$$= 2\int_0^1 \cos \pi x \cdot \sin n\pi x \, dx$$

$$= \int_0^1 \{\sin(n+1)\pi x + \sin(n-1)\pi x\}\, dx$$

$$= \left[-\frac{\cos(n+1)\pi x}{\pi(n+1)} - \frac{\cos(n-1)\pi x}{\pi(n-1)}\right]_0^1$$

$$= -\frac{1}{\pi}\left[\frac{\cos(n+1)\pi}{(n+1)} + \frac{\cos(n-1)\pi}{(n-1)} - \frac{1}{n+1} - \frac{1}{n-1}\right]$$

$$= -\frac{1}{\pi}\left[\frac{(-1)^{n+1}}{n+1} + \frac{(-1)^{n-1}}{n-1} - \frac{1}{n+1} - \frac{1}{n-1}\right] = \frac{2n}{\pi(n^2-1)}\left\{(-1)^n + 1\right\}$$

$$\Rightarrow \qquad \boxed{b_n = \begin{cases} 0, \text{ when } n \text{ is odd } (n \neq 1) \\ \dfrac{4n}{\pi(n^2-1)}, \text{ when } n \text{ is even} \end{cases}}$$

Now for $n = 1$

$$b_1 = 2\int_0^1 \cos \pi x \sin \pi x \, dx = \int_0^1 \sin 2\pi x \, dx = -\frac{1}{2\pi}[\cos 2\pi - \cos 0] = 0$$

Hence from (1), we get

$$f(x) = \sum_{n=2m(\text{even})}^{\infty} \frac{8m}{\pi(4m^2-1)} \sin 2m\pi x \qquad \Big|\, n = 2m$$

$$= \sum_{m=1}^{\infty} \frac{8m}{\pi(4m^2-1)} \sin 2m\pi x$$

$$\Rightarrow f(x) = \cos \pi x = \sum_{m=1}^{\infty} \frac{8m}{\pi(4m^2-1)} \sin 2m\pi x$$

In general

$$\cos \pi x = \frac{8}{\pi}\sum_{n=1}^{\infty} \frac{n}{4n^2-1} \sin 2n\pi x. \quad \textbf{Proved.}$$

Example 22: Show that

$$\frac{l}{2} - x = \frac{l}{\pi}\sum_{1}^{\infty} \frac{1}{n} \sin \frac{2n\pi x}{l}, \quad 0 < x < l.$$

Solution: This is Fourier half range sine series expansion of $\frac{l}{2} - x$; in the interval $0 < x < l$.

Let $$f(x) = \frac{l}{2} - x$$

$\therefore$ The Fourier half range sine series is given by

$$f(x) = \sum_{n=1}^{\infty} b_n \sin \frac{n\pi x}{l} \qquad ...(1)$$

$$b_n = \frac{2}{l}\int_0^l f(x) \cdot \sin \frac{n\pi x}{l}\, dx = \frac{2}{l}\int_0^l \left(\frac{l}{2} - x\right) \sin \frac{n\pi x}{l}\, dx$$

$$= \frac{2}{l}\left[-\left\{ \left(\frac{l}{2} - x\right) \frac{\cos \frac{n\pi x}{l}}{\frac{n\pi}{l}} \right\}_0^l - \frac{l}{n\pi}\int_0^l \cos \frac{n\pi x}{l}\, dx \right]$$

$$= \frac{2}{n\pi}\left[\frac{l}{2}\cos n\pi + \frac{l}{2} - \frac{l}{n\pi}\left(\sin \frac{n\pi x}{l} \right)_0^l \right]$$

$$= \frac{l}{n\pi}[\cos n\pi + 1 - \sin n\pi - \sin 0]$$

$\Rightarrow$ $$b_n = \frac{l}{n\pi}\left[(-1)^n + 1\right]$$

or $$b_n = \begin{cases} \dfrac{2l}{n\pi}, & n \text{ is even} \\ 0, & n \text{ is odd} \end{cases}$$

$\therefore$ From (1), we get

$$f(x) = \sum_{n=2\,(\text{even})}^{\infty} \frac{2l}{n\pi} \sin \frac{n\pi x}{l}$$

$\Rightarrow$ $$f(x) = \sum_{m=1}^{\infty} \frac{2l}{2m\pi} \sin 2m \frac{\pi x}{l} \qquad \Big| \text{ Let } n = 2m \text{ (even)}$$

$\Rightarrow$ $$\frac{l}{2} - x = \frac{l}{\pi}\sum_{m=1}^{\infty} \frac{1}{m} \sin \frac{2m\pi x}{l}$$

In general,

$$\frac{l}{2} - x = \frac{l}{\pi}\sum_{n=1}^{\infty} \frac{1}{n} \sin \frac{2n\pi x}{l}. \quad \textbf{Proved.}$$

Example 23: Express $f(x) = |x|$, $-\pi < x < \pi$, as Fourier series.

Hence show that

$$\frac{1}{1^2}+\frac{1}{3^2}+\frac{1}{5^2}+\ldots\ldots = \frac{\pi^2}{8}.$$

Solution: $f(-x) = |-x| = x = f(x) \quad \Rightarrow \quad f(-x) = f(x)$

$\therefore$ $f(x)$ is an even function and hence $b_n = 0$

$$\Rightarrow \qquad f(x) = \frac{a_0}{2} + \sum_{n=1}^{\infty} a_n \cos nx \qquad \ldots(1)$$

$$a_0 = \frac{2}{\pi}\int_0^{\pi} |x|\,dx = \frac{2}{\pi}\int_0^{\pi} x\,dx = \left[\frac{x^2}{\pi}\right]_0^{\pi} = \pi$$

$$\Rightarrow \qquad \boxed{a_0 = \pi}$$

and

$$a_n = \frac{2}{\pi}\int_0^{\pi} f(x)\cos nx\,dx = \frac{2}{\pi}\int_0^{\pi} |x|\cos nx\,dx = \frac{2}{\pi}\int_0^{\pi} x\cos nx\,dx$$

$$= \frac{2}{\pi}\left[x\left(\frac{\sin nx}{n}\right) - 1\left(-\frac{\cos nx}{n^2}\right)\right]_0^{\pi}$$

$$= \frac{2}{\pi}\left[\frac{\cos n\pi}{n^2} - \frac{1}{n^2}\right] = \frac{2}{\pi n^2}\left[(-1)^n - 1\right]$$

$$\Rightarrow \qquad \boxed{a_n = \begin{cases} 0, & \text{if } n \text{ is even} \\ -\dfrac{4}{\pi n^2}, & \text{if } n \text{ is odd} \end{cases}}$$

$\therefore$ From (1), we get

$$f(x) = \frac{\pi}{2} - \frac{4}{\pi}\left(\cos x + \frac{\cos 3x}{3^2} + \frac{\cos 5x}{5^2} + \ldots\ldots\right)$$

Putting $x = 0$ in above result, we get

$$0 = \frac{\pi}{2} - \frac{4}{\pi}\left(\frac{1}{1^2} + \frac{1}{3^2} + \frac{1}{5^2} + \ldots\ldots\right)$$

$$\Rightarrow \qquad \frac{1}{1^2}+\frac{1}{3^2}+\frac{1}{5^2}+\ldots\ldots = \frac{\pi^2}{8}. \qquad \textbf{Proved.}$$

EXERCISE 4.1

Find the Fourier series expansion of the following functions:

1. $f(x)=\begin{cases}0, & \text{when } -\pi\le x\le 0\\ x^2, & \text{when } 0\le x\le \pi.\end{cases}$

$$\left[\textbf{Ans. } f(x)=\frac{\pi^2}{6}+2\sum_{n=1}^{\infty}\frac{(-1)^n}{n^2}\cos nx+\sum_{n=1}^{\infty}\left[\frac{\pi}{n}(-1)^{n+1}+\frac{2}{\pi n^3}\{(-1)^n-1\}\right]\sin nx\right]$$

2. $f(x)=e^{ax}$ in $(0,\ 2\pi)$.

$$\left[\textbf{Ans. } f(x)=\left(\frac{e^{2a\pi}-1}{\pi}\right)\left\{\frac{1}{2}+\sum_{n=1}^{\infty}\left(\frac{1}{a^2+n^2}\right)(-n\sin nx)\right\}+\frac{ae^{2a\pi}-1}{\pi}\sum_{n=1}^{\infty}\frac{\cos nx}{(a^2+n^2)}\right]$$

3. $f(x)=\begin{cases}1, & \text{if } -\pi<x\le 0\\ -2, & \text{if } 0<x\le \pi.\end{cases}$ $\left[\textbf{Ans. } f(x)=-\frac{1}{2}-\frac{6}{\pi}\sum_{n=0}^{\infty}\frac{\sin(2n+1)x}{(2n+1)}\right]$

4. $f(x)=e^{-x}$ in $0<x<2\pi$.

$$\left[\textbf{Ans. } f(x)=\frac{1-e^{-2\pi}}{\pi}\left\{\frac{1}{2}+\left(\frac{1}{2}\cos x+\frac{1}{5}\cos 2x+\cdots\cdots\right)+\left(\frac{1}{2}\sin x+\frac{2}{5}\sin 2x+\frac{3}{10}\sin 3x+\cdots\right)\right\}\right]$$

5. $f(x)=\begin{cases}a\sin x, & \text{if } 0\le x\le \pi\\ 0, & \text{if } \pi\le x\le 2\pi\end{cases}$ and deduce $\frac{1}{2}=\frac{1}{1\cdot 3}+\frac{1}{3\cdot 5}+\cdots\cdots$

$$\left[\textbf{Ans. } f(x)=\frac{a}{\pi}+\frac{a}{2}\sin x-\frac{2a}{\pi}\sum_{n=1}^{\infty}\frac{\cos 2nx}{4n^2-1}\right]$$

6. $f(x)=\begin{cases}2x, & \text{when } 0\le x\le \pi\\ x, & \text{when } -\pi<x\le 0.\end{cases}$

$$\left[\textbf{Ans. } f(x)=\frac{\pi}{4}-\frac{2}{\pi}\left(\frac{\cos x}{1^2}+\frac{\cos 3x}{3^2}+\frac{\cos 5x}{5^2}+\cdots\cdot\right)+3\left(\frac{\sin x}{1}-\frac{\sin 2x}{2}+\cdots\cdot\right)\right]$$

7. $f(x)=x\sin x$, for $0<x<2\pi$. *(U.P.T.U. 2001)*

$$\left[\textbf{Ans. } -1+\pi\sin x-\frac{1}{2}\cos x+2\left(\frac{\cos 2x}{2^2-1}+\frac{\cos 3x}{3^2-1}+\frac{\cos 4x}{4^2-1}+\cdots\cdots\right)\right]$$

8. Expand the function $f(x)=\dfrac{\pi^2}{12}-\dfrac{x^2}{4}$ in a Fourier series in the interval $(-\pi, \pi)$.

$$\left[\textbf{Ans. } f(x)=\frac{\cos x}{1^2}-\frac{\cos 2x}{2^2}+\frac{\cos 3x}{3^2}-\frac{\cos 4x}{4^2}+\cdots\right]$$

9. Show that if m is neither zero nor an integer,

$$\sin mx = \frac{2\sin m\pi}{\pi}\left\{\frac{\sin x}{1-m^2}-\frac{2\sin 2x}{2^2-m^2}+\frac{3\sin 3x}{3^2-m^2}-\cdots\right\} \text{ in } (-\pi, \pi).$$

10. Obtain Fourier series expansion for the function $f(x)$ given by

$$f(x)=\begin{cases}1+\dfrac{2x}{\pi}, & -\pi\le x\le 0\\[2mm] 1-\dfrac{2x}{\pi}, & 0\le x\le \pi\end{cases}$$

Hence deduce that $\dfrac{1}{1^2}+\dfrac{1}{3^2}+\dfrac{1}{5^2}+\cdots\cdots=\dfrac{\pi^2}{8}$

$$\left[\textbf{Ans. } f(x)=\frac{8}{\pi^2}\left(\frac{\cos x}{1^2}+\frac{\cos 3x}{3^2}+\frac{\cos 5x}{5^2}+\cdots\right)\right]$$

11. Find the Fourier series of the function

$$f(x)=\begin{cases}-x, & \text{for } -2<x<0\\ x, & \text{for } 0<x<2\end{cases};\quad \text{period} = 4.$$

$$\left[\textbf{Ans. } f(x)=1-\frac{8}{\pi^2}\left\{\frac{1}{1^2}\cos\frac{\pi x}{2}+\frac{1}{3^2}\cos\frac{3\pi x}{2}+\frac{1}{5^2}\cos\frac{5\pi x}{2}+\cdots\right\}\right]$$

12. Expand the function $f(x) = \begin{cases}x, & \text{when } 0<x<1\\ 2-x, & \text{when } 1<x<2\end{cases}$ in the interval (0, 2) in a series of sines and also in a series of cosines.

$$\left[\textbf{Ans. } f(x)=1-\frac{8}{\pi^2}\sum_{n=0}^{\infty}\frac{(-1)^n}{(2n+1)^2}\sin\frac{(2n+1)\pi x}{2} \text{ and } f(x)=\frac{1}{2}-\frac{4}{\pi^2}\sum_{n=0}^{\infty}\frac{\cos(2n+1)\pi x}{(2n+1)^2}\right]$$

13. $f(x)=\begin{cases}-1, & \text{for } -1<x<0\\ 1, & \text{for } 0<x<1\end{cases}$; period = 2.

$$\left[\textbf{Ans. } f(x)=\frac{4}{\pi}\left(\sin\pi x+\frac{1}{3}\sin 3\pi x+\frac{1}{5}\sin 5\pi x+\cdots\right)\right]$$

14. $f(x)=\begin{cases}0, & \text{for } -2<x<0\\ 1, & \text{for } 0<x<2\end{cases}$; period = 4.

$$\left[\textbf{Ans. } f(x)=\frac{1}{2}+\frac{2}{\pi}\left(\sin\frac{\pi x}{2}+\frac{1}{3}\sin\frac{3\pi x}{2}+\frac{1}{5}\sin\frac{5\pi x}{2}+\cdots\right)\right]$$

15. $f(x)=[x]$ in $(-L, L)$.

$$\left[\textbf{Ans. } \text{even, } f(x)=\frac{L}{2}-\frac{4L}{\pi^2}\sum_{n=\text{odd}}\frac{\cos\left(\frac{n\pi x}{L}\right)}{n^2}\right]$$

16. $f(x)=\begin{cases}x, & \text{for } 0<x<1\\ 0, & \text{for } 1<x<2.\end{cases}$

$$\left[\textbf{Ans. } f(x)=\frac{1}{4}+\frac{1}{\pi^2}\sum_{n=1}^{\infty}\left(\frac{\cos n\pi-1}{n^2}\right)\cos n\pi x+\frac{1}{\pi}\sum_{n=1}^{\infty}\frac{(-1)^{n-1}}{n}\sin n\pi x\right]$$

17. $f(x)=\begin{cases}x, & \text{for } -1<x\le 0\\ x+2, & \text{for } 0<x\le 1\end{cases}$; Deduce that $\frac{\pi}{4}=1-\frac{1}{3}+\frac{1}{5}-\cdots$

$$\left[\textbf{Ans. } f(x)=1+\frac{2}{\pi}\sum_{n=1}^{\infty}\frac{1}{n}[1-2(-1)^n]\sin n\pi x\right]$$

18. $f(x)=\begin{cases}0, & \text{if } -2<x<-1\\ m, & \text{if } -1<x<1\\ 0, & \text{if } 1<x<2\end{cases}$ $\left[\textbf{Ans. } \text{even function, } f(x)=\frac{m}{2}+\frac{2m}{\pi}\sum_{n=1}^{\infty}\frac{\sin(n\pi/2)}{n}\cdot\cos\left(\frac{n\pi x}{2}\right)\right]$

19. $f(x)=x+x^2$ in $(-1, 1)$ $\left[\textbf{Ans. } f(x)=\frac{1}{3}+\frac{4}{\pi^2}\sum_{n=1}^{\infty}\frac{(-1)^n}{n^2}\cos n\pi x+\frac{2}{\pi}\sum_{n=1}^{\infty}\frac{(-1)^{n+1}}{n}\sin n\pi x\right]$

20. $f(x)=\begin{cases}-1, & \text{for } -\pi<x<0\\ 1, & \text{for } 0<x<\pi\end{cases}$ where $f(x+2\pi)=f(x)$. *(U.P.T.U., 2005)*

$$\left[\textbf{Ans. } f(x)=\frac{4}{\pi}\left[\sin x+\frac{1}{3}\sin 3x+\frac{1}{5}\sin 5x+\cdots\right]\right]$$

21. Obtain a Fourier series to represent the function $f(x)=|\sin x|$ for $-\pi<x<\pi$.

$$\left[\textbf{Ans. } f(x)=\frac{2}{\pi}-\frac{4}{\pi}\left[\frac{1}{3}\cos 2x+\frac{1}{15}\cos 4x+\frac{1}{35}\cos 6x+\cdots\right]\right]$$

22. Find the Fourier series of the function $f(x) = \begin{cases} -2, & \text{if } -4 < x < -2 \\ x, & \text{if } -2 < x < 2 \\ 2, & \text{if } 2 < x < 4. \end{cases}$

$$\left[\textbf{Ans. } f(x) = \frac{4}{\pi} + \frac{8}{\pi^2}\sin\frac{\pi x}{4} - \frac{2}{\pi}\sin\frac{2\pi x}{4} + \left(\frac{4}{3\pi} - \frac{8}{3^2\pi}\right)\sin\frac{3\pi x}{4} - \ldots\right]$$

23. Find the Fourier sine and cosine series of $f(x) = \begin{cases} x & \text{when } 0 < x < \dfrac{\pi}{2} \\ 0 & \text{when } \dfrac{\pi}{2} < x < \pi. \end{cases}$

$$\left[\textbf{Ans. } \text{Fourier sine series, } f(x) = \frac{2}{\pi}\sum_{n=1}^{\infty}\left[\frac{1}{n^2}\sin\frac{n\pi}{2} - \frac{\pi}{2n}\cos\frac{n\pi}{2}\right]\sin nx;\right.$$

$$\left.\text{Fourier cosine series, } f(x) = \frac{\pi}{8} + \sum_{n=1}^{\infty}\left[\frac{1}{n}\sin\frac{n\pi}{2} + \frac{2}{\pi n^2}\left(\cos\frac{n\pi}{2} - 1\right)\right]\cos nx\right]$$

24. Determine the half range sine series for $f(x) = e^x$ in $0 < x < 1$.

$$\left[\textbf{Ans. } f(x) = 2\pi\sum_{n=1}^{\infty}\frac{n[1 - e\cos n\pi]}{1 + n^2\pi^2}\sin n\pi x\right]$$

25. Expand $f(x) = \begin{cases} \dfrac{1}{4} - x & \text{if } 0 < x < 0.5 \\ x - \dfrac{3}{4} & \text{if } 0.5 < x < 1 \end{cases}$ as Fourier series of sine terms.

$$\left[\textbf{Ans. } f(x) = \sum_{n=1}^{\infty}\left\{\frac{1}{2n\pi}[1 - (-1)^n] - \frac{4\sin(n\pi/2)}{n^2\pi^2}\right\}\sin n\pi x\right]$$

26. Find the half range Fourier sine series of $f(x) = \begin{cases} \sin x & \text{when } 0 \le x \le \dfrac{\pi}{4} \\ \cos x & \text{when } \dfrac{\pi}{4} \le x \le \dfrac{\pi}{2}. \end{cases}$

$$\left[\textbf{Ans. } f(x) = \frac{8}{\pi}\frac{1}{\sqrt{2}}\left[\frac{\sin 2x}{1\cdot 3} - \frac{\sin 6x}{5\cdot 7} + \ldots\right]\right]$$

27. Find the half range sine series for the function $f(x) = x - x^2$ in the interval $0 < x < 1$.

(U.P.T.U. (C.O.) 2004)

$$\left[\textbf{Ans. } f(x) = \frac{8}{\pi^3}\left(\sin \pi x + \frac{1}{3^3}\sin 3\pi x + \frac{1}{5^3}\sin 5\pi x + \cdots\right)\right]$$

28. Expand the function $f(x) = \begin{cases} x, & \text{for } 0 \le x \le \pi/2 \\ \pi - x, & \text{for } \dfrac{\pi}{2} \le x \le \pi \end{cases}$

(*i*) In a series of sines (*ii*) In a series of cosines.

$$\left[\textbf{Ans. } (i)\ f(x) = \frac{4}{\pi}\left(\sin x - \frac{1}{3^2}\sin 3x + \frac{1}{5^2}\sin 5x - \cdots + \cdots\right);\right.$$

$$\left.(ii)\ f(x) = \frac{\pi}{4} - \frac{8}{\pi}\left(\frac{1}{2^2}\cos 2x + \frac{1}{6^2}\cos 6x + \frac{1}{10^2}\cos 10x + \cdots\right)\right]$$

29. Find the Fourier cosine series of the function $f(x) = x^2$ in $0 < x < l$.

$$\left[\textbf{Ans. } f(x) = \frac{l^2}{3} - \frac{4l^2}{\pi^2}\left(\cos\frac{\pi x}{l} - \frac{1}{4}\cos\frac{2\pi x}{l} + \frac{1}{9}\cos\frac{3\pi x}{l} - \frac{1}{16}\cos\frac{4\pi x}{l} + \cdots\right)\right]$$

30. Find the Fourier cosine series of $f(x) = \sin x$ in $0 < x < \pi$.

$$\left[\textbf{Ans. } f(x) = \frac{2}{\pi} - \frac{4}{\pi}\left(\frac{1}{1\cdot 3}\cos 2x + \frac{1}{3\cdot 5}\cos 4x + \frac{1}{5\cdot 7}\cos 6x + \cdots\right)\right]$$

31. Find the Fourier expansion for the function $f(x) = x - x^3$ in the interval $-1 < x < 1$.

$$\left[\textbf{Ans. } f(x) = \frac{12}{\pi^3}\left(\sin \pi x - \frac{\sin 2\pi x}{2^3} + \frac{\sin 3\pi x}{3^3} - \cdots\right)\right]$$

32. Find the two half range Fourier series of $f(x) = 1$ in $0 < x < l$.

$$\left[\textbf{Ans. } \text{sine series } f(x) = 1 = \frac{2}{\pi}\sum_{n=1}^{\infty}\frac{1}{n}\left[1 - \cos n\pi\right]\sin\frac{n\pi x}{l};\right.$$

$$\left.\text{cosine series } f(x) = 1 = \frac{2}{2} = 1\ (a_0 = 2,\ a_n = 0 \text{ for } n \ge 1)\right]$$

33. Find the Fourier series of $f(t)$

$$f(t) = \begin{cases} 0, & \text{if } -l < t < 0 \\ E\sin wt, & \text{if } 0 < t < l. \end{cases}$$

$$\left[\textbf{Ans. } f(t) = \frac{E}{\pi} + \frac{E}{2}\sin \omega t - \frac{2E}{\pi}\left(\frac{1}{1\cdot 3}\cos 2\omega t + \frac{1}{3\cdot 5}\cos 4\omega t + \cdots\right)\right]$$

34. Find the Fourier series of the function $f(x)=\begin{cases} 2x, & \text{when } 0 \le x \le \pi \\ x, & \text{when } -\pi < t \le 0. \end{cases}$

$$\left[\textbf{Ans. } f(x)=\frac{\pi}{4}-\frac{2}{\pi}\left(\frac{\cos x}{1^2}+\frac{\cos 3x}{3^2}+\frac{\cos 5x}{5^2}+\cdots\right) + 3\left(\frac{\sin x}{1}-\frac{\sin 2x}{2}+\frac{\sin 3x}{3}-\cdots\right)\right]$$

4.9 HARMONIC ANALYSIS

Harmonic analysis is the theory of expanding functions in Fourier series. Let $f(x)$ be a function with period 2π defined in the interval $[-\pi, \pi]$, then the Fourier series of $f(x)$ is given by

$$f(x) = \frac{a_0}{2}\sum_{n=1}^{\infty}(a_n \cos nx + b_n \sin nx)$$

Where
$$a_0 = \frac{1}{\pi}\int_{-\pi}^{\pi} f(x)dx \qquad ...(i)$$

$$a_n = \frac{1}{\pi}\int_{-\pi}^{\pi} f(x)\cos nx\, dx \qquad ...(ii)$$

$$b_n = \frac{1}{\pi}\int_{-\pi}^{\pi} f(x)\sin nx\, dx \qquad ...(iii)$$

Since the function $f(x)$ is in tabulated form. So harmonic analysis deals with the determination of the approximate values of a_0, a_n and b_n.

Divide the interval $(-\pi, \pi)$ into m equal parts with $(m + 1)$ points *i.e.*

$$-\pi = x_0, x_1, \ldots\ldots, x_m = \pi$$

and sub-interval size
$$h = \frac{\pi-(-\pi)}{m} = \frac{2\pi}{m}$$

Let
$$y_i = f(x_i);\ i = 0, 1, 2, \cdots$$

Then a_0, a_n and b_n are approximately evaluated using say rectangular formula (area = sum of n rectangles $= \sum_{i=1}^{n}(h \times y_i)$ where h is the width and y_i is ordinate.

∴ From (1), (2) and (3), we get

$$a_0 = \frac{1}{\pi}\int_{-\pi}^{\pi} f(x)dx = \frac{1}{\pi}\left[\sum_{i=1}^{n} hy_i\right] = \frac{1}{\pi}\cdot\frac{2\pi}{m}\sum_{i=1}^{n} y^i$$

or
$$\boxed{a_0 = \frac{2}{m}\sum_{i=1}^{n} y_i} \qquad ...(iv)$$

$$a_n = \frac{1}{\pi}\int_{-\pi}^{\pi} f(x)\cos nx dx = \frac{1}{\pi}\left[\sum_{i=1}^{n} h \cdot y_i \cos nx_i\right] = \frac{1}{\pi}\cdot\frac{2\pi}{m}\sum_{i=1}^{n} y_i \cos nx_i$$

or
$$\boxed{a_n = \frac{2}{m}\sum_{i=1}^{n} y_i \cos nx_i} \qquad ...(v)$$

Similarly,
$$\boxed{b_n = \frac{2}{m}\sum_{i=1}^{n} y_i \sin nx_i} \qquad ...(vi)$$

Note: Since the values of sine and cosine are repeated in four quadrants. So we choose here m as a number divisibly by 4. Usual choice m = 6, 12, 24 (in which (4), (5), (6) get simplified).

Note: The direct current part $= \frac{a_0}{2}$.

Example 24: Determine the first three coefficients of cosine and two coefficients of sine terms in the Fourier series expansion of the following tabulated function.

x :	0	$\frac{\pi}{6}$	$\frac{\pi}{3}$	$\frac{\pi}{2}$	$\frac{2\pi}{3}$	$\frac{5\pi}{6}$
y :	0	9.2	14.4	17.8	17.3	11.7

Solution: Here m = 6 (number of sub-intervals)

$\therefore$ sub-interval size $\qquad h = \frac{2\pi}{6} = \frac{\pi}{3}$

But in the problem sub-interval size is $= \frac{\pi}{6}$

$\Rightarrow$ Length of interval is $\qquad = \pi$

so $\qquad 2l = \pi \quad \text{or} \quad l = \frac{\pi}{2}$

In this case

$$f(x) = \frac{a_0}{2} + \sum_{n=1}^{\infty}(a_n \cos 2nx + b_n \sin 2nx) \qquad ...(1)$$

where
$$a_0 = \frac{2}{m}\sum_{i=1}^{n} y_i \qquad ...(2)$$

$$a_n = \frac{2}{m}\sum_{i=1}^{n} y_i \cos 2nx_i \qquad ...(3)$$

$$b_n = \frac{2}{m}\sum_{i=1}^{n} y_i \sin 2nx_i \qquad ...(4)$$

Now we construct the following table for determined the values of coefficients (2), (3) and (4).

x	y	cos 2x	cos 4x	sin 2x	sin 4x	y cos 2x	y cos 4x	y sin 2x	y sin 4x
0	0	1.0	1.0	0.0	0.0	0.0	0.0	0.0	0.0
$\frac{\pi}{6}$	9.2	0.5	– 0.5	0.8660	0.8660	4.6	– 4.6	7.9672	7.9672
$\frac{\pi}{3}$	14.4	– 0.5	– 0.5	0.8660	– 0.8660	– 7.2	– 7.2	12.470	– 12.470
$\frac{\pi}{2}$	17.8	– 1.0	1.0	0.00	0.0	– 17.8	17.8	0.0	0.0
$\frac{2\pi}{3}$	17.3	– 0.5	– 0.5	– 0.8660	0.8660	– 8.65	– 8.65	– 14.981	14.981
$\frac{5\pi}{6}$	11.7	0.5	– 0.5	– 0.8660	– 0.8660	5.85	– 5.85	– 10.132	– 10.132
Total Σ	70.4					– 23.20	– 8.50	– 4.675	0.3462

From (2), (3) and (4), we get

$$a_0 = \frac{2}{6}\times 70.4 = 23.466$$

$$a_1 = \frac{2}{6}\sum y_i \cos 2x_i = \frac{2}{6}\times(-23.20) = -7.733$$

$$a_2 = \frac{2}{6}\sum y_i \cos 4x_i = \frac{2}{6}\times(-8.50) = -2.833$$

$$b_1 = \frac{2}{6}\sum y_i \sin 2x_i = \frac{2}{6}\times(-4.675) = -1.566$$

$$b_2 = \frac{2}{6}\sum y_i \sin 4x_i = \frac{2}{6}\times(0.3462) = 0.116$$

And from (1), we get the Fourier series expansion.

$$f(x) = 11.733 - 7.733\cos 2x - 2.833\cos 4x - 1.566\sin 2x + 0.116\sin 4x.$$

Example 25: Find the Fourier coefficients a_0, a_1, a_2, a_3 and b_1, b_2, b_3 in the Fourier series expansion of function tabulated as follows. Find also Fourier series expansion and the amplitude of the first harmonic. Calculate $y(3)$.

x :	0	1	2	3	4	5
y :	9	18	24	28	26	20

Solution: Here $m = 6$.

$\therefore$ Sub-interval size $\qquad h = \dfrac{2\pi}{6} = \dfrac{\pi}{3} = 60°$

x	θ	y	cos θ	cos 2θ	cos 3θ	sin θ	sin 2θ	sin 3θ	y cos θ	y cos 2θ	y cos 3θ	y sin θ	y sin 2θ	y sin 3θ
0	0°	9	1.0	1.0	1	0	0	0	9	9	9	0	0	0
1	60°	18	0.5	– 0.5	– 1	0.866	0.866	0	9	– 9	– 18	15.588	15.588	0
2	120°	24	– 0.5	– 0.5	1	0.866	– 0.866	0	– 12	– 12	24	20.784	–20.784	0
3	180°	28	– 1.0	1.0	– 1	0.0	0.0	0	– 28	28	– 28	0	0	0
4	240°	26	– 0.5	– 0.5	1	– 0.866	0.866	0	– 13	– 13	26	– 22.516	22.516	0
5	300°	20	0.5	– 0.5	– 1	– 0.866	– 0.866	0	10	– 10	– 20	– 17.32	– 17.32	0
Total Σ		125							– 25	– 7	– 7	– 3.464	0	0

Now
$$a_0 = \frac{2}{m}\sum_{i=1}^{n} y_i = \frac{2}{6} \times 215 = 41.666$$

$$a_1 = \frac{2}{m}\sum_{i=1}^{n} y_i \cos x_i = \frac{2}{6} \times (-25) = -8.333$$

$$a_2 = \frac{2}{m}\sum_{i=1}^{n} y_i \cos 2x_i = \frac{2}{6} \times (-7) = -2.333$$

$$a_3 = \frac{2}{m}\sum_{i=1}^{n} y_i \cos 3x_i = \frac{2}{6}(-7) = -2.333$$

$$b_1 = \frac{2}{m}\sum_{i=1}^{n} y_i \sin x_i = \frac{2}{6}(-3.464) = -1.1547$$

$$b_2 = \frac{2}{m}\sum_{i=1}^{n} y_i \sin 2x_i = \frac{2}{6} \times 0 = 0$$

$$b_3 = 0$$

$\therefore$ Fourier series expansion is

$$f(x) = 20.833 - 8.333\cos x - 2.333\cos 2x - 2.333\cos 3x - 1.1547\sin x$$

And

Amplitude of the first harmonic $= \sqrt{a_1^2 + b_1^2} = \sqrt{(-8.333)^2 + (-1.1547)^2} = 8.4126$

At $x = 3$, $\theta = 180°$

$\therefore$
$$y(3) = y(\theta = 180°) = 20.833 - (8.333)(-1) - 2.333(1) - 2.333(-1) - 0 = 29.166$$

(Exact value : $y(3) = 28$).

Example 26: Find Fourier series expansion till first three coefficients of cosine and sine from the following data:

θ :	0	30	60	90	120	150	180	210	240	270	300	330
$f(\theta)$:	298	356	373	337	254	155	80	51	60	93	147	221

Solution: Here $m = 12$

θ	y	$\cos\theta$	$\cos 2\theta$	$\sin\theta$	$\sin 2\theta$	$\sin 3\theta$	$y\cos\theta$	$y\cos 2\theta$	$y\sin\theta$	$y\sin 2\theta$	$y\sin 3\theta$
0°	298	1	1	0	0	0	298	298	0	0	0
30°	356	0.866	0.5	0.5	0.866	1	308.30	178	178	308.30	356
60°	373	0.5	– 0.5	0.866	0.866	0	186.5	– 186.5	323.01	3230.1	0
90°	337	0.0	– 1	1	0	– 1	0.0	– 337	337	0	– 337
120°	254	– 0.5	– 0.5	0.866	– 0.866	0	– 127	– 127	219.97	– 219.97	0
150°	155	– 0.866	0.5	0.5	– 0.866	1	– 134.23	77.5	77.5	– 134.23	155
180°	80	– 1	1	0	0	0	– 80	80	0	0	0
210°	51	– 0.866	0.5	– 0.5	0.866	– 1	– 44.17	25.5	– 25.5	44.166	– 51
240°	60	– 0.5	– 0.5	– 0.866	0.866	0	– 30	– 30	– 51.96	51.96	0
270°	93	0.0	– 1	– 1	0	1	0.0	– 93	– 93	0	93
300°	147	0.5	– 0.5	– 0.866	– 0.866	0	73.5	– 73.5	– 127.30	– 127.30	0
330°	221	0.866	0.5	– 0.5	– 0.866	– 1	191.38	110.5	– 110.5	– 191.38	– 221
Total Σ	2425						642.28	–77.5	727.22	54.56	– 5

Now

$$a_0 = \frac{2}{m}\sum_{i=1}^{n} y_i = \frac{2}{12}\times 2425 = 404.16 \sim 404$$

$$a_1 = \frac{2}{m}\sum_{i=1}^{n} y_i \cos x_i = \frac{2}{12}\times 642.28 = 107.04 \sim 107$$

$$a_2 = \frac{2}{m}\sum_{i=1}^{n} y_i \cos 2x_i = \frac{2}{12}\times(-77.5) = -12.91 \sim -13$$

$$b_1 = \frac{2}{m}\sum_{i=1}^{n} y_i \sin x_i = \frac{2}{12}\times(727.22) = 121.20 \sim 121$$

$$b_2 = \frac{2}{m}\sum_{i=1}^{n} y_i \sin 2x_i = \frac{2}{12}(54.56) = 9.09 \sim 9$$

$$b_3 = \frac{2}{m}\sum_{i=1}^{n} y_i \sin 3x_i = \frac{2}{12}(-5) = -0.83 \sim -1$$

Thus

$$y = f(\theta) = 202 + 107\cos\theta - 13\cos 2\theta + 121\sin\theta + 9\sin 2\theta - \sin 3\theta$$

EXERCISE 4.2

1. Find a_0, a_1, a_2 and a_3 in the Fourier cosine series for y which is tabulated below:

x	0	1	2	3	4	5
y	4	8	15	7	6	2

[**Ans.** $a_0 = 14, a_1 = -2.8, a_2 = -1.5, a_3 = 2.7$]

2. A machine completes its cycle of operations every time as certain pulley completes a revolution. The displacement $f(x)$ of a point on a certain portion of the machine is given in the table given below for twelve positions of the pulley, x being the angle in degree turned through by the pulley. Find a Fourier series to represent $f(x)$ for all values of x.

x	30°	60°	90°	120°	150°	180°	210°	240°	270°	300°	330°	360°
$f(x)$	7.976	8.026	7.204	5.676	3.674	1.764	0.552	0.262	0.904	2.492	4.736	6.824

[**Ans.** $f(x) = 4.17 + 2.45\cos x + 0.12\cos 2x + 0.08\cos 3x + 3.16\sin x + \cdots$
$+ 3.16\sin x + 0.03\sin 2x + 0.01\sin 3x + \cdots$]

3. Compute $a_0, a_1, a_2, a_3, b_1, b_2, b_3$ of the Fourier series of the following data:

x	0	$\pi/3$	$2\pi/3$	π	$4\pi/3$	$5\pi/3$	2π
y	1.0	1.4	1.9	1.7	1.5	1.2	1.0

[**Ans.** $a_0 = 2.9, a_1 = -.37, a_2 = -0.1, a_3 = 0.03, b_1 = 0.17, b_2 = -0.06, b_3 = 0$]

4. Find $T(\theta = 75°)$ using the first four terms of the Fourier sine series representation of turning moment T given for a series of values of the crank angle $\theta° = 75°$.

θ°	0	30	60	90	120	150	180
T	0	5224	8097	7850	5499	2626	0

[**Ans.** 8332]

5. Compute the Fourier series, till first and second harmonic function given by table below:

x	0°	30°	60°	90°	120°	150°	180°	210°	240°	270°	300°	330°
$f(x)$	2.34	3.01	3.69	4.15	3.69	2.20	0.83	0.51	0.88	1.09	1.19	1.64

[**Hint:** First harmonic $= (a_1\cos x + b_1\sin x)$, second harmonic $= a_2\cos 2x + b_2\sin 2x$]

[**Ans.** $f(x) = 2.1015 + 0.559\cos x - 0.519\cos 2x + \cdots + 1.535\sin x - 0.091\sin 2x + \cdots$]

4.10 INTRODUCTION OF PARTIAL DIFFERENTIAL EQUATIONS (PDE)

Equations which contain one or more partial derivatives are called partial differential equations or the differential equation which involve two or more independent variables is called partial differential equation. PDE arise in geometry and physics when the number of independent variables is the problem under consideration is two or more, usually on time t and on one or several space variables. It is fair to say that only the simplest physical systems can be modeled by **ordinary** differential equations, whereas most problems in fluid mechanics elasticity, heat transfer, electromagnetic theory, quantum mechanics, and other areas of physics lead **partial** differential equations.

Indeed the range of application of the partial differential equation is tremendous compared to that of ordinary differential equations.

The independent variables will be denoted by x and y and the dependent variable by z. The partial differential **coefficients** are denoted as follows:

$$\frac{\partial z}{\partial x} = p, \frac{\partial z}{\partial y} = q, \frac{\partial^2 z}{\partial x^2} = r, \frac{\partial^2 z}{\partial x \partial y} = s, \frac{\partial^2 z}{\partial y^2} = t$$

4.10.1 Order and Degree of a Partial Differential Equation

The order of a partial differential equation to be the order of the highest order partial derivative occurring in the partial differential equation.

The degree of a partial differential equation is the degree of the highest order derivative which occurs in it after the partial differential equation has been rationalised. For example, $pz - qz = z^2 + (x + y)^2$ and $p \tan x + q \tan y = \tan z$ are the differential equations of first order and first degree and so on.

$p^2 + q^2 - 2px - 3qy + 3xy = 0$ and $p^2 + qy^2 = y^2 - x^2$ are the differential equation of first order and second degree.

$$r - 2s + t = \sin(2x + 3y) \text{ and } y^2 \frac{\partial^2 z}{\partial x^2} - 2xy \frac{\partial^2 z}{\partial x \partial y} + x^2 \frac{\partial^2 z}{\partial y^2} = \frac{y^2}{x}\frac{\partial z}{\partial x} + \frac{x^2}{y}\frac{\partial z}{\partial y}$$

are the differential equation of second order and first degree.

4.11 PARTIAL DIFFERENTIAL EQUATIONS OF FIRST ORDER

The general form a first order partial equation is

$$F\left(x, y, z, \frac{\partial z}{\partial x}, \frac{\partial z}{\partial y}\right) = 0$$

or

$$F(x, y, z, p, q) = 0 \qquad ...(i)$$

Complete solution: The solution

$$\phi(x, y, z, a, b) = 0$$

where a and b are arbitrarly constants satisfying the P.D.E. (1) is known as complete solution.

General solution: The solution $\phi(u, v) = 0$

where $u = u(x, y)$, $v = v(x, y)$ satisfying (*i*) is known as general solution.

4.12 DERIVATION OF A PARTIAL DIFFERENTIAL EQUATION

4.12.1 Elimination of Arbitrary Constants

Let there be any function

$$\phi(x, y, z, a, b) = 0 \qquad ...(i)$$

where a and b are arbitrary constants.

Differentiating equation (*i*) partially w.r.t. 'x' and 'y', we get

$$\frac{\partial \phi}{\partial x} + \frac{\partial \phi}{\partial z} \cdot \frac{\partial z}{\partial x} = 0 \quad \text{or} \quad \frac{\partial \phi}{\partial x} + \frac{\partial \phi}{\partial z} \cdot p = 0 \qquad ...(ii)$$

and $\dfrac{\partial \phi}{\partial y} + \dfrac{\partial \phi}{\partial z} \cdot \dfrac{\partial z}{\partial y} = 0$ or $\dfrac{\partial \phi}{\partial y} + \dfrac{\partial \phi}{\partial z} \cdot q = 0$ ...(*iii*)

We thus have three relations from which two quantities a and b can generally be eliminated. The resulting equation $\psi(x, y, z, p, q) = 0$ is a partial differential equation of order one. If there are more arbitrary constants than the number of independent variables the eleminant will give a differential equation of higher order than one.

Example 27: Find the partial differential equation by the elimination of a and b from

$$z = ax + by + ab.$$

Solution: $z = ax + by + ab$ ...(1)

Differentiating equation (1), partially w.r.t. 'x' and 'y', we get

$$\frac{\partial z}{\partial x} = a = p \quad ...(2)$$

and $$\frac{\partial z}{\partial y} = b = q \quad ...(3)$$

Now eliminating a, b from these equations, we have

$$z = \frac{\partial z}{\partial x} \cdot x + \frac{\partial z}{\partial y} \cdot y + \frac{\partial z}{\partial x} \cdot \frac{\partial z}{\partial y}$$

$$z = px + qy + pq. \quad \textbf{Ans.}$$

Example 28: Find the partial differential equation by the elimination of a and b from

$$z = (x + a)(y + b).$$

Solution: $z = (x + a)(y + b)$ ...(1)

Differentiating equation (1) w.r.t. 'x' and 'y', we get

$$\frac{\partial z}{\partial x} = y + b = p \quad ...(2)$$

and $$\frac{\partial z}{\partial y} = (x + a) = q \quad ...(3)$$

Now eliminating a, b from (1), (2), (3), we have

$$z = \frac{\partial z}{\partial x} \cdot \frac{\partial z}{\partial y} = pq$$

$$z = pq. \quad \textbf{Ans.}$$

4.12.2 Elimination of Arbitrary Functions of these Variables

Let u and v be two functions of x, y, z connected by the relation

$$\phi(u, v) = 0. \quad ...(i)$$

where ϕ is arbitrary. It is now to be shown, that on the elimination of the arbitrary function ϕ from (*i*), a partial differential equation will be formed and moreover, that this equation will be linear, *i.e.*, it will be of the first degree in p and q. Regarding z as dependent variable and differentiating (*i*) partially w.r.t. 'x' and 'y', we get

$$\frac{\partial \phi}{\partial u}\left(\frac{\partial u}{\partial x}+\frac{\partial u}{\partial z}\cdot\frac{\partial z}{\partial x}\right)+\frac{\partial \phi}{\partial v}\left(\frac{\partial v}{\partial x}+\frac{\partial v}{\partial z}\cdot\frac{\partial z}{\partial x}\right)=0$$

or
$$\frac{\partial \phi}{\partial u}\left(\frac{\partial u}{\partial x}+p\frac{\partial u}{\partial z}\right)+\frac{\partial \phi}{\partial v}\left(\frac{\partial v}{\partial x}+p\frac{\partial v}{\partial z}\right)=0 \qquad \text{...(ii)}$$

and
$$\frac{\partial \phi}{\partial u}\left(\frac{\partial u}{\partial y}+\frac{\partial u}{\partial z}\cdot\frac{\partial z}{\partial y}\right)+\frac{\partial \phi}{\partial v}\left(\frac{\partial v}{\partial y}+\frac{\partial v}{\partial z}\cdot\frac{\partial z}{\partial y}\right)=0$$

or
$$\frac{\partial \phi}{\partial u}\left(\frac{\partial u}{\partial y}+q\frac{\partial u}{\partial z}\right)+\frac{\partial \phi}{\partial v}\left(\frac{\partial v}{\partial y}+q\frac{\partial v}{\partial z}\right)=0 \qquad \text{...(iii)}$$

Eliminating $\frac{\partial \phi}{\partial u}, \frac{\partial \phi}{\partial v}$ between (*ii*) and (*iii*) by the determinant method, we get

$$\begin{vmatrix} \frac{\partial u}{\partial x}+p\frac{\partial u}{\partial z} & \frac{\partial v}{\partial x}+p\frac{\partial v}{\partial z} \\ \frac{\partial u}{\partial y}+q\frac{\partial u}{\partial z} & \frac{\partial v}{\partial y}+q\frac{\partial v}{\partial z} \end{vmatrix}=0$$

or
$$\left(\frac{\partial u}{\partial y}\cdot\frac{\partial v}{\partial z}-\frac{\partial u}{\partial z}\cdot\frac{\partial v}{\partial y}\right)p+\left(\frac{\partial u}{\partial z}\cdot\frac{\partial v}{\partial x}-\frac{\partial u}{\partial x}\cdot\frac{\partial v}{\partial z}\right)q = \frac{\partial u}{\partial x}\cdot\frac{\partial v}{\partial y}-\frac{\partial u}{\partial y}\cdot\frac{\partial v}{\partial x}$$

or
$$\frac{\partial(u,v)}{\partial(y,z)}p+\frac{\partial(u,v)}{\partial(z,x)}q = \frac{\partial(u,v)}{\partial(x,y)}$$

$$\boxed{Pp+Qq=R.}$$

where
$$P=\frac{\partial(u,v)}{\partial(y,z)}=\frac{\partial u}{\partial y}\cdot\frac{\partial v}{\partial z}-\frac{\partial u}{\partial z}\frac{\partial v}{\partial y}.$$

$$Q=\frac{\partial(u,v)}{\partial(z,x)}=\frac{\partial u}{\partial z}\cdot\frac{\partial v}{\partial x}-\frac{\partial u}{\partial x}\cdot\frac{\partial v}{\partial z}.$$

$$R=\frac{\partial(u,v)}{\partial(x,y)}=\frac{\partial u}{\partial x}\cdot\frac{\partial v}{\partial y}-\frac{\partial u}{\partial y}\cdot\frac{\partial v}{\partial x}.$$

If the given relation between x, y, z contains two arbitrary functions, then leaving a few exceptional cases, the partial differential equations of higher order than two will be formed.

Example 29: Form the partial differential equation by the elimination of ϕ from

$$lx + my + nz = \phi(x^2 + y^2 + z^2).$$

Solution: $$lx + my + nz = \phi(x^2 + y^2 + z^2) \quad ...(1)$$

Differentiating (1) partially w.r.t. 'x' and 'y', we get

$$l + np = \phi'(x^2 + y^2 + z^2)\cdot(2x + 2zp)$$

and $$m + nq = \phi'(x^2 + y^2 + z^2)\cdot(2y + 2zq)$$

Dividing

$$\frac{l + np}{m + nq} = \frac{(x + pz)}{(y + zq)}$$

or $$y(l + np) + z(lq - mp) = (m + nq)x. \quad \textbf{Ans.}$$

Example 30: Find the partial differential equation by eliminating the functions ϕ and ψ from

$$z = \phi(x + iy) + \psi(x - iy).$$

Solution: $$z = \phi(x + iy) + \psi(x - iy) \quad ...(1)$$

Here, we are having two functions, hence after eliminating ϕ and ψ we cannot get a partial differential equation of order one.

Differentiating (1), we get

$$\frac{\partial z}{\partial x} = \phi'(x + iy) + \psi'(x - iy)$$

$$\frac{\partial^2 z}{\partial x^2} = \phi''(x + iy) + \psi''(x - iy)$$

$$\frac{\partial z}{\partial y} = i\phi'(x + iy) - \psi'(x - iy)$$

$$\frac{\partial^2 z}{\partial y^2} = i^2\phi''(x + iy) + i^2\psi''(x - iy)$$

$$= -\{\phi''(x + iy) + \psi''(x - iy)\}$$

$$\Rightarrow \quad \frac{\partial^2 z}{\partial y^2} = -\frac{\partial^2 z}{\partial x^2} \quad \text{or} \quad \frac{\partial^2 z}{\partial x^2} + \frac{\partial^2 y}{\partial y^2} = 0.$$

4.13 SOLUTION OF FIRST ORDER LAGRANGE'S LINEAR PARTIAL DIFFERENTIAL EQUATIONS

The partial differential of the type

$$Pp + Qq = R \quad ...(i)$$

is called Lagrange's linear partial differential equation where P, Q and R are any functions of x, y and z.

Let the general solution of equation (*i*) is given by

$$\phi(u, v) = 0 \qquad ...(ii)$$

where ϕ is an arbitrary function and

$$u = u(x, y, z) = c_1 \qquad ...(iii)$$

$$v = v(x, y, z) = c_2 \qquad ...(iv)$$

c_1 and c_2 are arbitrary constants.

Differentiating (*iii*) and (*iv*), we get

$$\frac{\partial u}{\partial x}dx + \frac{\partial u}{\partial y}dy + \frac{\partial u}{\partial z}dz = 0$$

$$\frac{\partial v}{\partial x}dx + \frac{\partial v}{\partial y}dy + \frac{\partial v}{\partial z}dz = 0$$

Solving these for dx, dy and dz, we get

$$\frac{dx}{\frac{\partial u}{\partial y}\cdot\frac{\partial v}{\partial z} - \frac{\partial v}{\partial y}\cdot\frac{\partial u}{\partial z}} = \frac{dy}{\frac{\partial u}{\partial z}\cdot\frac{\partial v}{\partial x} - \frac{\partial u}{\partial x}\cdot\frac{\partial v}{\partial z}} = \frac{dz}{\frac{\partial u}{\partial x}\cdot\frac{\partial v}{\partial y} - \frac{\partial u}{\partial y}\cdot\frac{\partial v}{\partial x}} \qquad ...(v)$$

Using the values of P, Q, R as given in (4.11.2), (*v*) may be written as

$$\boxed{\frac{dx}{P} = \frac{dy}{Q} = \frac{dz}{R}} \qquad ...(vi)$$

The equation (*vi*) represents the differential equation whose solutions are $u = c_1$ and $v = c_2$. Having found u and v, we can say that $\phi(u, v) = 0$ is the solution of Lagrange's linear equation. Equation (*vi*) are called **Lagrange's auxiliary equations** or subsidiary equations.

Method of Solution:

Step 1. Put the given equation in the standard form

$$Pp + Qq = R \qquad ...(1)$$

Step 2. Write down Lagrange's auxiliary equations

$$\frac{dx}{P} = \frac{dy}{Q} = \frac{dz}{R} \qquad ...(2)$$

Step 3. Solve (2) by using well known methods

Let $u(x, y, z) = c_1$ and $v(x, y, z) = c_2$ be two independent solutions of (2).

Step 4. The general solution of (1) is then written in one of the following three equivalent forms:

$$\phi(u, v) = 0 \quad \text{or} \quad u = \phi(v) \quad \text{or} \quad v = \phi(u).$$

There are four cases for getting two independent solutions of equation (2).

Case I: If one of the variables is either absent or cancels out from any two fractions of given equation (1). Then an integral may be find by the usual methods. The same procedure can be repeated with another set of two fractions of given equation (1).

Case II: If one integral of (1) is known by using Case I and other integral cannot be obtained by Case I. Then one integral known to us is used to find another integral, the constant of integration of first integral should be removed later on.

Case III: After fail the Case I and Case II, we introduce Lagrange's multipliers P_1, Q_1, R_1 which are the functions of x, y, z or constants each fraction in (2) will be equal

$$\frac{P_1 dx + Q_1 dy + R_1 dz}{P_1 P + Q_1 Q + R_1 R} \quad ...(3)$$

If P_1, Q_1, R_1 choosen s.t. $P_1 P + Q_1 Q + R_1 R = 0$ then $P_1 dx + Q_1 dy + R_1 dz = 0$ which can be integrated. This method can be repeated to get another integral. Sometimes only one integral is possible by use of multipliers. In such cases second integral should be find by use Case I.

Case IV: In some problems multipliers may be choosen (more than once) *i.e.*, P_2, Q_2, R_2 then

$$\frac{P_2 dx + Q_2 dy + R_2 dz}{P_2 P + Q_2 Q + R_2 R} \quad ...(4)$$

is such that its numerator is exact differential of denominator. Fractions (3) and (4) are then combined to give an integral.

Example 31: Solve $(y^2 + z^2 - x^2)p - 2xyq + 2zx = 0.$

Solution: $Pp + Qq = R$

$$P = (y^2 + z^2 - x^2), Q = -2xy, R = -2zx$$

Auxiliary equations are

$$\frac{dx}{P} = \frac{dy}{Q} = \frac{dz}{R} \quad \text{or} \quad \frac{dx}{(y^2 + z^2 - x^2)} = \frac{dy}{-2xy} = \frac{dz}{-2zx}$$

$$\underset{(I)}{\frac{dx}{(y^2 + z^2 - x^2)}} = \underset{(II)}{\frac{dy}{-2xy}} = \underset{(III)}{\frac{dz}{-2zx}}$$

Taking *II* and *III*

$$\frac{dy}{-2xy} = \frac{dz}{-2zx} \Rightarrow \frac{dy}{y} = \frac{dz}{z}$$

$$\Rightarrow \quad \frac{y}{z} = c_1$$

Again $$\frac{dx}{(y^2 + z^2 - x^2)} = \frac{dy}{-2xy} = \frac{dz}{-2zx} = \frac{2xdx + 2ydy + 2zdz}{-2x(x^2 + y^2 + z^2)}$$

$$\Rightarrow \quad \frac{dz}{-2xz} = \frac{2xdx + 2ydy + 2zdz}{-2x(x^2 + y^2 + z^2)}$$

$$\Rightarrow \quad \frac{dz}{z} = \frac{2xdx + 2ydy + 2zdz}{x^2 + y^2 + z^2}$$

Integrating

$$\log c_2 z = \log(x^2 + y^2 + z^2)$$

$$zc_2 = x^2 + y^2 + z^2$$

Hence the general integral of the given partial differential equation is given by

$$\phi\left\{\frac{y}{z}, \frac{x^2 + y^2 + z^2}{z}\right\} = 0 \quad \text{or} \quad \frac{x^2 + y^2 + z^2}{z} = \psi\left(\frac{y}{z}\right)$$

Example 32: Solve $(mz - ny)p + (nx - lz)q = ly - mx$.

Solution: $(mz - ny)p + (nx - lz)q = ly - mx$ ..(1)

Now compare the equation (1) with

$$Pp + Qq = R$$

$P = mz - ny, \quad Q = nx - lz, \quad R = ly - mx.$

Auxiliary equations are

$$\frac{dx}{P} = \frac{dy}{Q} = \frac{dz}{R}$$

$$\Rightarrow \qquad \frac{dx}{mz - ny} = \frac{dy}{nx - lz} = \frac{dz}{ly - mx}$$

$$\Rightarrow \qquad \frac{dx}{mz - ny} = \frac{dy}{nx - lz} = \frac{dz}{ly - mx} = \frac{ldx + mdy + ndz}{0} = \frac{xdx + ydy + zdz}{0}$$

$$\Rightarrow \qquad ldx + mdy + ndz = 0 \quad \text{and} \quad xdx + ydy + zdz = 0$$

Integrating

$$\Rightarrow \qquad lx + my + nz = c_1 \quad \text{and} \quad x^2 + y^2 + z^2 = c_2$$

Hence the required general solution is given by

$$\phi(lx + my + nz, x^2 + y^2 + z^2) = 0$$

or

$$lx + my + nz = \psi(x^2 + y^2 + z^2).$$

Example 33: Solve $(x^2 - yz)p + (y^2 - zx)q = z^2 - xy$.

Solution: $(x^2 - yz)p + (y^2 - zx)q = z^2 - xy$ ...(1)

Compare equation (1) with $Pp + Qq = R$.

$P = x^2 - yz, \quad Q = y^2 - zx, \quad R = z^2 - xy$

Auxiliary equations are

$$\frac{dx}{P} = \frac{dy}{Q} = \frac{dz}{R}$$

$$\Rightarrow \qquad \frac{dx}{x^2 - yz} = \frac{dy}{y^2 - zx} = \frac{dz}{z^2 - xy}$$

Now each fraction

$$\underset{(I)}{\frac{dx - dy}{(x-y)(x+y+z)}} = \underset{(II)}{\frac{dy - dz}{(y-z)(y+z+x)}} = \underset{(III)}{\frac{dz - dx}{(z-x)(z+x+y)}}$$

Now (*I*) and (*II*)

$$\frac{dx - dy}{x - y} = \frac{dy - dz}{y - z} \quad \Rightarrow \quad \frac{x-y}{y-z} = c_1 \qquad ...(2)$$

Similarly, (*II*) and (*III*)

$$\frac{dy - dz}{(y-z)(y+z+x)} = \frac{dz - dx}{(z-x)(z+x+y)} \quad \Rightarrow \quad \frac{y-z}{z-x} = c_2 \qquad ...(3)$$

Hence the required general integral is given by

$$\phi\left\{\frac{x-y}{y-z}, \frac{y-z}{z-x}\right\} = 0 \quad \text{or} \quad \frac{x-y}{y-z} = \psi\left(\frac{y-z}{z-x}\right).$$

Example 34: Solve $(y^2 z/x)p + zxq = y^2$.

Solution: $(y^2 z/x)p + zxq = y^2$ ...(1)

$$P = y^2 z / x, Q = zx, R = y^2$$

Auxiliary equations are

$$\underset{(I)}{\frac{dx}{y^2 z/x}} = \underset{(II)}{\frac{dy}{zx}} = \underset{(III)}{\frac{dz}{y^2}}$$

Consider (*I*) and (*II*);

$$\frac{dx}{y^2 z/x} = \frac{dy}{zx} \quad \Rightarrow \quad \frac{xdx}{y^2 z} = \frac{dy}{zx}$$

$$\Rightarrow \quad \frac{xdx}{y^2} = \frac{dy}{x}$$

$$\Rightarrow \quad x^2 dx = y^2 dy$$

Integrating $\Rightarrow \quad x^3 - y^3 = c_1$

Now, consider (*I*) and (*III*);

$$\frac{xdx}{y^2 z} = \frac{dz}{y^2}$$

$$\Rightarrow \qquad xdx = zdz$$

$$\Rightarrow \qquad x^2 - z^2 = c_2$$

Hence the general integral is given by

$$\phi(x^3 - y^3, x^2 - z^2) = 0 \quad \text{or} \quad x^3 - y^3 = \psi(x^2 - z^2).$$

Example 35: Solve $(y^3x - 2y^4)p + (2y^4 - x^3y)q = 9z(x^3 - y^3)$.

Solution: $(y^3x - 2x^4)p + (2y^4 - x^3y)q = 9z(x^3 - y^3)$ ...(1)

$P = (y^3x - 2x^4), \qquad Q = 2y^4 - x^3y, \qquad R = 9z(x^3 - y^3)$

Auxiliary equations are

$$\underset{(I)}{\frac{dx}{y^3x - 2x^4}} = \underset{(II)}{\frac{dy}{2y^4 - x^3y}} = \underset{(III)}{\frac{dz}{9z(x^3 - y^3)}}$$

Consider (I) and (II)

$$\frac{dx}{y^3x - 2x^4} = \frac{dy}{2y^4 - x^3y}$$

$$\Rightarrow \qquad \frac{dy}{dx} = \frac{y(2y^3 - x^3)}{x(y^3 - 2x^3)} \qquad \text{...(2)}$$

Equation (2) is a homogeneous equation

Putting $y = vx$

$$\frac{dy}{dx} = v + x\frac{dv}{dx}$$

$$v + x\frac{dv}{dx} = \frac{vx(2v^3x^3 - x^3)}{x(v^3x^3 - 2x^3)} = \frac{vx^4(2v^3 - 1)}{x^4(v^3 - 2)}$$

$$x\frac{dv}{dx} = \frac{v(2v^3 - 1)}{(v^3 - 2)} - v = \frac{v(2v^3 - 1) - v^4 + 2v}{v^3 - 2}$$

$$x\frac{dv}{dx} = \frac{2v^4 - v - v^4 + 2v}{v^3 - 2} = \frac{v^4 + v}{v^3 - 2}$$

$$\frac{dv}{dx} = \left(\frac{v^4 + v}{v^3 - 2}\right)\frac{1}{x}$$

$$\left(\frac{v^3 - 2}{v^4 + v}\right)dv = \frac{dx}{x}$$

$$\frac{dx}{x} = \left\{\frac{-2}{v} + \frac{1}{v+1} + \frac{2v - 1}{v^2 - v + 1}\right\}dv$$

Integrating

$$\log c_1 x = -2\log v + \log(v+1) + \log(v^2 - v + 1)$$

$$= \log \frac{(v+1)(v^2 - v + 1)}{v^2}$$

$$c_1 x = \frac{(y/x+1)(y^2/x^2 - y/x + 1)}{y^2/x^2} = \frac{(y^2 - xy + x^2)(x+y)}{xy^2}$$

$$\Rightarrow \quad c_1 x^2 y^2 = x^3 + y^3 \quad \text{or} \quad c_1 = (x/y^2) + (y/x^2)$$

Now

$$\frac{dz}{9z(x^3 - y^3)} = \frac{(dx/x) + (dy/y) + (0 \cdot dz)}{3(y^3 - x^3)}$$

$$\Rightarrow \quad \frac{dx}{x} + \frac{dy}{y} = \frac{dz}{-3z}$$

Integrating, $\log x + \log y = -(1/3) \cdot \log z + \frac{1}{3}\log c_2$ or $3\log xy = -\log z + \log c_2 = \log z^{-1} + \log c_2$

or $x^3 y^3 z = c_2$.

Hence the required general integral is given by

$$x^3 y^3 z = \phi\left(\frac{x}{y^2} + \frac{y}{x^2}\right) \quad \text{or} \quad z = \frac{1}{x^3 y^3}\phi\left(\frac{x}{y^2} + \frac{y}{x^2}\right).$$

Example 36: Solve $x(y-z)p + y(z-x)q = z(x-y)$.

Solution: $x(y-z)p + y(z-x)q = z(x-y)$ ...(1)

$$P = x(y-z), \quad Q = y(z-x), \quad R = z(x-y)$$

Auxiliary equations are

$$\underset{(I)}{\frac{dx}{x(y-z)}} = \underset{(II)}{\frac{dy}{y(z-x)}} = \underset{(III)}{\frac{dz}{z(x-y)}}$$

$$\frac{\frac{dx}{x}}{y-z} = \frac{\frac{dy}{y}}{z-x} = \frac{\frac{dz}{z}}{x-y} = \frac{\frac{dx}{x} + \frac{dy}{y} + \frac{dz}{z}}{y - z + z - x + x - y}$$

$$\Rightarrow \quad \frac{dx}{x} + \frac{dy}{y} + \frac{dz}{z} = 0 \Rightarrow \log x + \log y + \log z = \log c_1$$

$$\Rightarrow xyz = c_1$$

$$\frac{dx}{x(y-z)} = \frac{dy}{y(z-x)} = \frac{dz}{z(x-y)} = \frac{dx+dy+dz}{xy-xz+yz-yx+zx-zy}$$

$$= \frac{dx+dy+dz}{0}$$

$$\Rightarrow \qquad dx+dy+dz = 0$$

$$\Rightarrow \qquad x+y+z = c_2$$

Hence the general integral is

$$\phi(x+y+z, xyz) = 0 \quad \text{or} \quad x+y+z = \psi(xyz).$$

Example 37: Solve $(y+z)p+(z+x)q=(x+y)$.

Solution: $(y+z)p+(z+x)q=(x+y)$ ...(1)

$$P = (y+z), \quad Q=(z+x), \quad R=(x+y)$$

Auxiliary equations are

$$\underset{(I)}{\frac{dx}{y+z}} = \underset{(II)}{\frac{dy}{z+x}} = \underset{(III)}{\frac{dz}{x+y}}$$

Obviously

$$(I) = (II) = (III) = \underset{(IV)}{\frac{dx-dy}{y-x}} = \underset{(V)}{\frac{dz-dx}{x-z}} = \underset{(VI)}{\frac{dx+dy+dz}{2(x+y+z)}}$$

Now taking (V) and (VI) and then integrating

$$-2\log(x-z) = \log(x+y+z) - \log c_1$$

or

$$\log c_1 = \log\{(x+y+z)\cdot(x-z)^2\}$$

$$c_1 = (x+y+z)\cdot(x-z)^2 \qquad \text{...(2)}$$

Also from (IV) and (V)

$$\frac{dx-dy}{y-x} = \frac{dz-dx}{x-z}$$

Integrating

$$\log(y-x) = \log\{c_2(x-z)\}$$

$$(y-x) = c_2(x-z)$$

$$c_2 = \frac{(y-x)}{(x-z)} \qquad \text{...(3)}$$

From (2) and (3)

$$\phi\left\{(x-z)^2(x+y+z), \frac{y-x}{x-z}\right\} = 0.$$ **Ans.**

Example 38: Solve $(2x^2+y^2+z^2-2yz-zx-xy)p+(x^2+2y^2+z^2-yz-2zx-xy)q$

$$= (x^2+y^2+2z^2-yz-zx-2xy).$$

Solution: $(2x^2+y^2+z^2-2yz-zx-xy)p+(x^2+2y^2+z^2-yz-2zx-xy)q$

$$=(x^2+y^2+2z^2-yz-zx-2xy) \qquad ...(1)$$

$$P = (2x^2+y^2+z^2-2yz-zx-xy)$$

$$Q = (x^2+2y^2+z^2-yz-2zx-xy)$$

$$R = (x^2+y^2+2z^2-yz-zx-2xy)$$

Auxiliary equations are

$$\frac{dx}{2x^2+y^2+z^2-2yz-zx-xy} = \frac{dy}{x^2+2y^2+z^2-yz-2zx-xy}$$

$$= \frac{dz}{x^2+y^2+2z^2-yz-zx-2xy}$$

$$\frac{dx-dy}{x^2-y^2-yz+zx} = \frac{dy-dz}{y^2-z^2-zx+xy} = \frac{dz-dx}{z^2-x^2-xy+yz}$$

$$\frac{dx-dy}{(x-y)(x+y+z)} = \frac{dy-dz}{(y-z)(x+y+z)} = \frac{dz-dx}{(z-x)(z+x+y)}$$

or
$$\underset{(I)}{\frac{dx-dy}{x-y}} = \underset{(II)}{\frac{dy-dz}{y-z}} = \underset{(III)}{\frac{dz-dx}{z-x}}$$

(*I*) and (*II*)

$$\frac{dx-dy}{x-y} = \frac{dy-dz}{y-z}$$

$$\Rightarrow \quad \log(x-y) = \log(y-z)+\log c_1$$

$$\Rightarrow \quad \log(x-y) = \log(y-z)c_1$$

$$\Rightarrow \quad (x-y) = (y-z)c_1$$

$$\Rightarrow \quad c_1 = \frac{(x-y)}{(y-z)} \qquad ...(2)$$

(*I*) and (*III*)

$$\frac{dx-dy}{x-y} = \frac{dz-dx}{z-x}$$

$$\Rightarrow \quad \log(x-y) = \log(z-x)+\log c_2$$

$$\Rightarrow \quad \log(x-y) = \log(z-x)c_2$$

$$\Rightarrow \quad c_2 = \left(\frac{x-y}{z-x}\right) \quad \text{...(3)}$$

Hence the general solution

$$\phi\left(\frac{x-y}{y-z}, \frac{x-y}{z-x}\right) = 0. \quad \textbf{Ans.}$$

Example 39: Solve $(x+2z)p+(4zx-y)q = 2x^2+y.$

Solution: $(x+2z)p+(4zx-y)q = 2x^2+y$ ...(1)

$$P = (x+2z), \quad Q = (4zx-y), \quad R = 2x^2+y$$

Auxiliary equations are

$$\frac{dx}{x+2z} = \frac{dy}{4zx-y} = \frac{dz}{2x^2+y}$$

or
$$\frac{y\,dx}{xy+2yz} = \frac{x\,dy}{4zx^2-yz} = \frac{2zdz}{4x^2z+2yz}$$

(Taking multipliers y, x, $-2z$)

$$\frac{ydx}{xy+2yz} = \frac{xdy}{4zx^2-yx} = \frac{2zdx}{4x^2z+2yz} = \frac{ydx+xdy-2zdz}{0}$$

$$\Rightarrow \quad y\,dx + x\,dy - 2z\,dz = 0$$

$$\Rightarrow \quad xy - z^2 = c_1 \quad \text{...(2)}$$

$$\frac{2x\,dx}{2x^2+4xz^2} = \frac{-dy}{-4zx+y} = \frac{-dz}{-2x^2-y} = \frac{2x\,dx-dy-dz}{0}$$

(Taking multipliers $2x$, -1, -1)

$$\Rightarrow \quad 2xdx - dy - dz = 0$$

$$\Rightarrow \quad x^2 - y - z = c_2 \quad \text{...(3)}$$

From (2) and (3)

$$\phi(xy-z^2, x^2-y-z) = 0$$

Example 40: Solve $p+3q = 5z+\tan(y-3x).$

Solution: $p+3q = 5z+\tan(y-3x)$ ...(1)

$$P = 1, \quad Q = 3, \quad R = 5z+\tan(y-3x).$$

Auxiliary equations are

$$\frac{dx}{1} = \frac{dy}{3} = \frac{dz}{5z+\tan(y-3x)}$$
$$(I) \quad (II) \quad (III)$$

From (*I*) and (*II*)

$$\frac{dx}{1} = \frac{dy}{3}$$

$$\Rightarrow \quad x = 3y + c_1$$

$$\Rightarrow \quad x - 3y = c_1 \qquad ...(2)$$

From (*I*) and (*III*)

$$\frac{dx}{1} = \frac{dz}{5z+\tan(y-3x)}$$

From (2), we have $y - 3x = c_1$

$$\frac{dx}{1} = \frac{dz}{5z+\tan c_1}$$

$$x = \frac{1}{5}\log(3z+\tan c_1) - \frac{1}{5}\log c_2$$

$$5x = \log(5z+\tan c_1) - \log c_2$$

$$5x = \log(5z+\tan(y-3x))c_2$$

$$(5z+\tan(y-3x))c_2 = e^{5x}$$

$$c_2 = e^{5x}(5z+\tan(y-3x))^{-1}.$$

Hence

$$\phi(y-3x, e^{5x}(5z+\tan(y-3x))^{-1}\} = 0.$$

Example 41: Solve $x(y^2+z)p - y(x^2+z)q = z(x^2-y^2)$. (*U.P.T.U. 2008*)

Solution: $x(y^2+z)p - y(x^2+z)q = z(x^2-y^2)$ ...(1)

$$P = x(y^2+z), \quad Q = -y(x^2+z), \quad R = z(x^2-y^2)$$

Auxiliary equations are

$$\frac{dx}{x(y^2+z)} = \frac{dy}{-y(x^2+z)} = \frac{dz}{z(x^2-y^2)}$$

(Taking multipliers x, y, -1)

$$\frac{xdx}{x^2y^2+x^2z} = \frac{ydy}{-y^2x^2-y^2z} = \frac{-dz}{-zx^2+zy^2} = \frac{xdx+ydy-dz}{0}$$

$$\Rightarrow \quad xdx + ydy - dz = 0$$

$$\Rightarrow \qquad x^2 + y^2 - 2z = c_2 \qquad \ldots(2)$$

Again using multipliers $\frac{1}{x}, \frac{1}{y}, \frac{1}{z}$, then we have

$$\frac{dx}{x} + \frac{dy}{y} + \frac{dz}{z} = 0$$

$$\Rightarrow \qquad \log xyz = \log c_2$$

$$\Rightarrow \qquad xyz = c_2$$

Hence

$$\phi(x^2 + y^2 - 2z, xyz) = 0. \quad \textbf{Ans.}$$

Example 42: Solve $y^2(x-y)p + x^2(y-x)q = z(x^2 + y^2)$.

Solution: We have $y^2(x-y)p + x^2(y-x)q = z(x^2 + y^2)$

$$P = y^2(x-y), Q = x^2(y-x), R = z(x^2 + y^2)$$

The Lagrange's auxiliary equations are

$$\frac{dx}{y^2(x-y)} = \frac{dy}{-x^2(x-y)} = \frac{dz}{z(x^2 + y^2)} \qquad \ldots(1)$$

Taking first two fractions

$$\frac{dx}{y^2(x-y)} = \frac{dy}{-x^2(x-y)} \Rightarrow -x^2 dx = y^2 dy$$

or $\qquad x^2 dx + 3y^2 dy = 0$

Integrating $\qquad x^3 + y^3 = c_1 \qquad \ldots(2)$

Choosing 1, –1, 0 as multipliers each fraction of (1)

$$\frac{dx}{y^2(x-y)} = \frac{dy}{-x^2(x-y)} = \frac{dz}{z(x^2 + y^2)} = \frac{dx - dy}{y^2(x-y) + x^2(x-y)} = \frac{dx - dy}{(x-y)(x^2 + y^2)} \qquad \ldots(3)$$

Taking

$$\frac{dz}{z(x^2 + y^2)} = \frac{dx - dy}{(x-y)(x^2 + y^2)}$$

or $$\frac{dx - dy}{x - y} = \frac{dz}{z}$$

or $$\frac{d(x-y)}{(x-y)} - \frac{dz}{z} = 0$$

Integrating

$$\log(x-y) - \log z = \log c_2$$

or $$\frac{x-y}{z} = c_2$$

Hence the required general solution is

$$\phi\left(x^3 + y^3, \frac{x-y}{z}\right) = 0.$$

EXERCISE 4.3

Solve the following equations:

1. $x^2(y-z)p + (z-x)y^2q = z^2(x-y)$. $\left[\textbf{Ans. } \phi\left(xyz, \frac{1}{x} + \frac{1}{y} + \frac{1}{z}\right) = 0\right]$

2. $xzp + yzq = xy$. $\left[\textbf{Ans. } \phi\left(\frac{x}{y}, \frac{y}{z}\right) = 0\right]$

3. $p \tan x + q \tan y = \tan z$. $\left[\textbf{Ans. } \frac{\sin z}{\sin y} = \phi\left(\frac{\sin x}{\sin y}\right)\right]$

4. $(x+2z)p + (4zx-y)q = x^2 + y$. $\left[\textbf{Ans. } \phi\left(\frac{1}{x} - \frac{1}{y}, \frac{1}{x} - \frac{1}{z}\right) = 0\right]$

5. $(x+2y-z)p + (3y-z)q = x+y$. $\left[\textbf{Ans. } (x-y)\,e^{(x-z)/(x-y)} = \phi\left\{\frac{(x-3y+z)}{(x-y)^2}\right\}\right]$

6. $(-a+x)p + (-b+y)q = (-c+z)$. $\left[\textbf{Ans. } \phi\left(\frac{y-b}{z-c}, \frac{x-a}{z-c}\right) = 0\right]$

7. $(y^2+z^2)p - xyq = -zx$. $[\textbf{Ans. } \phi(x^2+y^2+z^2, y/z) = 0]$

8. $pz - qz = z^2 + (x+y)^2$. $[\textbf{Ans. } \phi[x+y, e^{2y}\{(x+y)^2+z^2\}] = 0]$

9. $x\frac{\partial u}{\partial x} + y\frac{\partial u}{\partial y} + z\frac{\partial u}{\partial z} = xyz$. $\left[\textbf{Ans. } \phi\left\{\frac{x}{y}, \frac{y}{z}, xyz - 3u\right\} = 0\right]$

10. $(u+y+z)\frac{\partial u}{\partial x} + (u+z+x)\frac{\partial u}{\partial y} + (u+x+y)\frac{\partial u}{\partial z} = x+y+z$.

$$\left[\textbf{Ans. } \phi\left\{\frac{x-y}{y-z}, \frac{y-z}{z-u}, (x-y)(u+x+y+z)^{1/3}\right\} = 0\right]$$

4.14 HOMOGENEOUS LINEAR PARTIAL DIFFERENTIAL EQUATION OF HIGHER ORDER WITH CONSTANT COEFFICIENTS

A differential equation of the form

$$\frac{\partial^n z}{\partial x^n}+a_1\frac{\partial^n z}{\partial x^{n-1}\partial y}+a_2\frac{\partial^n z}{\partial x^{n-2}\partial y^2}+\ldots+a_n\frac{\partial^n z}{\partial y^n} = F(x,y) \qquad \ldots(i)$$

where $a_1, a_2, \cdots, a_n$ are constants, is called homogeneous linear partial differential equation of nth order with constant coefficients.

Let $D=\frac{\partial}{\partial x}$ and $D'=\frac{\partial}{\partial y}$, then equation (*i*) takes the form

$$(D^n+a_1D^{n-1}D'+a_2D^{n-2}D'^2+\cdots+a_nD'^n)z = F(x,y)$$

$$\Rightarrow \qquad f(D,D')z = F(x,y) \qquad \ldots(ii)$$

where $$f(D,D') = D^n+a_1D^{n-1}D'+a_2D^{n-2}D'^2+\cdots+a_nD'^n$$

is a homogeneous function of D and D'.

The method of solving equation (*i*) or (*ii*) is similar to that for ordinary differential equations and complete solution = C.F. + P.I.

4.14.1 Determination of C.F.

Since C.F. is the solution of $f(D,D')z=0$. $\qquad \ldots(i)$

So, we let $z=\phi(y+mx)$ be a solution of above equation

Now
$$Dz = \frac{\partial}{\partial x}\phi(y+mx) = m\phi'(y+mx)$$

$$D'z = \frac{\partial}{\partial y}\phi(y+mx) = \phi'(y+mx)$$

$$D^2z = \frac{\partial}{\partial x}\{m\cdot\phi'(y+mx)\} = m^2\phi''(y+mx)$$

$$D'^2z = \frac{\partial}{\partial y}\{\phi'(y+mx)\} = \phi''(y+mx)$$

... ...

... ...

$$D^rz = m^r\phi^{(r)}(y+mx)$$

$$D'^rz = \phi^{(r)}(y+mx)$$

... ...

$$D^rD'^sz = m^r\cdot\phi^{(r+s)}(y+mx)$$

Substituting these values in equation (*i*), we get

$$[m^n\phi^n(y+mx)+a_1m^{n-1}\phi^n(y+mx)+a_2m^{n-2}\phi^n(y+mx)+\cdots+a_n\phi^n(y+mx)]=0$$

$$\Rightarrow \quad [m^n+a_1m^{n-1}+a_2m^{n-2}+\cdots+a_n]\phi^n(y+mx)=0$$

$$\Rightarrow \quad f(m,1)\phi^n(y+mx)=0 \qquad \ldots(ii)$$

As $f(D,D')=D^n+a_1D^{n-1}D'+\cdots+D'_n$

$\Rightarrow f(m,1)=m^n+a_1m^{n-1}+\cdots+a_n$

Clearly equation (*ii*) is satisfied if

$$\boxed{f(m,1)=0.} \qquad \ldots(A)$$

This is called an auxiliary equation.

Thus we write the auxiliary equation (A.E.) by putting

$D=m$ and $D'=1$, in $f(D,D')=0$

Since (*A*) is an equation of *n*th degree in *m*, so there arise the following cases:

Case I: If $m_1 \neq m_2 \neq m_3 \cdots\cdots \neq m_n$, then

Corresponding to each root *m*, we have an arbitrary function

$\phi(y+m_rx)$ as a solution of $f(D,D')z=0$.

$$\therefore \qquad \text{C.F.} = \phi_1(y+m_1x)+\phi_2(y+m_2x)+\cdots+\phi_n(y+m_nx)$$

If there are only two roots *i.e.*, m_1 and m_2, then

$$\boxed{\text{C.F.}=\phi_1(y+m_1x)+\phi_2(y+m_2x).}$$

In this case factors of $f(D,D')$ are

$(D-m_1D')(D-m_2D')(D-m_3D')\cdots\cdots(D-m_nD')$.

Case II: Equal roots $(m_1=m_2=m)$

then $\qquad \text{C.F.} = \phi_1(y+mx)+x\phi_2(y+mx)$

Proof: Consider the equation $(D-mD')(D-mD')z=0$

Let $\qquad (D-mD')z = V$

then above equation takes the form

$$(D-mD')V=0$$

from case I, its solution is $V=\phi\ (y+mx)$

Hence $(D-mD')z=\phi(y+mx)$ which is of the form

$$p-mq=\phi(y+mx)$$

This is the Lagrange's form. Its subsidiary equations are

$$\frac{dx}{1}=\frac{dy}{-m}=\frac{dz}{\phi(y+mx)}$$

The first two members of these equations give

$$y + mx = c$$

Hence the first and third members gives

$$\phi(c)\,dx = dz \quad \Rightarrow \quad z = x\phi(c) + c_1$$

$$\Rightarrow \qquad z = x\phi(y+mx) + \phi_1(y+mx)$$

$$\therefore \qquad \boxed{\text{C.F.} = \phi_1(y+mx) + x\phi_2(y+mx).}$$

Similarly, if m repeated n times,

$$\text{C.F.} = \phi_1(y+mx) + x\phi_2(y+mx) + x^2\phi_3(y+mx) + \cdots + x^{n-1}\phi_n(y+mx).$$

Example 43: Solve $\dfrac{\partial^2 z}{\partial x^2} + \dfrac{\partial^2 z}{\partial y^2} = 0.$

Solution: Putting $\dfrac{\partial}{\partial x} = D$ and $\dfrac{\partial}{\partial y} = D'$, then the given equation may be written as

$$(D^2 + D'^2)z = 0$$

Again putting $D = m, \quad D' = 1$

$\therefore$ The A.E. is $\quad m^2 + 1 = 0 \;\Rightarrow\; m = \pm i$

hence the general solution is

$$z = \phi_1(y+ix) + \phi_2(y-ix). \quad \textbf{Ans.}$$

Example 44: Solve $25r - 40s + 16t = 0.$

Solution: The given equation can be written as

$$25\frac{\partial^2 z}{\partial x^2} - 40\frac{\partial^2 z}{\partial x \partial y} + 16\frac{\partial^2 z}{\partial y^2} = 0$$

or

$$(25D^2 - 40DD' + 16D'^2)z = 0$$

Putting $D = m, \quad D' = 1$

$\therefore$ The A.E. is $\quad 25m^2 - 40m + 16 = 0$

$$\Rightarrow \qquad (5m-4)^2 = 0 \;\Rightarrow\; m = \frac{4}{5}, \frac{4}{5}$$

$$\therefore \qquad z = \phi_1\left(y + \frac{4}{5}x\right) + x\phi_2\left(y + \frac{4}{5}x\right)$$

or

$$z = f_1(5y+4x) + x\,f_2(5y+4x). \quad \textbf{Ans.}$$

4.14.2 Determination of *P.I.* = $\frac{1}{f(D,D')} \times F(x,y)$

(*a*) If $F(x,y) = e^{ax+by}$

then P.I. $= \frac{1}{f(D,D')} e^{ax+by} = \frac{e^{ax+by}}{f(a,b)}$, provides $f(a,b) \neq 0, (D = a, D' = b)$.

Proof: $De^{ax+by} = ae^{ax+by}, D'e^{ax+by} = be^{ax+by}$

$D^2e^{ax+by} = a^2e^{ax+by}, D'^2e^{ax+by} = b^2e^{ax+by}$

and $DD'e^{ax+by} = abe^{ax+by}$

$$\therefore \quad (D^2 + a_1DD' + a_2D'^2)e^{ax+by} = (a^2 + a_1ab + a_2b^2)e^{ax+by}$$

$$\Rightarrow \quad f(D,D')e^{ax+by} = f(a,b)e^{ax+by}$$

$$\Rightarrow \quad \frac{1}{f(D,D')} f(D,D')e^{ax+by} = \frac{1}{f(D,D')} f(a,b)e^{ax+by}$$

$$\therefore \quad \boxed{\frac{1}{f(D,D')} e^{ax+by} = \frac{1}{f(a,b)} e^{ax+by}.}$$

(*b*) If $F(x,y) = \sin(ax+by)$ or $\cos(ax+by)$

then P.I. $= \frac{1}{f(D^2, DD', D'^2)} \sin(ax+by)$ or $\cos(ax+by)$

$= \frac{1}{f(-a^2, -ab, -b^2)} \sin(ax+by)$ or $\cos(ax+by)$

$$(D^2 = -a^2, DD' = -ab, D'^2 = -b^2)$$

(*c*) If $F(x,y) = x^h y^k$

$$\text{P.I.} = \frac{1}{f(D,D')} x^h y^k = [f(D,D')]^{-1} x^h y^k$$

where $[f(D,D')]^{-1}$ is to be expanded in powers of D and D'.

(*d*) When $F(x,y) = e^{ax+by} \cdot V$

$$\text{P.I.} = \frac{1}{f(D,D')} e^{ax+by} \cdot V = e^{ax+by} \cdot \frac{1}{f(D+a, D'+b)} V$$

(*e*) If $F(x, y) =$ any function of x and y

$$\text{P.I.} = \frac{1}{(D-m_1D')(D-m_2D')}F(x,y)$$

It can evaluate, if we can evaluate $\frac{1}{(D-mD')}F(x,y)$

For this, let $z = \frac{1}{(D-mD')}F(x,y)$

$\therefore\ p - mq = F(x, y).$

Its subsidiary equations are $\frac{dx}{1} = \frac{dy}{-m} = \frac{dz}{F(x,y)}$

One solution is $y + mx = c$ and other is given by

$$\int dz = \int F(x,y)\,dx$$

$$\Rightarrow \qquad z = \int F(x, c-mx)\,dx$$

Hence

$$\boxed{\text{P.I.} = \frac{1}{(D-mD')}F(x,y) = \int F(x,c-mx)\,dx.}$$

(*f*) When $f(a,b) = 0$ [failure case]

$$\text{P.I.} = \frac{1}{f(D,D')}F(x,y) = x\frac{1}{\frac{\partial}{\partial D}f(D,D')}F(x,y)$$

and so on.

Special case: If $F(x,y) = \phi(ax+by)$

then

$$\text{P.I.} = \frac{1}{f(D,D')}\phi(ax+by)$$

$$= \frac{1}{f(a,b)}\iint\cdots\int \phi(v)\,dv$$

where $v = ax + by.$

Solved Examples Related to '*a*'

Example 45: Solve $2r - s - 3t = 5e^{x-y}$.

Solution: The given equation can be written as

$$2\frac{\partial^2 z}{\partial x^2} - \frac{\partial^2 z}{\partial x \partial y} - 3\frac{\partial^2 z}{\partial y^2} = 5e^{x-y}$$

or $$(2D^2 - DD' - 3D'^2)z = 5e^{x-y} \qquad \left| \begin{array}{l} D = \partial/\partial x \\ D' = \partial/\partial y \end{array} \right.$$

Here the A.E. is

$$2m^2 - m - 3 = 0 \qquad |\text{Put } D = m, D' = 1$$

$$2m^2 - 3m + 2m - 3 = 0$$

$$\Rightarrow \quad (m+1)(2m-3) = 0$$

$$\Rightarrow \quad m = -1,\ 3/2$$

$$\therefore \quad \text{C.F.} = \phi_1(y-x) + \phi_2\left(y + \frac{3x}{2}\right)$$

and $$\text{P.I.} = \frac{1}{(2D^2 - DD' - 3D'^2)} 5e^{x-y}$$

$$= \frac{5e^{x-y}}{(2+1-3)} \quad \text{(case fail)}$$

$$= \frac{x}{(4D - D')} 5e^{x-y} = \frac{5xe^{x-y}}{4+1} = xe^{x-y}$$

Hence $z = \text{C.F.} + \text{P.I.} \Rightarrow z = \phi_1(y-x) + \phi_2\left(y + \frac{3x}{2}\right) + xe^{x-y}$. **Ans.**

Example 46: Solve $\dfrac{\partial^3 u}{\partial x^3} + \dfrac{\partial^3 u}{\partial x^2 \partial y} + 4\dfrac{\partial^3 u}{\partial y^3} = e^{x+2y}$. *(U.P.T.U. 2007)*

Solution: The given equation may be written as

$$(D^3 + D^2D' + 4D'^3)u = e^{x+2y}$$

Putting $D = m$, $D' = 1$ to obtain A.E.

$\therefore$ The A.E. is

$$m^3 + m^2 + 4 = 0$$

$$\Rightarrow \quad m^2(m+2) - m(m+2) + 2(m+2) = 0$$

$$\Rightarrow \quad (m+2)(m^2 - m + 2) = 0$$

$$\Rightarrow \quad m = -2, \frac{1 \pm i\sqrt{7}}{2}$$

$$\Rightarrow \quad m_1 = -2, m_2 = \frac{1+i\sqrt{7}}{2}, m_2 = \frac{1-i\sqrt{7}}{2}$$

$$\text{C.F.} = \phi_1(y-2x) + \phi_2\left[y + \frac{x}{2}(1+i\sqrt{7})\right] + \phi_3\left[y + \frac{x}{2}(1-i\sqrt{7})\right]$$

and $$\text{P.I.} = \frac{1}{(D^3 + D^2D' + 4D'^3)} e^{x+2y} = \frac{1}{1+1^2 \times 2 + 4 \times 2^3} e^{x+2y} \quad \left| \begin{matrix} D=1 \\ D'=2 \end{matrix} \right.$$

$$\text{P.I.} = \frac{e^{x+2y}}{35}$$

Hence the required solution of given equation is

$$u = \text{C.F.} + \text{P.I.} = \phi_1(y-2x) + \phi_2\left[y + \frac{x}{2}(1+i\sqrt{7})\right] + \phi_3\left[y + \frac{x}{2}(1-i\sqrt{7})\right] + \frac{e^{x+2y}}{7}$$

$$\Rightarrow \quad u = \phi_1(y-2x) + \phi_2\left[y + \frac{x}{2}(1+i\sqrt{7})\right] + \phi_3\left[y + \frac{x}{2}(1-i\sqrt{7})\right] + \frac{e^{x+2y}}{7}. \quad \textbf{Ans.}$$

Example 47: Solve $4r + 12s + 9t = e^{3x-2y}$.

Solution: The given equation may be written as

$$4\frac{\partial^2 z}{\partial x^2} + 12\frac{\partial^2 z}{\partial x \partial y} + 9\frac{\partial^2 z}{\partial y^2} = e^{3x-2y}$$

$$\Rightarrow \quad 4D^2 + 12DD' + 9D'^2 = e^{3x-2y} \quad \left| \begin{matrix} D = \partial/\partial x \\ D' = \partial/\partial y \end{matrix} \right.$$

Putting $D = m$ and $D' = 1$ to obtain A.E.

∴ A.E. is

$$4m^2 + 12m + 9 = 0$$

$$\Rightarrow \quad (2m+3)^2 = 0$$

$$\Rightarrow \quad m = -\frac{3}{2}, -\frac{3}{2}$$

$$\text{C.F.} = \phi_1\left(y - \frac{3}{2}x\right) + x\phi_2\left(y - \frac{3}{2}x\right)$$

or $$\text{C.F.} = f_1(2y-3x) + xf_2(2y-3x) \quad \left| \begin{matrix} \phi_1\left(\frac{2y-3x}{2}\right) \\ = f_1(2y-3x), \\ \phi_2\left(\frac{2y-3x}{2}\right) = f_2(2y-3x) \end{matrix} \right.$$

and $$\text{P.I.} = \frac{1}{(4D^2 + 12DD' + 9D'^2)} e^{3x-2y}$$

$$= \frac{1}{4\times 3^2 + 12\times 3(-2) + 9(-2)^2} e^{3x-2y} \quad \Big| D = 3, D' = -2$$

$$= \frac{1}{36 - 72 + 36} e^{3x-2y} \quad \text{(case fail)}$$

$$\therefore \quad \text{P.I.} = \frac{x}{\frac{\partial}{\partial D}\{4D^2 + 12DD' + 9D'^2\}} e^{3x-2y}$$

$$= \frac{x}{(8D + 12D')} e^{3x-2y} = \frac{xe^{3x-2y}}{8\times 3 + 12(-2)}$$

$$= \frac{xe^{3x-2y}}{24 - 24} \quad \text{(again case fail)}$$

$$\Rightarrow \quad \text{P.I.} = \frac{x^2}{\frac{\partial}{\partial D}(8D + 12D')} \cdot e^{3x-2y} = \frac{x^2 e^{3x-2y}}{8}$$

Hence $$z = f_1(2y - 3x) + xf_2(2y - 3x) + \frac{x^2 e^{3x-2y}}{8}. \quad \textbf{Ans.}$$

Solved Examples Related to '*b*'

Example 48: Solve $\frac{\partial^2 z}{\partial x^2} - 2\frac{\partial^2 z}{\partial x \partial y} + \frac{\partial^2 z}{\partial y^2} = \sin x.$ *(U.P.T.U. 2004, 2009)*

Solution: We have $(D^2 - 2DD' + D'^2)z = \sin x \quad \big| D = \partial/\partial x, D' = \partial/\partial y$

Putting $D = m$ and $D' = 1$

then A.E. is $m^2 - 2m + 1 = 0 \quad \Rightarrow \quad (m-1)^2 = 0 \quad \Rightarrow \quad m = 1, 1$

$$\therefore \quad \text{C.F.} = \phi_1(y + x) + x\phi_2(y + x)$$

and $$\text{P.I.} = \frac{1}{(D^2 - 2DD' + D'^2)} \sin x = \frac{1}{(D^2 - 2DD' + D'^2)} \sin(1\cdot x + 0\cdot y)$$

$$= \frac{1}{-1 - 2(0) + 0} \sin x = -\sin x \quad \Big| D^2 = -1, DD' = 0, D'^2 = 0$$

Hence $$z = \phi_1(y + x) + x\phi_2(y + x) - \sin x. \quad \textbf{Ans.}$$

Example 49: Solve $\frac{\partial^2 z}{\partial x^2} - 3\frac{\partial^2 z}{\partial x \partial y} + 2\frac{\partial^2 z}{\partial y^2} = e^{2x-y} + e^{x+y} + \cos(x+2y)$. *(U.P.T.U. 2006)*

Solution: The given equation may be written as

$$(D^2 - 3DD' + 2D'^2)z = e^{2x-y} + e^{x+y} + \cos(x+2y)$$

Putting $D = m, \quad D' = 1$

then A.E. is $m^2 - 3m + 2 = 0 \Rightarrow (m-1)(m-2) = 0 \Rightarrow m = 1, 2$

$$\therefore \quad \text{C.F.} = \phi_1(y+x) + \phi_2(y+2x)$$

$$\text{P.I.} = \frac{1}{(D^2 - 3DD' + 2D'^2)} e^{2x-y} + \frac{1}{(D^2 - 3DD' + 2D'^2)} e^{x+y} + \frac{1}{(D^2 - 3DD' + 2D'^2)} \cos(x+2y)$$

$$= \frac{1}{4 - 3\times 2\times(-1) + 2(-1)^2} e^{2x-y} + \frac{1}{1^2 - 3(1)(1) + 2(1)^2} e^{x+y} + \frac{\cos(x+2y)}{-1-3(-2)+2(-4)}$$

$$= \frac{e^{2x-y}}{12} + \frac{1}{0} e^{x+y} - \frac{\cos(x+2y)}{3} \quad \text{(Case fail)}$$

$$= \frac{e^{2x-y}}{12} + \frac{x}{(2D-3D')} e^{x+y} - \frac{\cos(x+2y)}{3}$$

$$= \frac{e^{2x-y}}{12} + \frac{xe^{x+y}}{(2-3)} - \frac{\cos(x+2y)}{3}$$

$$\Rightarrow \quad \text{P.I.} = \frac{e^{2x-y}}{12} - xe^{x+y} - \frac{\cos(x+2y)}{3}$$

$$\therefore \quad z = \phi_1(y+x) + \phi_2(y+2x) + \frac{e^{2x-y}}{12} - xe^{x+y} - \frac{\cos(x+2y)}{3}. \quad \textbf{Ans.}$$

Example 50: Solve $r - 2s = \sin x \cdot \cos 2y$. *(U.P.T.U. 2008)*

Solution: The given equation may be written as

$$\frac{\partial^2 z}{\partial x^2} - 2\frac{\partial^2 z}{\partial x \partial y} = \sin x \cos 2y$$

or $$(D^2 - 2DD')z = \frac{1}{2}(2\sin x \cos 2y) = \frac{1}{2}\left[\sin(x+2y) + \sin(x-2y)\right]$$

Putting $D = m$ and $D' = 1$

The A.E. is $m^2 - 2m = 0 \Rightarrow m(m-2) = 0 \Rightarrow m = 0,\ m = 2$

$$\text{C.F.} = \phi_1(y) + \phi_2(y + 2x)$$

and

$$\text{P.I.} = \frac{1}{2(D^2 - 2DD')} \cdot [\sin(x + 2y) + \sin(x - 2y)]$$

$$= \frac{1}{2(D^2 - 2DD')}\sin(x + 2y) + \frac{1}{2(D^2 - 2DD')}\sin(x - 2y)$$

$$= \frac{\sin(x+2y)}{2\{-1-2\times(-2)\}} + \frac{\sin(x-2y)}{2(-1-2\times 2)}$$

$D^2 = -1,\ DD' = -2$ in the first term and $D^2 = -1$, $DD' = -2$ in the second term

$$\text{P.I.} = \frac{\sin(x+2y)}{6} - \frac{\sin(x-2y)}{10}$$

Hence,

$$z = \phi_1(y) + \phi_2(y + 2x) + \frac{1}{6}\sin(x + 2y) - \frac{1}{10}\sin(x - 2y). \quad \textbf{Ans.}$$

EXERCISE 4.4

Solve the following partial differential equations:

1. $(D + 2D')(D - 3D')^2 z = 0.$ $\quad$ [**Ans.** $z = \phi_1(y - 2x) + \phi_2(y + 3x) + x\phi_3(y + 3x)$]

2. $r = a^2 t.$ $\quad$ [**Ans.** $z = \phi_1(y + ax) + \phi_2(y - ax)$]

3. $(D^3 - 4D^2D' + 4DD'^2)z = 0.$ $\quad$ [**Ans.** $z = \phi_1(y) + \phi_2(y + 2x) + x\phi_3(y + 2x)$]

4. $(D^2 - 3aDD' + 2a^2D'^2)z = 0.$ $\quad$ [**Ans.** $z = \phi_1(y + ax) + \phi_2(y + 2ax)$]

5. $(D^2 + 2DD' + D'^2)z = e^{2x+3y}.$ $\quad$ [**Ans.** $z = \phi_1(y - x) + x\phi_2(y - x) + \frac{1}{25}e^{2x+3y}$]

6. $\frac{\partial^2 z}{\partial x^2} + 5\frac{\partial^2 z}{\partial x \partial y} + 6\frac{\partial^2 z}{\partial y^2} = e^{x-y}.$ $\quad$ [**Ans.** $z = \phi_1(y - 2x) + \phi_2(y - 3x) + \frac{1}{2}e^{x-y}$]

7. $\frac{\partial^2 z}{\partial x^2} - 7\frac{\partial^2 z}{\partial x \partial y} + 12\frac{\partial^2 z}{\partial y^2} = e^{x-y}$. $\left[\textbf{Ans. } z = \phi_1(y+3x) + \phi_2(y+4x) + \frac{1}{20}e^{x-y}\right]$

8. $\frac{\partial^3 z}{\partial x^3} - 2\frac{\partial^3 z}{\partial x^2 \partial y} = 2e^{2x-y}$. $\left[\textbf{Ans. } z = \phi_1(y) + x\phi_2(y) + \phi_3(y+2x) + \frac{1}{8}e^{2x-y}\right]$

9. $(D^2 - 5DD' + 6D'^2)z = e^{3x-2y}$. $\left[\textbf{Ans. } z = \phi_1(y+2x) + \phi_2(y+3x) + \frac{1}{63}e^{3x-2y}\right]$

10. $\frac{\partial^2 z}{\partial x^2} + 2\frac{\partial^2 z}{\partial x \partial y} + \frac{\partial^2 z}{\partial y^2} = \sin(2x+3y)$. (*U.P.T.U. 2006*)

$\left[\textbf{Ans. } z = \phi_1(y-x) + x\phi_2(y-x) - \frac{1}{25}\sin(2x+3y)\right]$

11. $(D^3 - 4D^2D' + 4DD'^2)z = 4\sin(2x+y)$.

$\left[\textbf{Ans. } z = \phi_1(y) + \phi_2(y+2x) + x\phi_3(y+2x) - x^2\cos(2x+y)\right]$

12. $(D^2 + D'^2)z = \cos mx \cdot \cos ny$. $\left[\textbf{Ans. } z = \phi_1(y+ix) + \phi_2(y-ix) - \frac{1}{(m^2+n^2)} \cdot \cos mx \cos ny\right]$

13. $(D^3 - 7DD'^2 - 6D'^3)z = \sin(x+2y) + e^{2x+y}$.

$\left[\textbf{Ans. } z = \phi_1(y-x) + \phi_2(y-2x) + \phi_3(y+3x) - \frac{1}{75}\cos(x+2y) - \frac{1}{12}e^{2x+y}\right]$

14. $\frac{\partial^2 z}{\partial x^2} - 3\frac{\partial^2 z}{\partial x \partial y} + 2\frac{\partial^2 z}{\partial y^2} = e^{2x+3y} + \sin(x-2y)$.

$\left[\textbf{Ans. } z = \phi_1(y+x) + \phi_2(y+2x) + \frac{1}{4}e^{2x+3y} - \frac{1}{15}\sin(x-2y)\right]$

15. $(D^2 - 5DD' + 4D'^2)z = \sin(4x+y)$. $\left[\textbf{Ans. } z = \phi_1(y+x) + \phi_2(y+4x) - \frac{1}{2}x\cos(4x+y)\right]$

16. $(D^2 - DD')z = \cos 2y(\sin x + \cos x)$. (*U.P.T.U. 2003*)

$\left[\textbf{Ans. } z = \phi_1(y) + \phi_2(y+x) + \frac{1}{2}[\sin(x+2y) + \cos(x+2y)] - \frac{1}{6}[\sin(x-2y) + \cos[x-2y)]\right]$

Solved Examples Related to '*c*'

Example 51: Solve $(D^2+6DD'+9D'^2)z=6x+2y+e^{2x+y}$.

Solution: The A.E. is $m^2+6m+9 \quad (D = m,\ D' = 1)$

$$\Rightarrow \qquad (m+3)^2=0 \;\Rightarrow\; m=-3,-3$$

$$\therefore \qquad \text{C.F.} = \phi_1(y-3x)+x\phi_2(y-3x)$$

$$\text{P.I.} = \frac{1}{(D^2+6DD'+9D'^2)}[(6x+2y)+e^{2x+y}]$$

$$= \frac{1}{(D+3D')^2}(6x+2y)+\frac{1}{4+6\cdot 2\cdot 1+9\cdot 1^2}e^{2x+y}$$

$$= \frac{1}{D^2\left(1+\frac{3D'}{D}\right)^2}(6x+2y)+\frac{1}{25}\cdot e^{2x+y}$$

$$= \frac{1}{D^2}\left(1+\frac{3D'}{D}\right)^{-2}(6x+2y)+\frac{1}{25}e^{2x+y}$$

$$= \frac{1}{D^2}\left(1-\frac{6D'}{D}+\frac{9D'^2}{D^2}-\cdots\cdots\right)(6x+2y)+\frac{1}{25}e^{2x+y}$$

$$= \frac{1}{D^2}\left(6x+2y-\frac{12}{D}\right)+\frac{1}{25}e^{2x+y}$$

$$= \frac{6x^3}{6}+\frac{2x^2y}{2}-12\cdot\frac{x^3}{6}+\frac{1}{25}e^{2x+y}$$

$$\Rightarrow \qquad \text{P.I.} = -x^3+x^2y+\frac{1}{25}e^{2x+y}$$

Hence, $$z=\phi_1(y-3x)+x\phi_2(y-3x)-x^3+x^2y+\frac{1}{25}e^{2x+y}. \qquad \textbf{Ans.}$$

Example 52: Solve $(D^2+D'^2)z=x^2y^2$.

Solution: The A.E. is $m^2+1=0 \; (D=m,\ D'=1)$

$$\Rightarrow \qquad m=\pm i$$

so $$\text{C.F.} = \phi_1(y+ix)+\phi_2(y-ix)$$

$$\text{P.I.} = \frac{1}{(D^2+D'^2)}x^2y^2 = \frac{1}{D^2\left(1+\frac{D'^2}{D^2}\right)}x^2y^2 = \frac{1}{D^2}\left(1+\frac{D'^2}{D^2}\right)^{-1}x^2y^2$$

$$= \frac{1}{D^2}\left\{1-\frac{D'^2}{D^2}+\frac{D'^4}{D^4}-\cdots\right\}x^2y^2$$

$$= \frac{1}{D^2}\left\{x^2y^2-\frac{1}{D^2}(2x^2)+0-\cdots\right\}$$

$$= \frac{1}{D}\left(\frac{x^3y^2}{3}\right)-\frac{1}{D^3}\left(\frac{2x^3}{3}\right)=\frac{x^4y^2}{12}-\frac{1}{D^2}\left(\frac{2x^4}{12}\right)$$

$$= \frac{x^4y^2}{12}-\frac{2x^6}{360}=\frac{x^4y^2}{12}-\frac{x^6}{180}$$

$\Rightarrow$ $$\text{P.I.} = \frac{1}{180}(15x^4y^2-x^6)$$

Hence $$z = \phi_1(y+ix)+\phi_2(y-ix)+\frac{1}{180}(15x^4y^2-x^6). \quad \textbf{Ans.}$$

Example 53: Solve $\dfrac{\partial^3 z}{\partial x^3}-2\dfrac{\partial^3 z}{\partial x^2\partial y}=2e^{2x}+3x^2y.$

Solution: We have $\dfrac{\partial^3 z}{\partial x^3}-2\dfrac{\partial^3 z}{\partial x^2\partial y}=2e^{2x}+3x^2y$

or $$(D^3-2D^2D')z = 2e^{2x}+3x^2y$$

Putting $D=m$ and $D'=1$, then

A.E. is $m^3-2m^2=0 \Rightarrow m^2(m-2)=0$

$\Rightarrow$ $m = 0, 0, 2$

$\therefore$ $$\text{C.F.} = \phi_1(y)+x\phi_2(y)+\phi_3(y+2x)$$

and $$\text{P.I.} = \frac{1}{(D^3-2D^2D')}(2e^{2x}+3x^2y)$$

$$= \frac{2e^{2x}}{8-2\times4\times0}+\frac{3}{D^3\left(1-\frac{2D'}{D}\right)}(x^2y)$$

Put $D=2, D'=0$ in first term

$$= \frac{e^{2x}}{4} + \frac{3}{D^3}\left(1 + \frac{2D'}{D} + \frac{4D'^2}{D^2} + \cdots\right)(x^2 y)$$

$$= \frac{e^{2x}}{4} + \frac{3}{D^3}\left(x^2 y + \frac{2}{D}x^2 + 0 + \cdots\right)$$

$$= \frac{e^{2x}}{4} + \frac{3}{D^3}(x^2 y) + \frac{6}{D^3}\left(\frac{x^3}{3}\right) = \frac{e^{2x}}{4} + \frac{3}{D^2}\left(\frac{x^3 y}{3}\right) + \frac{2}{D^2}\left(\frac{x^4}{4}\right)$$

$$\text{P.I.} = \frac{e^{2x}}{4} + \frac{x^5 y}{20} + \frac{x^6}{60} = \frac{e^{2x}}{4} + \frac{x^5}{60}(3y + x).$$

Hence the complete solution of given problem is

$$z = \phi_1(y) + x\phi_2(y) + \phi_3(y + 2x) + \frac{e^{2x}}{4} + \frac{x^5}{60}(3y + x). \quad \textbf{Ans.}$$

Example 54: Solve $\dfrac{\partial^3 z}{\partial x^3} - \dfrac{\partial^3 z}{\partial y^3} = x^3 y^3$.

Solution: We have $\dfrac{\partial^3 z}{\partial x^3} - \dfrac{\partial^3 z}{\partial y^3} = x^3 y^3 \Rightarrow (D^3 - D'^3)z = x^3 y^3$

The A.E. is $m^3 - 1 = 0 \Rightarrow m = 1, w, w^2$ | w is the cube root of unity

$\therefore$ $$\text{C.F.} = \phi_1(y + x) + \phi_2(y + wx) + \phi_3(y + w^2 x)$$

and $$\text{P.I.} = \frac{1}{(D^3 - D'^3)} x^3 y^3 = \frac{1}{D^3\left(1 - \frac{D'^3}{D^3}\right)} x^3 y^3 = \frac{1}{D^3}\left(1 - \frac{D'^3}{D^3}\right)^{-1} x^3 y^3$$

$$= \frac{1}{D^3}\left\{1 + \frac{D'^3}{D^3} + \cdots\right\} x^3 y^3 = \frac{1}{D^3}\left\{x^3 y^3 + \frac{1}{D^3}(6x^3)\right\}$$

$$= \frac{1}{D^3}(x^3 y^3) + \frac{1}{D^6}(6x^3) = \frac{x^6 y^3}{120} + \frac{x^9}{10080}$$

Therefore the complete solution of given equation is

$$z = \phi_1(y + x) + \phi_2(y + wx) + \phi_3(y + w^2 x) + \frac{x^6 y^3}{120} + \frac{x^9}{10080}. \quad \textbf{Ans.}$$

EXERCISE 4.5

Solve the following partial differential equations:

1. $\frac{\partial^2 z}{\partial x^2} - 2\frac{\partial^2 z}{\partial x \partial y} = e^{2x} + x^3 \cdot y.$ $\left[\textbf{Ans. } z = \phi_1(y) + \phi_2(y+2x) + \frac{1}{4}e^{2x} + \frac{x^5}{15}(3y+x)\right]$

2. $\frac{\partial^2 y}{\partial x^2} = a^2 \frac{\partial^2 y}{\partial t^2} + t^2.$ $\left[\textbf{Ans. } y = \phi_1(x-at) + \phi_2(x+at) + \frac{1}{12}t^4\right]$

3. $\frac{\partial^2 z}{\partial x^2} + 2\frac{\partial^2 z}{\partial x \partial y} + \frac{\partial^2 z}{\partial y^2} = 2(y-x) + \sin(x-y).$

$\left[\textbf{Ans. } z = \phi_1(y-x) + x\phi_2(y-x) + x^2 y - x^3 + \frac{x^2}{2}\sin(x-y)\right]$

4. $(D^2 + 2DD' + D'^2)z = x^2 + xy + y^2.$

$\left[\textbf{Ans. } z = \phi_1(y-x) + x\phi_2(y-x) + \frac{1}{4}(x^4 - 2x^3 y + 2x^2 y^2)\right]$

5. $\frac{\partial^2 z}{\partial x^2} - 6\frac{\partial^2 z}{\partial x \partial y} + 9\frac{\partial^2 z}{\partial y^2} = 12x^2 + 36xy.$ $[\textbf{Ans. } z = \phi_1(y+3x) + x\phi_2(y+3x) + 6x^3 y + 10x^4]$

6. $(D^3 - 7DD'^2 - 6D'^3)z = \cos(x-y) + x^2 + xy^2 + y^3.$

$\left[\textbf{Ans. } z = \phi_1(y-x) + \phi_2(y-2x) + \phi_3(y+3x) + \frac{x}{6}\cos(x-y) + \frac{7}{72}x^6 + \frac{1}{60}x^5(21y+1) + \frac{1}{24}x^4 y^2 + \frac{1}{6}x^3 y^3\right]$

7. $(D^2 + 3DD' - 4D'^2)z = x + \cos y.$ $\left[\textbf{Ans. } z = \phi_1(y+x) + \phi_2(y-4x) + \frac{x^3}{6} + \frac{1}{4}\cos y\right]$

Solved Examples Related to '*d*'

Example 55: Solve $(D^3 - 3DD'^2 - 2D'^3)z = \cos(x+2y) - e^y(3+2x).$

Solution: The A.E. is $m^3 - 3m - 2 = 0 \Rightarrow (m+1)^2(m-2) = 0 \Rightarrow m = -1, -1, 2$

$\therefore$ C.F. $= \phi_1(y-x) + x\phi_2(y-x) + \phi_3(y+2x)$

and $$\text{P.I.} = \frac{1}{D^3 - 3DD'^2 - 2D'^3}[\cos(x+2y)$$

$$-e^y(3+2x)]\frac{1}{D^2 \cdot D - 3D \cdot D'D' - 2D'^2 \cdot D'}\cos(x+2y)$$

$$-e^y \cdot \frac{1}{\{D^3 - 3D(D'+1)^2 - 2(D'+1)^3\}}(3+2x)$$

$$= \frac{1}{(-D+6D'+8D')}\cos(x+2y) - e^y\{D^3 - 3DD'^2$$

$$-6DD' - 3D - 2D'^3 - 2 - 6D'^2 - 6D'\}^{-1}(3+2x)$$

$$= \frac{14D'+D}{(196D'^2 - D^2)}\cos(x+2y)$$

$$-\frac{e^y}{(-2)}\left\{1 - \left(3D' + \frac{3D}{2} + 3DD' + 3D'^2 + \frac{3DD'^2}{2} + D'^3 - \frac{D^3}{2}\right)\right\}(3+2x)$$

$$= \frac{(14D'+D)}{196(-4)-(-1)}\cos(x+2y) + \frac{e^y}{2}\left\{(3+2x) - \left(\frac{3}{2}\times 2\right)\right\}$$

$$= \frac{29}{783}\sin(x+2y) + \frac{e^y}{2}\{3+2x-3\}$$

$$\Rightarrow \quad \text{P.I.} = \frac{1}{27}\sin(x+2y) + xe^y$$

Hence the required solution of given equation is

$$z = \phi_1(y-x) + x\phi_2(y-x) + \phi_3(y+2x) + \frac{1}{27}\sin(x+2y) + xe^y. \quad \textbf{Ans.}$$

Example 56: Solve $\dfrac{\partial^2 z}{\partial x^2} + \dfrac{\partial^2 z}{\partial x \partial y} - 2\dfrac{\partial^2 z}{\partial y^2} = (y-1)e^x.$

Solution: The given equation may be written as

$$(D^2 + DD' - 2D'^2)z = (y-1)e^x$$

$\therefore$ The A.E. is $m^2 + m - 2 = 0 \quad (D = m, D' = 1)$

$\Rightarrow \quad (m-1)(m+2) = 0 \quad \Rightarrow \quad m = 1, -2$

$$\therefore \quad \text{C.F.} = \phi(y+x) + \phi_2(y-2x)$$

$$\text{P.I.} = \frac{1}{(D^2 + DD' - 2D'^2)}(y-1)e^x$$

$$= e^x \cdot \frac{1}{\{(D+1)^2 + (D+1)D' - 2D'^2\}}(y-1)$$

$$= e^x \cdot \frac{1}{\{1+(2D+DD'+D'+D^2-2D'^2)\}}(y-1)$$

$$= e^x \cdot \{1+(2D+DD'+D'+D^2-2D'^2)\}^{-1}(y-1)$$

$$= e^x \cdot \{1-(2D+DD'+D'+D^2-2D'^2)\}(y-1) = e^x\{(y-1)-(1)\}$$

$\Rightarrow$ $$\text{P.I.} = e^x(y-2)$$

Hence $$z = \phi_1(y+x)+\phi_2(y-2x)+e^x(y-2). \quad \textbf{Ans.}$$

Solved Examples Related to '*e*'

Example 57: Solve $\frac{\partial^2 z}{\partial x^2}+\frac{\partial^2 z}{\partial x \partial y}-6\frac{\partial^2 z}{\partial y^2} = y\cos x.$ *(U.P.T.U. 2003)*

Solution: The given equation may written as

$$(D^2+DD'-6D'^2)z = y\cos x$$

The A.E. is

$$m^2+m-6 = 0 \quad |D=m, D'=1$$

$\Rightarrow$ $$(m-2)(m+3) = 0$$

$\Rightarrow$ $$m = 2, -3$$

$\therefore$ $$\text{C.F.} = \phi_1(y+2x)+\phi_2(y-3x)$$

Now $$\text{P.I.} = \frac{1}{(D^2+DD'-6D'^2)}y\cos x = \frac{1}{(D+3D')(D-2D')}y\cos x$$

$$= \frac{1}{(D+3D')}\left[\int (c-2x)\cos x\,dx\right] \quad |y=c-2x$$

$$= \frac{1}{(D+3D')}[(c\sin x-2x\sin x-2\cos x]$$

$$= \frac{1}{(D+3D')}[(c-2x)\sin x-2\cos x]$$

$$= \frac{1}{(D+3D')}[y\sin x-2\cos x] = \frac{1}{(D+3D')}y\sin x - \frac{2}{(D+3D')}\cos x$$

$$= \int (c_1+3x)\sin x\,dx - \frac{2(D-3D')}{(D^2-9D'^2)}\cos x \quad |y=c_1+3x$$

$$= -c_1\cos x - 3x\cos x + 3\sin x - \frac{2(D-3D')\cos x}{-1-9\times 0} \quad |D^2=-1^2, D'^2=0$$

$$= -(c_1 + 3x)\cos x + 3\sin x + 2(-\sin x - 0)$$
$$= -y\cos x + 3\sin x - 2\sin x$$

$\Rightarrow$ P.I. $= -y\cos x + \sin x$

Hence, $z = \phi_1(y + 2x) + \phi_2(y - 3x) - y\cos x + \sin x.$ **Ans.**

Example 58: Solve $\dfrac{\partial^2 z}{\partial x^2} + \dfrac{\partial^2 z}{\partial x \partial y} - 6\dfrac{\partial^2 z}{\partial y^2} = x^2 \sin(x + y).$

Solution: We have $(D^2 + DD' - 6D'^2)z = x^2 \sin(x + y)$

$\therefore$ The A.E. is $m^2 + m - 6 = 0$ $(D = m, D' = 1)$

$\Rightarrow$ $(m - 2)(m + 3)$ $\Rightarrow$ $m = 2, -3$

$\Rightarrow$ C.F. $= \phi_1(y + 2x) + \phi_2(y - 3x)$

Now P.I. $= \dfrac{1}{(D^2 + DD' - 6D'^2)} x^2 \sin(x + y) = \dfrac{1}{(D + 3D')(D - 2D')} x^2 \sin(x + y)$

$$= \frac{1}{(D + 3D')} \int x^2 \sin\{x + c - 2x\}\, dx \quad |\text{As } y = c - 2x$$

$$= \frac{1}{(D + 3D')} \int x^2 \sin(c - x)\, dx$$

$$= \frac{1}{(D + 3D')} \{(x^2 - 2)\cos(c - x) + 2x\sin(c - x)\}$$

$$= \frac{1}{(D + 3D')} \{(x^2 - 2)\cos(y + x) + 2x\sin(y + x)\} \quad |\text{Put, } c = y + 2x$$

$$= \int (x^2 - 2)\cos(c_1 + 3x + x)\, dx + 2\int x\sin(c_1 + 3x + x)\, dx \quad |y = c_1 + 3x$$

$$= \int (x^2 - 2)\cos(c_1 + 4x)\, dx + 2\int x\sin(c_1 + 4x)\, dx$$

$$= \frac{1}{4}(x^2 - 2)\sin(c_1 + 4x) - \frac{1}{4} \times 2\int x\sin(c_1 + 4x)\, dx + 2\int x\sin(c_1 + 4x) dx$$

$$= \frac{1}{4}(x^2 - 2)\sin(c_1 + 4x) + \frac{3}{2}\int x\sin(c_1 + 4x)\, dx$$

$$= \frac{1}{4}(x^2 - 2)\sin(c_1 + 4x) - \frac{3x}{8}\cos(c_1 + 4x) + \frac{3}{32}\sin(c_1 + 4x)$$

$\Rightarrow$ P.I. $= \dfrac{1}{4}(x^2 - 2)\sin(y + x) - \dfrac{3x}{8}\cos(y + x) + \dfrac{3}{32}\sin(y + x)$ $\quad |\text{Put } c_1 = y - 3x$

$$\Rightarrow \quad \text{P.I.} = \frac{1}{32}\{8x^2 - 13\}\sin(y+x) - \frac{3x}{8}\cos(y+x)$$

$$\therefore \quad z = \text{C.F.} + \text{P.I.}$$

$$\Rightarrow \quad z = \phi_1(y+2x) + \phi_2(y-3x) + \frac{1}{32}(8x^2-13)\sin(y+x) - \frac{3}{8}x\cos(y+x). \qquad \textbf{Ans.}$$

Example 59: Solve $(D^2 - D'^2)z = \tan^3 x \tan y - \tan x \tan^3 y$.

Solution: We have $(D^2 - D'^2)z = \tan^3 x \tan y - \tan x \tan^3 y$

$\therefore$ The A.E. is $m^2 - 1 = 0 \;\Rightarrow\; m = \pm 1$

$$\Rightarrow \quad \text{C.F.} = \phi_1(y+x) + \phi_2(y-x)$$

Now

$$\text{P.I.} = \frac{1}{(D^2 - D'^2)} \tan x \tan y\,(\tan^2 x - \tan^2 y)$$

$$= \frac{1}{(D+D')(D-D')} \tan x \tan y(\tan^2 x - \tan^2 y)$$

$$= \frac{1}{(D+D')(D-D')} \tan x \tan y(\sec^2 x - \sec^2 y) \qquad |\tan^2\theta = \sec^2\theta - 1$$

$$= \frac{1}{(D+D')}\int [\tan x \sec^2 x \cdot \tan(c-x) - \tan x \tan(c-x)\sec^2(c-x)]dx \qquad |y = c - x$$

$$= \frac{1}{(D+D')}\left[\frac{1}{2}\tan^2 x \tan(c-x) + \int \frac{1}{2}\tan^2 x \sec^2(c-x)dx\right.$$

$$\left. + \frac{1}{2}\tan x \tan^2(c-x) - \int \frac{1}{2}\tan^2(c-x)\sec^2 x\, dx\right]$$

$$= \frac{1}{2(D+D')}\left[\tan^2 x \tan(c-x) + \tan x \tan^2(c-x) + \int\{(\sec^2 x - \sec^2(c-x)\}dx\right]$$

$$= \frac{1}{2(D+D')}\left[\tan^2 x \tan(c-x) + \tan x \tan^2(c-x) + \tan x + \tan(c-x)\right]$$

$$= \frac{1}{2(D+D')}\left[\tan^2 x \tan y + \tan x \tan^2 y + \tan x + \tan y\right] \qquad |\text{Put } y = c - x$$

$$= \frac{1}{2(D+D')}[\tan x \sec^2 y + \tan y \sec^2 x]$$

$$= \frac{1}{2}\int \tan x \sec^2(c_1 + x) + \tan(c_1 + x)\sec^2 x]dx \quad |y = c_1 + x$$

$$= \frac{1}{2}[\tan x \tan(c_1 + x) - \int \sec^2 x \tan(c_1 + x)dx + \int \tan(c_1 + x)\sec^2 x\, dx$$

$$= \frac{1}{2}\tan x \tan(c_1 + x) = \frac{1}{2}\tan x \tan y \quad |\text{Put } y = c_1 + x$$

Hence the complete solution is

$$z = \phi_1(y + x) + \phi_2(y - x) + \frac{1}{2}\tan x \tan y. \quad \textbf{Ans.}$$

Solved Examples Related to Special Case:

Example 60: Solve $(D^3 - 3DD'^2 + 2D'^3)z = (x + 2y)^{1/2}$.

Solution: Then A.E. is $m^3 - 3m + 2 = 0 \quad (D = m, D' = 1)$

$\Rightarrow \quad (m-1)^2\ (m + 2) = 0 \quad \Rightarrow \quad m = 1, 1, -2$

$\therefore \quad \text{C.F.} = \phi_1(y + x) + x\phi_2(y + x) + \phi_3(y - 2x)$

and

$$\text{P.I.} = \frac{1}{(D^3 - 3DD'^2 + 2D'^3)}(x + 2y)^{1/2}$$

$$= \frac{1}{(1^3 - 3\times1\times2^2 + 2\times2^3)}\iiint v^{1/2}dv\,dv\,dv$$

$$= \frac{1}{5}\iint \frac{2}{3}v^{3/2}dv\,dv \quad \begin{array}{|l} D = 1 \text{ and } D' = 2, \text{ Let} \\ v = x + 2y \\ \text{Here } n = 3, \text{ so we integrate 3 times} \end{array}$$

$$= \frac{2}{15}\int \frac{2}{5}v^{5/2}dv = \frac{4}{75}\times\frac{2}{7}v^{7/2} = \frac{8}{525}v^{7/2}$$

$$\Rightarrow \quad \text{P.I.} = \frac{8}{525}(x + 2y)^{7/2} \quad |\text{Put } v = x + 2y$$

Hence the complete solution of given equation is

$$z = \phi_1(y + x) + x\phi_2(y + x) + \phi_3(y - 2x) + \frac{8}{525}(x + 2y)^{7/2}. \quad \textbf{Ans.}$$

Example 61: Solve $6\frac{\partial^2 z}{\partial x^2}+5\frac{\partial^2 z}{\partial x\,\partial y}-6\frac{\partial^2 z}{\partial y^2}=132\log(x+3y)$.

Solution: We have $(6D^2+5DD'-6D'^2)\,z=132\log(x+3y)$

The A.E. is

$$6m^2+5m-6=0 \qquad \left|\, D=m,\ D'=1\right.$$

$$\Rightarrow \quad (3m-2)(2m+3)=0 \quad\Rightarrow\quad m=\frac{2}{3},\frac{-3}{2}$$

$$\therefore \quad \text{C.F.}=\phi_1\left(y+\frac{2}{3}x\right)+\phi_2\left(y-\frac{3}{2}x\right)=\phi_1\left(\frac{3y+2x}{3}\right)+\phi_2\left(\frac{2y-3x}{2}\right)$$

$$\Rightarrow \quad \text{C.F.}=f_1(3y+2x)+f_2(2y-3x) \qquad \left|\begin{array}{l} f_1(3y+2x)=\phi_1\left(\dfrac{3y+2x}{3}\right)\\ f_2(2y-3x)=\phi_2\left(\dfrac{2y-3x}{2}\right)\end{array}\right.$$

and

$$\text{P.I.}=\frac{1}{(6D^2+5DD'-6D'^2)}132\log(x+3y)$$

$$=\frac{132}{6\times1^2\times5\times1\times3-6\times3^2}\iint\log v\,dv\,dv \qquad \left|\, x+3y=v\right.$$

$$=-4\int(v\log v-v)\,dv=4\left[\frac{v^2}{2}-\int v\log v\,dv\right]$$

$$=4\left[\frac{v^2}{2}-\int\log v\,d\left(\frac{v^2}{2}\right)\right] \qquad \left|\,\text{As } d\left(\frac{v^2}{2}\right)=v\,dv\right.$$

$$=4\left[\frac{v^2}{2}-\frac{v^2}{2}\cdot\log v+\frac{1}{2}\frac{v^2}{2}\right]$$

$$\Rightarrow \quad \text{P.I.}=4\left[\frac{3v^2}{4}-\frac{v^2}{2}\cdot\log v\right]=v^2[3-2\log v]$$

$$\Rightarrow \quad \text{P.I.}=(x+3y)^2[3-2\log(x+3y)] \qquad \left|\,\text{Put } v=x+3y\right.$$

Hence, the complete solution is

$$z=f_1(3y+2x)+f_2(2y-3x)+(x+3y)^2[3-2\log(x+3y)].$$ **Ans.**

Example 62: Solve $r - t = \dfrac{1}{x^2}$.

Solution: The given equation may be written as

$$\frac{\partial^2 z}{\partial x^2} - \frac{\partial^2 z}{\partial y^2} = \frac{1}{x^2} \Rightarrow (D^2 - D'^2)z = \frac{1}{x^2}$$

The A.E. is

$$m^2 - 1 = 0 \qquad (D = m,\ D' = 1)$$

$\Rightarrow$ $$m = \pm 1$$

$\therefore$ $$\text{C.F.} = \phi_1(y+x) + \phi_2(y-x)$$

and $$\text{P.I.} = \frac{1}{(D^2 - D'^2)} \cdot \frac{1}{x^2} = \frac{1}{(D^2 - D'^2)} \frac{1}{(1 \cdot x + 0 \cdot y)^2}$$

$$= \frac{1}{1-0} \iint \frac{1}{v^2}\, dv\, dv \qquad \left| \begin{array}{l} D = 1,\ D' = 0 \\ x = v \end{array} \right.$$

$$= -\int \frac{1}{v}\, dv = -\log v$$

$\Rightarrow$ $$\text{P.I.} = -\log x \quad (\text{Put } v = x)$$

Hence the complete solution of given equation is

$$z = \phi_1(y+x) + \phi_2(y-x) - \log x. \quad \textbf{Ans.}$$

EXERCISE 4.6

Solve the following partial differential equations:

1. $(D^3 + D^2 D' - DD'^2 - D'^3)z = e^x \cos 2y.$

$$\left[\textbf{Ans. } z = \phi_1(y-x) + \phi_2(y-x) + \phi_3(y+x) + \frac{1}{25} e^x (\cos 2y + 2\sin 2y)\right]$$

2. $(D^2 + DD' - 2D'^2)z = e^x(y+1).$ $\left[\textbf{Ans. } z = \phi_1(y+x) + \phi_2(y-2x) + ye^x\right]$

3. $(D^3 + 2D^2 D' - DD'^2 - 2D'^3)z = e^x(y+2).$

$$\left[\textbf{Ans. } z = \phi_1(y+x) + \phi_2(y-x) + \phi_3(y-2x) + ye^x\right]$$

4. $(D^2 + 5DD' + 5D'^2)z = x \cdot \sin(3x - 2y).$

$$\left[\textbf{Ans. } z = \phi_1\left\{y + \frac{x}{2}(-5+\sqrt{5})\right\} + \phi_2\left\{y + \frac{x}{2}(-5-\sqrt{5})\right\} + x\sin(3x-2y) + 4\cos(3x-2y)\right]$$

5. $\dfrac{\partial^2 z}{\partial x^2}+3\dfrac{\partial^2 z}{\partial x\,\partial y}+2\dfrac{\partial^2 z}{\partial y^2}=x+y.$ $\left[\textbf{Ans. } z=\phi_1(y-x)+\phi_2(y-2x)+\dfrac{x^2y}{2}-\dfrac{x^3}{3}\right]$

6. $\dfrac{\partial^2 z}{\partial x^2}+3\dfrac{\partial^2 z}{\partial x\,\partial y}+2\dfrac{\partial^2 z}{\partial y^2}=12xy.$ $\left[\textbf{Ans. } z=\phi_1(y-x)+\phi_2(y-2x)+2x^3y-\dfrac{3x^4}{2}\right]$

7. $(D^2-DD'-2D'^2)z=(2x^2+xy-y^2)\sin xy-\cos xy.$

$[\textbf{Ans. } z=\phi_1(y+2x)+\phi_2(y-x)+\sin xy]$

8. $(D^3-7DD'^2-6D'^3)z=\sin(x+2y)+e^{2x+y}.$

$\left[\textbf{Ans. } z=\phi_1(y-x)+\phi_2(y-2x)+\phi_3(y+3x)-\dfrac{1}{75}\cos(x+2y)-\dfrac{1}{12}e^{2x+y}\right]$

9. $r+(a+b)s+abt=xy.$ $\left[\textbf{Ans. } z=\phi_1(y-ax)+\phi_2(y-bx)+\dfrac{1}{6}x^3y-\dfrac{(a+b)}{24}x^4\right]$

10. $\dfrac{\partial^2 z}{\partial x^2}+\dfrac{\partial^2 z}{\partial x\partial y}-2\dfrac{\partial^2 z}{\partial y^2}=8\log(x+5y).$

$\left[\textbf{Ans. } z=f_1(2x-y)+f_2(x+y)+\dfrac{1}{22}[3-2\log(x+5y)](x+5y)^2\right]$

11. $(D^2+2DD'-8D'^2)z=\sqrt{2x+3y}.$ $\left[\textbf{Ans. } z=\phi_1(y+2x)+\phi_2(y-4x)-\dfrac{1}{210}(2x+3y)^{5/2}\right]$

12. $\left(\dfrac{\partial^2 z}{\partial x^2}+2\dfrac{\partial^2 z}{\partial x\,\partial y}+\dfrac{\partial^2 z}{\partial y^2}\right)=2\cos y-x\sin y.$ $[\textbf{Ans. } z=\phi_1(y-x)+x\,\phi_2(y-x)+x\sin y]$

13. $(D^2-2DD'+D'^2)z=e^{x+2y}+x^3.$ $[\textbf{Ans. } z=\phi_1(y+x)+x\,\phi_2(y+x)+e^{x+2y}+x^5/20]$

14. $(D^3-4D^2D'+5DD'^2-2D'^3)z=e^{y+2x}+(y+x)^{1/2}.$

$\left[\textbf{Ans. } z=\phi_1(y+x)+x\phi_2(y+x)+\phi_3(y+2x)+xe^{y+x}-\dfrac{x^2}{3}(y+x)^{3/2}\right]$

15. $(D^3-7DD'^2-6D'^3)z=\sin(x+2y)+e^{3x+y}.$

$\left[\textbf{Ans. } z=\phi_1(y-x)+\phi_2(y-2x)+\phi_3(y+3x)-\dfrac{1}{75}\cos(x+2y)+\dfrac{x}{20}e^{3x+y}\right]$

4.15 NON-HOMOGENEOUS LINEAR EQUATIONS

A linear partial differential equation which is not homogeneous is called 'Non-homogeneous linear equation.'

Let $f(D, D')z = F(x, y)$, is a non-homogeneous equation then the complete solution of above equation is

$$z = \text{C.F.} + \text{P.I.}$$

4.15.1 Determination of C.F.

Consider a non-homogeneous equation

$$(D - mD' - a)z = 0 \quad \Rightarrow \quad \frac{\partial z}{\partial x} - m\frac{\partial z}{\partial y} - az = 0$$

$$\Rightarrow \qquad p - mq = az$$

Then by Lagrange's method, we get

$$\frac{dx}{1} = \frac{dy}{-m} = \frac{dz}{az}$$

From first two term, we have

$$-m\,dx = dy$$

$$\Rightarrow \qquad dy + m\,dx = 0$$

On integrating, we get

$$y + mx = c_1$$

Again taking first and last term we have

$$dx = \frac{dz}{az} \quad \Rightarrow \quad x = \frac{1}{a}\log z + c_2 \quad \Rightarrow \quad z = c_2 e^{ax}$$

$$\Rightarrow \qquad z = e^{ax}\phi(y + mx)$$

There are following cases to obtain C.F.

(a) If the factor of equation are

$$(D - m_1D' - a_1)(D - m_2D' - a_2)\cdots\cdots(D - m_nD' - a_n)z = 0$$

then

$$\text{C.F.} = e^{a_1x}\phi_1(y + m_1x) + e^{a_2x}\phi_2(y + m_2x) + \cdots\cdots$$

(b) If there are repeated factors, *i.e.*,

$$(D - mD' - a)^n z = 0$$

$$\text{C.F.} = e^{ax}\phi_1(y + mx) + xe^{ax}\phi_2(y + mx) + x^2e^{ax}\phi_3(y + mx) + \cdots\cdots$$

(c) If $f(D, D')$ not factorized into linear form consider the equation

$$f(D, D')z = 0 \qquad \ldots(i)$$

Let the solution of equation (*i*) is

$$z = Ae^{hx + ky}$$

$$\therefore \qquad f(h, k)\,Ae^{hx + ky} = 0$$

which is satisfied if

$$f(h, k) = 0 \qquad ...(ii)$$

This is an auxiliary equation of the equation (*i*), we solve (*ii*) to find *k* in terms of *h*.

Let $\qquad k = \phi(h),$

then $\qquad z = Ae^{hx + \phi(h)y}$

or The general solution is $z = \Sigma Ae^{hx+\phi(h)y}$

Clearly the sum of all such solutions will give C.F. of the given differential equation.

$$\therefore \qquad \text{C.F.} = \sum A_1 e^{hx+\phi_1(h)y} + \sum A_2 e^{hx+\phi_2(h)y} + \ldots\ldots$$

Usually $\sum Ae^{h(ax+by)}$ is written as $\phi(ax+by)$ in its final form.

4.15.2 Determination of *P.I.*

The P.I. can be obtained as the P.I. to homogeneous linear partial differential equations.

Example 63: Solve $(D^2 + 2DD' + D'^2 - 2D - 2D')z = \sin(x+2y)$. *(U.P.T.U. 2004)*

Solution: We have

$$(D^2 + 2DD' + D'^2 - 2D - 2D')z = \sin(x+2y)$$

for C.F. we consider

$$(D^2 + 2DD' + D'^2 - 2D - 2D')z = 0$$

$$\Rightarrow \qquad \{(D+D')^2 - 2(D+D')\}z = 0$$

$$\Rightarrow \qquad (D+D')(D+D'-2)z = 0$$

$$\therefore \qquad \text{C.F.} = e^{0\cdot x}\phi_1(y-x) + e^{2\cdot x}\phi_2(y-x) \qquad \left| \begin{array}{l} m_1 = -1,\ m_2 = -1,\ a_1 = 0 \\ \text{and } a_2 = 2 \end{array} \right.$$

$$\text{C.F.} = \phi_1(y-x) + e^{2x}\phi_2(y-x)$$

and

$$\text{P.I.} = \frac{1}{(D^2 + 2DD' + D'^2 - 2D - 2D')}\sin(x+2y)$$

$$\text{P.I.} = \frac{1}{-1+2(-2)+(-4)-2D-2D'}\sin(x+2y) \qquad \left| \begin{array}{l} D^2 = -1^2,\ D'^2 = -2^2 \\ DD' = -2 \end{array} \right.$$

$$= \frac{1}{-9-2D-2D'}\sin(x+2y) = -\frac{1}{(2D+2D'+9)}\sin(x+2y)$$

$$= \frac{9-2(D+D')}{81-4(D+D')^2}\sin(x+2y) = -\frac{9-2(D+D')}{81-4(D^2+2DD'+D'^2)}\sin(x+2y)$$

$$= -\frac{9\sin(x+2y) - 2\cos(x+2y) - 4\cos(x+2y)}{81-4(-1-4-4)}$$

$$= \frac{6\cos(x+2y) - 9\sin(x+2y)}{117}$$

$$\Rightarrow \qquad \text{P.I.} = \frac{2\cos(x+2y) - 3\sin(x+2y)}{39}$$

$\therefore$ Complete solution is

$$z = \phi_1(y-x) + e^{2x}\phi_2(y-x) + \frac{1}{39}\left[2\cos(x+2y) - 3\sin(x+2y)\right]. \quad \textbf{Ans.}$$

Example 64: Solve the partial differential equation

$$(D^2 - D'^2 - 3D + 3D')z = xy + e^{x+2y}. \qquad \textit{(U.P.T.U. 2002)}$$

Solution: We have

$$(D^2 - D'^2 - 3D + 3D')z = xy + e^{x+2y}$$

For C.F., we consider

$$(D^2 - D'^2 - 3D + 3D')z = 0$$

$$\Rightarrow \quad \{(D-D')(D+D') - 3(D-D')\}z = 0$$

$$\Rightarrow \quad (D-D')(D+D'-3)z = 0$$

$$\therefore \quad \text{C.F.} = e^{0.x}\phi_1(y+x) + e^{3x}\phi_2(y-x) \qquad \left| \begin{array}{l} m_1 = 1, m_2 = -1, a_1 = 0 \\ a_2 = 3 \end{array} \right.$$

$$\Rightarrow \quad \text{C.F.} = \phi_1(y+x) + e^{3x}\phi_2(y-x)$$

and

$$\text{P.I.} = \frac{1}{(D-D')(D+D'-3)}[xy + e^{x+2y}]$$

$$= \frac{1}{(D-D')(D+D'-3)}xy + \frac{1}{(D-D')(D+D'-3)}e^{x+2y}$$

$$= -\frac{1}{3D}\left(1 - \frac{D'}{D}\right)^{-1}\left(1 - \frac{D}{3} - \frac{D'}{3}\right)^{-1} xy + \frac{1}{(1-2)(1+2-3)}e^{x+2y} \text{(case fail)}$$

$$= -\frac{1}{3D}\left(1 + \frac{D}{3} + \frac{D'}{3} + \frac{D'}{D} + \frac{D'}{3} + \frac{2DD'}{9} + \ldots\right)xy + \frac{x}{(2D-3)}e^{x+2y}$$

$$= -\frac{1}{3D}\left(xy + \frac{y}{3} + \frac{2}{3}x + \frac{x^2}{2} + \frac{2}{9}\right) + \frac{x\,e^{x+2y}}{(2\times1-3)}$$

$$= -\frac{1}{3}\left(\frac{x^2 y}{2} + \frac{xy}{3} + \frac{x^2}{3} + \frac{x^3}{6} + \frac{2x}{9}\right) - xe^{x+2y}$$

Hence the complete solution is

$$z = \phi_1(y+x) + e^{3x}\phi_2(y-x) - \frac{1}{3}\left[\frac{x^2 y}{2} + \frac{xy}{3} + \frac{x^2}{3} + \frac{x^3}{6} + \frac{2x}{9}\right] - xe^{x+2y}. \quad \textbf{Ans.}$$

Example 65: Solve $(D-3D'-2)^2 z = 2e^{2x}\tan(y+3x)$.

Solution: Consider the equation

$$(D-3D'-2)^2 z = 0$$

$$\therefore \quad \text{C.F.} = e^{2x}\phi_1(y+3x)+xe^{2x}\phi_2(y+3x) \quad \Big| \, m=3, a=2$$

and

$$\text{P.I.} = \frac{1}{(D-3D'-2)^2} 2e^{2x}\cdot\tan(y+3x)$$

$$= 2e^{2x}\frac{1}{(D+2-3D'-2)^2}\tan(y+3x)$$

$$= 2e^{2x}\frac{1}{(D-3D')^2}\tan(y+3x) \quad \Big| \, f(a,b)=0$$

$$= 2xe^{2x}\cdot\frac{1}{2(D-3D')}\tan(y+3x) \quad \Big| \, f'(a,b)=0$$

$$= x^2e^{2x}\cdot\tan(y+3x).$$

Hence, the complete solution of given equation is

$$z = e^{2x}\phi_1(y+3x)+xe^{2x}\phi_2(y+3x)+x^2e^{2x}\tan(y+3x). \quad \textbf{Ans.}$$

Example 66: Solve $(D^3-3DD'+D'+4)z = e^{2x+y}$.

Solution: Consider the equation

$$(D^3-3DD'+D'+4)z = 0 \qquad \ldots(1)$$

The equation (1) cannot be resolved into linear factors in D and D'.

Here the A.E. is

$$h^3-3hk+k+4 = 0 \quad (D=h,\ D'=k) \qquad \ldots(2)$$

$$\therefore \quad \text{C.F.} = \sum Ae^{hx+ky}, \text{ where } h^3-3hk+k+4=0$$

Let $k = \phi(h)$ (function of h only from (2))

then $\text{C.F.} = \sum Ae^{hx+\phi(h)y}$

and

$$\text{P.I.} = \frac{1}{(D^3-3DD'+D'+4)}e^{2x+y} = \frac{e^{2x+y}}{2^3-3(2)(1)+1+4} = \frac{1}{7}e^{2x+y}$$

Hence the complete, solution of given equation is

$$z = \sum Ae^{hx+\phi(h)y}+\frac{1}{7}e^{2x+y}. \quad \textbf{Ans.}$$

Example 67: Solve $s+p-q=z+xy$.

Solution: The given equation may be written as

$$\frac{\partial^2 z}{\partial x\,\partial y}+\frac{\partial z}{\partial x}-\frac{\partial z}{\partial y} = z+xy$$

$$\Rightarrow \quad \frac{\partial^2 z}{\partial x\,\partial y}+\frac{\partial z}{\partial x}-\frac{\partial z}{\partial y}-z = xy$$

or $$(DD'+D-D'-1)z = xy$$

Now consider the equation $(DD'+D-D'-1)z=0$

$$\Rightarrow \quad (D-1)(D'+1)z = 0$$

$$\therefore \quad \text{C.F.} = e^x\phi_1(y)+e^{-y}\phi_2(x) \qquad \Big| m_1=0,\ a_1=1,\ a_2=-1, m_2=0$$

and $$\text{P.I.} = \frac{1}{(D-1)(D'+1)}xy = -\left\{(1-D)^{-1}(1+D')^{-1}\right\}xy$$

$$= -\left\{(1+D+D^2+\cdots)\ (1-D'+D'^2-\cdots)\right\}xy$$

$$= -\{1-D'+D-DD'\}xy = -\{xy-x+y-1\}$$

$$\Rightarrow \quad \text{P.I.} = 1-xy+x-y$$

Hence $$z = e^x\phi_1(y)+e^{-y}\phi_2(x)+1-xy+x-y. \qquad \textbf{Ans.}$$

EXERCISE 4.7

Solve the following non-homogeneous partial differential equations:

1. $(2D+D'+1)(D^2+3DD'-3D)z=0.$ $\left[\textbf{Ans.}\ z=\phi_1(y)+e^{-y}\phi_2(2y-x)+e^{2x}\phi_3(y-3x)\right]$

2. $(D^2-a^2D'^2+2abD+2abD')z=0.$ $\left[\textbf{Ans.}\ z=\phi_1(y-ax)+e^{-2abx}\phi_2(y+ax)\right]$

3. $(D-2D'+5)^2z=0.$ $\left[\textbf{Ans.}\ z=e^{-5x}\phi_1(y+2x)+xe^{-5x}\phi_2(y+2x)\right]$

4. $D(D+D'-1)(D+3D'-2)z=0.$ $\left[\textbf{Ans.}\ z=\phi_1(y)+e^x\phi_2(y-x)+e^{2x}\phi_3(y-3x)\right]$

5. $(D-2D'-1)(D-2D'^2-1)z=0.$ $\left[\textbf{Ans.}\ z=e^x\phi_1(y+2x)+\sum Ae^{hx+\frac{\sqrt{h-1}}{2}\cdot y}\right]$

6. $2s+t-3q=0.$ $\left[\textbf{Ans.}\ z=\phi_1(x)+e^{\left(\frac{3x}{2}\right)}\phi_2(2y-x)\right]$

7. $(D^2 - D'^2 + D + 3D' - 2)z = e^{x-y} - x^2 y.$

$$\left[\textbf{Ans. } z = e^x \phi_1(y-x) + e^{-2x}\phi_2(y+x) - \frac{1}{4}e^{x-y} + \frac{1}{2}x^2 y + \frac{3}{4}x^2 + \frac{1}{2}xy + \frac{3x}{2} + \frac{3y}{4} + \frac{21}{8}\right]$$

8. $(D^2 + DD' + D' - 1)z = e^x + \dfrac{1}{e^x}.$ $\left[\textbf{Ans. } z = e^{-x}\phi_1(y) + e^x \phi_2(y-x) + \dfrac{x}{2}(e^x - e^{-x})\right]$

9. $(2DD' + D'^2 - 3D')z = \cos(3x - 2y)\cdot$

$$\left[\textbf{Ans. } z = \phi_1(x) + e^{3y}\phi_2(x-2y) + \frac{1}{50}[4\cos(3x-2y) + 3\sin(3x-2y)]\right]$$

10. $r - pq - 2t + 2p + 2q = e^{2x+3y} + \sin(2x+y) + xy.$ *[U.P.T.U., (C.O.) 2003]*

$$\left[\textbf{Ans. } z = \phi_1(y-x) + e^{-2x}\phi_2(y+2x) - \frac{1}{10}e^{2x+3y} - \frac{1}{6}\cos(y+2x) + \frac{1}{24}x(6xy - 6y - 2x^2 + 9x - 12)\right]$$

11. $(D^2 - D')z = xe^{ax + a^2 y}.$ $\left[\textbf{Ans. } P.I. = e^{ax+a^2 y}\left(\dfrac{x^2}{4a} - \dfrac{x}{4a^2}\right)\right]$

12. $D(D - 2D')(D + D')z = e^{x+2y}(x^2 + 4y^2).$

$$\left[\textbf{Ans. } z = \phi_1(y) + \phi_2(y+2x) + \phi_3(y-x) - \frac{1}{81}(9x^2 + 36y^2 - 18x - 72y + 76)e^{x+2y}\right]$$

13. $(D^2 - DD' + D' - 1)z = \cos(x+2y) + e^y.$

$$\left[\textbf{Ans. } z = e^x\phi_1(y) + e^{-x}\phi_2(y+x) + \frac{1}{2}\sin(x+2y) - xe^y\right]$$

14. $(D + D' - 1)(D + 2D' - 3)z = 2x + 3y.$

$$\left[\textbf{Ans. } z = e^x\phi_1(y+x) + e^{2x}\phi_2(y+x) + \frac{1}{2}e^{2x-y} + \frac{1}{4}(2x+3)\right]$$

15. $(D - 3D' - 2)^2 z = 2e^{2x}\sin(y+3x).$

$$\left[\textbf{Ans. } z = e^{2x}\phi_1(y+3x) + xe^{2x}\phi_2(y+3x) + x^2 e^{2x}\sin(y+3x)\right]$$

16. $D(D + D' - 1)(D + 3D' - 2)z = x^2 - 4xy + y^2.$

$$\left[\textbf{Ans. } z = \phi_1(y) + e^x\phi_2(y-x) + e^{2x}\phi_3(y-3x) + \frac{1}{6}x^3 - x^2 y + xy^2 - \frac{7}{4}x^2 + 2xy + \frac{x}{4}\right]$$

17. $(D^2D' + D'^2 - 2)z = e^{2y}\sin 3x - e^x \cos y.$

$$\left[\textbf{Ans. } z = \sum Ae^{hx+\phi(h)y} - \frac{1}{16}e^{2y}\sin 3x + \frac{1}{20}e^x(3\cos 2y - \sin 2y)\right]$$

18. $(2DD' + D'^2 - 3D')z = 5\cos(3x - 2y).$

$$\left[\textbf{Ans. } z = \phi_1(x) + e^{3y}\phi_2(x-2y) + \frac{2}{5}\cos(3x-2y) + \frac{3}{10}\sin(3x-2y)\right]$$

Indicate true or false of the following statement:

19. If $(D - m_1D' - a_1)(D - m_2D' - a_2)Z = 0$ then C.F. = $e^{a_1x}\phi_1(y + m_1x) + e^{a_2x}\phi_2(y + m_2x)$

True/False [**Ans.** True]

20. If $(D^2 + 2DD' + D'^2 - 2D - 2D')Z = 0$ then C.F. = $\phi_1(y + x) + e^{-2x}\phi_2(y - x)$

True/False [**Ans.** False]

4.16 CLASSIFICATION OF PARTIAL DIFFERENTIAL EQUATIONS OF SECOND ORDER

[U.P. T.U. (C.O.) 2005]

The most general second order P.D.E. may be written as

$$A(x,y)\frac{\partial^2 u}{\partial x^2} + B(x,y)\frac{\partial^2 u}{\partial x \partial y} + C(x,y)\frac{\partial^2 u}{\partial y^2} + f\left(x, y, u, \frac{\partial u}{\partial x}, \frac{\partial u}{\partial y}\right) = 0 \quad ...(i)$$

This equation is linear in second order terms.

The equation (*i*) is elliptic, parabolic and hyperbolic if

(*a*) $B^2 - 4AC < 0$ (Elliptic)

(*b*) $B^2 - 4AC = 0$ (Parabolic)

(*c*) $B^2 - 4AC > 0$ (Hyperbolic).

Example 68: Characterize the following partial differential equations into elliptic, parabolic and hyperbolic equations.

$$A\frac{\partial^2 u}{\partial x^2} + 2B\frac{\partial^2 u}{\partial x \partial y} + C\frac{\partial^2 u}{\partial y^2} = f(x, y, u, u_x, u_y)$$

where *A*, *B*, *C* may be functions of *x* and *y*. *(U.P.T.U. 2007)*

Solution: Let the most general form of P.D.E. is

$$A_1\frac{\partial^2 u}{\partial x^2} + B_1\frac{\partial^2 u}{\partial x \partial y} + C_1\frac{\partial^2 u}{\partial y^2} + f(x, y, u, u_x\ x_y) = 0 \text{ [Art. 5.2 equation (i)]} \quad ...(1)$$

Comparing the given equation with equation (1), we get

$A_1 = A, \quad B_1 = 2B \quad \text{and} \quad C_1 = C$

For elliptic $\quad B_1^2 - 4A_1C_1 < 0 \Rightarrow 4B^2 - 4AC < 0 \Rightarrow \boxed{B^2 - AC < 0.}$

For parabolic $\quad B_1^2 - 4A_1C_1 = 0 \Rightarrow 4B^2 - 4AC = 0 \Rightarrow \boxed{B^2 - AC = 0.}$

For hyperbolic $\quad B_1^2 - 4A_1C_1 > 0 \Rightarrow 4B^2 - 4AC > 0 \Rightarrow \boxed{B^2 - AC > 0.}$

Example 69: Show that the equation

$\frac{\partial^2 u}{\partial t^2} = c^2 \frac{\partial^2 u}{\partial x^2}$ is hyperbolic. *(U.P.T.U. 2003)*

Solution: The given equation may be written as

$$\frac{\partial^2 u}{\partial t^2} - c^2 \frac{\partial^2 u}{\partial x^2} = 0$$

Here $A = 1, \quad B = 0, \quad C = -c^2$

Now $\quad B^2 - 4AC = 0 - 4 \times 1(-c^2) = 4c^2$

$\Rightarrow \quad B^2 - 4AC = (2c)^2$ is +ve always

$\Rightarrow \quad B^2 - 4AC > 0$

$\therefore$ The given equation is hyperbolic.

Example 70: Classify the equation

$$(1 - x^2)\frac{\partial^2 z}{\partial x^2} - 2xy\frac{\partial^2 z}{\partial x \partial y} + (1 - y^2)\frac{\partial^2 z}{\partial y^2} + x\frac{\partial z}{\partial x} + 3x^2 y\frac{\partial z}{\partial y} - 2z = 0.$$

Solution: Here $A = 1 - x^2, \quad B = -2xy, \quad C = 1 - y^2$

Now $\quad B^2 - 4AC = 4x^2y^2 - 4(1 - x^2)(1 - y^2)$

$= 4x^2y^2 - 4 + 4x^2 + 4y^2 - 4x^2y^2 = 4(x^2 + y^2 - 1)$

$\Rightarrow \quad B^2 - 4AC = 4(x^2 + y^2 - 1)$

If $\quad x^2 + y^2 > 1 \Rightarrow B^2 - 4AC > 0$ (Hyperbolic)

If $\quad x^2 + y^2 = 1 \Rightarrow B^2 - 4AC = 0$ (Parabolic)

and if $\quad x^2 + y^2 < 1 \Rightarrow B^2 - 4AC < 0$ (Elliptic).

Example 71: Classify the following differential equation in the second quadrant of xy-plane

$$(x^2+y^2)^{1/2}\frac{\partial^2 u}{\partial x^2}+2(x-y)\frac{\partial^2 u}{\partial x\partial y}+(x^2+y^2)^{1/2}\frac{\partial^2 u}{\partial y^2}=0.$$

Solution: Here

$$A=(x^2+y^2)^{1/2}$$
$$B=2(x-y)$$
$$C=(x^2+y^2)^{1/2}$$

$$\therefore \quad B^2-4AC=4(x-y)^2-4(x^2+y^2)^{1/2}(x^2+y^2)^{1/2}$$
$$=4(x-y)^2-4(x^2+y^2)$$
$$=4(x^2-2xy+y^2-x^2-y^2)$$
$$B^2-4AC=-8xy \qquad \ldots(i)$$

In second quadrant, x always negative and y positive.

$\therefore$ xy will be negative always in second quadrant

Hence from (i) $B^2-4AC>0$

The given differential represents hyperbolic curve.

Example 72: Classify the equation $(1+x^2)\frac{\partial^2 u}{\partial x^2}+(5+2x^2)\frac{\partial^2 u}{\partial x\partial t}+(4+x^2)\frac{\partial^2 u}{\partial t^2}=0.$

Solution: Here $A=(1+x^2), B=(5+2x^2), C=(4+x^2)$

$$\therefore \quad B^2-4AC=(5+2x^2)^2-4(1+x^2)(4+x^2)$$
$$=25+20x^2+4x^4-16-20x^2-4x^4$$
$$B^2-4AC=9>0$$

Equation is elliptical.

EXERCISE 4.8

1. Classify the partial differential equation $x^2\frac{\partial^2 u}{\partial t^2}+3\frac{\partial^2 u}{\partial x\partial t}+x\frac{\partial^2 u}{\partial x^2}+17\frac{\partial u}{\partial t}=100u.$

   [**Ans.** $9-4x^3<0$ (elliptic), $9-4x^3=0$ (parabolic), $9-4x^3>0$ (hyperbolic)]

2. Classify the following:

   (*a*) $\frac{\partial^2 z}{\partial x^2}=\frac{\partial z}{\partial y}$ (*U.P.T.U. 2003*) (*b*) $\frac{\partial^2 z}{\partial x^2}+\frac{\partial^2 z}{\partial y^2}=0.$ [**Ans.** (*a*) parabolic (*b*) elliptic]

3. Classify the following differential equation:

$$\sqrt{x^2+y^2}\frac{\partial^2 u}{\partial x^2}+2(x-y)\frac{\partial^2 u}{\partial x \partial y}+\sqrt{x^2+y^2}\frac{\partial^2 y}{\partial y^2}=0$$ **[Ans.** hyperbolic]

4. Show that the equation $\frac{\partial^2 u}{\partial x^2}+x\frac{\partial^2 u}{\partial y^2}+\frac{\partial u}{\partial y}=0$ is elliptic for $x > 0$ and hyperbolic for $x < 0$.

5. Classify the partial differential equation $\frac{\partial^2 u}{\partial t^2}+t\frac{\partial^2 u}{\partial x \partial t}+x\frac{\partial^2 u}{\partial x^2}+2\frac{\partial u}{\partial t}+\frac{\partial u}{\partial x}+6u=0$

[**Ans.** $t^2-4x<0$ (elliptic), $t^2-4x>0$ (hyperbolic), $t^2-4x=0$ (parabolic)]

6. Classify the equation $(1-x^2)\frac{\partial^2 z}{\partial x^2}-2xy\frac{\partial^2 z}{\partial x \partial y}+(1-y^2)\frac{\partial^2 z}{\partial y^2}+x\frac{\partial z}{\partial x}+3x^2y\frac{\partial z}{\partial y}-2z=0$

[**Ans.** $x^2+y^2>1$ (hyperbolic), $x^2+y^2<1$ (elliptic), $x^2+y^2=1$ (parabolic)]

OBJECTIVE TYPE QUESTIONS

A. Pick the correct answers of the given choices below:

1. If $f(x)=x+x^2$ for $-\pi<x<\pi$ then a_0 is equal to

(*i*) $\frac{2\pi^3}{3}$ (*ii*) $\frac{2\pi^2}{3}$ (*iii*) $\frac{\pi^2}{3}$ (*iv*) $\frac{3\pi^2}{4}$

2. If $f(x)=\frac{\pi-x}{2}$ in the interval $(0, 2\pi)$ then a_n is equal to

(*i*) $\frac{1}{12}$ (*ii*) $\frac{(-1)^n}{2}$ (*iii*) 0 (*iv*) None of these

3. If $f(x)=x^2$ in the interval $0<x<3$ then a_0 is equal to

(*i*) $\frac{3}{2}$ (*ii*) $-\frac{3}{2}$ (*iii*) $\frac{5}{2}$ (*iv*) 0

4. If $f(x)=x\cos x, -\pi<x<\pi,$ then a_0 is equal to

(*i*) $-\frac{3}{2}$ (*ii*) $\frac{5}{2}$ (*iii*) 0 (*iv*) $\frac{1}{2}$

5. P.I. of $(D^2 - 2DD' + D'^2)z = e^{x+2y}$ is

(*i*) e^{x+2y} (*ii*) e^{x-2y} (*iii*) $-e^{x+2y}$ (*iv*) None of these

6. Solution of $(D^3 - 6D^2D' + 11DD'^2 - 6D'^3)z = 0$ is

(*i*) $z = \phi_1(y-x) + \phi_2(y+x) + \phi_3(y+2x)$

(*ii*) $z = \phi_1(y+x) + \phi_2(y+2x) + \phi_3(y+3x)$

(*iii*) $z = (y+2x)\phi_1(x-y) + \phi_2(y) + \phi_3(y+3x)$

(*iv*) None of these.

7. The solution of equation $\dfrac{\partial^2 z}{\partial x^2} + \dfrac{\partial^2 z}{\partial y^2} = 0$ is

(*i*) $z = \phi_1(y+ix) + \phi_2(y-ix)$ (*ii*) $z = \phi_1(y+x) + \phi_2(y-x)$

(*iii*) $z = \phi_1(y) - \phi_2(y-ix)$ (*iv*) $z = \phi_1(y+ix) - \phi_2(y-ix)$

8. The general solution of $r = 6x$ is

(*i*) $z = x + y\phi_1(y) + \phi_2(y)$ (*ii*) $z = x^3 + y\phi_1(y) + \phi_2(y)$

(*iii*) $z = -x - y\phi_1(y) + \phi_2(y)$ (*iv*) $z = x^4 + y\phi_1(y) + \phi_2(y)$

9. The general solution of $\dfrac{\partial z}{\partial x} + \dfrac{\partial z}{\partial y} = \sin x$ is

(*i*) $z = f(y-x) - \cos x$ (*ii*) $z = f(y-x) + \cos x$

(*iii*) $z = f(y-x) - \sin x$ (*iv*) $z = f(y-x) - \operatorname{cosec} x$

10. C.F. of $4r + 12s + 9t = e^{3x-2y}$ is

(*i*) $\phi_1\left(y + \frac{3}{2}x\right) + x\phi_2\left(y + \frac{3}{2}x\right)$ (*ii*) $\phi_1(y+3x) + \phi_2(y-3x)$

(*iii*) $\phi_1\left(y - \frac{3}{2}x\right) + x\phi_2\left(y - \frac{3}{2}x\right)$ (*iv*) $\phi_1(y) + \phi_2(y-x)$

B. Fill in the blanks:

1. The general solution of first order linear equation $Pp + Qq = R$ is given by Lagrange's equation as ________ . [*U.P.T.U. (Special) 2009*]

2. The solution of $\dfrac{\partial^2 z}{\partial y^2} = \sin(xy)$ is ________ .

3. P.I. of $(D^2 + DD')z = \sin(x+y)$ is ________ .

4. In the half range Fourier sine series the values of constants a_0 and a_n always ________ .

5. If $f(x) = k$ for $0 < x < 2$ in a half range cosine series then $a_0 =$ ________ .

6. If $f(x) = |x|, -\pi < x < \pi$ then $a_0 =$ ________ .

7. Solution of $\dfrac{\partial^2 z}{\partial x^3} = 0$ is ________ .

C. Indicate True or False for the following statements:

1. (*i*) A function *f*(*x*) is even if $f(-x) = -f(x)$. (*U.P.T.U. 2009*)

(*ii*) A function *f*(*x*) is odd if $f(-x) = f(x)$. (*U.P.T.U. 2009*)

(*iii*) Most functions are neither even nor odd. (*U.P.T.U. 2009*)

(*iv*) A function *f*(*x*) can always be expressed as an arithmetic mean of an even and odd function as $f(x) = \frac{1}{2}[f(x) + f(-x)] + \frac{1}{2}[f(x) - f(-x)]$. (*U.P.T.U. 2009*)

2. (*i*) Fourier series is an infinite series representation of periodic function in terms of Trigonometric sine and cosine functions.

(*ii*) A function *f*(*x*) is not periodic if $f(x+T) = f(x)$ for all real *x* and some positive number *T*.

(*iii*) The equation $u\dfrac{\partial u}{\partial x} + \dfrac{\partial^2 u}{\partial x^2} = 0$ is a linear partial differential equation.

(*iv*) The equation $2\dfrac{\partial^2 u}{\partial x^2} + u\dfrac{\partial^2 u}{\partial x \partial y} + 3\dfrac{\partial^2 u}{\partial y^2} = 0$ represents an elliptic curve.

D. Match the following:

1. A second order P.D.E. in the function '*u*' of two independent variables *x*, *y* given with usual symbols. $Au_{xx} + Bu_{xy} + Cu_{yy} + F(u) = 0$ then (*U.P.T.U. 2009*)

(*i*) Hyperbolic	(*a*) $B^2 - 4AC = 0$
(*ii*) Parabolic	(*b*) $B^2 - 4AC < 0$
(*iii*) Elliptic	(*c*) $B^2 - 4AC > 0$
(*iv*) Not classified	(*d*) $A = B = C = 0$

2.

(*i*) a_0	(*a*) $\dfrac{dx}{P} = \dfrac{dy}{Q} = \dfrac{dz}{R}$
(*ii*) b_n	(*b*) $B^2 - 4AC > 0$

(*iii*) Lagrange's equations (*c*) $\frac{1}{\pi}\int_{-\pi}^{\pi} f(x)dx$

(*iv*) Parabola (*d*) $\frac{1}{\pi}\int_{-\pi}^{\pi} f(x)\sin nx\, dx$

ANSWERS TO OBJECTIVE TYPE QUESTIONS

A. Pick the correct answers:

1. (*ii*)	**2.** (*iii*)	**3.** (*i*)	**4.** (*iii*)	**5.** (*i*)
6. (*ii*)	**7.** (*i*)	**8.** (*ii*)	**9.** (*i*)	**10.** (*iii*)

B. Fill in the blanks:

1. $\phi(u, v) = 0$; $u = c_1$ and $v = c_2$ **2.** $\phi_1(x) + y\phi_2(x) - \frac{1}{x^2}\sin(xy)$

3. $-\frac{1}{2}\sin(x+y)$ **4.** Zero

5. $2k$ **6.** π

7. $z = \phi_1(y) + x\phi_2(y) + x^2\phi_3(y)$

C. True or False:

1. (*i*) F	(*ii*) F	(*iii*) T	(*iv*) T
2. (*i*) T	(*ii*) F	(*iii*) F	(*iv*) T

D. Match the following:

1. (*i*) $\rightarrow c$	(*ii*) $\rightarrow a$	(*iii*) $\rightarrow b$	(*iv*) $\rightarrow d$
2. (*i*) $\rightarrow c$	(*ii*) $\rightarrow d$	(*iii*) $\rightarrow a$	(*iv*) $\rightarrow b$

UNIT V

Applications of Partial Differential Equations

5.1 INTRODUCTION

The problems related to fluid mechanics, solid mechanics, heat transfer, wave equation and other areas of physics are designed as Initial Boundary Value Problems consisting of partial differential equations and initial conditions.

These problems can be solved by 'Method of separation of variables,' in this unit we derive and solve one dimensional heat equation, wave equation, Laplace's equation in two dimensions etc. by separation of variables method. The general solution of partial differential equation consists arbitrary functions which can be obtained by Fourier Series.

5.2 METHOD OF SEPARATION OF VARIABLES

In this method, we assume that the required solution is the product of two functions *i.e.*,

$$u(x, y) = X(x)Y(y) \qquad ...(i)$$

Then we substitute the value of $u(x, y)$ from (*i*) and its derivatives reduces the *P.D.E.* to the form

$$f_1(X, X', \cdots\cdots) = f_2(Y, Y', \cdots\cdots) \qquad ...(ii)$$

which is separable in X and Y. Since $f_2(Y, Y', \cdots)$ is function Y only and $f_1(X, X', \cdots)$ is function of X only, then equation (*ii*) must be equal to a common constant say k. Thus (*ii*) reduces to

$$f_1(X, X', \cdots) = f_2(X, X', \cdots) = k.$$

Example 1: Using, the method of separation of variables solve $\dfrac{\partial u}{\partial x} = 2\dfrac{\partial u}{\partial t} + u$, where $u(x, 0) = 6e^{-3x}$. (*U.P.T.U. 2006*)

Solution: We have

$$\frac{\partial u}{\partial x} = 2\frac{\partial u}{\partial t} + u \qquad ...(1)$$

Let

$$u = X(x) \cdot T(t) \qquad ...(2)$$

then $$\frac{\partial u}{\partial x} = \frac{\partial X}{\partial x} T = \frac{dX}{dx} T \qquad \left| \frac{\partial X}{\partial x} = \frac{dX}{dx}, \text{ As } X \text{ is the function of } x \text{ alone} \right.$$

and $$\frac{\partial u}{\partial t} = X \frac{\partial T}{\partial t} = X \frac{dT}{dt}$$

Putting these values in equation (1), we get

$$T \frac{dX}{dx} = 2X \frac{dT}{dt} + XT$$

$$\Rightarrow \quad \frac{1}{X}\frac{dX}{dx} = \frac{2}{T}\frac{dT}{dt} + 1 \qquad \text{(On dividing by } XT\text{)}$$

$$\Rightarrow \quad \frac{1}{X}\frac{dX}{dx} = \frac{2}{T}\frac{dT}{dt} + 1 = k$$

Now $$\frac{1}{X}\frac{dX}{dx} = k \Rightarrow \frac{dX}{X} = k\,dx \qquad ...(3)$$

On integrating, we get

$$\log_e X = kx + \log_e c_1$$

$$\Rightarrow \quad \log_e \frac{X}{c_1} = kx \Rightarrow \boxed{X = c_1 e^{kx}.}$$

and again from (3), we get

$$\frac{2}{T}\frac{dT}{dt} + 1 = k$$

$$\Rightarrow \quad \frac{2}{T}\frac{dT}{dt} = k - 1 \Rightarrow \frac{dT}{T} = \frac{k-1}{2} dt$$

On integrating, we obtain

$$\log_e T = \frac{k-1}{2} t + \log_e c_2 \Rightarrow \log_e \frac{T}{c_2} = \frac{k-1}{2} t$$

$$\Rightarrow \quad \boxed{T = c_2 e^{\left(\frac{k-1}{2}\right)t}.}$$

Putting the values of X and T is equation (2), we get

$$u = c_1 e^{kx} \cdot c_2 e^{\left(\frac{k-1}{2}\right)t}$$

$$\Rightarrow \quad u = c_1 c_2 e^{kx + \frac{1}{2}(k-1)t} \qquad ...(4)$$

On putting $t = 0$ and $u = 6e^{-3x}$ in (4), we have

$$6e^{-3x} = c_1 c_2 e^{kx} \Rightarrow c_1 c_2 = 6 \text{ and } k = -3$$

Hence the required solution of given equation is

$$u = 6e^{-3x + \frac{1}{2}(-3-1)t} = 6e^{-3x-2t}$$

$$\Rightarrow \qquad u = 6e^{-3x}e^{-2t}. \quad \textbf{Ans.}$$

Example 2: Solve the following equation by the method of separation of variables.

$$\frac{\partial u}{\partial x} = 4\frac{\partial u}{\partial y} \text{ where } u(0, y) = 8e^{-3y} \qquad (U.P.T.U.\ 2008)$$

Solution: We have

$$\frac{\partial u}{\partial x} = 4\frac{\partial u}{\partial y} \qquad \ldots(1)$$

Let $u = Xy$ ...(2)

then $\dfrac{\partial u}{\partial x} = \dfrac{dX}{dx}Y, \dfrac{\partial u}{\partial y} = X\dfrac{dY}{dy}$

Putting these values in equation 1, we get

$$\frac{dX}{dx}Y = 4X\frac{dY}{dy}$$

or $$\frac{1}{X}\cdot\frac{dX}{dx} = \frac{4}{Y}\cdot\frac{dY}{dy} = k$$

Now $$\frac{1}{X}\frac{dX}{dx} = K \Rightarrow \frac{dX}{X} = kdx \Rightarrow \log_e X = kx + \log_e c_1$$

or $$\boxed{X = c_1e^{kx}}$$

And $$\frac{4}{Y}\frac{dY}{dy} = k \Rightarrow \frac{dY}{Y} = \frac{k}{4}dy \Rightarrow \log_e Y = \frac{k}{4}y + \log_e c_2$$

or $$\boxed{Y = c_2e^{\frac{ky}{4}}}$$

$\therefore$ From (2), we get

$$u = c_1c_2e^{\left(x+\frac{y}{4}\right)k} \qquad \ldots(3)$$

Putting $x = 0$, in the equation (3), we get

$$8e^{-3y} = c_1c_2e^{\frac{yk}{4}}$$

Comparing on both sides

$$c_1c_2 = 8 \text{ and } \frac{k}{4} = -3 \Rightarrow k = -12$$

Hence

$$u = 8e^{-(12x+3y)} \qquad \textbf{Ans.}$$

Example 3: Solve by the method of separation of variables

$$\frac{\partial^2 z}{\partial x^2} - 2\frac{\partial z}{\partial x} + \frac{\partial z}{\partial y} = 0. \qquad (U.P.T.U.\ 2005, 2009)$$

Solution: We have

$$\frac{\partial^2 z}{\partial x^2} - 2\frac{\partial z}{\partial x} + \frac{\partial z}{\partial y} = 0 \qquad ...(1)$$

Let $\qquad z = X(x)Y(y) \qquad ...(2)$

$$\Rightarrow \quad \frac{\partial z}{\partial x} = \frac{dX}{dx}Y \Rightarrow \frac{\partial^2 z}{\partial x^2} = \frac{d^2X}{dx^2}Y$$

and

$$\frac{\partial z}{\partial y} = \frac{dY}{dy}\cdot X$$

Putting these values in equation (1), we get

$$Y\frac{d^2X}{dx^2} - 2Y\frac{dX}{dx} + X\cdot\frac{dY}{dy} = 0$$

$$\Rightarrow \frac{1}{X}\frac{d^2X}{dx^2} - \frac{2}{X}\frac{dX}{dx} + \frac{1}{Y}\frac{dY}{dy} = 0 \qquad \text{(divide by } XY)$$

$$\Rightarrow \frac{1}{X}\frac{d^2X}{dx^2} - \frac{2}{X}\frac{dX}{dx} = -\frac{1}{Y}\frac{dY}{dy} = k \qquad ...(3)$$

Now $\qquad \frac{1}{X}\frac{d^2X}{dx^2} - \frac{2}{X}\frac{dX}{dx} = k \Rightarrow (D^2 - 2D - k)X = 0$

The A.E. is $m^2 - 2m - k = 0$

$$\Rightarrow \quad m = \frac{2 \pm \sqrt{4+4k}}{2} \Rightarrow m = 1 \pm \sqrt{1+k}$$

$$\therefore \quad X = c_1e^{\{1+\sqrt{1+k}\}x} + c_2e^{\{1-\sqrt{1+k}\}x}$$

and again from (3), we have

$$\frac{1}{Y}\frac{dY}{dy} = -k \quad \Rightarrow \quad \frac{dY}{Y} = -k\,dy$$

On integrating, we get

$$\log_e Y = -ky + \log c_3$$

$$\Rightarrow \qquad \log_e \frac{Y}{c_3} = -ky \Rightarrow Y = c_3 e^{-ky}$$

Hence the required solution of equation (1) is

$$z = \left[c_1 e^{\{1+\sqrt{1+k}\}x} + c_2 e^{\{1+\sqrt{1-k}\}x}\right] c_3 e^{-ky}. \qquad \textbf{Ans.}$$

Example 4: Solve by the method of separation of variables $\frac{\partial u}{\partial x} = 2\frac{\partial u}{\partial y} + u$, where $u(x, 0) = 3e^{-5x} - 2e^{-3x}$.

Solution: We have

$$\frac{\partial u}{\partial x} = 2\frac{\partial u}{\partial y} + u \qquad \text{...(1)}$$

Let $\qquad u = XY \qquad$ ...(2)

Putting this value in equation (1), we get

$$\frac{\partial}{\partial x}(XY) = 2\frac{\partial}{\partial y}(XY) + XY$$

$$\Rightarrow \qquad Y\frac{dX}{dx} = 2X\frac{dY}{dy} + XY$$

$$\Rightarrow \qquad \frac{1}{X}\frac{dX}{dx} = \frac{2}{Y}\frac{dY}{dy} + 1 = k \qquad \text{...(3)}$$

Now $\qquad \frac{1}{X}\frac{dX}{dx} = k \quad \Rightarrow \quad \frac{dX}{X} = k\,dx$

On integrating, we get

$$\log_e X = kx + \log_e c_1$$

$$\Rightarrow \qquad \boxed{X = c_1 e^{kx}.}$$

And taking last two terms of equation (3), we have

$$\frac{2}{Y}\frac{dY}{dy} + 1 = k \quad \Rightarrow \quad \frac{2}{Y}\frac{dY}{dy} = k - 1$$

$$\Rightarrow \qquad \frac{dY}{Y} = \frac{(k-1)}{2} dy$$

On integrating, $\log_e Y = \frac{(k-1)}{2} y + \log_e c_2$

$$\Rightarrow \qquad \boxed{Y = c_2 e^{(k-1) \cdot \frac{y}{2}}}.$$

From (2), we get

$$u = c_1 c_2 e^{kx} \cdot e^{(k-1)y/2}$$

$$\Rightarrow \qquad u = \sum_{n=1}^{\infty} b_n e^{k_n x} e^{(k_n - 1)y/2} \qquad \ldots(4)$$

$(b_n = c_1 c_2)$ and $(k = k_n)$

which is the most general solution of equation (1).

Putting $y = 0$ and $u = 3e^{-5x} - 2e^{-3x}$ in equation (4), we get

$$3e^{-5x} - 2e^{-3x} = \sum_{n=1}^{\infty} b_n e^{k_n x} = b_1 e^{k_1 x} + b_2 e^{k_2 x}$$

Comparing the terms on both sides, we get

$b_1 = 3, \quad k_1 = -5, \quad b_2 = -2, \quad k_2 = -3$

Hence the required solution of given equation is from (4), we have

$$u = 3e^{-5x} \cdot e^{-3y} + (-2)e^{-3x} \cdot e^{-2y}$$

$$\Rightarrow \qquad u = 3e^{-(5x+3y)} - 2e^{-(3x+2y)}. \qquad \textbf{Ans.}$$

EXERCISE 5.1

Solve the following P.D.E. by the method of separation of variables:

1. $\frac{\partial u}{\partial x} = 2\frac{\partial u}{\partial y} + u, u(x, 0) = 4e^{-3x}$. [**Ans.** $u = 4e^{-(3x+2y)}$]

2. $4\frac{\partial u}{\partial x} + \frac{\partial u}{\partial y} = 3u,$ and $u(0, y) = e^{-5y}$. [**Ans.** $u = e^{2x-5y}$]

3. $\frac{\partial^2 u}{\partial x^2} = \frac{\partial u}{\partial y} + 2u$, with $u(0, y) = 0, \frac{\partial u(0, y)}{\partial x} = 1 + e^{-3y}$. *(U.P.T.U. 2008)*

$$\left[\textbf{Ans. } u = \frac{1}{\sqrt{2}} \sin h\sqrt{2}x + e^{-3y} \sin x\right]$$

4. $4\frac{\partial u}{\partial t} + \frac{\partial u}{\partial x} = 3u, u(x, 0) = 3e^{-x} - e^{-5x}$. [**Ans.** $u = 3e^{t-x} - e^{2t-5x}$]

5. $2x\dfrac{\partial z}{\partial x} - 3y\dfrac{\partial z}{\partial y} = 0.$ **[Ans.** $z = Ax^{3b}y^{2b}$, where $A = c_1c_2, k = 6b$]

6. $\dfrac{\partial^2 u}{\partial x^2} = \dfrac{\partial u}{\partial t}$ if $u(x, 0) = x^2(25 - x^2)$. **[Ans.** $u = x^2(25 - x^2)e^{-p^2 t}$]

7. $\dfrac{\partial^2 z}{\partial x^2} = \dfrac{\partial z}{\partial t}$ given that $z = 0$ when $t \to \infty$ as well as $z = 0$ at $x = 0$ and $x = l$.

$$\left[\textbf{Ans. } z = \sum_{n=1}^{\infty} b_n e^{-\left(\frac{n^2\pi^2}{l^2}\right)t} \cdot \sin\frac{n\pi x}{l}\right]$$

8. $\dfrac{\partial u}{\partial x} = \dfrac{\partial u}{\partial y}$. **[Ans.** $u = Ae^{x^2} \cdot y^k$]

9. $\dfrac{\partial^2 u}{\partial x \partial t} = e^{-t}\cos x$ with $u(x, 0)$ and $\dfrac{\partial u(0, t)}{\partial t} = 0.$ **[Ans.** $u = \sin x - e^{-t}\sin x$]

5.3 ONE DIMENSIONAL WAVE EQUATION

(U.P.T.U. 2007)

The one dimensional wave equation arises in the study of transverse vibrations of an elastic string. Consider an elastic string, stretched to its length 'l' between two points O and A fixed. Let the function $y(x, t)$ denote the displacement of string at any point x and at any time $t > 0$ from the equilibrium position (x-axis). When the string released after stretching then it vibrates and therefore the transverse vibrations formed a one dimensional wave equation.

Let the string is perfectly flexible and does not offer resistance to bending. Let T_1 and T_2 be tensions at the end points P and Q of the portion of the string. Since there is no motion in the horizontal direction. Thus the sum of the forces in the horizontal direction must be zero *i.e.*,

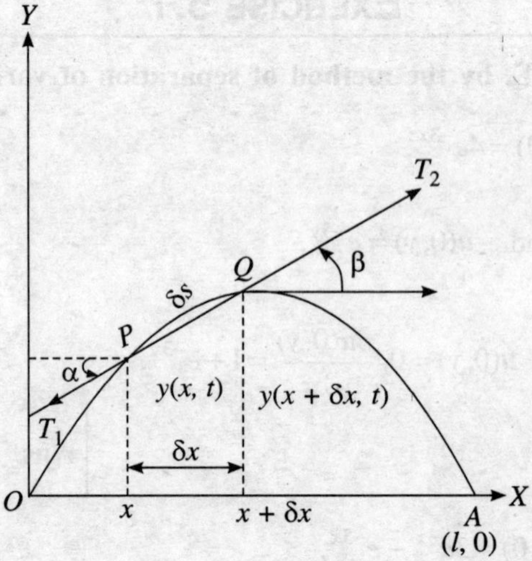

$$-T_1\cos\alpha + T_2\cos\beta = 0$$

$$\Rightarrow \qquad T_1\cos\alpha = T_2\cos\beta = T \quad \text{(constant)} \qquad ...(i)$$

Let m be the mass of the string per unit length then the mass of portion $PQ = m\delta s$.

Now by Newton's second law of motion, the equation of motion in the vertical direction is

mass × acceleration = resultant of forces

$$m\delta s\frac{\partial^2 y}{\partial t^2} = T_2\sin\beta - T_1\sin\alpha \qquad ...(ii)$$

Dividing (*ii*) by (*i*), we have

$$\frac{m\delta s}{T}\frac{\partial^2 y}{\partial t^2} = \frac{T_2\sin\beta}{T_2\cos\beta} - \frac{T_1\sin\alpha}{T_1\cos\alpha}$$

$$\Rightarrow \qquad \frac{m\delta s}{T}\frac{\partial^2 y}{\partial t^2} = \tan\beta - \tan\alpha$$

$$\Rightarrow \qquad \frac{\partial^2 y}{\partial t^2} = \frac{T}{m\delta s}(\tan\beta - \tan\alpha)$$

$$= \frac{T}{m}\left\{\frac{\left(\frac{\partial y}{\partial x}\right)_{x+\delta x} - \left(\frac{\partial y}{\partial x}\right)_x}{\delta x}\right\} \qquad \Big|\ \delta s \to \delta x$$

Taking $\delta x \to 0$

$$\Rightarrow \qquad \frac{\partial^2 y}{\partial t^2} = \frac{T}{m}\lim_{\delta x\to 0}\left\{\frac{\left(\frac{\partial y}{\partial x}\right)_{x+\delta x} - \left(\frac{\partial y}{\partial x}\right)_x}{\delta x}\right\}$$

$$\Rightarrow \qquad \frac{\partial^2 y}{\partial t^2} = \frac{T}{m}\frac{\partial^2 y}{\partial x^2}$$

Thus

$$\boxed{\frac{\partial^2 y}{\partial t^2} = c^2\frac{\partial^2 y}{\partial x^2}.} \qquad ...(iii)$$

where $c^2 = \frac{T}{m}$. Equation (*iii*) is known as one dimensional wave equation.

5.3.1 Solution of One Dimensional Wave Equation

The wave equation is

$$\frac{\partial^2 y}{\partial t^2} = c^2 \frac{\partial^2 y}{\partial x^2} \qquad \text{...(i)}$$

Let $$y = X(x)\, T(t). \qquad \text{...(ii)}$$

where X is the function of x only and T is the function of t only.

Then $$\frac{\partial^2 y}{\partial t^2} = X\frac{d^2 T}{dt^2} \quad \text{and} \quad \frac{\partial^2 y}{\partial x^2} = T\frac{d^2 X}{dx^2}$$

Putting these values in equation (*i*), we get

$$X\frac{d^2T}{dt^2} = c^2 T\frac{d^2X}{dx^2}$$

$$\Rightarrow \quad \frac{1}{X}\frac{d^2X}{dx^2} = \frac{1}{c^2 T}\frac{d^2T}{dt^2} = k \qquad \text{...(iii)}$$

Now $$\frac{1}{X}\frac{d^2X}{dx^2} = k$$

$$\Rightarrow \quad (D^2 - k)X = 0; \quad \frac{d}{dx} = D$$

The A.E. is $m^2 - k = 0$

$$m = \pm\sqrt{k}$$

$$\therefore \quad \boxed{X = c_1 e^{\sqrt{k}x} + c_2 e^{-\sqrt{k}x}}$$

and, again from (*iii*), we get

$$\frac{d^2T}{dt^2} = kc^2 T \quad \Rightarrow \quad (D^2 - kc^2)T = 0\,; \quad D = \frac{d}{dt}$$

$\Rightarrow$ The A.E. is $m^2 - kc^2 = 0 \Rightarrow m = \pm c\sqrt{k}$

$$\therefore \quad \boxed{T = c_3 e^{c\sqrt{k}t} + c_4 e^{-c\sqrt{k}t}.}$$

Thus, from equation (*ii*), we get

$$y = (c_1 e^{\sqrt{k}x} + c_2 e^{-\sqrt{k}x})(c_3 e^{c\sqrt{k}t} + c_4 e^{-c\sqrt{k}t})$$

There are arise following cases:

***Case I*:** If $k > 0$ let $k = p^2$

then $$\boxed{y = (c_1 e^{px} + c_2 e^{-px})(c_3 e^{cpt} + c_4 e^{-cpt}).} \qquad \text{...(A)}$$

***Case II*:** If $k < 0$, let $k = -p^2$

then $$m^2 = -p^2 \Rightarrow m = \pm pi$$

$\therefore$ $$X = c_1 \cos px + c_2 \sin px$$

and $$m^2 = -p^2c^2 \Rightarrow m = \pm ipc$$

$\therefore$ $$T = c_3 \cos cpt + c_4 \sin cpt$$

then $$\boxed{y = (c_1 \cos px + c_2 \sin px)(c_3 \cos cpt + c_4 \sin cpt).} \quad ...(B)$$

***Case III*:** If $k = 0$

then $$D^2X = 0 \Rightarrow m = 0, 0$$

$\therefore$ $$X = (c_1 + c_2x)$$

And $$D^2 T = 0 \Rightarrow m = 0, 0$$

$\therefore$ $$T = (c_3 + c_4t)$$

Then $$\boxed{y = (c_1 + c_2x)(c_3 + c_4t).} \quad ...(C)$$

Of these three solutions, we have choose the solution which is consistent with the physical nature of the problem.

Since the physical nature of one dimension wave equation is periodic so we consider the solution which has periodic nature.

Here only the solution in equation (B) is periodic (as both sine and cosine are periodic).

Thus, the desired solution, for one dimensional wave equation is

$$\boxed{y(x, t) = (c_1 \cos px + c_2 \sin px)(c_3 \cos cpt + c_4 \sin cpt).} \quad ...(iv)$$

Now using the boundary conditions

At $x = 0$ (origin), $y = 0$ from (iv), we get

$$0 = c_1(c_3 \cos cpt + c_4 \sin cpt) \Rightarrow c_1 = 0$$

using the value of c_1 in (iv), we get

$$y(x, t) = c_2 \sin px(c_3 \cos cpt + c_4 \sin cpt) \quad ...(v)$$

At $x = l$ (at A), $y = 0$ from (v), we get

$$0 = c_2 \sin pl(c_3 \cos cpt + c_4 \sin cpt)$$

$$\Rightarrow \quad \sin pl = 0 = \sin n\pi \Rightarrow pl = n\pi$$

$$\Rightarrow \quad \boxed{p = \frac{n\pi}{l}}, \quad \text{where } n = 1, 2, 3, \ldots\ldots$$

Hence, the solution of wave equation satisfying the boundary conditions is, from (v)

$$y(x, t) = c_2 \sin\frac{n\pi x}{l}\left(c_3 \cos\frac{n\pi ct}{l} + c_4 \sin\frac{n\pi ct}{l}\right) \qquad \left| \; p = \frac{n\pi}{l} \right.$$

$$= \left(a_n \cos\frac{n\pi ct}{l} + b_n \sin\frac{n\pi ct}{l}\right)\sin\frac{n\pi x}{l} \qquad \left| \; \begin{array}{l} c_2c_3 = a_n \\ c_2c_4 = b_n \end{array} \right.$$

$\therefore$ The general solution of wave equation is

$$\boxed{y(x,t) = \sum_{n=1}^{\infty}\left(a_n \cos\frac{n\pi ct}{l} + b_n \sin\frac{n\pi ct}{l}\right)\sin\frac{n\pi x}{l}} \quad ...(vi)$$

Remark: We can apply initial conditions on above equation (*vi*) in time domain *i.e.*, at $t = 0$.

Example 5: A string of length l is fastened at both ends A and C. At a distance 'b' from the end A, the string is transversely displaced to a distance 'd' and is released from rest when it is in this position. Find the equation of the subsequent motion.

Solution: Let $y(x, t)$ is the displacement of the string.

Now, by the one dimensional wave equation, we have

$$\frac{\partial^2 y}{\partial t^2} = c^2 \frac{\partial^2 y}{\partial x^2} \quad ...(1)$$

Let $\qquad y = XT \qquad ...(2)$

where X is the function of x and T is the function of t.

From equation (2), we get

$$\frac{\partial y}{\partial x} = \frac{\partial X}{\partial x}T = \frac{dX}{dx}T \Rightarrow \frac{\partial^2 y}{\partial x^2} = \frac{d^2 X}{\partial x^2}T$$

and
$$\frac{\partial y}{\partial t} = X\frac{\partial T}{\partial t} = X\frac{dT}{dt} = \Rightarrow \frac{\partial^2 y}{\partial t^2} = X\frac{d^2 T}{\partial t^2}$$

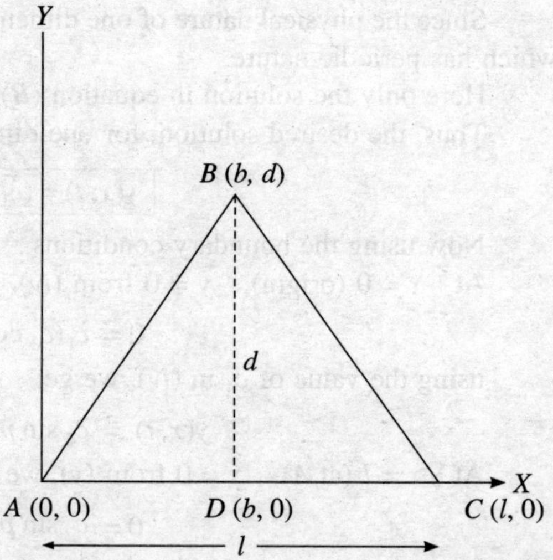

Using these valves in equation (1), we get

$$X\frac{d^2 T}{dt^2} = C^2 T\frac{d^2 X}{dx^2}$$

$$\Rightarrow \quad \frac{1}{C^2 T}\frac{d^2 T}{dt^2} = \frac{1}{X}\frac{d^2 X}{dx^2} = -p^2$$

$$\Rightarrow \quad \frac{d^2 X}{dx^2} + p^2 X = 0$$

$$\Rightarrow \quad (D^2 + p^2)X = 0$$

The A.E. is $\qquad m^2 + p^2 = 0$

$$\Rightarrow \quad m = \pm ip$$

$$\therefore \quad X = C_1 \cos px + C_2 \sin px \quad ...(3)$$

and
$$\frac{d^2 T}{dt^2} + p^2 C^2 T = 0$$

$\Rightarrow \qquad (D^2 + p^2C^2)T = 0$

The A.E. is $\qquad m^2 + p^2C^2 = 0$

$\Rightarrow \qquad m = \pm ipC$

$\therefore \qquad T = C_3 \cos cpt + C_4 \sin cpt \qquad \text{...(4)}$

Using equation (3) and (4) in (2), we get the solution of equation (1)

$$y(x,t) = (c_1 \cos px + c_2 \sin px)(c_3 \cos cpt + c_4 \sin cpt) \qquad \text{...(5)}$$

Now using the boundary conditions as follows:

The boundary conditions are

(*i*) At $x = 0$(at A), $y = 0 \Rightarrow y(0, t) = 0$

and (*ii*) At $x = l$ (at C), $y = 0 \Rightarrow y(l, t) = 0$

From (2), we have

$$0 = c_1(c_3 \cos cpt + c_4 \sin cpt) \quad \Rightarrow \quad c_1 = 0$$

Using $c_1 = 0$, in equation (2), we get

$$y(x,t) = c_2 \sin px(c_3 \cos cpt + c_4 \sin cpt) \qquad \text{...(6)}$$

Using (*ii*) boundary condition, from (6), we have

$$0 = c_2 \sin pl(c_3 \cos cpt + c_4 \sin cpt)$$

$$\Rightarrow \qquad \sin pl = 0 \Rightarrow \sin pl = \sin n\pi \Rightarrow \boxed{p = \frac{n\pi}{l}}.$$

Using the value of p in (6), we obtain

$$y(x, t) = c_2 \sin\frac{n\pi x}{l}\left(c_3 \cos\frac{n\pi ct}{l} + c_4 \sin\frac{n\pi ct}{l}\right) \qquad \text{...(7)}$$

Next, the initial conditions are as follows:

(*iii*) velocity $\dfrac{\partial y}{\partial t} = 0$ at $t = 0$

and displacement at $t = 0$ is

$$\text{(iv)}\ y(x, 0) = \begin{cases} \dfrac{d \cdot x}{b}, & 0 \le x \le b \\ \dfrac{d(x-l)}{(b-l)}, & b \le x \le l \end{cases}$$

Equation of AB is $y = \dfrac{d \cdot x}{b}$ and equation of BC is $y = \dfrac{d(x-l)}{(b-l)}$

From (4),

$$\frac{\partial y}{\partial t} = c_2 \sin\frac{n\pi x}{l}\left(-\frac{n\pi c c_3}{l}\sin\frac{n\pi ct}{l} + \frac{n\pi c c_4}{l}\cos\frac{n\pi ct}{l}\right)$$

Using (*iii*) in above equation, we get

$$0 = c_2 c_4 \frac{n\pi c}{l} \cdot \sin \frac{n\pi x}{l} \quad \Rightarrow \quad c_4 = 0$$

Using $c_4 = 0$ in equation (7), we get

$$y(x, t) = c_2 c_3 \sin \frac{n\pi x}{l} \cdot \cos \frac{n\pi ct}{l}$$

$\therefore$ The general solution of given problem is

$$y(x, t) = \sum_{n=1}^{\infty} b_n \sin \frac{n\pi x}{l} \cdot \cos \frac{n\pi ct}{l} \qquad ...(8)$$

Using initial condition (*iv*) in equation (8), we get

$$y(x, 0) = \sum_{n=1}^{\infty} b_n \sin \frac{n\pi x}{l}$$

which is half range Fourier sine series, so we have

$$b_n = \frac{2}{l} \int_0^l y(x, 0) \cdot \sin \frac{n\pi x}{l} \, dx$$

$$= \frac{2}{l} \int_0^b \frac{d}{b} x \cdot \sin\left(\frac{n\pi x}{l}\right) dx + \frac{2}{l} \frac{d}{(b-l)} \int_b^l (x - l) \sin \frac{n\pi x}{l} \, dx$$

$$= \frac{2d}{bl} \left[x \left(\frac{-l}{n\pi} \right) \cdot \cos \frac{n\pi x}{l} - \left(\frac{-l^2}{n^2 \pi^2} \right) \sin \frac{n\pi x}{l} \right]_0^b$$

$$+ \frac{2d}{l(b-l)} \left[(x - l) \left(\frac{-l}{n\pi} \right) \cos \frac{n\pi x}{l} - \left(\frac{l^2}{n^2 \pi^2} \right) \cdot \sin \frac{n\pi x}{l} \right]_b^l$$

$$\Rightarrow \quad b_n = -\frac{2d}{n\pi} \cos \frac{n\pi b}{l} + \frac{2dl^2}{b l n^2 \pi^2} \cdot \sin \frac{n\pi b}{l} + \frac{2d}{n\pi} \cdot \cos \frac{n\pi b}{l}$$

$$- \frac{2dl^2}{l(b-l) n^2 \pi^2} \sin \frac{n\pi b}{l}$$

$$\Rightarrow \quad b_n = \frac{2dl^2}{b(l-b) n^2 \pi^2} \sin \frac{n\pi b}{l}$$

$\therefore$ From (8), we get

$$y(x, t) = \frac{2dl^2}{b(l-b)\pi^2} \sum_{n=1}^{\infty} \frac{1}{n^2} \sin \frac{n\pi b}{l} \times \sin \frac{n\pi x}{l} \cdot \cos \frac{n\pi ct}{l}. \qquad \textbf{Ans.}$$

Example 6: A string is stretched and fastened to two points l apart. Motion is started by displacing the string in the form $y = a \sin \dfrac{\pi x}{l}$ from which it is released at a time $t = 0$. Show that the displacement of any point at a distance x from one end at time t is given by

$$y(x, t) = a \sin\left(\frac{\pi x}{l}\right)\cos\left(\frac{\pi ct}{l}\right). \qquad \textit{(U.P.T.U. 2004)}$$

Solution: Let $y(x, t)$ be the displacement at any point $P(x, y)$ at any time.
Then by the wave equation, we have

$$\frac{\partial^2 y}{\partial t^2} = c^2 \frac{\partial^2 y}{\partial x^2} \qquad \text{...(1)}$$

The solution of equation (1) is of the form

$$y(x, t) = (c_1 \cos px + c_2 \sin px)(c_3 \cos cpt + c_4 \sin cpt) \qquad \text{...(2)}$$

Now using the boundary conditions

(*i*) At $x = 0$, the displacement $y = 0 \Rightarrow y(0, t) = 0$

(*ii*) At $x = l$, the displacement $y = 0 \Rightarrow y(l, t) = 0$

Using (*i*) boundary condition in (2), we get

$$y(0, t) = 0 = c_1(c_3 \cos cpt + c_4 \sin cpt) \Rightarrow \boxed{c_1 = 0.}$$

$\therefore$ From (2), we get

$$y(x, t) = c_2 \sin px(c_3 \cos cpt + c_4 \sin cpt) \qquad \text{...(3)}$$

Using (*ii*) boundary condition in equation (3), we get

$$y(l, t) = 0 = c_2 \sin pl(c_3 \cos cpt + c_4 \sin cpt)$$

$$\Rightarrow \qquad \sin pl = 0 = \sin n\pi \Rightarrow \boxed{p = \frac{n\pi}{l}.}$$

Now using the initial conditions

(*iii*) At $t = 0$, the velocity $\dfrac{\partial y}{\partial t} = 0 \Rightarrow \left(\dfrac{\partial y}{\partial t}\right)_{t=0} = 0$

(*iv*) At $t = 0$, the displacement $y = a \sin \dfrac{\pi x}{l} \Rightarrow y(x, 0) = a \sin \dfrac{\pi x}{l}$

$\therefore$ From (3), we have

$$\frac{\partial y}{\partial t} = c_2 \sin px[c_3(-cp) \sin cpt + c_4 (cp) \cos cpt]$$

Using (*iii*) initial condition in above equation, we get

$$0 = c_2 c_4 cp \sin px \Rightarrow c_2 c_4 cp = 0$$

$$\Rightarrow \boxed{c_4 = 0.} \qquad \left|\ \begin{array}{l} c_2 \neq 0, \text{ otherwise there} \\ \text{is trivial solution} \end{array}\right.$$

Using $p = \frac{n\pi}{l}$ and $c_4 = 0$, in equation (3), we get

$$y(x, t) = c_2 c_3 \sin\frac{n\pi x}{l}\cos\frac{n\pi ct}{l}$$

$\therefore$ The general solution is

$$y(x, t) = \sum_{n=1}^{\infty} b_n \sin\frac{n\pi x}{l}\cos\frac{n\pi ct}{l} \quad (b_n = c_2 c_3) \qquad ...(4)$$

Finally using *(iv)* initial condition in equation (4), we get

$$y(x, 0) = a\sin\frac{\pi x}{l} = \sum_{n=1}^{\infty} b_n \sin\frac{n\pi x}{l}$$

or

$$a\sin\frac{\pi x}{l} = b_1 \sin\frac{\pi x}{l} + b_2 \sin\frac{2\pi x}{l} + \ldots\ldots$$

Equating the coefficient of $\sin\frac{\pi x}{l}$, we get

$$b_1 = a, \quad b_2 = b_3 = \cdots = 0$$

Hence the required solution of given problem is

$$y(x, t) = a\sin\frac{\pi x}{l}\cos\frac{\pi ct}{l}; \quad n = 1. \qquad \textbf{Proved.}$$

Example 7: Find the displacement of a string stretched between two fixed points at a distance $2l$ apart when the string is initially at rest in equilibrium position and points of the string are given initial velocity v where

$$v = \begin{cases} \frac{x}{l}, & \text{when } 0 < x < l \\ \frac{2l - x}{l}, & \text{when } l < x < 2l \end{cases}$$

x being the distance measured from one end.

Solution: The displacement $y(x, t)$ is given by wave equation

$$\frac{\partial^2 y}{\partial t^2} = c^2 \frac{\partial^2 y}{\partial x^2} \qquad ...(1)$$

The solution of equation is given by

$$y(x, t) = (c_1 \cos px + c_2 \sin px)(c_3 \cos cpt + c_4 \sin cpt) \qquad ...(2)$$

Now, the boundry conditions are

(i) At $x = 0$, $y = 0 \Rightarrow y(0, t) = 0$

(ii) At $x = 2l$, $y = 0 \Rightarrow y(2l, t) = 0$

Using (*i*) boundary condition in (2), we get

$$0 = c_1(c_3 \cos cpt + c_4 \sin cpt) \quad \Rightarrow \quad c_1 = 0$$

$\therefore$ From (2), we get

$$y(x, t) = c_2 \sin px(c_3 \cos cpt + c_4 \sin cpt) \quad \text{...(3)}$$

Using (*ii*) condition in (3), we get

$$0 = c_2 \sin 2pl(c_3 \cos cpt + c_4 \sin cpt)$$

$$\Rightarrow \qquad \sin 2pl = 0 = \sin n\pi \quad \Rightarrow \quad \boxed{p = \frac{n\pi}{2l}}.$$

Now, the initial conditions are

(*iii*) At $t = 0$ the displacement $y(x, 0) = 0$.

(*iv*) At $t = 0$, $\frac{\partial y}{\partial t} = v$.

Making use of initial condition (*iii*) in (3), we get

$$y(x, 0) = 0 = c_2 \sin px(c_3) \Rightarrow c_3 = 0$$

$\therefore$ From (3), we get

$$y(x, t) = c_2 c_4 \sin\frac{n\pi x}{2l} \sin\frac{n\pi c}{2l}t \quad \Big| \; p = \frac{n\pi}{2l}$$

The general solution is

$$y(x, t) = \sum_{n=1}^{\infty} b_n \sin\frac{n\pi x}{2l} \cdot \sin\frac{n\pi c}{2l}t \quad \text{...(4)}$$

$$\Rightarrow \qquad \frac{\partial y}{\partial t} = \frac{\pi c}{2l}\sum_{n=1}^{\infty} nb_n \sin\frac{n\pi x}{2l} \cos\frac{n\pi c}{2l}t$$

Using initial condition (*iv*) in above equation, we get

$$v = \frac{\pi c}{2l}\sum_{n=1}^{\infty} nb_n \sin\frac{n\pi x}{2l}$$

which represents half range Fourier sine series

$$\therefore \qquad \frac{\pi c}{2l} nb_n = \frac{2}{2l}\int_0^{2l} v \sin\frac{n\pi x}{2l}\,dx$$

$$= \frac{1}{l}\int_0^{l} \frac{x}{l}\sin\frac{n\pi x}{2l}\,dx + \frac{1}{l}\int_l^{2l}\left(\frac{2l - x}{l}\right)\sin\frac{n\pi x}{2l}\,dx$$

$$= \frac{1}{l}\left[\frac{x}{l}(-1)\left(\frac{2l}{n\pi}\right)\cos\frac{n\pi x}{2l} - \frac{1}{l}(-1)\frac{4l^2}{n^2\pi^2}\sin\frac{n\pi x}{2l}\right]_0^l$$

$$+\frac{1}{l}\left[\left(\frac{2l-x}{l}\right)\left(\frac{-2l}{n\pi}\right)\cdot\cos\frac{n\pi x}{2l}-\left(\frac{-1}{l}\right)\left(\frac{-4l^2}{n^2\pi^2}\right)\sin\frac{n\pi x}{2l}\right]_l^{2l}$$

$$= -\frac{2}{n\pi}\cos\frac{n\pi}{2}+\frac{4}{n^2\pi^2}\sin\frac{n\pi}{2}+\frac{2}{n\pi}\cos\frac{n\pi}{2}+\frac{4}{n^2\pi^2}\sin\frac{n\pi}{2}$$

$$\Rightarrow \qquad b_n = \frac{2l}{n\pi c}\left[\frac{8}{n^2\pi^2}\cdot\sin\frac{n\pi}{2}\right]$$

$$\Rightarrow \qquad b_n = \frac{16l}{n^3\pi^3 c}\cdot\sin\frac{n\pi}{2}$$

Hence the displacement function is given by, from equation (4), we get

$$y(x,t) = \frac{16l}{\pi^3 c}\sum_{n=1}^{\infty}\frac{1}{n^3}\sin\frac{n\pi}{2}\times\sin\frac{n\pi x}{2l}\cdot\sin\frac{n\pi ct}{2l}. \quad \textbf{Ans.}$$

Example 8: A string is stretched and fastened to two points l apart. Motion is started by displacing the string into the form $y = k(lx - x^2)$ from which it is released at time $t = 0$. Find the displacement of any point on the string at a distance of x from one end at time t. *(U.P.T.U. 2002)*

Solution: Let the displacement $y(x, t)$ given by the wave equation

$$\frac{\partial^2 y}{\partial t^2} = c^2\frac{\partial^2 y}{\partial x^2} \qquad ...(1)$$

$$\therefore \qquad y(x,t) = (c_1\cos px + c_2\sin px)(c_3\cos cpt + c_4\sin cpt) \qquad ...(2)$$

Using the boundary conditions (*i*) $y(0,t)=0$ (*ii*) $y(l,t)=0$

Using (*i*) in equation (2), we get

$$y = c_2\sin px(c_3\cos cpt + c_4\sin cpt) \qquad ...(3)$$

Using (*ii*), in equation (3), we get

$$0 = c_2\sin pl(c_3\cos cpt + c_4\sin cpt) \quad \Rightarrow \quad \sin pl = 0$$

$$\Rightarrow \qquad \sin pl = \sin n\pi \quad \Rightarrow \quad \boxed{p = \frac{n\pi}{l}}.$$

and the initial conditions are

(*iii*) $\left(\frac{\partial y}{\partial t}\right)_{t=0} = 0$ (*iv*) $y(x,0) = k(lx - x^2)$

$\therefore$ From (3), we get

$$\frac{\partial y}{\partial t} = c_2\sin px\{-c_3 c\sin cpt + c_4 c\cos cpt\}$$

Using (*iii*) in above relation, we get

$$0 = c_2(c_4 c)\sin px \quad \Rightarrow \quad \boxed{c_4 = 0.}$$

Using the values of c and p in equation (3), we get

$$y = c_2 c_3 \sin\frac{n\pi x}{l}\cos\frac{n\pi ct}{l}$$

$\therefore$ The general solution is

$$y = \sum_{n=1}^{\infty} b_n \sin\frac{n\pi x}{l}\cos\frac{n\pi ct}{l} \qquad \text{...(4)}$$

Making use of (*iv*) in (4), we get

$$k(lx - x^2) = \sum_{n=1}^{\infty} b_n \sin\frac{n\pi x}{l}$$

which represents half range Fourier sine series

$$\therefore \qquad b_n = \frac{2k}{l}\int_0^l (lx - x^2)\sin\frac{n\pi x}{l}\,dx$$

$$\Rightarrow \qquad b_n = \frac{2k}{l}\left[(lx - x^2)\left(-\cos\frac{n\pi x}{l}\right)\frac{l}{\pi x} - (l - 2x)\left(-\sin\frac{n\pi x}{l}\right)\frac{l^2}{n^2\pi^2} + (-2)\left(\cos\frac{n\pi x}{l}\right)\frac{l^3}{n^3\pi^3}\right]_0^l$$

$$\Rightarrow \qquad b_n = \frac{2k}{l}\left[(-1)^{n+1}\frac{2l^3}{n^3\pi^3} + \frac{2l^3}{n^3\pi^3}\right] = \frac{8kl^2}{n^3\pi^3} \text{ when } n \text{ is odd.}$$

$$= 0 \quad \text{when } n \text{ is even.}$$

Hence the required displacement is, from (4), we get

$$y = \sum_{n=1}^{\infty} \frac{8kl^2}{n^3\pi^3}\sin\frac{n\pi x}{l}\cos\frac{n\pi c}{l}t, \quad \text{when } n \text{ is odd.} \quad \textbf{Ans.}$$

Example 9: A tightly stretched string with fixed end points $x = 0$ and $x = l$ is initially at rest in its equilibrium position. If it is set vibrating by giving to each of its points on initial velocity $\lambda x(l - x)$, find the displacement of the string at any distance x from one end at any time t.

[*U.P.T.U. 2002 (C.O.) 2005*]

Solution: We know that the solution of one dimensional wave equation with boundary conditions

$$y(0, t) = y(l, t) = 0 \quad \text{is}$$

$$y(x,\ t) = c_2 \sin px(c_3 \cos cpt + c_4 \sin cpt) \qquad ...(1)$$

where $p = \dfrac{n\pi}{l}$.

Now the initial conditions are

$(a)\ \ y(x,0) = 0 \qquad (b)\ \ \left(\dfrac{\partial y}{\partial t}\right)_{t=0} = \lambda x(l-x)$

Making use of (a) in (1), we get

$$0 = c_3 c_2 \sin px \quad \Rightarrow \quad \boxed{c_3 = 0.}$$

$\therefore$ From (1), we get

$$y(x,t) = c_2 c_4 \sin\frac{n\pi x}{l}\sin\frac{n\pi ct}{l}$$

The general solution of wave equation is

$$y(x,t) = \sum_{n=1}^{\infty} b_n \sin\frac{n\pi x}{l}\cdot\sin\frac{n\pi ct}{l} \qquad ...(2)$$

From (2), $\dfrac{\partial y}{\partial t} = \displaystyle\sum_{n=1}^{\infty}(n\pi c/l)\cdot b_n \sin\frac{n\pi x}{l}\cos\frac{n\pi ct}{l}$

$$\Rightarrow \qquad \left(\frac{\partial y}{\partial t}\right)_{t=0} = \lambda x(l-x) = \sum_{n=1}^{\infty}\left(\frac{n\pi c}{l}\right)b_n \sin\frac{n\pi x}{l}$$

$$\therefore \qquad \frac{n\pi c}{l}b_n = \frac{2}{l}\int_0^l \lambda x(l-x)\sin\frac{n\pi x}{l}\,dx$$

$$= \frac{2\lambda}{l}\left[x(l-x)\left(-\frac{l}{n\pi}\cos\frac{n\pi x}{l}\right) - (l-2x)\left(-\frac{l^2}{n^2\pi^2}\sin\frac{n\pi x}{l}\right) + (-2)\left(\frac{l^3}{n^3\pi^3}\cos\frac{n\pi x}{l}\right)\right]_0^l$$

$$= \frac{4\lambda l^2}{n^3\pi^3}(1-\cos n\pi) = \frac{4\lambda l^2}{n^3\pi^3}\left[1-(-1)^n\right]$$

$$\Rightarrow \qquad b_n = \begin{cases} \dfrac{8\lambda l^3}{cn^4\pi^4}, & \text{when } n \text{ is odd} \\ 0, & \text{when } n \text{ is even} \end{cases}$$

∴ From (2) the required solution is

$$y(x, t) = \frac{8\lambda l^3}{c\pi^4}\sum_{n=1}^{\infty}\frac{1}{n^4}\sin\frac{n\pi x}{l}\sin\frac{n\pi ct}{l}, \quad n \text{ is odd.} \quad \textbf{Ans.}$$

Example 10: If a string of length l is initially at rest in equilibrium position and each of its points is given the velocity $\left(\frac{\partial y}{\partial t}\right)_{t=0} = b\sin^3\frac{\pi x}{l}$, find the displace $y(x, t)$. *(U.P.T.U. 2001, 2006)*

Solution: The equation for vibrations of the string is

$$\frac{\partial^2 y}{\partial t^2} = c^2\frac{\partial^2 y}{\partial x^2}$$

The solution of above equation is

$$y(x, t) = (c_1\cos px + c_2\sin px)(c_3\cos cpt + c_4\sin cpt) \quad \text{...(1)}$$

At $x = 0$, the displacement $y = 0$

$$\Rightarrow \quad 0 = c_1(c_3\cos cpt + c_4\sin cpt) \quad \Rightarrow \quad \boxed{c_1 = 0.}$$

From (1), we get $\quad y(x, t) = c_2\sin px(c_3\cos cpt + c_4\sin cpt) \quad$ ...(2)

Again at $x = l$, the displacement $y = 0$

∴ From (2), we get

$$0 = c_2\sin pl(c_3\cos cpt + c_4\sin cpt)$$

$$\Rightarrow \quad \sin pl = 0 = \sin n\pi \quad \Rightarrow \quad \boxed{p = \frac{n\pi}{l}.}$$

Next making the use of initial conditions *i.e.*, at $t = 0$, the displacement $y(x, 0) = 0$.

Again from (2), we get

$$0 = c_2 c_3\sin px \quad \Rightarrow \quad \boxed{c_3 = 0}$$

Using $c_3 = 0$ and $p = \frac{n\pi}{l}$, equation (2) takes the form

$$y(x, t) = c_2 c_4\sin\frac{n\pi x}{l}\sin\frac{n\pi ct}{l}$$

∴ The general solution is

$$y(x, t) = \sum_{n=1}^{\infty} b_n\cdot\sin\frac{n\pi x}{l}\sin\frac{n\pi ct}{l} \quad \text{...(3)}$$

$$\Rightarrow \quad \frac{\partial y}{\partial t} = \sum_{n=1}^{\infty} b_n\cdot\frac{n\pi c}{l}\sin\frac{n\pi x}{l}\cos\frac{n\pi ct}{l}$$

At $t = 0$, $\left(\dfrac{\partial y}{\partial t}\right)_{t=0} = b\sin^3 \dfrac{\pi x}{l}$, we get

$$b\sin^3 \frac{\pi x}{l} = \sum_{n=1}^{\infty} b_n \cdot \frac{n\pi c}{l} \sin\frac{n\pi x}{l}$$

$$= b_1 \frac{\pi c}{l}\sin\frac{\pi x}{l} + b_2 \cdot \frac{2\pi c}{l}\sin\frac{2\pi x}{l} + b_3 \frac{3\pi c}{l}\sin\frac{3\pi x}{l} + \cdots\cdots$$

$$b\sin^3 \frac{\pi x}{l} = (b_1 + 9b_3)\frac{\pi c}{l}\sin\frac{\pi x}{l} + b_2 \cdot \frac{2\pi c}{l}\sin\frac{2\pi x}{l} - \frac{12\pi c}{l} b_3 \sin^3\frac{\pi x}{l} + \cdots\cdots$$

Equating the coefficient of like powers of $\sin \dfrac{\pi x}{l}$, we get

$$b_1 + 9b_3 = 0 \qquad \ldots(i)$$

and $$\frac{-12\pi c}{l} b_3 = b \;\Rightarrow\; b_3 = -\frac{bl}{12\pi c}$$

from (i), $$b_1 = \frac{3bl}{4\pi c}, \quad b_2 = b_4 = b_5 = \cdots = 0$$

Hence from (3), the required displacement is

$$y(x, t) = \frac{3bl}{4\pi c}\sin\frac{\pi x}{l}\sin\frac{\pi ct}{l} + \left(-\frac{bl}{12\pi c}\right)\sin\frac{3\pi x}{l}\sin\frac{3\pi ct}{l}. \quad \textbf{Ans.}$$

Example 11: Solve the equation $\dfrac{\partial^2 u}{\partial x^2} + \dfrac{\partial^2 u}{\partial x \partial y} - 2\dfrac{\partial^2 u}{\partial y^2} = 0$ using the transformation $v = x + y$, $z = 2x - y$. [*U.P.T.U. (C.O.) 2005*]

Solution: We have

$$\frac{\partial^2 u}{\partial x^2} + \frac{\partial^2 u}{\partial x \partial y} - 2\frac{\partial^2 u}{\partial y^2} = 0 \qquad \ldots(1)$$

Given that

$$v = x + y \qquad \ldots(2)$$

and $$z = 2x - y \qquad \ldots(3)$$

Now $$\frac{\partial u}{\partial x} = \frac{\partial u}{\partial v}\cdot\frac{\partial v}{\partial x} + \frac{\partial u}{\partial z}\cdot\frac{\partial z}{\partial x} = \frac{\partial u}{\partial v} + 2\frac{\partial u}{\partial z} \qquad \text{using, (2) and (3)}$$

$$\Rightarrow \quad \frac{\partial}{\partial x} = \frac{\partial}{\partial v} + 2\frac{\partial}{\partial z}$$

$$\therefore \quad \frac{\partial^2 u}{\partial x^2} = \left(\frac{\partial}{\partial v} + 2\frac{\partial}{\partial z}\right)\left(\frac{\partial u}{\partial v} + 2\frac{\partial u}{\partial z}\right) = \frac{\partial^2 u}{\partial v^2} + 4\frac{\partial^2 u}{\partial z \partial v} + 4\frac{\partial^2 u}{\partial z^2}$$

and
$$\frac{\partial u}{\partial y} = \frac{\partial u}{\partial v}\cdot\frac{\partial v}{\partial y} + \frac{\partial u}{\partial z}\cdot\frac{\partial z}{\partial y} = \frac{\partial u}{\partial v} - \frac{\partial u}{\partial z}$$

$$\Rightarrow \quad \frac{\partial}{\partial y} = \frac{\partial}{\partial v} - \frac{\partial}{\partial z}$$

$$\therefore \quad \frac{\partial^2 u}{\partial y^2} = \left(\frac{\partial}{\partial v} - \frac{\partial}{\partial z}\right)\left(\frac{\partial u}{\partial v} - \frac{\partial u}{\partial z}\right) = \frac{\partial^2 u}{\partial v^2} - 2\frac{\partial^2 u}{\partial z \partial v} + \frac{\partial^2 u}{\partial z^2}$$

Also
$$\frac{\partial^2 u}{\partial x \partial y} = \left(\frac{\partial}{\partial v} + 2\frac{\partial}{\partial z}\right)\left(\frac{\partial u}{\partial v} - \frac{\partial u}{\partial z}\right)$$

$$= \frac{\partial^2 u}{\partial v^2} - \frac{\partial^2 u}{\partial v \partial z} + 2\frac{\partial^2 u}{\partial z \partial v} - 2\frac{\partial^2 u}{\partial z^2}$$

$$= \frac{\partial^2 u}{\partial v^2} + \frac{\partial^2 u}{\partial v \partial z} - 2\frac{\partial^2 u}{\partial z^2}$$

Putting the values of $\frac{\partial^2 u}{\partial x^2}, \frac{\partial^2 u}{\partial x \partial y}$ and $\frac{\partial^2 u}{\partial y^2}$ in equation (1), we get

$$\frac{\partial^2 u}{\partial v^2} + 4\frac{\partial^2 u}{\partial z \partial v} + 4\frac{\partial^2 u}{\partial z^2} + \frac{\partial^2 u}{\partial v^2} + \frac{\partial^2 u}{\partial v \partial z} - 2\frac{\partial^2 u}{\partial z^2} - 2\frac{\partial^2 u}{\partial v^2} + 4\frac{\partial^2 u}{\partial z \partial v} - 2\frac{\partial^2 u}{\partial z^2} = 0$$

$$\Rightarrow \quad 2\frac{\partial^2 u}{\partial v^2} - 2\frac{\partial^2 u}{\partial v^2} + 9\frac{\partial^2 u}{\partial z \partial v} + 4\frac{\partial^2 u}{\partial z^2} - 4\frac{\partial^2 u}{\partial z^2} = 0$$

$$\Rightarrow \quad 9\frac{\partial^2 u}{\partial z \partial v} = 0$$

$$\Rightarrow \quad \frac{\partial^2 u}{\partial z \partial v} = 0$$

Integrating the above equation w.r.t. 'z' partially, we get

$$\frac{\partial u}{\partial v} = f_1(v)$$

Again integrate w.r.t. 'v' partially, we get

$$u = \int f_1(v)\partial v + F_2(z)$$

$$= F_1(v) + F_2(z)$$

$$\Rightarrow \quad u = F_1(x+y) + F_2(2x-y). \qquad \textbf{Ans.}$$

Example 12: The points of trisection of a string are pulled aside through the same distance on opposite sides of the position of equilibrium and the string is released from rest. Derive an expression for the displacement of the string at subsequent time and show that the mid-point of the string always remains at rest. *(A.M.I.E.T.E., 1999)*

Solution: Let the string *OA* be trisected at *B* and *C*.

Let the equation of vibrating string is

$$\frac{\partial^2 y}{\partial t^2} = c^2 \frac{\partial^2 y}{\partial x^2} \qquad \text{...(1)}$$

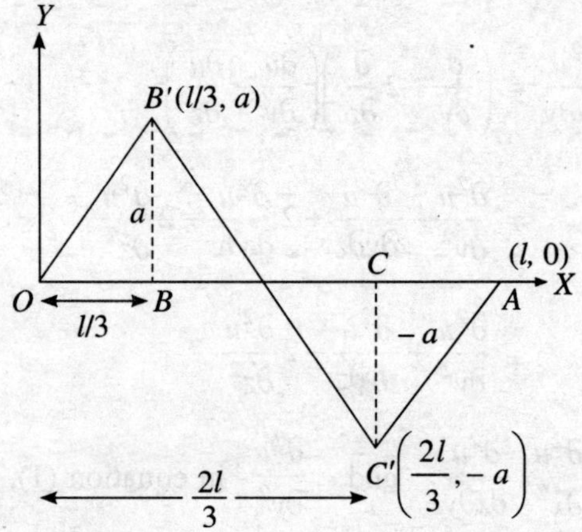

∴ The solution of equation (1) is

$$y(x, t) = (c_1 \cos px + c_2 \sin px)(c_3 \cos cpt + c_4 \sin cpt) \qquad \text{...(2)}$$

Now using the boundary conditions

(*i*) $y(0, t) = 0$ (*ii*) $y(l, t) = 0$

Making use of (*i*) and (*ii*) in equation (2), we get

$$y(x, t) = c_2 \sin px(c_3 \cos cpt + c_4 \sin cpt) \qquad \text{...(3)}$$

where $p = \dfrac{n\pi}{l}$.

Next equation of *OB′* is $y = \dfrac{ax}{l/3} \Rightarrow y = \dfrac{3a}{l}x$

Equation of *B′C′* is $y - a = \dfrac{a + a}{\dfrac{l}{3} - \dfrac{2l}{3}}\left(x - \dfrac{l}{3}\right)$ $\quad\left|\; y - y_1 = \dfrac{y_2 - y_1}{x_2 - x_1}(x - x_1)\right.$

$$\Rightarrow \qquad y = \frac{3a}{l}(l - 2x)$$

and equation of $C'A$ is $\quad y - 0 = \dfrac{-a-0}{\dfrac{2l}{3}-l}(x-l)$

$$\Rightarrow \qquad y = \frac{3a}{l}(x-l)$$

$\therefore$ The initial conditions of given problem are

(*iii*) $\left(\dfrac{\partial y}{\partial t}\right)_{t=0} = 0$

(*iv*) $y(x, 0) = \begin{cases} \dfrac{3a}{l}x, & 0 \le x \le l/3 \\ \dfrac{3a}{l}(l-2x), & \dfrac{l}{3} \le x \le \dfrac{2l}{3} \\ \dfrac{3a}{l}(x-l), & \dfrac{2l}{3} \le x \le l \end{cases}$

From equation (3), we get

$$\frac{\partial y}{\partial t} = c_2 \sin px(-c_3 cp \sin cpt + c_4 cp \cos cpt)$$

Using (*iii*) initial condition in above, we obtain

$$0 = c_2 \sin px(c_4 cp) \quad \Rightarrow \quad \boxed{c_4 = 0.}$$

Again from (3), we have

$$y(x, t) = c_2 c_3 \sin\frac{n\pi x}{l} \cdot \cos\frac{n\pi ct}{l}$$

$\therefore$ The general solution of equation (1) is

$$y(x, t) = \sum_{n=1}^{\infty} b_n \sin\frac{n\pi x}{l} \cos\frac{n\pi ct}{l} \qquad \text{...(4)}$$

Using (*iv*) condition in equation (4), we get

$$y(x, 0) = \sum_{n=1}^{\infty} b_n \sin\frac{n\pi x}{l}$$

$$\therefore \qquad b_n = \frac{2}{l}\int_0^l y(x, 0) \cdot \sin\frac{n\pi x}{l}\, dx$$

$$= \frac{2}{l}\left[\int_0^{\frac{l}{3}} \frac{3ax}{l} \sin\frac{n\pi x}{l}\, dx + \int_{\frac{l}{3}}^{\frac{2l}{3}} \frac{3a}{l}(l-2x)\sin\frac{n\pi x}{l}\, dx + \int_{\frac{2l}{3}}^{l} \frac{3a}{l}(x-l)\sin\frac{n\pi x}{l}\, dx\right]$$

$$= \frac{6a}{l^2}\left[x\left(-\frac{l}{n\pi}\cos\frac{n\pi x}{l}\right)-(1)\left(-\frac{l^2}{n^2\pi^2}\sin\frac{n\pi x}{l}\right)\right]_0^{\frac{l}{3}} + \frac{6a}{l^2}\left[(l-2x)\left(\frac{-l}{n\pi}\cos\frac{n\pi x}{l}\right)\right.$$

$$\left. -(-2)\left(\frac{-l^2}{n^2\pi^2}\sin\frac{n\pi x}{l}\right)\right]_{\frac{l}{3}}^{\frac{2l}{3}} + \frac{6a}{l^2}\left[(x-l)\left(\frac{-l^2}{n\pi}\cos\frac{n\pi x}{l}\right)-(1)\left(\frac{-l^2}{n^2\pi^2}\sin\frac{n\pi x}{l}\right)\right]_{\frac{2l}{3}}^{l}$$

$$= \frac{6a}{l^2}\left[\left(-\frac{l^2}{3n\pi}\cos\frac{n\pi}{3}+\frac{l^2}{n^2\pi^2}\sin n\pi\right)+\frac{l^2}{3n\pi}\cos\frac{2n\pi}{3}-\frac{2l^2}{n^2\pi^2}\sin\frac{2n\pi}{3}\right.$$

$$\left. +\frac{l^2}{3n\pi}\cos\frac{n\pi}{3}+\frac{2l^2}{n^2\pi^2}\sin\frac{n\pi}{3}-\left(\frac{l^2}{3n\pi}\cos\frac{2n\pi}{3}+\frac{l^2}{n^2\pi^2}\sin\frac{2n\pi}{3}\right)\right]$$

$$= \frac{6a}{l^2}\cdot\frac{3l^2}{n^2\pi^2}\left(\sin\frac{n\pi}{3}-\sin\frac{2n\pi}{3}\right)$$

$$= \frac{18a}{n^2\pi^2}\sin\frac{n\pi}{3}[1+(-1)^n] \qquad \sin\frac{2n\pi}{3} = \sin\left(n\pi-\frac{n\pi}{3}\right) = -(-1)^n\sin\frac{n\pi}{3}$$

$$\Rightarrow \qquad b_n = \begin{cases} 0, \text{ when } n \text{ is odd} \\ \dfrac{36a}{n^2\pi^2}\sin\dfrac{n\pi}{3}, \text{ when } n \text{ is even} \end{cases}$$

Putting the value of b_n in equation (4), we get

$$y(x, t) = \sum_{n=2\,(\text{even})}^{\infty} \frac{36a}{n^2\pi^2}\sin\frac{n\pi}{3}\sin\frac{n\pi x}{l}\cos\frac{n\pi ct}{l}. \quad \textbf{Ans.}$$

Example 13: A tightly stretched string with fixed end points $x = 0$ and $x = \pi$ is initially at rest in its equilibrium position. If it is set vibrating by giving to each of its points an initial velocity

$$\left(\frac{\partial y}{\partial t}\right)_{t=0} = 0.05\sin x - 0.06\sin 2x$$

then find the displacement $y\ (x, t)$ at any point of string at any time t.

Solution: We know that

$$\frac{\partial^2 y}{\partial t^2} = c^2\frac{\partial^2 y}{\partial x^2} \qquad \ldots(1)$$

and its solution is

$$y(x, t) = (c_1 \cos px + c_2 \sin px)(c_3 \cos cpt + c_4 \sin cpt) \quad ...(2)$$

The boundary conditions are

(a) $y(0, t) = 0$ (b) $y(\pi, t) = 0$

Using (a) in equation (2), we get $c_1 = 0$

$$\Rightarrow \quad y(x, t) = c_2 \sin px(c_3 \cos cpt + c_4 \sin cpt) \quad ...(3)$$

Using (b) in equation (3), we get

$$0 = c_2 \sin p\pi(c_3 \cos cpt + c_4 \sin cpt)$$

$$\Rightarrow \quad \sin pn = 0 = \sin n\pi \quad \Rightarrow \quad \boxed{p = n.}$$

Now, the initial conditions are

(c) $y(x, 0) = 0$ (d) $\left(\frac{\partial y}{\partial t}\right)_{t=0} = 0.05 \sin x - 0.06 \sin 2x$

At $t = 0$, from (3), we have

$$0 = c_2 c_3 \sin px \quad \Rightarrow \quad \boxed{c_3 = 0.}$$

$\therefore$ Again from (3), we obtain

$$y(x, t) = c_2 c_4 \sin nx \sin nct \quad |\text{As } p = n$$

The general solution is

$$y(x, t) = \sum_{n=1}^{\infty} b_n \sin nx \sin nct \quad ...(4)$$

$$\Rightarrow \quad \frac{\partial y}{\partial t} = \sum_{n=1}^{\infty} b_n \sin nx \cdot (nc) \cos nct$$

At $t = 0$

$$\left(\frac{\partial y}{\partial t}\right)_{t=0} = 0.05 \sin x - 0.06 \sin 2x = \sum_{n=1}^{\infty} nc \cdot b_n \sin nx$$

$$\Rightarrow \quad 0.05 \sin x - 0.06 \sin 2x = cb_1 \sin x + 2cb_2 \sin 2x + \cdots$$

$$\Rightarrow \quad cb_1 = 0.05 \quad \Rightarrow \quad b_1 = \frac{0.05}{c} \quad \text{and} \quad 2cb_2 = -0.06 \quad \Rightarrow \quad b_2 = -\frac{0.03}{c}$$

$\therefore$ From (4), we get

$$y(x, t) = \frac{0.05}{c} \sin x \sin ct - \frac{0.03}{c} \sin 2x \sin 2ct. \quad \textbf{Ans.}$$

Example 14: If the string of length l is initially at rest in equilibrium position and each of its points is given the velocity.

$v_0 \sin\left(\frac{3\pi x}{l}\right)\cos\left(\frac{2\pi x}{l}\right)$ where $0 < x < l$ at $t = 0$ determine the displacement function $y(x, t)$.

Solution: The displacement $y(x, t)$ given by wave equation

$$\frac{\partial^2 y}{\partial t^2} = c^2 \frac{\partial^2 y}{\partial x^2} \qquad \text{...(1)}$$

we know that the solution of (1) is

$$y(x, t) = (c_1 \cos px + c_2 \sin px)(c_3 \cos cpt + c_4 \sin cpt) \qquad \text{...(2)}$$

Using the boundary conditions

(*i*) $y(0, t) = 0$ (*ii*) $y(l, t) = 0$

We get from (2)

$$y(x, t) = c_2 \sin px(c_3 \cos cpt + c_4 \sin cpt) \qquad \text{...(3)}$$

where $p = \frac{n\pi}{l}$.

and the initial conditions are

(*iii*) $y(x, 0) = 0$ (*iv*) $\left(\frac{\partial y}{\partial t}\right)_{t=0} = v_0 \sin\left(\frac{3\pi x}{l}\right)\cos\left(\frac{2\pi x}{l}\right)$

Using (*iii*) in equation (3), we get

$$0 = c_2 c_3 \sin px \cos cpt \quad \Rightarrow \quad \boxed{c_3 = 0.}$$

Making use $c_3 = 0$ in equation (3), we get

$$y(x, t) = c_2 c_4 \sin px \sin cpt$$

where $p = \frac{n\pi}{l}$.

or the general form of solution is

$$y(x, t) = \sum_{n=1}^{\infty} b_n \sin\frac{n\pi x}{l} \sin\frac{n\pi ct}{l} \qquad \text{...(4)}$$

Differentiating partially w.r.t. 't' equation (4), we get

$$\frac{\partial y}{\partial t} = \sum_{n=1}^{\infty} b_n \left(\frac{n\pi c}{l}\right) \sin\frac{n\pi x}{l} \cdot \cos\frac{n\pi ct}{l}$$

$$\therefore \quad \left(\frac{\partial y}{\partial t}\right)_{t=0} = v_0 \sin\frac{3\pi x}{l} \cos\frac{2\pi x}{l} = \sum_{n=1}^{\infty} b_n \left(\frac{n\pi c}{l}\right) \sin\frac{n\pi x}{l}$$

$$\Rightarrow \quad \frac{v_0}{2}\left[\sin\frac{5\pi x}{l}+\sin\frac{\pi x}{l}\right] = \sum_{n=1}^{\infty} b_n\left(\frac{n\pi c}{l}\right)\sin\frac{n\pi x}{l}$$

Equating the coefficient of like terms, we have

$$\frac{v_0}{2} = b_1\left(\frac{\pi c}{l}\right) \quad \Rightarrow \quad b_1 = \frac{l v_0}{2c\pi}$$

$$\frac{v_0}{2} = b_5\left(\frac{5\pi c}{l}\right) \quad \Rightarrow \quad b_5 = \frac{l v_0}{5c\pi}$$

and $\qquad b_2 = b_3 = b_4 = b_5 = b_6 = \cdots = 0$

Using these values in equation (4), we get the required solution

$$y(x,t) = \left(\frac{l v_0}{2c\pi}\right)\sin\left(\frac{\pi x}{l}\right)\sin\frac{\pi ct}{l} + \left(\frac{l v_0}{5c\pi}\right)\sin\left(\frac{5\pi x}{l}\right)\sin\frac{5\pi ct}{l}. \qquad \textbf{Ans.}$$

5.4 D'ALEMBERT'S SOLUTION OF WAVE EQUATION

Transform the equation $\frac{\partial^2 y}{\partial t^2} = c^2\frac{\partial^2 y}{\partial x^2}$ to its normal form using the transformation $u = x+ct,\ v = x-ct$

and hence solve it. Show that the solution may be put in the form $y = \frac{1}{2}\left[f(x+ct)+f(x-ct)\right]$.

Assume initial conditions $y = f(x)$ and $\frac{\partial y}{\partial t} = 0$ at $t = 0$. *(U.P.T.U., 2003)*

Proof: Consider one dimensional wave equation

$$\frac{\partial^2 y}{\partial t^2} = c^2\frac{\partial^2 y}{\partial x^2} \qquad \ldots(1)$$

Let $u = x+ct$ and $v = x-ct$, be a transformation of x and t into u and v.

then $$\frac{\partial y}{\partial x} = \frac{\partial y}{\partial u}\frac{\partial u}{\partial x}+\frac{\partial y}{\partial v}\frac{\partial v}{\partial x} = \frac{\partial y}{\partial u}+\frac{\partial y}{\partial v} \qquad \left|\ \frac{\partial u}{\partial x} = 1, \frac{\partial v}{\partial x} = 1\right.$$

or $$\frac{\partial}{\partial x} = \frac{\partial}{\partial u}+\frac{\partial}{\partial v}$$

$$\Rightarrow \quad \frac{\partial^2 y}{\partial x^2} = \frac{\partial}{\partial x}\left(\frac{\partial y}{\partial x}\right) = \left(\frac{\partial}{\partial u}+\frac{\partial}{\partial v}\right)\left(\frac{\partial y}{\partial u}+\frac{\partial y}{\partial v}\right)$$

$$= \frac{\partial^2 y}{\partial u^2}+\frac{\partial^2 y}{\partial u\partial v}+\frac{\partial^2 y}{\partial v\partial u}+\frac{\partial^2 y}{\partial v^2}$$

$$\Rightarrow \qquad \frac{\partial^2 y}{\partial x^2} = \frac{\partial^2 y}{\partial u^2} + 2\frac{\partial^2 y}{\partial u \partial v} + \frac{\partial^2 y}{\partial v^2} \qquad \text{...(2)}$$

and
$$\frac{\partial y}{\partial t} = \frac{\partial y}{\partial u}\frac{\partial u}{\partial t} + \frac{\partial y}{\partial v}\frac{\partial v}{\partial t} = c\frac{\partial y}{\partial u} - c\frac{\partial y}{\partial v} \qquad \left| \frac{\partial u}{\partial t} = c, \frac{\partial v}{\partial t} = -c \right.$$

or
$$\frac{\partial}{\partial t} = c\left(\frac{\partial}{\partial u} - \frac{\partial}{\partial v}\right)$$

$$\therefore \qquad \frac{\partial^2 y}{\partial t^2} = \frac{\partial}{\partial t}\left(\frac{\partial y}{\partial t}\right) = c\left(\frac{\partial}{\partial u} - \frac{\partial}{\partial v}\right)\left(c\frac{\partial y}{\partial u} - c\frac{\partial y}{\partial v}\right)$$

$$= c^2\left(\frac{\partial^2 y}{\partial u^2} - \frac{\partial^2 y}{\partial u \partial v} - \frac{\partial^2 y}{\partial v \partial u} + \frac{\partial^2 y}{\partial v^2}\right)$$

$$\Rightarrow \qquad \frac{\partial^2 y}{\partial t^2} = c^2\left(\frac{\partial^2 y}{\partial u^2} - 2\frac{\partial^2 y}{\partial u \partial v} + \frac{\partial^2 y}{\partial v^2}\right) \qquad \text{...(3)}$$

Making use of equations (2) and (3) in equation (1), we get

$$c^2\left(\frac{\partial^2 y}{\partial u^2} - 2\frac{\partial^2 y}{\partial u \partial v} + \frac{\partial^2 y}{\partial v^2}\right) = c^2\left(\frac{\partial^2 y}{\partial u^2} + 2\frac{\partial^2 y}{\partial u \partial v} + \frac{\partial^2 y}{\partial v^2}\right)$$

$$\Rightarrow \qquad 4c^2\frac{\partial^2 y}{\partial u \partial v} = 0 \quad \Rightarrow \quad \frac{\partial^2 y}{\partial u \partial v} = 0 \qquad \text{...(4)}$$

Integrating equation (4), w.r.t. 'v', we get

$$\frac{\partial^2 y}{\partial u} = \phi(u) \qquad \text{...(5)}$$

where $\phi(u)$ is a constant in respect of v.

Again integrate equation (5) w.r.t. 'u', we get

$$y = \int \phi(u)\,du + \phi_2(v)$$

$$\Rightarrow \qquad y = \phi_1(u) + \phi_2(v)$$

$$\Rightarrow \qquad \boxed{y(x,t) = \phi_1(x+ct) + \phi_2(x-ct).} \qquad \text{...(6)}$$

The solution (6) is D'Alembert's solution of wave equation

Now, we applying initial conditions $y = f(x)$ and $\frac{\partial y}{\partial t} = 0$ at $t = 0$

From (6), we get at $t = 0$

$$f(x) = \phi_1(x) + \phi_2(x) \qquad \text{...(7)}$$

and $$\frac{\partial y}{\partial t} = c\phi_1'(x+ct) - c\phi_2'(x-ct)$$

$$\Rightarrow \quad \left(\frac{\partial y}{\partial t}\right)_{t=0} = 0 = c\phi_1'(x+0) - c\phi_2'(x-0)$$

$$\Rightarrow \quad \phi_1'(x) - \phi_2'(x) = 0$$

$$\Rightarrow \quad \phi_1'(x) = \phi_2'(x)$$

On integrating, we get

$$\phi_1(x) = \phi_2(x) + c_1 \qquad \text{...(8)}$$

Using equation (8) in equation (7), we get

$$f(x) = \phi_2(x) + c_1 + \phi_2(x) = 2\phi_2(x) + c_1$$

$$\Rightarrow \quad \phi_2(x) = \frac{1}{2}\left[f(x) - c_1\right] \quad \Rightarrow \quad \phi_2(x-ct) = \frac{1}{2}\left[f(x-ct) - c_1\right]$$

and $$\phi_1(x) = \frac{1}{2}\left[f(x) + c_1\right] \quad \Rightarrow \quad \phi_1(x+ct) = \frac{1}{2}\left[f(x+ct) + c_1\right]$$

Putting the values of $\phi_1(x+ct)$ and $\phi_2(x-ct)$ in equation (6), we get

$$y(x, t) = \frac{1}{2}\left[f(x+ct) + f(x-ct)\right]. \quad \textbf{Proved.}$$

EXERCISE 5.2

1. Solve completely the equation $\frac{\partial^2 y}{\partial t^2} = c^2 \frac{\partial^2 y}{\partial x^2}$, representing the vibrations of a string of length l, fixed at both ends, given that $y(0, t) = 0$, $y(l, t) = 0$; $y(x, 0) = f(x)$ and $\left(\frac{\partial y}{\partial t}\right)_{t=0} = 0$, $0 < x < l$.

   $\left[\textbf{Ans. } y(x, t) = \sum_{n=1}^{\infty} b_n \sin\frac{n\pi x}{l} \cos\frac{n\pi ct}{l};\ b_n = \frac{2}{l}\int_0^l f(x)\sin\frac{n\pi x}{l}\right]$ (*U.P.T.U. 2005*)

2. A taut string of length $2l$ is fastened at both ends. The midpoint of the string is taken to a height h and then released from rest in that position. Find the displacement of the string.

   $\left[\textbf{Ans. } y(x, t) = \frac{8h}{\pi^2}\sum_{n=1}^{\infty}\frac{(-1)^{n-1}}{(2n-1)^2}\sin\frac{(2n-1)\pi x}{2l}.\cos\frac{(2n-1)\pi ct}{2l}\right]$

3. A uniform elastic string of length 60 cm is subjected to a constant tension of 2 kg. If the ends are fixed and the initial displacement is $y(x, 0) = 60x - x^2$ for $0 < x < 60$ while the initial velocity is zero, find $y(x, t)$.

$$\left[\textbf{Ans. } y(x,t) = \frac{8\times 60^2}{\pi^3}\left(\sum_{n=1}^{\infty}\frac{1}{(2n-1)^3}\sin\frac{(2n-1)\pi x}{60}\times\cos\frac{(2n-1)\pi ct}{60}\right)\right]$$

4. A taut string of length 20 cms fastened at both ends is displaced from its position of equilibrium by imparting to each of its points an initial velocity is given by

$$v = \begin{cases} x & \text{in } 0 < x < 10 \\ 20 - x & \text{in } 10 < x < 20 \end{cases}$$

x being the distance from one end. Determine the displacement at any subsequent time.

$$\left[\textbf{Ans. } y(x,t) = \frac{1600}{\pi^3 c}\sum_{n=1}^{\infty}\frac{1}{n^3}\sin\left(\frac{n\pi}{2}\right)\sin\frac{n\pi x}{20}\cdot\sin\frac{n\pi ct}{20}\right]$$

5. A tightly stretched string with fixed end points $x = 0$ and $x = \pi$ is initially at rest in the equilibrium position. If it is set vibrating by giving each point a velocity $\left(\frac{\partial y}{\partial t}\right)_{t=0} = 0.03$ $\sin x - 0.04 \sin 3x$ then find the displacement $y(x, t)$ at any point of the string at any time t.

$$\left[\textbf{Ans. } y(x,t) = \frac{1}{c}[0.03\sin x\sin ct - 0.133\sin 3x\sin 3ct]\right]$$

6. A tightly stretched string with fixed end points $x = 0$ and $x = l$ is initially in a position given by $y = y_0 \sin^3\left(\frac{\pi x}{l}\right)$. If it is released from rest from this position, find the displacement $y(x, t)$.

$$\left[\textbf{Ans. } y(x,t) = \frac{y_0}{4}\left(3\sin\frac{\pi x}{l}\cos\frac{c\pi t}{l} - \sin\frac{3\pi x}{l}\cos\frac{3c\pi t}{l}\right)\right]$$

7. A tightly stretched flexible string has its ends fixed at $x = 0$ and $x = l$. At time $t = 0$ the string is given a shape defined by $y = \mu x(l - x)$, μ is a constant and then released. Find the displacement $y(x, t)$ of any point x of the string at any time $t > 0$.

$$\left[\textbf{Ans. } y(x,t) = \frac{4\mu l^2}{\pi^3}\sum_{n=1}^{\infty}\frac{1-(-1)^n}{n^3}\cdot\cos\frac{n\pi ct}{l}\cdot\sin\frac{n\pi x}{l}\right]$$

8. A uniform string of length l is struck in such a way that an initial velocity of v_0 is imparted to the portion of the string $\frac{l}{4}$ and $\frac{3l}{4}$ while the string is in equilibrium position. Find the subsequent displacement of the string as a function of x and t.

$$\left[\textbf{Ans. } y(x,t)=\frac{4lv_0}{\sqrt{2}\pi^2 c}\left(\frac{1}{1^2}\sin\frac{\pi x}{l}.\sin\frac{\pi ct}{l}-\frac{1}{3^2}\sin\frac{3\pi x}{l}\sin\frac{3\pi ct}{l}+\frac{1}{5^2}\sin\frac{5\pi x}{l}\sin\frac{5\pi ct}{l}-\ldots\right)\right]$$

9. Find the displacement if a string of length a is vibrating between fixed end points with initial velocity zero and initial displacement given by $y(x, 0) = \begin{cases} \dfrac{2px}{a} & \text{for } 0 < x < \dfrac{a}{2} \\ 2p - \dfrac{2px}{a} & \text{for } \dfrac{a}{2} < x < a. \end{cases}$

$$\left[\textbf{Ans. } y(x,t)=\frac{8p}{\pi^2}\sum_{n=1}^{\infty}\frac{(-1)^{n+1}}{(2n-1)^2}\sin\frac{(2n-1)\pi x}{a}\cdot\cos\frac{(2n-1)\pi ct}{a}\right]$$

10. A taut string of length l has its ends $x = 0$ and $x = l$ fixed. The mid-point is taken to a small height h and released from rest at time $t = 0$. Find the displacement $y(x, t)$.

$$\left[\textbf{Ans. } y(x,t)=\sum_{n=1}^{\infty}\frac{8h}{n^2\pi^2}\sin\frac{n\pi}{2}\cdot\sin\frac{n\pi x}{l}\cdot\cos\frac{n\pi ct}{l}\right]$$

11. If a string of length l is released from rest in the position $y = \dfrac{4kx(l-x)}{l^2}$ show that the motion is described by the equation.

$$\left[\textbf{Ans. } y(x,t)=\frac{32k}{\pi^2}\sum_{n=1}^{\infty}\frac{1}{(2n+1)^3}\sin\frac{(2n+1)\pi x}{l}\cos\frac{(2n+1)\pi ct}{l}\right]$$

5.5 ONE DIMENSIONAL FLOW

In this section we set up the mathematical model for one dimensional heat flow and derive the corresponding partial differential equation.

Consider a bar or a rod of equal thickness at every point.

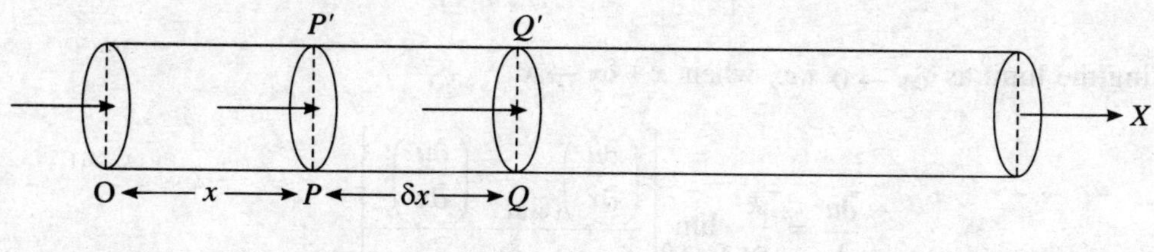

Let the area of cross-sectional = A cm^2.

and density of material of rod = ρ gr/cm^3

Here we consider a small element PQ of length δx.

$\therefore$ The mass of the element $PQ = A\rho\delta x$

Let $u(x, t)$ is the temperature of the rod at a distance x at time t.

We know that the amount of heat in a body is always proportional to the mass of the body and to the temperature change.

Thus the rate of increase of heat in element

$$= sA\rho\delta x \frac{\partial u}{\partial t} \quad (s \text{ is specific heat}) \qquad \ldots(i)$$

Since the direction of heat flow in a body becomes always toward decreasing temperature. Physical experiment shows that the rate of flow is proportional to the area and to the temperature gradient normal to the area. If we suppose Q_1 and Q_2 are the quantities of heat flowing at the points P and Q respectively,

then $$Q_1 = -kA\left(\frac{\partial u}{\partial x}\right)_x \text{ per second}$$

and $$Q_2 = -kA\left(\frac{\partial u}{\partial x}\right)_{x+\delta x} \text{ per second}$$

The negative sign shows the direction of heat flow towards lower temperature

where k is a constant known as thermal conductivity.

$\therefore$ Total amount of heat in the element $= Q_1 - Q_2 = kA\left[\left(\frac{\partial u}{\partial x}\right)_{x+\delta x} - \left(\frac{\partial u}{\partial x}\right)_x\right]$ per second $\qquad \ldots(ii)$

From (i) and (ii)

$$sA\rho\,\delta x \frac{\partial u}{\partial t} = kA\left[\left(\frac{\partial u}{\partial x}\right)_{x+\delta x} - \left(\frac{\partial u}{\partial x}\right)_x\right]$$

$$\Rightarrow \quad \frac{\partial u}{\partial t} = \frac{k}{\rho s}\left[\frac{\left(\frac{\partial u}{\partial x}\right)_{x+\delta x} - \left(\frac{\partial u}{\partial x}\right)_x}{\delta x}\right]$$

Taking the limit as $\delta x \to 0$ *i.e.*, when $x + \delta x \to x$

$$\therefore \quad \frac{\partial u}{\partial t} = \frac{k}{\rho s} \lim_{\delta x \to 0}\left[\frac{\left(\frac{\partial u}{\partial x}\right)_{x+\delta x} - \left(\frac{\partial u}{\partial x}\right)_x}{\delta x}\right]$$

$$= \frac{k}{\rho s}\frac{\partial}{\partial x}\left(\frac{\partial u}{\partial x}\right) = \frac{k}{\rho s}\frac{\partial^2 u}{\partial x^2}$$

Let $\dfrac{k}{\rho s} = c^2$ is called diffusivity of the substance

Thus
$$\boxed{\frac{\partial u}{\partial t} = c^2 \frac{\partial^2 u}{\partial x^2}.}$$

Alternate: Since the rate of flow is proportional to the gradient of the temperature

$$\therefore \qquad \vec{V} = -k\nabla u \qquad ...(i)$$

where $\vec{V}$ is the velocity of the heat flow and $u(x, y, z, t)$ is temperature and k is thermal conductivity.

Let R be the region in the body and let s be its boundary surface.

Then amount of heat leaving R per unit time $= \iint_S \vec{V}\cdot\hat{n}\,dA$

where $\hat{n}$ is the outward unit normal of s.

From (i) and by Gauss divergence theorem, we obtain

$$\iint_S \vec{V}\cdot\hat{n}dA = -k\iiint_R \nabla\cdot(\nabla u)\,dx\,dy\,dz$$

$$= -k\iiint_R \nabla^2 u\,dx\,dy\,dz \qquad ...(ii)$$

And the amount of heat in R is given by

$$Q = \iiint_R s\rho u\,dx\,dy\,dz$$

where s is the specific heat of the material of the body. Here the rate of decrease of Q is given by

$$-\frac{\partial Q}{\partial t} = -\iiint_R s\rho\frac{\partial u}{\partial t}\,dx\,dy\,dz \qquad ...(iii)$$

Since the rate of decrease of Q must be equal to the amount of heat leaving R, from (ii) and (iii), we get

$$-\iiint_R s\rho\frac{\partial u}{\partial t}\,dx\,dy\,dz = -k\iiint_R \nabla^2 u\,dx\,dy\,dz$$

$$\Rightarrow \qquad \iiint_R \left(s\rho\frac{\partial u}{\partial t} - k\nabla^2 u\right)dx\,dy\,dz = 0$$

Since this holds for any region R in the body the integrand must be zero everywhere.

$$\therefore \qquad s\rho\frac{\partial u}{\partial t} = k\nabla^2 u$$

$$\Rightarrow \qquad \frac{\partial u}{\partial t} = \frac{k}{s\rho}\nabla^2 u \qquad ...(iv)$$

Let $\frac{k}{s\rho} = c^2$, then equation (*iv*) reduces in the form

$$\frac{\partial u}{\partial t} = c^2 \nabla^2 u \Rightarrow \boxed{\frac{\partial u}{\partial t} = c^2 \frac{\partial^2 u}{\partial x^2}.}$$

$\nabla^2 u = \nabla \cdot \nabla u = \frac{\partial^2 u}{\partial x^2} + \frac{\partial^2 u}{\partial y^2} + \frac{\partial^2 u}{\partial z^2}$ in one dimension *i.e.*, along X. $\nabla^2 u = \frac{\partial^2 u}{\partial x^2}$

5.5.1 Solution of One Dimensional Heat Equation (*U.P.T.U. 2007*)

We know that the heat equation

$$\frac{\partial u}{\partial t} = c^2 \frac{\partial^2 u}{\partial x^2} \quad \text{...(1)}$$

Let $u(x, t) = X(x)\ T(t)$ ...(2)

$$\Rightarrow \quad \frac{\partial u}{\partial x} = T\frac{dX}{dx} \quad \text{or} \quad \frac{\partial^2 u}{\partial x^2} = T\frac{d^2 X}{dx^2}$$

and $$\frac{\partial u}{\partial t} = X\frac{dT}{dt}$$

Using these values in equation (1), we get

$$X\frac{dT}{dt} = c^2 T\frac{d^2 X}{dx^2}$$

$$\Rightarrow \quad \frac{1}{c^2 T}\frac{dT}{dt} = \frac{1}{X}\frac{d^2 X}{dx^2} = k \quad \text{...(3)}$$

Taking 2nd and 3rd terms

$$\therefore \quad \frac{1}{X}\frac{d^2 X}{dx^2} = k \Rightarrow \frac{d^2 X}{dx^2} - kX = 0$$

$$\Rightarrow \quad (D^2 - k)X = 0$$

Hence $$X = c_1 e^{\sqrt{k}x} + c_2 e^{-\sqrt{k}x}$$

and $$\frac{1}{c^2 T}\frac{dT}{dt} = k \Rightarrow \frac{dT}{T} = kc^2 dt$$

On integrating, $$\log_e T = kc^2 t + \log c_3$$

$$\Rightarrow \quad T = c_3 e^{kc^2 t}$$

$\therefore$ From (2), we get

$$\boxed{u(x, t) = (c_1 e^{\sqrt{k}x} + c_2 e^{-\sqrt{k}x}) c_3 e^{kc^2 t}}$$

There are arise following cases:

***Case I*: If $k > 0$, let $k = p^2$**

then $$\boxed{u(x, t) = (c_1 e^{px} + c_2 e^{-px}) c_3 e^{p^2 c^2 t}.} \quad ...(A)$$

***Case II*: If $k < 0$, Let $k = -p^2$**

then A.E. is $$m^2 = -p^2 \quad \Rightarrow \quad m = \pm\, pi$$

$\therefore$ $$X = c_1 \cos px + c_2 \sin px$$

then $$\boxed{u(x, t) = (c_1 \cos px + c_2 \sin px) c_3 e^{-p^2 c^2 t}.} \quad ...(B)$$

***Case III*: If $k = 0$, then**

$$\boxed{u(x, t) = (c_1 + c_2 x) c_3.} \quad ...(C)$$

Note 1: The solution (temperature) in case I tends to infinity when $t \rightarrow \infty$ it is not possible because the temperature always reduces with time.

Note 2: The solution (temperature) in case case III is independent of time t which is not possible without any external source of heat.

Result: Here the solution (temperature) is given by case II reduces with time which is the physical nature of heat flow.

Hence, the required solution of heat equation is

$$\boxed{u(x, t) = (c_1 \cos px + c_2 \sin px) c_3 e^{-p^2 c^2 t}.} \quad ...(4)$$

The boundary conditions are

(*i*) $u(0, t) = 0$ and (*ii*) $u(l, t) = 0$

and initial condition is (*iii*) $u(x, 0) = f(x)$.

Using (*i*) boundary condition in (4), we get

$$0 = c_1 c_3 e^{-p^2 c^2 t} \quad \Rightarrow \quad c_1 = 0$$

$\therefore$ From (4), we get

$$u(x, t) = c_2 c_3 \sin px \cdot e^{-p^2 c^2 t} \quad ...(5)$$

Using (*ii*) boundary condition in (5), we get

$$0 = c_2 c_3 \sin pl \quad \Rightarrow \quad \sin pl = 0 = \sin n\pi \quad \Rightarrow \quad \boxed{p = \frac{n\pi}{l}}$$

$\therefore$ $$u(x, t) = c_2 c_3 \sin \frac{n\pi x}{l} e^{-\frac{n^2 \pi^2 c^2 t}{l^2}}$$

The general form of above solution is

$$u(x, t) = \sum_{n=1}^{\infty} b_n \sin\frac{n\pi x}{l} \cdot e^{-n^2\pi^2c^2t/l} \qquad (b_n = c_2c_3) \qquad ...(6)$$

Again using initial condition (*iii*) in equation (6), we get

$$f(x) = \sum_{n=1}^{\infty} b_n \sin\frac{n\pi x}{l}$$

which represents Fourier half range sine series so, we have

$$b_n = \frac{2}{l}\int_0^l f(x)\sin\frac{n\pi x}{l}\,dx$$

Thus the required solution is

$$u(x, t) = \sum_{n=1}^{\infty} b_n \sin\frac{n\pi x}{l}\, e^{-\frac{n^2\pi^2c^2t}{l}}$$

$$\text{where } b_n = \frac{2}{l}\int_0^l f(x)\sin\frac{n\pi x}{l}\,dx.$$

Remark: In steady state $\frac{\partial u}{\partial t} = 0$, so $\boxed{\frac{\partial^2 u}{\partial x^2} = 0.}$

Example 15: Determine the solution of one dimension heat equation

$$\frac{\partial u}{\partial t} = c^2\frac{\partial^2 u}{\partial x^2}$$

Under the conditions $u(0, t) = u(l, t) = 0$ and $u(x,0) = \begin{cases} x & \text{if } 0 \le x \le l/2 \\ l - x & \text{if } \frac{l}{2} \le x \le l. \end{cases}$

Solution: We have

$$\frac{\partial u}{\partial t} = c^2\frac{\partial^2 u}{\partial x^2} \qquad ...(1)$$

Let the solution of equation (1) is

$$u = XT \qquad ...(2)$$

$$\therefore \qquad \frac{\partial u}{\partial x} = \frac{dX}{dx}T \quad \Rightarrow \frac{\partial^2 u}{\partial x^2} = \frac{d^2X}{dx^2}T$$

and $$\frac{\partial u}{\partial t} = X\frac{dT}{dt}$$

Using these values in equation (1), we get

$$X\frac{\partial T}{dt} = c^2 T\frac{d^2X}{dx^2}$$

$$\Rightarrow \quad \frac{1}{c^2T}\frac{dT}{dt} = \frac{1}{X}\frac{d^2X}{dx^2} = -p^2$$

$$\therefore \quad \frac{d^2X}{dx^2} + p^2X = 0$$

$$\Rightarrow \quad (D^2 + p^2)X = 0$$

The A.E. is $m^2 + p^2 = 0$

$$\Rightarrow \quad m = \pm ip$$

So $$X = c_1 \cos px + c_2 \sin px \qquad ...(3)$$

and $$\frac{1}{T}\frac{dT}{dt} = -p^2c^2$$

$$\Rightarrow \quad \frac{dT}{T} = -p^2c^2dt$$

integrating, we get

$$\log T = -p^2c^2t + \log c_3$$

$$\Rightarrow \quad T = c_3e^{-p^2c^2t} \qquad ...(4)$$

Putting the value of X and T from (3) and (4) in equation (2), we get

$$u(x, t) = (c_1 \cos px + c_2 \sin px)c_3e^{-p^2c^2t} \qquad ...(5)$$

At $x = 0$, we get

$$0 = c_1c_3e^{-p^2c^2t} \Rightarrow c_1 = 0.$$

$\therefore$ From equation (5), we get

$$u(x,t) = c_2c_3 \sin px \cdot e^{-p^2c^2t} \qquad ...(6)$$

Again at $x = l$ from (6), we get

$$0 = c_2c_3 \sin pl \cdot e^{-p^2c^2t} \Rightarrow \sin pl = 0 = \sin n\pi$$

$$\Rightarrow \quad p = \frac{n\pi}{l}$$

From (6), we get

$$u(x,\ t) = c_2 c_3 \sin\frac{n\pi x}{l} e^{-\frac{n^2\pi^2c^2}{l^2}t}$$

⇒ Therefore the general form of solution can be written as

$$u(x,\ t) = \sum_{n=1}^{\infty} b_n \sin\frac{n\pi x}{l} \cdot e^{-\frac{n^2\pi^2c^2t}{l^2}} \qquad ...(7)$$

At $t = 0$, from equation (7), we get

$$u(x,\ 0) = \sum_{n=1}^{\infty} b_n \sin\frac{n\pi x}{l}$$

$$\therefore \qquad b_n = \frac{2}{l}\int_0^l u(x,0)\sin\frac{n\pi x}{l}\,dx$$

$$= \frac{2}{l}\left[\int_0^{l/2} x\sin\frac{n\pi x}{l}\,dx + \int_{l/2}^{l}(l-x)\sin\frac{n\pi x}{l}\,dx\right]$$

$$= \frac{2}{l}\left\{\left[-x\frac{l}{n\pi}\cos\frac{n\pi x}{l} + \frac{l^2}{n^2\pi^2}\sin\frac{n\pi x}{l}\right]_0^{l/2}\right.$$

$$\left. + \left[-(l-x)\frac{l}{n\pi}\cos\frac{n\pi x}{l} - \frac{l^2}{n^2\pi^2}\sin\frac{n\pi x}{l}\right]_{l/2}^{l}\right\}$$

$$= \frac{2}{l}\left[-\frac{l^2}{2n\pi}\cos\frac{n\pi}{2} + \frac{l^2}{n^2\pi^2}\sin\frac{n\pi}{2} + \frac{l^2}{2n\pi}\cos\frac{n\pi}{2} + \frac{l^2}{n^2\pi^2}\sin\frac{n\pi}{2}\right]$$

$$= \frac{4l}{n^2\pi^2}\sin\left(\frac{n\pi}{2}\right).$$

∴ From equation (7), we get

$$u(x,\ t) = \frac{4l}{\pi^2}\sum_{n=1}^{\infty}\frac{1}{n^2}\sin\frac{n\pi}{2}\cdot\sin\frac{n\pi x}{l}\cdot e^{-\frac{n^2\pi^2c^2}{l^2}t}. \quad \textbf{Ans.}$$

Example 16: Find the temperature $u(x, t)$ in a homogeneous bar of heat conducting material of length L can with its end kept at zero temperature and initial temperature is given by $\dfrac{dx(L-x)}{L^2}$.

[*U.P.T.U. (Summer) 2009*]

Solution: From heat equation, we have

$$\frac{\partial u}{\partial t} = c^2 \frac{\partial^2 u}{\partial x^2} \quad ...(1)$$

We know that the solution of equation (*i*) is

$$u(x, t) = (c_1 \cos px + c_2 \sin px) c_3 e^{-p^2 c^2 t} \quad ...(2)$$

The boundary conditions are

(*a*) $u(0, t) = 0$

(*b*) $u(L, t) = 0$

and the initial condition is

$$u(x, 0) = \frac{dx(L-x)}{L^2}$$

Using condition (*a*) in (*ii*), we get

$$0 = c_1 c_3 e^{-c^2 p^2 t} \Rightarrow c_1 = 0$$

Putting $c_1 = 0$ in (*ii*), we get

$$u(x, t) = c_2 c_3 \sin px \cdot e^{-p^2 c^2 t} \quad ...(3)$$

Using condition (*b*) in (*iii*), we get

$$0 = c_2 c_3 \sin pL \cdot e^{-p^2 c^2 t}$$

$$\Rightarrow \quad \sin pL = 0 = \sin n\pi$$

$$\therefore \quad p = \frac{n\pi}{L}$$

from (*iii*), we get

$$u(x, t) = c_2 c_3 \cdot \sin \frac{n\pi x}{L} e^{-\frac{n^2 \pi^2 c^2 t}{L^2}}$$

The general solution is

$$u(x, t) = \sum_{n=1}^{\infty} b_n \cdot \sin \frac{n\pi x}{L} \cdot e^{-\frac{n^2 \pi^2 c^2 t}{L^2}} \quad ...(4)$$

Using initial condition in (*iv*), we get

$$\frac{dx(L-x)}{L} = \sum_{n=1}^{\infty} b_n \cdot \sin \frac{n\pi x}{L}$$

$$\therefore \quad b_n = \frac{2}{L}\int_0^L \frac{dx(L-x)}{L^2}\sin\frac{n\pi x}{L}\,dx$$

$$= \frac{2d}{L^3}\left[x(L-x)\left(-\frac{L}{n\pi}\right)\cos\frac{n\pi x}{L} - (L-2x)\left(-\frac{L^2}{n^2\pi^2}\right)\sin\frac{n\pi x}{L}\right.$$

$$\left. +(-2)\left(-\frac{L^3}{n^3\pi^3}\right)\left(-\cos\frac{n\pi x}{L}\right)\right]_0^L$$

$$= \frac{2d}{L^3}\left[0+0-\frac{2L^3}{n^3\pi^3}\{\cos n\pi - 1\}\right]$$

$$b_n = -\frac{2d}{L^3}\left[\frac{2L^3}{n^3\pi^3}\{(-1)^n - 1\}\right] = -\frac{4d}{n^3\pi^3}\left\{(-1)^n - 1\right\}$$

$$\Rightarrow \quad b_n = \begin{cases} \dfrac{8d}{n^3\pi^3} & \text{if } n \text{ is odd} \\ 0 & \text{if } n \text{ is even} \end{cases}$$

Hence from (*iv*)

$$u(x, t) = \frac{8d}{\pi^3}\sum_{n=1(odd)}^{\infty}\frac{1}{n^3}\sin\frac{n\pi x}{L}\cdot e^{-\frac{n^2\pi^2c^2t}{L^2}}. \quad \textbf{Ans.}$$

Example 17: An insulated rod of length l has its ends A and B maintained 0ºC and 100ºC respectively until steady state conditions prevail. If B is suddenly reduced to 0ºC and maintained at 0ºC find the temperature at a distance x from A at time t. (*U.P.T.U., 2004, 2005*)

Solution: From one dimensional heat equation, we have

$$\frac{\partial u}{\partial t} = c^2\frac{\partial^2 u}{\partial x^2} \quad \text{...(1)}$$

The boundary conditions are

(*i*) $u(0, t)$ = 0ºC and (*ii*) $u(l, t)$ = 100ºC

In steady state condition $\frac{\partial u}{\partial t} = 0$ here from (1), we get

$$\frac{\partial^2 u}{\partial x^2} = 0$$

On integrating, we get $u(x) = c_1x + c_2$ ...(2)

where c_1 and c_2 are constants to be determined

At $x = 0$, from equation (2), we have

$$\boxed{0 = c_2.}$$

and $$\text{at } x = l, \quad 100 = c_1 l + 0 \quad \Rightarrow \quad \boxed{c_1 = \frac{100}{l}.}$$

∴ From (2)

$$u(x) = \frac{100}{l}x \qquad \text{...(3)}$$

Now the temperature at B is suddenly changed we have again transient state. If $u(x, t)$ is the subsequent temperature function, the boundary conditions are

(*iii*) $u(0, t) = 0°C$, (*iv*) $u(l, t) = 0°C$ and the initial condition (*v*) $u(x, 0) = \frac{100}{l}x$

Since the subsequent steady state function $u_s(x)$ satisfies the equation

$$\frac{\partial^2 u_s}{\partial x^2} = 0$$

or $$\frac{d^2 u_s}{dx^2} = 0 \quad \Rightarrow \quad u_s(x) = c_3 x + c_4$$

at $x = 0$, we get $$0 = c_4$$

and at $x = l$, we get $$0 = c_3 l + 0 \quad \Rightarrow \quad c_3 = 0$$

Thus $$u_s(x) = 0 \qquad \text{...(4)}$$

If $u_T(x, t)$ is the temperature in transient state then the temperature distribution in the rod $u(x, t)$ can be expressed in the form

$$u(x, t) = u_s(x) + u_T(x, t)$$

$$\Rightarrow \quad u(x, t) = u_T(x, t) \quad \Big| \text{ As } u_s(x) = 0 \qquad \text{...(5)}$$

Again from heat equation, we have

$$\frac{\partial u_T}{\partial t} = c^2 \frac{\partial^2 u_T}{\partial x^2} \qquad \text{...(6)}$$

The solution of equation (6) is

$$u_T(x, t) = (c_1 \cos px + c_2 \sin px) c_3 e^{-c^2 p^2 t}$$

$$\Rightarrow \quad u(x, t) = (c_1 \cos px + c_2 \sin px) c_3 e^{-c^2 p^2 t} \qquad \text{...(7)}$$

At $$x = 0, \ u(0, t) = 0$$

$$\Rightarrow \quad 0 = c_3 c_1 e^{-c^2 p^2 t} \quad \Rightarrow \quad c_1 = 0.$$

From (7), we get

$$u(x, t) = c_2 \sin px \cdot c_3 e^{-c^2 p^2 t} \qquad \text{...(8)}$$

Again at $$x = l, \quad u(l, t) = 0$$

$$\Rightarrow \quad 0 = c_2 c_3 \sin pl \cdot e^{-c^2 p^2 t} \Rightarrow \sin pl = 0 = \sin n\pi$$

$$\Rightarrow \quad p = \frac{n\pi}{l}$$

From (8), we get

$$u(x, t) = c_2 c_3 \sin \frac{n\pi x}{l} \cdot e^{-\frac{n^2 c^2 \pi^2 t}{l^2}}$$

$$\Rightarrow \quad u(x, t) = \sum_{n=1}^{\infty} b_n \sin \frac{n\pi x}{l} \cdot e^{-\frac{n^2 \pi^2 c^2 t}{l^2}} \quad \text{...(9)}$$

Using initial condition *i.e.*, at $t = 0$, $u = \frac{100}{l} x$, we get

$$u(x, 0) = \frac{100}{l} x = \sum_{n=1}^{\infty} b_n \sin \frac{n\pi x}{l}$$

$$\therefore \quad b_n = \frac{2}{l} \int_0^l \frac{100x}{l} \cdot \sin \frac{n\pi x}{l} \, dx$$

$$= \frac{200}{l^2} \int_0^l x \cdot \sin \frac{n\pi x}{l} \, dx = \frac{200}{l^2} \left[-\frac{xl}{n\pi} \cos \frac{n\pi x}{l} + \frac{l^2}{n^2 \pi^2} \sin \frac{n\pi x}{l} \right]_0^l$$

$$\Rightarrow \quad b_n = \frac{200}{l^2} \left[\frac{-l^2}{n\pi} \cos n\pi \right] = \frac{200}{n\pi} (-1)^{n+1}$$

Hence from equation (9), we get

$$u(x, t) = \frac{200}{\pi} \sum_{n=1}^{\infty} \frac{(-1)^{n+1}}{n} \sin \frac{n\pi x}{l} e^{-\frac{n^2 \pi^2 c^2}{l^2} t} \cdot \textbf{ Ans.}$$

Example 18: Determine the solution of one dimensional heat equation $\frac{\partial u}{\partial t} = c^2 \frac{\partial^2 u}{\partial x^2}$ subject to the boundary conditions $u(0, t) = 0$, $u(l, t) = 0$ $(t > 0)$ and the initial condition $u(x, 0) = x$, l being the length of the bar. *(U.P.T.U. 2006)*

Solution: We have

$$\frac{\partial u}{\partial t} = c^2 \frac{\partial^2 u}{\partial x^2} \quad \text{...(1)}$$

We know that the solution of equation (1) is given by

$$u(x,\ t) = (c_1 \cos px + c_2 \sin px) c_3 e^{-p^2c^2t} \qquad ...(2)$$

At $\qquad x = 0,\ u = 0$

$\Rightarrow \qquad 0 = c_1 c_3 e^{-p^2c^2t} \Rightarrow c_1 = 0$

From (2), we get

$$u(x,\ t) = c_2 c_3 \sin px\, e^{-p^2c^2t} \qquad ...(3)$$

Again at $\qquad x = l, \quad u = 0$

$\Rightarrow \qquad 0 = c_2 c_3 \sin pl \cdot e^{-p^2c^2t}$

$\Rightarrow \qquad \sin pl = 0 = \sin n\pi$

$\Rightarrow \qquad p = \dfrac{n\pi}{l}.$

From (3) the general solution of equation (1) is

$$u(x,\ t) = \sum_{n=1}^{\infty} b_n \sin\frac{n\pi x}{l} \cdot e^{-\frac{n^2\pi^2c^2t}{l^2}} \qquad ...(4)$$

At $\qquad t = 0, \quad u = x$

$$\Rightarrow \qquad x = \sum_{n=1}^{\infty} b_n \sin\frac{n\pi x}{l}$$

$$\therefore \qquad b_n = \frac{2}{l}\int_0^l x \sin\frac{n\pi x}{l}\,dx = \frac{2}{l}\left[x \cdot \frac{l}{n\pi}\left(-\cos\frac{n\pi x}{l}\right) - \left(\frac{-l^2}{n^2\pi^2}\sin\frac{n\pi x}{l}\right)\right]_0^l$$

$$= \frac{2}{l}\left[\left(l \cdot \frac{l}{n\pi}(-\cos n\pi) + \frac{l^2}{n^2\pi^2}\sin n\pi\right) - 0\right]$$

$$\Rightarrow \qquad b_n = \frac{2}{l}\left[\frac{-l^2}{n\pi}(-1)^n\right] = (-1)^{n+1} \cdot \frac{2l}{n\pi}$$

Putting the value of b_n in equation (4), we get

$$u(x,\ t) = \frac{2l}{\pi}\sum_{n=1}^{\infty}\frac{(-1)^{n+1}}{n}\sin\frac{n\pi x}{l}\, e^{-\frac{n^2\pi^2c^2t}{l^2}}. \quad \textbf{Ans.}$$

Example 19: Find the temperature in a bar of length 2 whose end are kept at zero and lateral surface is insulated if the initial temperature is $\sin\dfrac{nx}{2} + 3\sin 5\pi\dfrac{x}{2}$. *(U.P.T.U. 2009)*

Solution: From heat equation

$$\frac{\partial u}{\partial t} = c^2\frac{\partial^2 u}{\partial x^2} \qquad ...(1)$$

Let $$u = XT \quad \text{...(2)}$$

from (*ii*), $$\frac{\partial u}{\partial x} = \frac{\partial X}{\partial x}T \Rightarrow \frac{\partial^2 u}{\partial x^2} = \frac{\partial^2 X}{\partial x^2}T = \frac{d^2 X}{dx^2}T$$

and $$\frac{\partial u}{\partial t} = X\frac{\partial T}{\partial t} = X\frac{dT}{dt}$$

using these values in equation (*i*), we get

$$X\frac{dT}{dt} = c^2 T\frac{d^2 X}{dx^2}$$

$$\Rightarrow \quad \frac{1}{c^2 T}\frac{dT}{dt} = \frac{1}{X}\frac{d^2 X}{dx^2} = -p^2$$

Now, $$\frac{d^2 X}{dx^2} + p^2 X = 0 \Rightarrow (D^2 + p^2)X = 0$$

The A.E. is $m^2 + p^2 = 0 \Rightarrow m = \pm pi$

$$\therefore \quad X = c_1 \cos px + c_2 \sin px$$

And $$\frac{dT}{dt} + c^2 p^2 T = 0 \Rightarrow \frac{dT}{T} = -c^2 p^2 dt$$

On integrating, we get

$$\log_e T = -c^2 p^2 t + \log_e c_3$$

$$\Rightarrow \quad T = c_3 e^{-c^2 p^2 t}$$

From (*ii*), the solution of equation (*i*) is

$$u(x, t) = (c_1 \cos px + c_2 \sin px) c_3 e^{-c^2 p^2 t} \quad \text{...(3)}$$

The boundary conditions are

(*i*) $u(0, t) = 0$ and (*ii*) $u(2, t) = 0$

The initial condition is

$$u(x, 0) = \sin\frac{nx}{2} + 3\sin 5\pi\frac{x}{2}$$

Using (*i*) boundary condition in (*iii*), we get

$$0 = c_1 c_3 e^{-c^2 p^2 t} \Rightarrow c_1 = 0$$

Using in (*iii*) $c_1 = 0$, we get

$$u(x, t) = c_2 c_3 \sin px \cdot e^{-c^2 p^2 t} \quad \text{...(4)}$$

Using (*ii*) boundary condition in (*iv*), we get

$$0 = c_2 c_3 \sin 2p \cdot e^{-c^2 p^2 t}$$

$$\Rightarrow \qquad \sin 2p = 0 = \sin n\pi$$

$$\Rightarrow \qquad p = \frac{n\pi}{2}$$

putting $p = \frac{n\pi}{2}$ in (*iv*), we get

$$u(x, t) = c_2 c_3 \sin\frac{n\pi x}{2} \cdot e^{-\frac{n^2\pi^2c^2}{4}t}$$

$\therefore$ The general solution is

$$u(x, t) = \sum_{n=1}^{\infty} b_n \sin\frac{n\pi x}{2} \cdot e^{-\frac{n^2\pi^2c^2}{4}t} \qquad ...(5)$$

Using the initial condition ($t = 0$) in (v), we get

$$\sin\frac{nx}{2} + 3\sin 5\pi\frac{x}{2} = \sum_{1}^{\infty} b_n \sin\frac{n\pi x}{2}$$

which is Fourier half range sine series

$$\therefore \qquad b_n = \int_0^2 \left(\sin\frac{nx}{2} + 3\sin 5\pi\frac{x}{2}\right)\sin\frac{n\pi x}{2}\,dx \qquad ...(6)$$

Hence, the equation (v) gives required temperature in wire with equation (vi).

Example 20: A homogeneous rod of conducting material of length l has its ends kept at zero temperature. The temperature at the centre is T and falls uniformly to zero at the two ends. Find the temperature distribution.

Solution: From heat equation in one dimension, we have

$$\frac{\partial u}{\partial t} = c^2\frac{\partial^2 u}{\partial x^2}. \qquad ...(1)$$

The boundary conditions are (*i*) $u(0, t) = 0$ (*ii*) $u(l, t) = 0$.

Since the temperature at the centre is T and falls uniformly to zero at the two ends, its distribution at $t = 0$ is given in the figure.

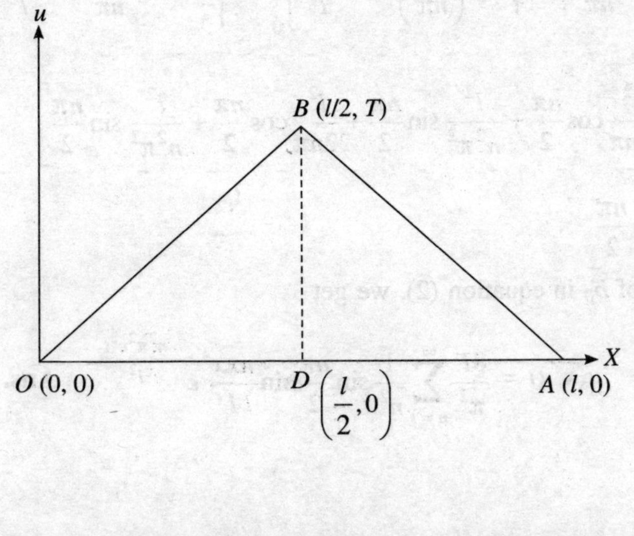

The equation of line segment OB is $u = \dfrac{2T}{l}x;\ 0 \le x \le \dfrac{l}{2}$ and the equation of line segment BA is given by

$$u = \frac{2T(l-x)}{l};\ \frac{l}{2} \le x \le l \qquad \left|\ \text{equation of line } y - y_1 = \frac{y_2 - y_1}{x_2 - x_1}(x - x_1)\right.$$

Thus initial condition *i.e.*, at $t = 0$ is given by

$$u(x,\ 0) = \begin{cases} \dfrac{2Tx}{l}, & \text{when } 0 \le x \le \dfrac{l}{2} \\ \dfrac{2T(l-x)}{l}, & \text{when } \dfrac{l}{2} \le x \le l \end{cases}$$

We know that the solution of equation (1) satisfying the boundary conditions (*i*) and (*ii*) is given by

$$u(x,\ t) = \sum_{n=1}^{\infty} b_n \sin\frac{n\pi x}{l}\, e^{-\frac{n^2\pi^2c^2t}{l^2}} \qquad \text{...(2)}$$

At $t = 0$, from (2), we get

$$u(x,\ 0) = \sum_{n=1}^{\infty} b_n \sin\frac{n\pi x}{l}$$

$$\therefore\quad b_n = \frac{2}{l}\int_0^l u(x, 0)\cdot\sin\frac{n\pi x}{l}\,dx$$

$$b_n = \frac{2}{l}\left[\int_0^{l/2} \frac{2Tx}{l}\sin\frac{n\pi x}{l}\,dx + \int_{l/2}^{l} \frac{2T(l-x)}{l}\cdot\sin\frac{n\pi x}{l}\,dx\right]$$

$$= \frac{4T}{l^2}\left[\left\{-x\frac{l^2}{n\pi}\cos\frac{n\pi x}{l} + \left(\frac{l}{n\pi}\right)^2\sin\frac{n\pi x}{l}\right\}_0^{l/2} + \left\{-(l-x)\frac{l}{n\pi}\cos\frac{n\pi x}{l} + \left(\frac{l}{n\pi}\right)^2\sin\frac{n\pi x}{l}\right\}_{l/2}^{l}\right]$$

$$= \frac{4T}{l^2}\left[-\frac{l^2}{2n\pi}\cos\frac{n\pi}{2} + \frac{l^2}{n^2\pi^2}\sin\frac{n\pi}{2} + \frac{l^2}{2n\pi}\cos\frac{n\pi}{2} + \frac{l^2}{n^2\pi^2}\sin\frac{n\pi}{2}\right]$$

$$\Rightarrow\quad b_n = \frac{8T}{n^2\pi^2}\sin\frac{n\pi}{2}$$

Putting the value of b_n in equation (2), we get

$$u(x,\ t) = \frac{8T}{\pi^2}\sum_{n=1}^{\infty}\frac{1}{n^2}\sin\frac{n\pi}{2}\sin\frac{n\pi x}{l}\, e^{-\frac{n^2\pi^2c^2t}{l^2}}. \quad \textbf{Ans.}$$

Example 21: A rod of length 10 cm has its ends *A* and *B* kept at 50ºC and 100ºC until steady state conditions prevail. The temperature at *A* suddenly raised to 90ºC and that at *B* is lowered to 60ºC and they are maintained. Find the temperature at *a* distance *x* from one end at time *t*.

Solution: From one dimensional heat equation

$$\frac{\partial u}{\partial t} = c^2 \frac{\partial^2 u}{\partial x^2} \qquad ...(1)$$

The boundary conditions are (*i*) $u(0, t) = 50°$ and (*ii*) $u(10, t) = 100°C$

In steady state condition $\frac{\partial u}{\partial t} = 0$, so from (1), we get

$$\frac{\partial^2 u}{\partial x^2} = 0$$

On integrating, we get $u(x) = C_1 x + C_2$ ...(2)

Making use of boundary conditions in above equation, we get

$$50 = C_2$$

and $$100 = 10C_1 + C_2 \Rightarrow C_1 = 5$$

∴ From equation (2)

$$u(x) = 5x + 50 \qquad ...(3)$$

Now, the temperatures at *A* and *B* are suddenly changed we have again transient state. If $u_1(x, t)$ is subsequent temperature function then the boundary conditions are

$$u_1(0, t) = 90°C, \quad u_1(10, t) = 60°C$$

and the initial condition

$$u_1(x, 0) = 5x + 50$$

If $u_s(x)$ is the subsequent steady state function then

$$\frac{\partial^2 u_s}{\partial x^2} = 0 \Rightarrow \frac{d^2 u_s}{dx^2} = 0$$

$$\Rightarrow \quad u_s(x) = C_3 x + C_4 \qquad ...(4)$$

At $x = 0$, we get

$$90 = C_4 \text{ and at } x = 10, \text{ we get}$$

$$60 = 10\,C_3 + 90 \Rightarrow C_3 = -3$$

$$\therefore \quad u_s(x) = -3x + 90$$

$u_1 \to u_s$
$\Rightarrow u_s(0) = 90$
$u_s(10) = 60$

Thus the total temperature distribution in the rod at a distance *x* at time *t* is given by

$$u(x, t) = u_s(x) + u_T(x, t)$$

where $u_T(x, t)$ is the temperature in transient state

$$\Rightarrow \quad u(x, t) = (-3x + 90) + u_T(x, t) \qquad ...(5)$$

Now again by heat equation, we have

$$\frac{\partial u_T}{\partial t} = c^2 \frac{\partial^2 u_T}{\partial x^2}$$

We know that the solution of above equation $u_T(x, t)$ is given by

$$u_T(x, t) = (C_1 \cos px + C_2 \sin px) C_3 e^{-c^2 p^2 t} \quad \text{...(6)}$$

Now the boundary conditions which satisfied by $u_T(x, t)$ are as follows:

$$u_T(0, t) = u_1(0, t) - u_s(0) = 90 - 90 = 0$$

$$u_T(l, t) = u_1(10, t) - u_s(10) = 60 - 60 = 0$$

and $$u_T(x, 0) = u_1(x, 0) - u_s(x)$$

$\Rightarrow$ $$u_T(x, 0) = 5x + 50 - (-3x + 90) = 8x - 40$$

The general solution for $u_T(x, t)$ is given by

$$u_T(x, t) = \sum_{n=1}^{\infty} b_n \sin \frac{n\pi x}{10} e^{-\frac{n^2 \pi^2 c^2 t}{100}} \quad \text{...(7)}$$

At $t = 0, \quad u_T = 8x - 40$

$\Rightarrow$ $$8x - 40 = \sum_{n=1}^{\infty} b_n \sin \frac{n\pi x}{10}$$

$\therefore$ $$b_n = \frac{2}{10} \int_0^{10} (8x - 40) \sin \frac{n\pi x}{10} dx$$

$$= \frac{1}{5} \left[-\frac{10}{n\pi} (8x - 40) \cos \frac{n\pi x}{10} + \frac{800}{n^2 \pi^2} \sin \frac{n\pi x}{10} \right]_0^{10}$$

$$= \frac{1}{5} \left[-\frac{400}{n\pi} \cos n\pi + \frac{800}{n^2 \pi^2} \sin n\pi - \frac{400}{n\pi} \right]$$

$\Rightarrow$ $$b_n = \frac{1}{5} \left[-\frac{400}{n\pi} (-1)^n - \frac{400}{n\pi} \right] = \frac{80}{n\pi} \left\{ (-1)^{n+1} - 1 \right\}$$

$\Rightarrow$ $$b_n = \begin{cases} 0, \text{ when } n \text{ is odd} \\ -\dfrac{160}{n\pi}, \text{ when } n \text{ is even} \end{cases}$$

From (7), we get

$$u_T(x, t) = \sum_{m=1}^{\infty} -\frac{160}{2m\pi} \sin \frac{2m\pi x}{10} e^{-\frac{4m^2 \pi^2 c^2 t}{100}} \quad | n = 2m \text{(even)}$$

or $$u_T(x, t) = -\frac{80}{\pi}\sum_{m=1}^{\infty}\frac{1}{m}\sin\frac{m\pi x}{5}\, e^{-\frac{m^2\pi^2c^2t}{25}} \quad ...(8)$$

Hence from equation (5), we get

$$u(x, t) = -3x + 90 - \frac{80}{\pi}\sum_{m=1}^{\infty}\frac{1}{m}\sin\frac{m\pi x}{5}\cdot e^{-\frac{m^2\pi^2c^2t}{25}}. \quad \textbf{Ans.}$$

Example 22: The temperature of a bar 50 cm long with insulated sides is kept at 0ºC at one end and 100ºC at the other end until steady conditions prevail. The two ends are then suddenly insulated so that the temperature gradient is zero at each end thereafter. Find the temperature distribution.

Solution: The temperature function $u(x, t)$ is the solution of the one dimensional heat equation

$$\frac{\partial u}{\partial t} = c^2\frac{\partial^2 u}{\partial x^2} \quad ...(1)$$

When the steady state condition prevails $\frac{\partial u}{\partial t} = 0$ and hence from (1), we get

$$\frac{\partial^2 u}{\partial x^2} = 0$$

On integrating, we get

$$u(x) = c_1x + c_2 \quad ...(2)$$

At $x = 0,\ u = 0$

$\therefore$ $0 = c_2$

and at $x = 50$, $u = 100$, from (2), we get

$$100 = 50\,c_1 + 0 \Rightarrow c_1 = 2$$

Hence $u(x) = 2x \Rightarrow u(x, 0) = 2x \quad ...(3)$

and the subsequent temperature function $u_1(x, t)$ satisfy the boundary conditions

$$u_1(0, t) = 0, \quad u_1(50, t) = 0$$

Under these conditions, we find the steady state function $u_s(x)$ vanishes

i.e., $u_s(x) = 0 \qquad \big| u_1 \to u_s$

$\Rightarrow$ $u(x, t) = u_s(x) + u_T(x, t) = 0 + u_T(x, t)$

$\Rightarrow$ $u(x, t) = u_T(x, t) \quad ...(4)$

where $u_T(x, t)$ is the temperature in transient state which satisfied the boundary conditions

$$u_T(0, t) = 0 = u_T(50, t)$$

$\therefore$ The temperature $u_T(x, t)$ can be obtained by the solution of one dimensional heat equation

$\Rightarrow$ $u_T(x, t) = (c_1\cos px + c_2\sin px)c_3e^{-c^2p^2t} \quad ...(5)$

At $x = 0,\ u_T = 0$

$$\Rightarrow \qquad 0 = c_1 c_3 e^{-c^2p^2t} \quad \Rightarrow \quad c_1 = 0 \text{ (otherwise } u_T(x, t) = 0)$$

From (5), we get

$$u_T(x, t) = c_2 c_3 \sin px \cdot e^{-c^2p^2t} \qquad \text{...(6)}$$

And at $x = 50, \quad u_T = 0$

$$\Rightarrow \qquad 0 = c_2 c_3 \sin 50p \cdot e^{-c^2p^2t}$$

$$\Rightarrow \qquad \sin 50p = 0 = \sin n\pi \quad \Rightarrow \quad p = \frac{n\pi}{50}$$

$\therefore$ From (6), we get

$$u_T(x, t) = c_2 c_3 \sin \frac{n\pi x}{50} e^{-\frac{n^2\pi^2c^2t}{2500}}$$

The general form of solution is

$$u_T(x, t) = \sum_{n=1}^{\infty} b_n \sin \frac{n\pi x}{50} e^{-\frac{n^2\pi^2c^2t}{2500}} \qquad \text{...(7)}$$

At $t = 0, u_T = 2x$ (from equation 3)

$$\Rightarrow \qquad 2x = \sum_{n=1}^{\infty} b_n \sin \frac{n\pi x}{50}$$

$$\therefore \qquad b_n = \frac{2}{50}\int_0^{50} 2x \sin \frac{n\pi x}{50} dx = \frac{2}{25}\left[-\frac{50}{n\pi} x \cos \frac{n\pi x}{50} + \frac{2500}{n^2\pi^2} \sin \frac{n\pi x}{50}\right]_0^{50}$$

$$b_n = \frac{2}{25}\left[-\frac{2500}{n\pi}\cos n\pi + \frac{2500}{n^2\pi^2}\sin n\pi + 0\right] = \frac{200}{n\pi}(-1)^{n+1}$$

Putting the value of b_n in equation (7), we get

$$u_T(x, t) = \sum_{n=1}^{\infty} \frac{200}{n\pi}(-1)^{n+1} \cdot \sin \frac{n\pi x}{50} \cdot e^{-\frac{n^2\pi^2c^2t}{2500}} \qquad \text{...(8)}$$

Hence from (4) and (8), we get

$$u(x, t) = \frac{200}{\pi} \sum_{n=1}^{\infty} \frac{(-1)^{n+1}}{n} \sin \frac{n\pi x}{50} \cdot e^{-\frac{n^2\pi^2c^2t}{2500}}. \quad \textbf{Ans.}$$

Example 23: A rod of length l with insulated sides is initially at a uniform temperature u. Its ends are suddenly cooled 0ºC and are kept that temperature. Prove that the temperature function $u(x, t)$ is given by

$$u(x, t) = \sum_{n=1}^{\infty} b_n \sin \frac{n\pi x}{l} \cdot e^{-\frac{n^2\pi^2c^2t}{l^2}}$$

where $$b_n = \frac{2}{l}\int_0^l u_0(x)\sin\frac{n\pi x}{l}\,dx.$$

Solution: Since the temperature distribution $u(x, t)$ is the solution of heat equation

$$\therefore \quad \frac{\partial u}{\partial t} = c^2\frac{\partial^2 u}{\partial x^2} \qquad ...(1)$$

The solution of equation (1) is as follows

$$u(x, t) = (c_1\cos px + c_2\sin px)\,c_3 e^{-p^2c^2t} \qquad ...(2)$$

The boundary conditions are

(*i*) $u(0, t) = 0$ (*ii*) $u(l, t) = 0$

Making use of these boundary conditions the equation (2) takes the form

$$u(x, t) = c_2c_3\sin\frac{n\pi x}{l}\,e^{-\frac{n^2\pi^2c^2t}{l^2}}$$

The general form of the solution is

$$u(x, t) = \sum_{n=1}^{\infty} b_n\sin\frac{n\pi x}{l}\cdot e^{-\frac{n^2\pi^2c^2t}{l^2}} \qquad ...(3)$$

and initial condition *i.e.*, at $t = 0$, $u = u_0(x)$ (say)

$$u_0(x) = \sum_{n=1}^{\infty} b_n\cdot\sin\frac{n\pi x}{l}$$

which is half range Fourier sine series

$$b_n = \frac{2}{l}\int_0^l u_0(x)\cdot\sin\frac{n\pi x}{l}\,dx$$

Hence

$$u(x, t) = \sum_{n=1}^{\infty} b_n\sin\frac{n\pi x}{l}\cdot e^{-\frac{n^2\pi^2c^2t}{l^2}}$$

where $$b_n = \frac{2}{l}\int_0^l u_0(x)\cdot\sin\frac{n\pi x}{l}\,dx.$$ **Proved.**

Example 24: Two ends A and B of a rod 20 cm long have the temperature at 30ºC and 80ºC respectively until steady state prevails. The temperature at the end are changed to 40ºC and 60ºC respectively find the temperature distribution in the rod.

Solution: The heat equation in one dimensional is

$$\frac{\partial u}{\partial t} = c^2\frac{\partial^2 u}{\partial x^2} \qquad ...(1)$$

The boundary conditions are

(*i*) $u(0, t) = 30°C$ (*ii*) $u(20, t) = 80°C$

In steady condition $\frac{\partial u}{\partial t} = 0$

$\therefore$ From (1), we get $\frac{\partial^2 u}{\partial x^2} = 0$, on integrating, we get

$$u(x) = c_1 x + c_2 \quad \text{...(2)}$$

at $x = 0$, $u = 30$, so $30 = 0 + c_2 \Rightarrow c_2 = 30$

and at $x = 20$ $u = 80$ so $80 = c_1 \times 20 + 30 \Rightarrow c_1 = \frac{5}{2}$

From equation (2), we get

$$u(x) = \frac{5x}{2} + 30 \quad \text{...(3)}$$

Now the temperatures at *A* and *B* are suddenly changed we have again gain transient state. If $u_1(x, t)$ is subsequent temperature function then the boundary conditions are

$$u_1(0, t) = 40°C \quad \text{and} \quad u_1(20, t) = 60°C$$

and the initial condition *i.e.*, at $t = 0$, is given by equation (3)

Since the subsequent steady state function $u_s(x)$ satisfies the equation

$$\frac{\partial^2 u_s}{\partial x^2} = 0 \quad \text{or} \quad \frac{d^2 u_s}{dx^2} = 0$$

The solution of above equation is

$$u_s(x) = c_3 x + c_4 \quad \text{...(4)}$$

At $x = 0$, $u_s = 40 \Rightarrow 40 = 0 + c_4 \Rightarrow c_4 = 40$

and at $x = 20$, $u_s = 60 \Rightarrow 60 = 20\,c_3 + 40 \Rightarrow c_3 = 1$

$\left| \begin{array}{l} u_s(0) = 40°C \\ u_s(20) = 60°C \end{array} \right.$

$\therefore$ From (4), we get

$$u_s(x) = x + 40 \quad \text{...(5)}$$

Thus the temperature distribution in the rod at time t is given by

$$u(x, t) = u_s(x) + u_T(x, t)$$

$$\Rightarrow \quad u(x, t) = (x + 40) + u_T(x, t) \quad \text{...(6)}$$

where $u_T(x, t)$ is the transient state function which satisfying the conditions

$$\left. \begin{array}{l} u_T(0, t) = u_1(0, t) - u_s(0) = 40 - 40 = 0 \\ u_T(20, t) = u_1(20, t) - u_s(20) = 60 - 60 = 0 \\ \text{and} \quad u_T(x, 0) = u_1(x, 0) - u_s(x) = \frac{5x}{2} + 30 - x - 40 = \frac{3x}{2} - 10 \end{array} \right\}$$

The general solution for $u_T(x, t)$ is given by

$$u_T(x, t) = \sum_{n=1}^{\infty} b_n \sin\frac{n\pi x}{20}\, e^{-\frac{n^2\pi^2c^2t}{400}} \qquad ...(7)$$

At $t = 0$, from (7), we get

$$\frac{3x}{2} - 10 = \sum_{n=1}^{\infty} b_n \sin\frac{n\pi x}{20}$$

$$\therefore \qquad b_n = \frac{2}{20}\int_0^{20}\left(\frac{3x}{2} - 10\right)\sin\frac{n\pi x}{20}\,dx$$

$$= \frac{1}{10}\left[\left(\frac{3x}{2} - 10\right)\left(-\frac{20}{n\pi}\cos\frac{n\pi x}{20}\right) - \frac{3}{2}\left(-\frac{400}{n^2\pi^2}\sin\frac{n\pi x}{20}\right)\right]_0^{20}$$

$$= \frac{1}{10}\left[-20\left(\frac{20}{n\pi}\right)(-1)^n - (-10)\left(\frac{20}{n\pi}\right)\right] = -\frac{20}{n\pi}\left[2(-1)^n + 1\right]$$

Putting the value of b_n in equation (7), we get

$$u_T(x, t) = -\frac{20}{\pi}\sum_{n=1}^{\infty}\frac{2(-1)^n + 1}{n}\sin\frac{n\pi x}{20}\cdot e^{-\frac{n^2\pi^2c^2t}{400}} \qquad ...(8)$$

From (6) and (8), we get

$$u(x, t) = (x + 40) - \frac{20}{\pi}\sum_{n=1}^{\infty}\frac{2(-1)^n + 1}{n}\cdot\sin\frac{n\pi x}{20}\cdot e^{-\frac{n^2\pi^2c^2t}{400}}. \quad \textbf{Ans.}$$

EXERCISE 5.3

1. Solve the boundary value problem $\frac{\partial u}{\partial t} = c^2\frac{\partial^2 u}{\partial x^2}$ with the conditions $u(l, t) = 0$ for all $t \geq 0$

$\frac{\partial u}{\partial t}(0, t) = 0$ and $u(x, 0) = 20x$ for $0 < x < l$.

$$\left[\textbf{Ans. } u(x,t) = \frac{40l}{\pi}\sum_{n=1}^{\infty}\frac{(-1)^{n+1}}{n}\sin\left(\frac{n\pi x}{l}\right)e^{-\frac{n^2\pi^2c^2t}{l^2}}\right]$$

2. Solve $\frac{\partial u}{\partial t} = c^2\frac{\partial^2 u}{\partial x^2}$ for $0 < x < \pi$, $t > 0$.

$u_x(0, t) = u_x(\pi, t) = 0$ and $u(x, 0) = \sin x$. $\left[\textbf{Ans. } u(x,t) = \frac{2}{\pi} - \frac{4}{\pi}\sum_{n=1}^{\infty}\frac{1}{4n^2 - 1}\cos 2nx\cdot e^{-4n^2c^2t}\right]$

3. Solve $\frac{\partial u}{\partial t} = c^2 \frac{\partial^2 u}{\partial x^2}$ with the conditions $\frac{\partial u}{\partial t}(0, t) = 0, u(0, t) = 0$, $u(l, t) = 0$ and $u(x, 0) = lx - x^2$, $0 \le x \le l$.

$$\left[\textbf{Ans. } u(x, t) = \frac{8l^2}{\pi^3} \sum_{n=1}^{\infty} \frac{1}{(2n-1)^3} \sin \frac{(2n-1)\pi x}{l} \cdot e^{-\frac{c^2(2n-1)^2\pi^2 t}{l^2}}\right]$$

4. Solve $\frac{\partial u}{\partial t} = k \frac{\partial^2 u}{\partial x^2}$ under the conditions (*i*) $u \neq \infty$ if $t \to \infty$

(*ii*) $\frac{\partial u}{\partial x} = 0$ for $x = 0$ and $x = l$ (*iii*) $u = lx - x^2$ for $t = 0$; $0 \le x \le l$.

$$\left[\textbf{Ans. } u(x, t) = \frac{l^2}{6} - \frac{l^2}{\pi^2} \sum_{m=1}^{\infty} \frac{1}{m^2} \cos \frac{2m\pi x}{l} e^{-\frac{4m^2\pi^2 kt}{l^2}}\right]$$

5. The temperature distribution in a bar of length π, which is prefectly insulated at the ends $x = 0$ and $x = \pi$ is governed by the partial differential equation $\frac{\partial u}{\partial t} = \frac{\partial^2 u}{\partial x^2}$. Assuming the initial temperature as $u(x, 0) = \cos 2x$, find the temperature distribution at any instant of time. $\left[\textbf{Ans. } u(x, t) = e^{-4t} \cdot \cos 2x\right]$

6. The ends A and B of a rod 30 cms long have their temperatures kept at 20ºC and 80ºC respectively until steady state conditions prevail. The temperature at the end B is then suddenly reduced to 60ºC and at the end A is raised to 40ºC and maintained so. Find $u(x, t)$ at any time t.

$$\left[\textbf{Ans. } u(x, t) = \frac{2x}{3} + 40 - \frac{40}{\pi} \sum_{n=1}^{\infty} \frac{1}{n} \{(-1)^n + 1\} \sin \frac{n\pi x}{30} \cdot e^{-\frac{n^2 c^2 \pi^2 t}{900}}\right]$$

7. A bar 40 cm long has originally a temperature of 0ºC along all its length. At $t = 0$, the temperature at the end $x = 0$ is raised to 50ºC while that at other end is raised to 100ºC. Determine the resulting temperature function.

$$\left[\textbf{Ans. } u(x,t) = \frac{5x}{4} + 50 - \frac{2}{\pi} \sum_{n=1}^{\infty} \frac{50 + 100(-1)^{n+1}}{n} \cdot \sin \frac{n\pi x}{40} \cdot e^{-\frac{c^2 n^2 \pi^2 t}{1600}}\right]$$

8. An insulated rod of length l has its ends A and B maintained at 0ºC and 100ºC respectively until steady state conditions prevails. If B is suddenly reduced to 0ºC and maintained 0ºC, find the temperature at a distance x from A at time t. *(U.P.T.U. 2004)*

$$\left[\textbf{Ans. } u(x, t) = \frac{200}{\pi} \sum_{n=1}^{\infty} \frac{(-1)^{n+1}}{n} \sin \frac{n\pi x}{l} \cdot e^{-\frac{n^2\pi^2 c^2 t}{l^2}}\right]$$

9. Solve $\frac{\partial u}{\partial t} = c^2 \frac{\partial^2 u}{\partial x^2}$ under the conditions (*i*) $\frac{\partial u}{\partial x}(0, t) = 0$, (*ii*) $\frac{\partial u}{\partial x}(5, t) = 0$ and

(*iii*) $u(x, 0) = x$. $\left[\textbf{Ans. } u(x, t) = \frac{5}{2} - \frac{20}{\pi^2} \sum_{n=1}^{\infty} \frac{1}{(2n-1)^2} \cos \frac{(2n-1)\pi x}{5} \cdot e^{-(2n-1)^2 \pi^2 t/25} \right]$

10. A rod of length l has its ends A and B kept 0ºC and 120ºC until steady state conditions prevail. If the temperature at B is reduced to 0ºC and kept so while that of A is maintained find $u(x, t)$. $\left[\textbf{Ans. } u(x, t) = \frac{240}{\pi} \sum_{n=1}^{\infty} \frac{(-1)^{n+1}}{n} \sin \frac{n\pi x}{l} e^{-\frac{n^2 \pi^2 c^2}{l^2}} \right]$

11. Solve $\frac{\partial u}{\partial t} = c^2 \frac{\partial^2 u}{\partial x^2}$ under the boundary conditions
(*i*) u is finite when $t \to \infty$ (*ii*) $u = 0$ when $x = 0$ and $x = \pi$ for all values of t, and (*iii*) $u = x$ from $x = 0$ to $x = \pi$ when $t = 0$. $\left[\textbf{Ans. } u(x, t) = 2 \sum_{n=1}^{\infty} \frac{(-1)^{n+1}}{n} \sin nx \cdot e^{-n^2 c^2 t} \right]$

5.6 TWO DIMENSIONAL HEAT EQUATION

We have

$$\frac{\partial u}{\partial t} = c^2 (\nabla^2 u) \quad \text{| In Section (5.5) (Alternate method)} \quad \ldots(i)$$

where $c^2 = \frac{k}{s\rho}$, k is the thermal conductivity of the body, s is the specific heat of the material of the body and ρ is the density.

In case of two dimensional, we may suppose that z-coordinate is constant.

$$\therefore \quad \nabla^2 u = \frac{\partial^2 u}{\partial x^2} + \frac{\partial^2 u}{\partial y^2} + 0 \quad \left| \begin{array}{l} \text{As } u(z) = \text{constant} \\ \therefore \ \frac{\partial^2 u}{\partial z^2} = 0 \end{array} \right.$$

$$\Rightarrow \quad \nabla^2 u = \frac{\partial^2 u}{\partial x^2} + \frac{\partial^2 u}{\partial y^2} \quad \ldots(ii)$$

From (*i*) and (*ii*), we get

$$\boxed{\frac{\partial u}{\partial t} = c^2 \left(\frac{\partial^2 u}{\partial x^2} + \frac{\partial^2 u}{\partial y^2} \right)}. \quad \ldots(A)$$

In steady state u always independent of t so that $\frac{\partial u}{\partial t} = 0$.

Hence from equation (*A*), we get

$$\boxed{\frac{\partial^2 u}{\partial x^2}+\frac{\partial^2 u}{\partial y^2}=0.} \qquad \ldots(B)$$

The equation (*B*) is known as Laplace's equation.

5.6.1 Solution of Two Dimensional Heat Equation

We have

$$\frac{\partial u}{\partial t}=c^2\left(\frac{\partial^2 u}{\partial x^2}+\frac{\partial^2 u}{\partial y^2}\right) \qquad \ldots(i)$$

Let $u(x, y, t) = XYT$ ...(*ii*)

Putting the value of $u(x, y, t)$ from (*ii*) in equation (*i*), we get

$$XY\frac{dT}{dt}=Tc^2\left(Y\frac{d^2X}{dx^2}+X\frac{d^2Y}{dy^2}\right)$$

$$\Rightarrow \quad \frac{1}{C^2T}\frac{dT}{dt}=\frac{1}{X}\frac{d^2X}{dx^2}+\frac{1}{Y}\frac{d^2Y}{dy^2}$$

There are three possibilities

$$(a)\quad \frac{1}{X}\frac{d^2X}{dx^2}=0,\ \frac{1}{Y}\frac{d^2y}{dy^2}=0,\ \frac{1}{c^2T}\frac{dT}{dt}=0.$$

$$(b)\quad \frac{1}{X}\frac{d^2X}{dx^2}=p_1^2,\ \frac{1}{Y}\frac{d^2Y}{dy^2}=p_2^2,\ \frac{1}{c^2T}\frac{dT}{dt}=p^2.$$

$$(c)\quad \frac{1}{X}\frac{d^2X}{dx^2}=-p_1^2,\ \frac{1}{Y}\frac{d^2Y}{dy^2}=-p_2^2,\ \frac{1}{c^2T}\frac{dT}{dt}=-p^2.$$

where $p^2=p_1^2+p_2^2$.

Out of these three possibilities, we have to select that solution which suits the physical nature of the problem and the given boundary conditions.

Here $u(x, y, t) = (c_1x+c_2)(c_3y+c_4)c_5$ (For *a*)

$u(x, y, t) = (c_1e^{p_1x}+c_2e^{-p_1x})(c_3e^{p_2y}+c_4e^{-p_2y})c_5e^{p^2c^2t}$ (For *b*)

and $u(x, y, t) = (c_1\cos p_1x+c_2\sin p_1x)(c_3\cos p_2y+c_4\sin p_2y)c_5e^{-p^2c^2t}$ (For *c*)

5.6.2 Solution of Laplace Equation

The Laplace equation is

$$\frac{\partial^2 u}{\partial x^2}+\frac{\partial^2 u}{\partial y^2}=0$$

Let $u(x,y) = X(x)Y(y)$. ...(i)

$$\therefore \quad \frac{\partial^2 u}{\partial x^2} = \frac{d^2 X}{dx^2} Y \text{ and } \frac{\partial^2 u}{\partial y^2} = X \frac{d^2 y}{dy^2}$$

Putting these values in given equation, we get

$$Y \frac{d^2 X}{dx^2} + X \frac{d^2 Y}{dy^2} = 0$$

$$\Rightarrow \quad Y \frac{d^2 X}{dx^2} = -X \frac{d^2 Y}{dy^2} \Rightarrow \frac{1}{X} \frac{d^2 X}{dx^2} = -\frac{1}{Y} \frac{d^2 Y}{dy^2} = -p^2 \text{ (say)} \quad ...(ii)$$

Taking first two terms, we get

$$\frac{d^2 X}{dx^2} + p^2 X = 0$$

$$\Rightarrow \quad X = c_1 \cos px + c_2 \sin px$$

And taking last two terms, we get

$$\frac{d^2 Y}{dy^2} - p^2 y = 0$$

$$\Rightarrow \quad Y = c_3 e^{py} + c_4 e^{-py}$$

From (i)

$$u(x,y) = (c_1 \cos px + c_2 \sin px)(c_3 e^{py} + c_4 e^{-py}) \quad ...(iii)$$

which is the solution of Laplace equation.

Example 25: A thin rectangular plate whose surface is impervious to heat flow has $t = 0$ an arbitrary distribution of temperature $f(x, y)$. Its four edges $x = 0$, $x = a$, $y = 0$ and $y = b$ are kept at zero temperature. Determine the temperature at a point of the plate as t increases. *(U.P.T.U. 2002)*

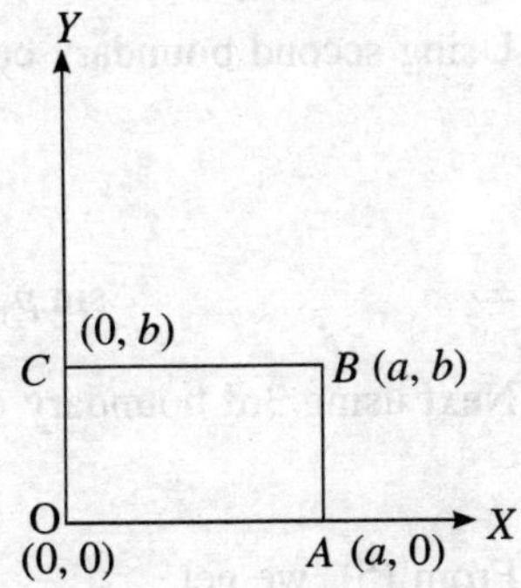

Solution: The heat equation in two dimensional is

$$\frac{\partial u}{\partial t} = c^2 \left(\frac{\partial^2 u}{\partial x^2} + \frac{\partial^2 u}{\partial y^2} \right) \quad ...(1)$$

Let $\quad u = XYT \quad ...(2)$

From (1) $\quad XY \frac{dT}{dt} = Tc^2 \left(Y \frac{d^2 X}{dx^2} + X \frac{d^2 Y}{dy^2} \right)$

$$\Rightarrow \quad \frac{1}{c^2 T} \frac{dT}{dt} = \frac{1}{X} \frac{d^2 X}{dx^2} + \frac{1}{Y} \frac{d^2 Y}{dy^2} \quad ...(3)$$

Since the physical nature of problem is periodic so we choosen the constant as follows:

From (3), we get

$$\frac{1}{X} \frac{d^2 X}{dx^2} = -p_1^2 \Rightarrow \frac{d^2 X}{dx^2} + p_1^2 X = 0 \Rightarrow (D^2 + p_1^2) X = 0$$

$$\therefore \text{ The A.E. is } \quad m^2 + p_1^2 = 0 \Rightarrow m^2 = i^2 p_1^2 \Rightarrow m = \pm i p_1$$

$$\Rightarrow \qquad X = c_1 \cos p_1 x + c_2 \sin p_1 x$$

Again from (3), we get

$$\frac{1}{Y}\frac{d^2Y}{dy^2} = -p_2^2 \Rightarrow (D^2 + p_2^2)X = 0$$

so
$$Y = (c_3 \cos p_2 y + c_4 \sin p_2 y)$$

and
$$\frac{1}{c^2 T}\frac{dT}{dt} = -p^2, \quad \text{where} \quad p^2 = p_1^2 + p_2^2.$$

$$\Rightarrow \qquad \frac{dT}{T} = -p^2c^2 dt \Rightarrow \log_e T = -p^2c^2 t + \log_e c_5$$

$$\Rightarrow \qquad T = c_5 e^{-p^2c^2t}$$

Hence from equation (2), we get

$$u = (c_1 \cos p_1 x + c_2 \sin p_1 x)(c_3 \cos p_2 y + c_4 \sin p_2 y) c_5 e^{-p^2c^2t} \qquad ...(4)$$

Now the boundary conditions are:

(*i*) $u(0, y, t) = 0$, (*ii*) $u(a, y, t) = 0$, (*iii*) $u(x, 0, t) = 0$; and (*iv*) $u(x, b, t) = 0$

Using first boundary condition in (4), we get

$$0 = c_1(c_3 \cos p_2 y + c_4 \sin p_2 y) c_5 e^{-p^2c^2t}$$

$$\Rightarrow \qquad c_1 = 0 \qquad (\text{otherwise } u(x, y, t) = 0)$$

From (4), we get

$$u = c_2 c_5 \sin p_1 x (c_3 \cos p_2 y + c_4 \sin p_2 y) e^{-p^2c^2t} \qquad ...(5)$$

Using second boundary condition in (5), we get

$$0 = c_2 c_5 \sin p_1 a (c_3 \cos p_2 y + c_4 \sin p_2 y) e^{-p^2c^2t}$$

$$\Rightarrow \qquad \sin p_1 a = 0 = \sin m\pi \Rightarrow \boxed{p_1 = \frac{m\pi}{a}}$$

Next using 3rd boundary condition in (5), we get

$$0 = c_3 c_2 c_5 \sin p_1 x \cdot e^{-p^2c^2t} \Rightarrow c_3 = 0 \quad (\text{otherwise } u(x, y, t) = 0)$$

From (5), we get

$$u = c_2 c_4 c_5 \sin p_1 x \cdot \sin p_2 y \cdot e^{-p^2c^2t} \qquad ...(6)$$

And using 4th boundary condition in (6), we get

$$0 = c_2 c_4 c_5 \sin p_1 x \cdot \sin p_2 b \cdot e^{-p^2c^2t}$$

$$\Rightarrow \qquad \sin p_2 b = 0 = \sin n\pi \Rightarrow \boxed{p_2 = \frac{n\pi}{b}.}$$

Since $$p^2 = p_1^2 + p_2^2$$

so $$p^2 = \pi^2\left(\frac{m^2}{a^2}+\frac{n^2}{b^2}\right)$$

Putting the values of p_1, p_2 and p in equation (6), we get

$$u(x, y, t) = c_2c_4c_5 \sin\frac{m\pi x}{a}\cdot\sin\frac{n\pi y}{b} e^{-\pi^2\left(\frac{m^2}{a^2}+\frac{n^2}{b^2}\right)c^2t}$$

The general form of solution is

$$u(x, y, t) = \sum_{m=1}^{\infty}\sum_{n=1}^{\infty} A_{mn} \sin\frac{m\pi x}{a}\cdot\sin\frac{n\pi y}{b} e^{-\pi^2\left(\frac{m^2}{a^2}+\frac{n^2}{b^2}\right)c^2t} \quad ...(7)$$

and the initial condition is

$$u(x, y, 0) = f(x, y)$$

From (7), we get

$$f(x, y) = \sum_{m=1}^{\infty}\sum_{n=1}^{\infty} A_{nm}\cdot\sin\frac{m\pi x}{a}\cdot\sin\frac{n\pi x}{b}$$

which is the double Fourier since series of $f(x, y)$.

$$\therefore \quad A_{mn} = \frac{2}{a}\cdot\frac{2}{b}\int_{x=0}^{a}\int_{y=0}^{b} f(x, y)\sin\frac{m\pi x}{a}\sin\frac{n\pi x}{b}\,dxdy \quad ...(8)$$

Hence the equation (7) is required temperature distribution with the equation (8). **Ans.**

Example 26: Solve the following Laplace equation:

$$\frac{\partial^2 u}{\partial x^2}+\frac{\partial^2 u}{\partial y^2} = 0$$

in a rectangle with $u(0, y) = 0, u(a, y) = 0, u(x, b) = 0$ and $u(x, 0) = f(x)$ along x-axis

(U.P.T.U. 2008)

Solution: Let $u(x, y) = X(x)Y(y)$. ...(1)

$$\therefore \quad \frac{\partial^2 u}{\partial x^2} = \frac{d^2X}{dx^2}Y \text{ and } \frac{\partial^2 u}{\partial y^2} = X\frac{d^2y}{\partial y^2}$$

Putting these values in given equation, we get

$$Y\frac{d^2X}{dx^2}+X\frac{d^2Y}{dy^2} = 0$$

$$\Rightarrow \quad Y\frac{d^2X}{dx^2} = -X\frac{d^2Y}{dy^2} \Rightarrow \frac{1}{X}\frac{d^2X}{dx^2} = -\frac{1}{Y}\frac{d^2Y}{dy^2} = -p^2 \text{ (say)} \quad ...(2)$$

Taking first two terms, we get

$$\frac{d^2X}{dx^2} + p^2X = 0$$

$$\Rightarrow \qquad X = c_1 \cos px + c_2 \sin px$$

And taking last two terms, we get

$$\frac{d^2Y}{dy^2} - p^2y = 0$$

$$\Rightarrow \qquad Y = c_3e^{py} + c_4e^{-py}$$

From (*i*)

$$u(x, y) = (c_1 \cos px + c_2 \sin px)(c_3e^{py} + c_4e^{-py}) \qquad ...(3)$$

Putting $x = 0$, we get

$$0 = c_1(c_3e^{py} + c_4e^{-py}) = 0 \Rightarrow c_1 = 0 \qquad \left| \begin{array}{l}\text{otherwise} \\ u = 0\end{array}\right.$$

From (*iii*) $\qquad u(x, y) = c_2 \sin px(c_3e^{py} + c_4e^{-py}) \qquad ...(4)$

Putting $x = a$, in (*iv*), we get

$$0 = c_2 \sin pa(c_3e^{py} + c_4e^{-py}) \Rightarrow \sin pa = 0$$

or $\qquad \sin pa = \sin n\pi \Rightarrow p = \dfrac{n\pi}{a}$

From (*iv*), $\qquad u(x, y) = c_2 \sin \dfrac{n\pi x}{a}\left(c_3e^{\frac{n\pi}{a}y} + c_4e^{-\frac{n\pi}{a}y}\right) \qquad ...(5)$

Putting $y = b$ in (*v*), we get

$$0 = c_2 \sin \frac{n\pi x}{a}\left(c_3e^{\frac{n\pi b}{a}} + c_4e^{-\frac{n\pi b}{a}}\right)$$

$$\Rightarrow \qquad c_3e^{\frac{n\pi b}{a}} + c_4e^{-\frac{n\pi b}{a}} = 0 \Rightarrow c_4 = -c_3e^{2\frac{n\pi b}{a}}$$

From (*v*), we get

$$u(x, y) = c_2c_3 \cdot \sin \frac{n\pi x}{a} \cdot \left\{e^{\frac{n\pi}{a}y} - e^{-\frac{n\pi}{a}(y-2b)}\right\}$$

or $\qquad u(x, y) = \displaystyle\sum_{n=1}^{\infty} b_n \cdot \sin \frac{n\pi x}{a} \cdot \left\{e^{\frac{n\pi y}{a}} - e^{-\frac{n\pi}{a}(y-2b)}\right\} \qquad ...(6)$

Putting $y = 0$, we get

$$f(x) = \sum_{n=1}^{\infty} b_n \cdot \sin \frac{n\pi x}{a} \cdot \left(1 - e^{+\frac{2n\pi b}{a}}\right) \qquad ...(7)$$

which is Fourier sine series

$$\therefore \qquad b_n\left(1-e^{\frac{2n\pi b}{a}}\right) = \frac{2}{a}\int_0^a f(x)\cdot\sin\frac{n\pi x}{a}\,dx$$

$$\Rightarrow \qquad b_n = \frac{2}{a\left(1-e^{\frac{2n\pi b}{a}}\right)}\int_0^a f(x)\cdot\sin\frac{n\pi x}{a}\,dx \qquad ...(8)$$

Hence equation (*vi*) is required solution with coefficient b_n (equations (*viii*)).

Example 27: An infinitely long uniform plate is bounded by two parallel edges and an end at right angles to them. The breadth is π, the end is maintained at a temperature 100ºC at all points and other edges are at 0ºC. Show that the steady state temperature is given by

$$u(x, y) = \frac{400}{\pi}\left[e^{-y}\sin x + \frac{1}{3}e^{-3y}\sin 3x + \frac{1}{5}e^{-5y}\sin 5x + \cdots\right].$$

Solution: The steady state temperature $u(x, y)$ is the solution of

$$\frac{\partial^2 u}{\partial x^2} + \frac{\partial^2 u}{\partial y^2} = 0 \qquad ...(1)$$

The boundary conditions are

(*i*) $u(0, y) = 0 = u(\pi, y)$ for all $y \geq 0$

(*ii*) $u(x, \infty) = \lim_{y\to\infty}(x, y) = 0$ for $0 \leq x \leq \pi$

and initial condition is (*iii*) $u(x, 0) = 100$

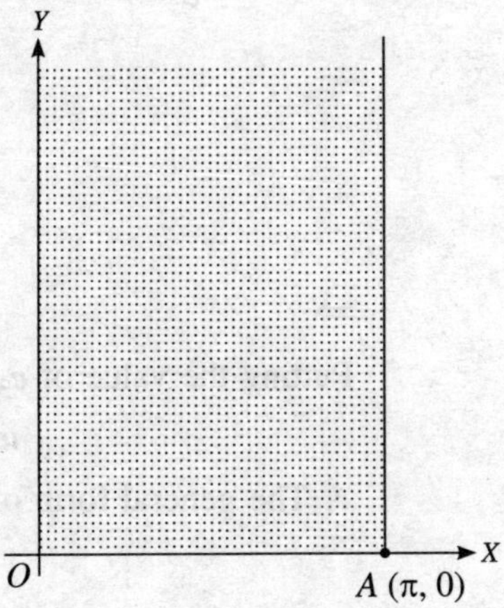

Let $\qquad u(x, y) = XY \qquad ...(2)$

Putting the value of u in equation (1), we get

$$Y\frac{d^2X}{dx^2} + X\frac{d^2Y}{dy^2} = 0$$

$$\Rightarrow \qquad Y\frac{d^2X}{dx^2} = -X\frac{d^2Y}{dy^2} \Rightarrow \frac{1}{X}\frac{d^2X}{dx^2} = -\frac{1}{Y}\frac{d^2Y}{dy^2} = -p^2 \text{ (say)} \qquad ...(3)$$

Taking first two member, we get

$$\frac{d^2X}{dx^2} + p^2X = 0$$

$$\Rightarrow \qquad X = c_1\cos px + c_2\sin px$$

And now taking last two member of equation (3), we get

$$\frac{d^2Y}{dy^2} - p^2Y = 0 \Rightarrow Y = c_3e^{py} + c_4e^{-py}$$

Putting the values of X and Y in equation (2), we get

$$u(x, y) = (c_1 \cos px + c_2 \sin px)(c_3e^{py} + c_4e^{-py}) \quad ...(4)$$

At $\quad x = 0, \quad u = 0$

$$\Rightarrow \quad 0 = c_1(c_3e^{py} + c_4e^{-py}) \Rightarrow \boxed{c_1 = 0.}$$

From equation (4), we have

$$u(x, y) = c_2 \sin px(c_3e^{py} + c_4e^{-py}) \quad ...(5)$$

Again at $\quad x = \pi, \quad u = 0$

$$\Rightarrow \quad 0 = c_2 \sin p\pi(c_3e^{py} + c_4e^{-py})$$

$$\Rightarrow \quad \sin p\pi = 0 = \sin n\pi \Rightarrow p = \frac{n\pi}{\pi} = n \quad \left| \begin{array}{l} \text{otherwise} \\ u(x, y) = 0 \end{array} \right.$$

Putting the value of p in equation (5), we get

$$u(x, y) = c_2 \sin nx(c_3e^{ny} + c_4e^{-ny}) \quad ...(6)$$

Now using (*ii*) boundary condition in equation (6) *i.e.*, when $y \to \infty$, $u = 0$, we get

$$0 = c_2 \sin nx \cdot \lim_{y\to\infty}(c_3e^{ny} + c_4e^{-ny})$$

$$0 = c_2 \sin nx \cdot \left\{c_3 \lim_{y\to\infty}(e^{ny}) + 0\right\} \quad \left| \text{ As } \lim_{y\to\infty} e^{-ny} \to 0 \right.$$

$$\Rightarrow \quad \boxed{c_3 = 0} \quad \text{(otherwise } u(x, y) = 0)$$

Putting the value of c_3 in equation (6), we get

$$u(x, y) = c_2c_4 \sin nx \cdot e^{-ny}$$

∴ The general form of solution may be written as

$$u(x, y) = \sum_{n=1}^{\infty} b_n \sin nx \cdot e^{-ny} \quad ...(7)$$

Using (*iii*) condition (*i.e.*, at $y = 0$) in equation (7), we get

$$100 = \sum_{n=1}^{\infty} b_n \sin nx$$

which is Fourier half range sine series in $0 \le x \le \pi$

$$\therefore \quad b_n = \frac{2}{\pi}\int_0^{\pi} 100 \cdot \sin nx dx = \frac{200}{n\pi}\left[-\cos nx\right]_0^{\pi} = \frac{200}{n\pi}\left[-\cos n\pi + \cos 0\right]$$

$$b_n = \frac{200}{n\pi}\left[-(-1)^n + 1\right] = \begin{cases} \dfrac{400}{n\pi} & \text{if } n \text{ is odd} \\ 0 & \text{if } n \text{ is even} \end{cases}$$

Hence from equation (7), we get

$$u(x, y) = \frac{400}{\pi}\left[e^{-y}\sin x + \frac{1}{3}e^{-3y}\sin 3x + \frac{1}{5}e^{-5y}\sin 5x + \cdots\right].$$ **Proved.**

Example 28: Solve $\dfrac{\partial^2 u}{\partial x^2} + \dfrac{\partial^2 u}{\partial y^2} = 0$ which satisfies the conditions

$$u(0, y) = u(l, y) = u(x, 0) = 0 \text{ and } u(x, a) = \sin\left(\frac{n\pi x}{l}\right).$$ *(U.P.T.U. 2004)*

Solution: We have

$$\frac{\partial^2 u}{\partial x^2} + \frac{\partial^2 u}{\partial y^2} = 0 \qquad ...(1)$$

The solution of equation (1) can be obtained as in (Example 27, equation (4) on page no. 494) is given below:

$$u(x, y) = (c_1 \cos px + c_2 \sin px)\,(c_3 e^{py} + c_4 e^{-py}) \qquad ...(2)$$

At $x = 0, \quad u = 0$

$$0 = c_1(c_3 e^{py} + c_4 e^{-py}) \quad \Rightarrow \quad c_1 = 0 \quad [\text{otherwise } u(x, y) = 0]$$

Putting the value of c_1 in equation (2), we get

$$u(x, y) = c_2 \sin px(c_3 e^{py} + c_4 e^{-py}) \qquad ...(3)$$

at $x = l, \quad u = 0$

$$0 = c_2 \sin pl(c_3 e^{py} + c_4 e^{-py})$$

$$\Rightarrow \quad \sin pl = 0 = \sin n\pi \quad \Rightarrow \quad \boxed{p = \frac{n\pi}{l}}.$$

From equation (3), we get

$$u(x, y) = c_2 \sin\frac{n\pi x}{l}\left(c_3 e^{\frac{n\pi y}{l}} + c_4 e^{-\frac{n\pi y}{l}}\right) \qquad ...(4)$$

At $y = 0, \quad u = 0,$ we get

$$0 = 2c_2 \sin\frac{n\pi x}{l}(c_3 + c_4) \quad \Rightarrow \quad c_3 + c_4 = 0 \quad \Rightarrow \quad \boxed{c_4 = -c_3}$$

Putting the value of c_4 in equation (4), we get

$$u(x, y) = c_2c_3 \sin\frac{n\pi x}{l}\left(e^{\frac{n\pi y}{l}} - e^{-\frac{n\pi y}{l}}\right) = 2c_2c_3 \sin\frac{n\pi x}{l}\cdot\sin h\frac{n\pi y}{l} \quad \left| \sin h\theta = \frac{e^{\theta} - e^{-\theta}}{2}\right.$$

The general form of above solution is given as

$$u(x, y) = \sum_{n=1}^{\infty} b_n \sin\frac{n\pi x}{l} \sin h\frac{n\pi y}{l} \qquad (2c_1c_2 = b_n) \qquad ...(5)$$

Putting $y = a$ and $u = \sin\frac{n\pi x}{l}$ in equation (5), we get

$$\sin\frac{n\pi x}{l} = \sum_{n=1}^{\infty} b_n \sin\frac{n\pi x}{l} \sin h\frac{n\pi a}{l}$$

Equating the coefficient of $\sin\frac{n\pi x}{l}$ on both sides, we get

$$1 = b_n \sin h\frac{n\pi a}{l} \Rightarrow b_n = \frac{1}{\sin h\frac{n\pi a}{l}}$$

and $$b_1 = b_2 = b_3 = ... = b_{n-1} = 0$$

Hence, from equation (5), we get

$$u(x, y) = \frac{\sin(n\pi x/l)}{\sin h\left(\frac{n\pi a}{l}\right)}\cdot\sin h\frac{n\pi y}{l}. \quad \textbf{Ans.}$$

Example 29: A thin rectangular plate whose surface is impervious to heat flow has $t = 0$ an arbitrary distribution of temperature $f(x, y)$. Its four edges $x = 0$, $x = a$, $y = 0$, $y = b$ are kept at 0°C temperature. Determine the temperature at a point of a plate as t increases. Discuss when

(*U.P.T.U. 2002*)

$$f(x, y) = \beta \sin\left(\frac{\pi x}{a}\right)\sin\left(\frac{\pi y}{b}\right).$$

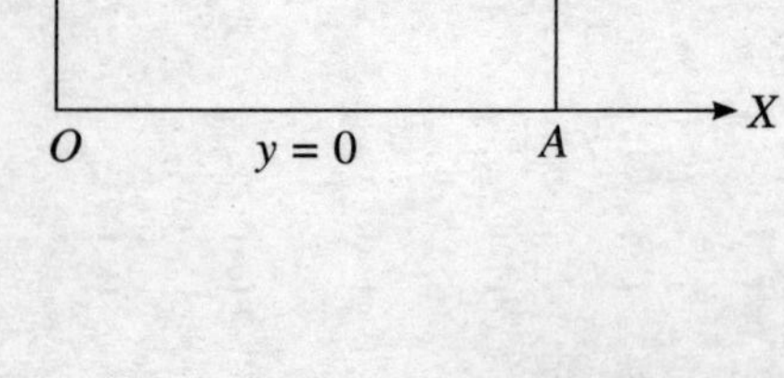

Solution: Two dimensional heat flow equation is

$$\frac{\partial^2 u}{\partial x^2} + \frac{\partial^2 u}{\partial y^2} = \frac{1}{c^2}\frac{\partial u}{\partial t} \qquad ...(1)$$

Boundary conditions are

$$u(0, y, t) = 0$$
$$u(a, y, t) = 0$$

$$u(x,\ 0,\ t) = 0$$
$$u(x,\ b,\ t) = 0$$

and the initial condition is

$$u(x,\ y,\ t) = f(x,\ y) \text{ at } t = 0.$$

The solution of equation (1) is given below (Example 25, equation (4) on page 514)

$$u = (c_1 \cos p_1 x + c_2 \sin p_1 x)(c_3 \cos p_2 y + c_4 \sin p_2 y)\, c_5 e^{-p^2 c^2 t} \qquad \text{...(2)}$$

At $\quad x = 0, \quad u = 0$, then we get

$$u(0,\ y,\ t) = 0 = c_1(c_3 \cos p_2 y + c_4 \sin p_2 y) c_5 e^{-c^2 p^2 t}$$

$\Rightarrow \quad c_1 = 0$

$\therefore$ from (2), $\quad u(x,\ y,\ t) = c_2\, c_5 \sin p_1 x(c_3 \cos p_2 y + c_4\, \mathit{sin}\, p_2 y)\ (e^{-c^2 p^2 t}) \qquad \text{...(3)}$

from (3), $\quad u(a,\ y,\ t) = 0 = c_2 c_5 \sin p_1 a(c_3 \cos p_2 y + c_4 \sin p_2 y) e^{-c^2 p^2 t}$

$\Rightarrow \quad \sin p_1 a = 0 = \sin m\pi\ (m \in I)$

$$\therefore \quad p_1 = \frac{m\pi}{a}.$$

From (3), $\quad u(x,\ y,\ t) = c_2 c_5 \sin \dfrac{m\pi x}{a}(c_3 \cos p_2 y + c_4 \sin p_2 y)(e^{-c^2 p^2 t}) \qquad \text{...(4)}$

$$u(x,\ 0,\ t) = 0 = c_2 c_5 \sin \frac{m\pi x}{a} \cdot c_3 e^{-c^2 p^2 t}$$

$\Rightarrow \quad c_3 = 0.$

$\therefore$ from (4), $\quad u(x,\ y,\ t) = c_2 c_5 c_4 \sin \dfrac{m\pi x}{a} \sin p_2 y \cdot e^{-c^2 p^2 t} \qquad \text{...(5)}$

$$u(x,\ b,\ t) = 0 = c_2 c_5 c_4 \sin \frac{m\pi x}{a} \sin p_2 b \cdot e^{-c^2 p^2 t}$$

$\Rightarrow \quad \sin p_2 b = 0 = \sin n\pi\ (n \in I)$

$$p_2 b = n\pi$$

$$\Rightarrow \quad p_2 = \frac{n\pi}{b}$$

$\therefore$ from (5), $\quad u(x,\ y,\ t) = c_2 c_5 c_4 \sin \dfrac{m\pi x}{a} \sin \dfrac{n\pi y}{b} e^{-c^2 p^2 t}$

$$= A_{mn} \sin \frac{m\pi x}{a} \sin \frac{n\pi y}{b} e^{-c^2 p^2 t} \qquad \text{...(6)} \quad \Big|\ \text{where } c_2 c_5 c_4 = A_{mn}$$

But, $\quad p^2 = p_1^2 + p_2^2 = \dfrac{m^2 \pi^2}{a^2} + \dfrac{n^2 \pi^2}{b^2}$

or $\quad p_{mn}^2 = \pi^2 \left(\dfrac{m^2}{a^2} + \dfrac{n^2}{b^2} \right)$

By using p_{mn}, equation (6) becomes,

$$u(x, y, t) = \sum_{m=1}^{\infty}\sum_{n=1}^{\infty} A_{mn} \sin\frac{m\pi x}{a} \sin\frac{n\pi x}{a} e^{-c^2 p_{mn}^2 t} \quad ...(7)$$

which is the most general solution.

$$\therefore \quad u(x, y, 0) = f(x, y) = \sum_{m=1}^{\infty}\sum_{n=1}^{\infty} A_{mn} \sin\frac{m\pi x}{a} \sin\frac{n\pi y}{b}$$

which is the double Fourier half range sine series for $f(x, y)$.

$$A_{mn} = \frac{2}{a}\cdot\frac{2}{b}\int_{x=0}^{a}\int_{y=0}^{b} \sin\frac{m\pi x}{a} \sin\frac{n\pi y}{b} f(x, y)dx\,dy \qquad \textbf{Ans.}$$

when $$f(x, y) = \beta \sin\left(\frac{\pi x}{a}\right)\sin\left(\frac{\pi y}{b}\right)$$

$$\therefore \quad A_{mn} = \frac{2}{a}\cdot\frac{2}{b}\int_{x=0}^{a}\int_{y=0}^{b} \sin\frac{m\pi x}{a} \sin\frac{n\pi y}{b} \beta \sin\frac{\pi x}{a} \sin\frac{\pi y}{b}\, dx\,dy$$

$$= \frac{4\beta}{ab}\int_0^a \sin\frac{m\pi x}{a} \sin\frac{\pi x}{a} dx \int_0^b \sin\frac{n\pi y}{b} \sin\frac{\pi y}{b} dy$$

$$= \frac{\beta}{ab}\int_0^a \left[\cos(m-1)\frac{\pi x}{a} - \cos(m+1)\frac{\pi x}{a}\right]dx \times \int_0^b \left[\cos(n-1)\frac{\pi y}{b} - \cos(n+1)\frac{\pi y}{b}\right]dy$$

$$= \frac{\beta}{ab}\left[\frac{\sin(m-1)\frac{\pi x}{a}}{(m-1)\frac{\pi}{a}} - \frac{\sin(m+1)\frac{\pi x}{a}}{(m+1)\frac{\pi}{a}}\right]_0^a \times \left[\frac{\sin(n-1)\frac{\pi y}{b}}{(n-1)\frac{\pi}{b}} - \frac{\sin(n+1)\frac{\pi y}{b}}{(n+1)\frac{\pi}{b}}\right]_0^b$$

$$= \frac{\beta}{ab}\cdot 0 = 0 \qquad (\because \sin m\pi = 0 \text{ and } \sin n\pi = 0)$$

From (7), $\quad u(x, y, t) = 0$ when $f(x, y) = \beta \sin\frac{\pi x}{a} \sin\frac{\pi y}{b}$. **Proved.**

Example 30: The edges of a thin plate of side π cm are kept at temperature zero and faces insulated. The initial temperature is $u(x, y, 0) = f(x, y) = xy(\pi - x)(\pi - y)$. Determine temperature in the plate at time t.

Solution: Referred to solution of Example 4, where, $a = \pi$ and $b = \pi$ and $f(x, y) = xy(\pi - x)(\pi - y)$

$$p_{mn}^2 = \pi^2\left(\frac{m^2}{\pi^2}+\frac{n^2}{\pi^2}\right) = m^2+n^2$$

$$u(x, y, t) = \sum_{m=1}^{\infty}\sum_{n=1}^{\infty} A_{mn} \sin mx \sin ny \, e^{-c^2(m^2+n^2)t} \qquad \text{...(1)}$$

$$u(x, y, 0) = \sum_{m=1}^{\infty}\sum_{n=1}^{\infty} A_{mn} \sin mx \sin ny \qquad \text{...(2)}$$

$$A_{mn} = \frac{4}{\pi^2}\int_0^{\pi}\int_0^{\pi} xy(\pi - x)(\pi - y)\sin mx \sin ny \, dx\, dy$$

$$= \frac{4}{\pi^2}\int_0^{\pi}(\pi x - x^2)\sin mx\, dx \int_0^{\pi}(\pi y - y^2)\sin ny\, dy \qquad \text{...(3)}$$

Now $$\int_0^{\pi}(\pi x - x^2)\sin mx\, dx = \left[(\pi x - x^2)\left(\frac{-\cos mx}{m}\right) - (\pi - 2x)\left(\frac{-\sin mx}{m^2}\right) + (-2)\left(\frac{\cos mx}{m^3}\right)\right]_0^{\pi}$$

$$= \frac{2}{m^3}[1-(-1)^m]$$

and similarly

$$\int_0^{\pi}(\pi y - y^2)\sin ny\, dy = \frac{2}{n^3}[1-(-1)^n]$$

$$\therefore \qquad A_{mn} = \frac{16}{\pi^2 m^3 n^3}[1-(-1)^m][1-(-1)^n]$$

or $$A_{mn} = \begin{cases} 0, & \text{when } m = 2p \text{ and } n = 2q \\ \dfrac{64}{\pi^2(2p-1)^3(2q-1)^3}, & \text{when } m = 2p-1 \text{ and } n = 2q-1 \end{cases}$$

Hence from (2)

$$u(x, y, t) = \sum_{p=1}^{\infty}\sum_{q=1}^{\infty} A_{pq} \sin(2p-1)x \sin(2q-1)y e^{-\mu_{pq}^2 t}. \qquad \textbf{Ans.}$$

where $$\mu_{pq}^2 = c^2[(2p-1)^2 + (2q-1)^2]$$

and $$A_{pq} = \frac{64}{\pi^2 (2p-1)^3 (2q-1)^3}.$$

Example 31: A square plate is bounded by the lines $x = 0$, $y = 0$, $x = 10$ and $y = 10$. Its faces are insulated. The temperature along the upper horizontal edge is given by $u(x, 10) = x(10 - x)$, while the other three edges are kept at 0°C. Find the steady state temperature in the plate.

Solution: The steady state temperature $u(x, y)$ is the solution of the equation

$$\frac{\partial^2 u}{\partial x^2} + \frac{\partial^2 u}{\partial y^2} = 0 \quad \text{...(1)}$$

The solution of equation (1) is given below

$$u(x, y) = (c_1 \cos px + c_2 \sin px)(c_3 e^{py} + c_4 e^{-py}) \quad \text{...(2)}$$

[See eqn. (4) on page no. 490]

The boundary conditions are

$$\left.\begin{aligned} &u(0, y) = u(10, y) = 0, \; 0 \le y \le 10 \\ &u(x, 0) = 0; \qquad 0 \le x \le 10 \\ \text{and} \quad &u(x, 10) = x(10 - x); \; 0 \le x \le 10 \end{aligned}\right\} \quad \text{...(A)}$$

Putting $x = 10$ and $u = 0$ in (2), we get

$$0 = (c_1 c_3 e^{py} + c_4 e^{-py}) \Rightarrow c_1 = 0$$

∴ From (2), we have

$$u(x, y) = c_2 \sin px (c_3 e^{py} + c_4 e^{-py}) \quad \text{...(3)}$$

Next putting $x = 10$ in (3), we get

$$0 = c_2 \sin 10p (c_3 e^{py} + c_4 e^{-py}) \Rightarrow \sin 10p = 0 = \sin n\pi \Rightarrow \boxed{p = \frac{n\pi}{10}}.$$

Again putting $y = 0$ and $u = 0$ in equation (3), we get

$$0 = c_2 \sin px (c_3 + c_4) \Rightarrow c_3 + c_4 = 0 \Rightarrow \boxed{c_4 = -c_3}$$

Putting the values of c_4 and p in equation (3), we get

$$u(x, y) = c_2 c_3 \sin \frac{n\pi x}{10} \left(e^{\frac{n\pi y}{10}} - e^{-\frac{n\pi y}{10}} \right) = 2c_2 c_3 \sin \frac{n\pi x}{10} \sin h \frac{n\pi y}{10}$$

The general solution is given by

$$u(x, y) = \sum_{n=1}^{\infty} b_n \sin \frac{n\pi x}{10} \cdot \sin h \frac{n\pi y}{10} \quad \text{...(4)}$$

Putting $y = 10$ and $u = x(10 - x)$ in equation (4), we get

$$x(10-x) = \sum_{n=1}^{10} b_n \sin\frac{n\pi x}{10} \sin h n\pi$$

$$\therefore \quad b_n \sin hn\pi = \frac{2}{10}\int_0^{10} x(10-x)\sin\frac{n\pi x}{10} \qquad \Big|\ \sin hn\pi \text{ is a constant}$$

$$= \frac{1}{5}\Bigg[-(10x-x^2)\cdot\frac{10}{n\pi}\cos\frac{n\pi x}{10} + (10-2x)\ \frac{100}{n^2\pi^2}\sin\frac{n\pi x}{10}$$

$$-2\frac{1000}{n^3\pi^3}\cos\frac{n\pi x}{10}\Bigg]_0^{10}$$

$$= \frac{1}{5}\left[-2\times\frac{1000}{n^3\pi^3}\cos n\pi + 2\times\frac{1000}{n^3\pi^3}\right]$$

$$= \frac{2000}{5n^3\pi^3}[-(-1)^n+1] = \begin{cases} \dfrac{800}{n^3\pi^3}, & \text{if } n \text{ is odd} \\ 0, & \text{if } n \text{ is even} \end{cases}$$

$$\therefore \quad b_n = \begin{cases} \dfrac{800}{n^3\pi^3 \sin hn\pi}, & \text{if } n \text{ is odd} \\ 0, & \text{if } n \text{ is even} \end{cases}$$

Putting the value of b_n in equation (4), we get

$$u(x, y) = \frac{800}{\pi^3}\sum_{m=1}^{\infty} \frac{\sin\dfrac{(2m-1)\pi x}{10}\cdot \sin h\dfrac{(2m-1)\pi y}{10}}{(2m-1)^3 \sin h(2m-1)\pi}; \text{ where } n = 2m-1. \quad \textbf{Ans.}$$

Example 32: A rectangular metal plate is bounded by $x = 0$, $x = a$, $y = 0$ and $y = b$. The three sides $x = 0$, $x = a$ and $y = b$ are insulated and the edge $y = 0$ is kept at $u_0 \cos \dfrac{\pi x}{a}\cdot$ Show that the temperature in the steady state is

$$u(x, y) = u_0 \operatorname{sec} h\left(\frac{(b-a)\pi}{a}\right)\cos h\left(\frac{\pi(b-y)}{a}\right)\cos\left(\frac{\pi x}{a}\right).$$

Solution: The steady state temperature $u(x, y)$ is the solution of the equation

$$\frac{\partial^2 u}{\partial x^2} + \frac{\partial^2 u}{\partial y^2} = 0 \qquad \text{...(1)}$$

Since sides $x = 0$, $x = a$ and $y = b$ are insulated

$$\left(\frac{\partial u}{\partial x}\right)_{x=0} = 0 \qquad \text{...(2)}$$

$$\left(\frac{\partial u}{\partial x}\right)_{x=a} = 0 \quad ...(3)$$

$$\left(\frac{\partial u}{\partial x}\right)_{y=b} = 0 \quad ...(4)$$

Also $$u(x, a) = u_0 \cos\left(\frac{\pi x}{a}\right) \quad ...(5)$$

The solution of (1) is

$$u(x, y) = (c_1 \cos px + c_2 \sin px)(c_3 e^{py} + c_4 e^{-py}) \quad ...(6)$$

Now $$\frac{\partial u}{\partial x} = (-c_1 p \sin px + c_2 p \cos px)(c_3 e^{py} + c_4 e^{-py})$$

$$\left(\frac{\partial u}{\partial x}\right)_{x=0} = 0 \Rightarrow c_2 p(c_3 e^{py} + c_4 e^{-py}) = 0$$

$$\Rightarrow \quad c_2 = 0 \quad (\because \; c_3 e^{py} + c_4 e^{-py} \neq 0)$$

$$\left(\frac{\partial u}{\partial x}\right)_{x=a} = 0 \Rightarrow -c_1 p \sin pa(c_3 e^{py} + c_4 e^{-py}) = 0$$

$$\Rightarrow \quad \sin pa = 0$$ ($\because$ $c_1 \neq 0$ otherwise $u(x, y)$ would be a trivial solution)

$$\Rightarrow \quad pa = n\pi \quad (n \text{ is an integer})$$

$$\Rightarrow \quad p = \frac{n\pi}{a}$$

Now $$\frac{\partial u}{\partial y} = c_1 \cos px(c_3 p e^{py} - c_4 p e^{-py}).$$

$$\left(\frac{\partial u}{\partial y}\right)_{y=b} = 0 \Rightarrow c_1 \cos px(c_3 p e^{pb} - c_4 p e^{-pb}) = 0$$

$$\Rightarrow \quad c_3 p e^{pb} - c_4 p e^{-pb} = 0$$

$$\Rightarrow \quad p(c_3 e^{pb} - c_4 e^{-pb}) = 0$$

$$\Rightarrow \quad c_3 e^{pb} = c_4 e^{-pb}$$

$$\Rightarrow \quad c_4 = c_3 e^{2pb}$$

$\therefore$ From (6) $$u(x, y) = c_1 \cos px(c_3 e^{py} + c_3 e^{2pb} e^{-py}) \quad \text{where } p = \frac{n\pi}{a}$$

$$= c_1 c_3 \cos px \left(e^{py} + \frac{e^{pb} e^{-py}}{e^{-pb}} \right)$$

$$= \frac{c_1 c_3}{e^{-pb}} \cos px (e^{-pb} e^{py} + e^{pb} e^{-py})$$

$$= c_1 c_3 e^{pb} \cos px [e^{-p(b-y)p} + e^{(b-y)p}]$$

$$= 2c_1 c_3 e^{pb} \cos px \cosh (b-y)p$$

$$= K \cos\left(\frac{n\pi x}{a}\right) \cosh\left(\frac{(b-y)n\pi}{a}\right) \quad \text{where} \quad K = 2c_1 c_3 e^{pb}$$

$\therefore$ The most general solution of (1) is

$$u(x, y) = \sum_{n=1}^{\infty} A_n \cos\left(\frac{n\pi x}{a}\right) \cosh\left(\frac{(b-y)n\pi}{a}\right)$$

where A_n's a are to be determined.

Since $u(x, a) = u_0 \cos\left(\frac{\pi x}{a}\right)$ for $0 < x < a$,

$$u(x, a) = \sum_{n=1}^{\infty} A_n \cos\left(\frac{n\pi x}{a}\right) \cosh\left(\frac{(b-a)n\pi}{a}\right) = u_0 \cos\left(\frac{\pi x}{a}\right)$$

$$\therefore \quad u_0 \cos\left(\frac{\pi x}{a}\right) = \sum_{n=1}^{\infty} A_n \cos\left(\frac{n\pi x}{a}\right) \cosh\left(\frac{(b-a)n\pi}{a}\right)$$

$$= A_1 \cos\left(\frac{\pi x}{a}\right) \cosh\left(\frac{(b-a)\pi}{a}\right) + A_2 \cos\left(\frac{2\pi x}{a}\right) \cosh\left(\frac{(b-a)2\pi}{a}\right) + \ldots.$$

Comparing both sides, we get

$$A_1 = u_0 \operatorname{sech}\left(\frac{(b-a)\pi}{a}\right)$$

$$A_2 = A_3 = \cdots = 0$$

$$\therefore \quad u(x, y) = u_0 \operatorname{sech}\left(\frac{(b-a)\pi}{a}\right) \cos\left(\frac{\pi x}{a}\right) \cosh\left(\frac{(b-y)\pi}{a}\right)$$

$$u(x, y) = u_0 \operatorname{sech}\left(\frac{(b-a)\pi}{a}\right) \cosh\left(\frac{(b-y)\pi}{a}\right) \cos\left(\frac{\pi x}{a}\right).$$ **Proved.**

Exercise 5.4

1. An infinite long plane uniform plate is bounded by two parallel edges and an end at right angles to them. The breadth is π. This end is maintained at temperature u_0 at all points and other are at zero temperature. Determine the temperature at any point of the plate in the steady state. *[U.P.T.U. (C.O.) 2003]*

$$\left[\textbf{Ans. } u(x, y) = \frac{4u_0}{\pi}\sum_{n=1}^{\infty}\frac{1}{2n-1}\sin(2n-1)xe^{-(2n-1)y}\right]$$

2. A square plate is bounded by the lines $x = 0$, $y = 0$, $x = 20$ and $y = 20$. Its faces are insulated. The temperature along the upper horizontal edge is given by $u(x, 20) = x(20 - x)$ when $0 < x < 20$, while the other three edges are kept at 0°C. Find the steady temperature in the rod.

$$\left[\textbf{Ans. } u(x,y) = \frac{3200}{\pi^3}\sum_{m=1}^{\infty}\frac{\sin\frac{(2m-1)\pi x}{20}\sin h\frac{(2m-1)\pi y}{20}}{\sin A(2m-1)\pi}; 2m-1 = n\right]$$

3. A rectangular plate with insulated surfaces is 10 cm wide and so long compared to its width that it may be considered infinite in length without introducing an appreciable error. If the temperature along the short edge $y = 0$ is given by

$$u(x, 0) = 20x, \quad 0 < x \le 5$$
$$= 20(10 - x), \quad 5 < x < 10.$$

while the two long edges $x = 0$ and $x = 10$ as well as other short edges are kept at 0°C. Find the steady state temperature at any point (x, y) of the plate.

$$\left[\textbf{Ans. } u(x, y) = \frac{800}{\pi^2}\sum_{m=1}^{\infty}\frac{(-1)^{m+1}}{(2m-1)^2}\sin\frac{(2m-1)\pi x}{10}e^{-\frac{(2m-1)\pi y}{10}}\right]$$

4. A long rectangular plate of width lam with insulated surfaces has its temperature u equal to zero on both the long sides and one of the shorter sides so that $u(0, y) = u(l, y) = u(x, \infty) = 0$ and $u(x, 0) = kx$. Find $u(x, y)$.

$$\left[\textbf{Ans. } u(x, y) = \frac{2kl}{\pi}\sum_{n=1}^{\infty}\frac{(-1)^{n+1}}{n}\sin\frac{n\pi x}{l}e^{-\frac{n\pi y}{l}}\right]$$

5. A long rectangular plate has its surface insulated and the two long sides as well one of the short sides are kept at 0°C while the other sides is kept at $u(x, 0) = 3x$ and the length being 5 cm. Find $u(x, y)$.

$$\left[\textbf{Ans. } u(x, y) = \frac{30}{\pi}\sum_{n=1}^{\infty}\frac{(-1)^{n+1}}{n}\sin\frac{n\pi x}{5}e^{-\frac{n\pi y}{5}}\right]$$

6. A square plate has its faces and the edge $y = 0$ insulated. Its edges $x = 0$ and $x = \pi$ are kept at 0°C and the edge $y = \pi$ is kept at temperature $f(x)$. Find the steady state temperature.

$$\left[\textbf{Ans. } u(x, y) = \sum_{n=1}^{\infty} b_n\frac{\sin nx \cos h\, ny}{\cos h\, n\pi}, \text{ where } b_n = \frac{2}{\pi}\int_0^{\pi} f(x)\sin nx\, dx\right]$$

7. A long rectangular plate has its surface insulated and the two long sides as well as one of the short sides are maintained at 0°C. Find an expression for the steady state temperature $u(x, y)$ if the short side $y = 0$ is 30 cm long and kept at 40°C.

$$\left[\textbf{Ans. } u(x, y) = \frac{160}{\pi}\sum_{m=1}^{\infty}\frac{1}{2m-1}\sin\frac{(2m-1)\pi x}{30}\cdot e^{-\frac{(2m-1)\pi y}{30}}\right]$$

5.7 WAVE EQUATION IN TWO DIMENSIONS

Consider a tightly stretched uniform membrane in xy-plane having the same tension T at every point.

Since the deflection $u(x, y, t)$ is in the perpendicular direction to xy-plane.

The vertical component of $T\,\delta y = T\,\delta y \sin\beta - T\,\delta y \sin\alpha$

$$= T\,\delta y(\tan\beta - \tan\alpha) \quad \left(\begin{array}{l}\text{approximately, since } \alpha \text{ and } \beta \text{ are very small}\\ \text{so } \sin\beta \approx \tan\beta, \sin\alpha \approx \tan\alpha\end{array}\right)$$

$$= T\,\delta y\left\{\left(\frac{\partial u}{\partial x}\right)_{x+\delta x} - \left(\frac{\partial u}{\partial x}\right)_x\right\} = T\,\delta y\frac{\partial^2 u}{\partial x^2}\delta x$$

Similarly, the vertical component due to the force $T\,\delta x$

$$= T\,\delta x\frac{\partial^2 u}{\partial y^2}\delta y$$

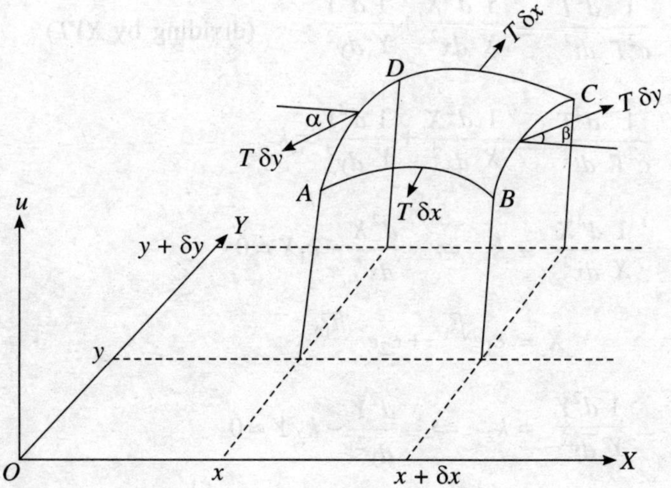

Hence the equation of motion of the element $ABCD$ is

$$m\,\delta x\,\delta y\frac{\partial^2 u}{\partial t^2} = T\left(\frac{\partial^2 u}{\partial x^2} + \frac{\partial^2 u}{\partial y^2}\right)\delta x\,\delta y$$

m is the mass of membrane per unit area.

or
$$\frac{\partial^2 u}{\partial t^2} = \frac{T}{m}\left(\frac{\partial^2 u}{\partial x^2}+\frac{\partial^2 u}{\partial y^2}\right)$$

$$\Rightarrow \quad \boxed{\frac{\partial^2 u}{\partial t^2} = c^2\left(\frac{\partial^2 u}{\partial x^2}+\frac{\partial^2 u}{\partial y^2}\right);\ c^2 = \frac{T}{m}.}$$

This is two dimensional wave equation.

5.7.1 Solution of Two Dimensional Wave Equation

We have

$$\frac{\partial^2 u}{\partial t^2} = c^2\left(\frac{\partial^2 u}{\partial x^2}+\frac{\partial^2 u}{\partial y^2}\right) \quad ...(i)$$

Let
$$u(x, y, t) = X(t)\ Y(y)\ T(t) \quad ...(ii)$$

$$\Rightarrow \quad \frac{\partial^2 u}{\partial t^2} = XY\frac{d^2T}{dt^2}, \frac{\partial^2 u}{\partial x^2} = YT\frac{d^2X}{dx^2} \text{ and } \frac{\partial^2 u}{\partial y^2} = XT\frac{d^2y}{dy^2}$$

Putting these values in equation (*i*), we get

$$XY\frac{d^2T}{dt^2} = c^2\left(YT\frac{d^2X}{dx^2}+XT\frac{d^2Y}{dy^2}\right)$$

$$\Rightarrow \quad \frac{1}{c^2T}\frac{d^2T}{dt^2} = \frac{1}{X}\frac{d^2X}{dx^2}+\frac{1}{Y}\frac{d^2Y}{dy^2} \quad \text{(dividing by } XYT)$$

$$\Rightarrow \quad \frac{1}{c^2T}\frac{d^2T}{dt^2} = \frac{1}{X}\frac{d^2X}{dx^2}+\frac{1}{Y}\frac{d^2Y}{dy^2} = k \quad ...(iii)$$

Let
$$\frac{1}{X}\frac{d^2X}{dx^2} = k_1 \Rightarrow \frac{d^2X}{dx^2} - k_1X = 0$$

$$\therefore \quad X = c_1e^{\sqrt{k_1}x} + c_2e^{-\sqrt{k_1}x}$$

and again let
$$\frac{1}{Y}\frac{d^2Y}{dy^2} = k_2 \Rightarrow \frac{d^2Y}{dy^2} - k_2\,Y = 0$$

$$\therefore \quad Y = c_3e^{\sqrt{k_2}y} + c_4e^{-\sqrt{k_2}y}$$

Finally
$$\frac{1}{c^2T}\frac{d^2T}{dt^2} = k \Rightarrow \frac{d^2T}{dt^2} = kc^2T \Rightarrow \frac{d^2T}{dt^2} - kc^2T = 0$$

$$\therefore \qquad T = c_5 e^{c\sqrt{k}t} + c_6 e^{-c\sqrt{k}t}$$

From equation (*ii*), we get

$$u(x, y, t) = (c_1 e^{\sqrt{k_1}x} + c_2 e^{-\sqrt{k_1}x})(c_3 e^{\sqrt{k_2}y} + c_4 e^{-\sqrt{k_2}y})(c_5 e^{c\sqrt{k}t} + c_6 e^{-c\sqrt{k}t})$$

where $k = k_1 + k_2$.

There arise the following cases:

***Case I*:** If $k = 0$, then

$$u(x, y, t) = (c_1 x + c_2)(c_3 y + c_4)(c_5 t + c_6)$$

***Case II*:** If $k = p^2$, then

$$u(x, y, t) = (c_1 e^{p_1 x} + c_2 e^{-p_1 x})(c_3 e^{p_2 y} + c_4 e^{-p_2 y})(c_5 e^{cpt} + c_6 e^{-cpt})$$

where $p^2 = p_1^2 + p_2^2$.

***Case III*:** If $k = -p^2$, then

$$u(x, y, t) = (c_1 \cos p_1 x + c_2 \sin p_1 x)(c_3 \cos p_2 y + c_4 \sin p_2 y)(c_5 \cos cpt + c_6 \sin cpt)$$

where $p^2 = p_1^2 + p_2^2$.

Since the physical nature of the problem for a rectangular membranes (0, 0), (*a*, 0), (*a*, *b*), (0, *b*) is periodic so the solution in case III, will be consistent with the physical nature of the problem. Here we will consider case III solution.

Example 33: Find the deflection $u(x, y, t)$ of a rectangular membranes $(0 \le x \le a,\ 0 \le y \le b)$ whose boundary is fixed given that it starts from rest and $u(x, y, 0) = xy\,(a - x)\,(b - y)$.

Solution: The vibrations of the membrane are governed by the wave equation

$$\frac{\partial^2 u}{\partial t^2} = c^2 \left(\frac{\partial^2 u}{\partial x^2} + \frac{\partial^2 u}{\partial y^2} \right) \qquad \text{...(1)}$$

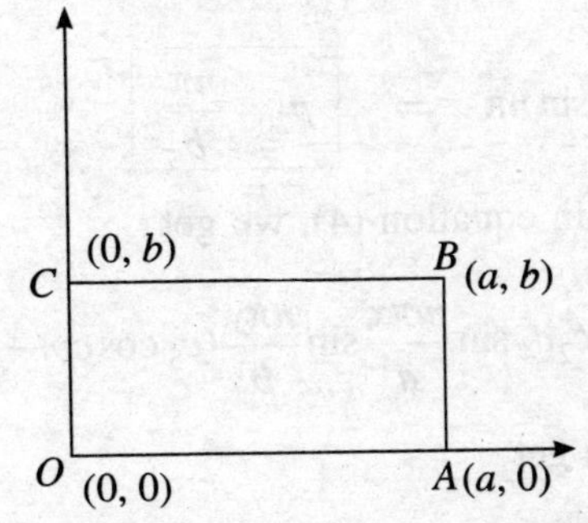

The physical nature of the problem is periodic so, we consider solution of equation (1) in case III.

$$\therefore\ u(x, y, t) = (c_1 \cos p_1 x + c_2 \sin p_1 x)(c_3 \cos p_2 y + c_4 \sin p_2 y)\ (c_5 \cos cpt + c_6 \sin cpt) \qquad \text{...(2)}$$

where $p^2 = p_1^2 + p_2^2$.

The boundary conditions are

(*i*) $u(0, y, t) = u(a, y, t) = 0$ (*ii*) $u(x, 0, t) = u(x, b, t) = 0$

and the initial condition are (*iii*) $u(x, y, 0) = xy(a - x)(b - y)$ (*iv*) $\frac{\partial u}{\partial t} = 0$, at $t = 0$

Using boundary conditions in (2), we get

At $x = 0$, the temperature the deflection $u = 0$

$$\Rightarrow \quad 0 = c_1(c_3 \cos p_2 y + c_4 \sin p_2 y)(c_5 \cos cpt + c_6 \sin cpt)$$

$$\Rightarrow \quad c_1 = 0 \text{ [(otherwise } u(x, y, t) = 0) \text{ which is impossible]}$$

From (2), we obtain

$$u(x, y, t) = c_2 \sin p_1 x(c_3 \cos p_2 y + c_4 \sin p_2 y)(c_5 \cos cpt + c_6 \sin cpt) \quad ...(3)$$

At $x = a$, $u = 0$

$$\Rightarrow \quad 0 = c_2 \sin p_1 a(c_3 \cos p_2 y + c_4 \sin p_2 y)(c_5 \cos cpt + c_6 \sin cpt)$$

$$\Rightarrow \quad \sin p_1 a = 0 \Rightarrow \sin p_1 a = \sin m\pi \Rightarrow \boxed{p_1 = \frac{m\pi}{a}}.$$

At $y = 0$, $u = 0$

(3) gives $0 = c_2 \sin p_1 x(c_3 + 0)(c_5 \cos cpt + c_6 \sin cpt)$

$$\Rightarrow \quad c_3 = 0 \text{ [(otherwise, } u = 0) \text{ which is impossible]}$$

Putting $c_3 = 0$ in equation (3), we get

$$u(x, y, t) = c_2 c_4 \sin p_1 x \sin p_2 y(c_5 \cos cpt + c_6 \sin cpt) \quad ...(4)$$

At $y = b$, $u = 0$

Equation (4) gives

$$0 = c_2 c_4 \sin p_1 x \cdot \sin p_2 b(c_5 \cos cpt + c_6 \sin cpt)$$

$$\Rightarrow \quad \sin p_2 b = 0 \text{ [otherwise } u(x, y, t) = 0, \text{ which is impossible]}$$

$$\Rightarrow \quad \sin p_2 b = \sin n\pi \Rightarrow \boxed{p_2 = \frac{n\pi}{b}}.$$

Putting the values of p_1 and p_2 in equation (4), we get

$$u(x, y, t) = c_2 c_4 \sin \frac{m\pi x}{a} \sin \frac{n\pi y}{b}(c_5 \cos cpt + c_6 \sin cpt) \quad ...(5)$$

Putting $\partial u/\partial t = 0$ and $u = 0$, we get

$$\Rightarrow \quad \left(\frac{\partial u}{\partial t}\right)_{t=0} = 0 = c_2 c_4 \sin \frac{m\pi x}{a} \sin \frac{n\pi y}{b}(-c_5 cp \sin cp(0) + c_6 cp \cos cp(0))$$

$$\Rightarrow \quad 0 = c_2 c_4 c_6 cp \sin \frac{m\pi x}{a} \sin \frac{n\pi y}{b} \Rightarrow c_6 = 0 \quad \Big| \text{ otherwise } u = 0$$

$\therefore$ Equation (5), gives

$$u(x, y, t) = c_2c_4c_5 \sin\frac{m\pi x}{a} \sin\frac{n\pi y}{b} \cos cpt$$

The general form of solution is

$$u(x, y, t) = \sum_{m=1}^{\infty}\sum_{n=1}^{\infty} A_{mn} \sin\frac{m\pi x}{a} \sin\frac{n\pi y}{b} \cos cp_{mn}t \qquad ...(6)$$

where $A_{mn} = c_2c_4c_5$ and $p_{mn} = \pi\sqrt{\frac{m^2}{a^2}+\frac{n^2}{b^2}}$

At $t = 0$, from (6), we get $xy(a-x)(b-y) = \sum_{m=1}^{\infty}\sum_{n=1}^{\infty} A_{mn} \cdot \sin\frac{m\pi x}{a} \sin\frac{n\pi y}{b}$

which is double half range Fourier sine series.

$$\therefore \qquad A_{mn} = \frac{2}{a}\cdot\frac{2}{b}\int_{x=0}^{a}\int_{y=0}^{b} xy(a-x)(b-y)\sin\frac{m\pi x}{a}\sin\frac{n\pi y}{b}\,dx\,dy$$

$$= \frac{4}{ab}\int_0^a x(a-x)\sin\frac{m\pi x}{a}\,dx\int_0^b y(b-y)\sin\frac{n\pi y}{b}\,dy$$

$$= \frac{4}{ab}\left[-x(a-x)\frac{\cos\frac{m\pi x}{a}}{\frac{m\pi}{a}} + (a-2x)\frac{a^2}{m^2\pi^2}\sin\frac{m\pi x}{a} - \frac{2a^3}{m^3\pi^3}\cos\frac{m\pi x}{a}\right]_0^a$$

$$\times\left[-y(b-y)\frac{b}{n\pi}\cos\frac{n\pi y}{b} + (b-2y)\frac{b^2}{n^2\pi^2}\sin\frac{n\pi y}{b} - \frac{2b^3}{n^3\pi^3}\cos\frac{n\pi y}{b}\right]_0^b$$

$$= \frac{4}{ab}\left(\frac{-2a^3}{m^3\pi^3}\cos m\pi + \frac{2a^3}{m^3\pi^3}\right)\left(\frac{-2b^3}{n^3\pi^3}\cos n\pi + \frac{2b^3}{n^3\pi^3}\right)$$

$$= \frac{4}{ab}\cdot\frac{2a^3}{m^3\pi^3}\cdot\frac{2b^3}{n^3\pi^3}\left\{(-1)^{m+1}+1\right\}\left\{(-1)^{n+1}+1\right\}$$

$$= \frac{64a^2b^2}{m^3n^3\pi^6}; \; m \text{ and } n \text{ both are odd.}$$

Putting the value of A_{mn} in equation (6), we get

$$u(x, y, t) = \sum_{m=(1,3,5,..)}^{\infty}\;\sum_{n=(1,3,5,..)}^{\infty} \frac{64a^2b^2}{m^3n^3\pi^6}\cdot\sin\frac{m\pi x}{a}\cdot\sin\frac{n\pi y}{b}\cos cp_{mn}t$$

$$= \frac{64a^2b^2}{\pi^6} \sum_{m=(1,3,5,..)}^{\infty} \sum_{n=(1,3,5,..)}^{\infty} \frac{1}{m^3n^3} \sin\frac{m\pi x}{a} \sin\frac{n\pi y}{b} \cos cp_{mn}t \quad \textbf{Ans.}$$

where $p_{mn} = \pi\sqrt{\frac{m^2}{a^2} + \frac{n^2}{b^2}}$.

EXERCISE 5.5

1. The initial displacement of the rectangular membrane $(0 < x < 1, 0 < y < 2)$ where boundary is fixed given that it starts from rest and $u(x, y, 0) = xy(1 - x)(2 - y)$. Find $u(x, y, t)$.

$$\left[\begin{array}{l} \textbf{Ans.}\, u(x,y,t) = \sum\sum \frac{256}{m^3n^3\pi^6} \cos cp_{mn}t \sin m\pi x \sin\frac{n\pi y}{2} \\ \text{where } m \text{ and } n \text{ both are odd and } p_{mn} = \pi\sqrt{m^2 + \frac{n^2}{4}} \end{array}\right]$$

2. The initial displacement of a rectangular membrane $(0 < x < a,\ 0 < y < b)$ whose boundary is fixed is given by $f(x, y) = \lambda(ax - x^2)(by - y^2)$ show that

$$u(x,y,t) = \sum_{m=1}^{\infty} \sum_{n=1}^{\infty} A_{mn} \cos cp_{mn}t \sin\frac{m\pi x}{a} \sin\frac{n\pi y}{b}$$

where $A_{mn} = \frac{16a^2b^2\lambda}{m^3n^3\pi^6}(1 - \cos m\pi)(1 - \cos n\pi)$;

and $p_{mn}^2 = \pi^2\left(\frac{m^2}{a^2} + \frac{n^2}{b^2}\right)$.

3. Find the deflection $u(x, y, t)$ of the square membrane with $a = b = c = 1$, if the initial velocity is zero and the initial deflection $f(x, y) = A \sin\pi x \sin 2\pi y$.

$$[\textbf{Ans.}\, u(x,y,t) = A\cos\sqrt{5}\pi t . \sin\pi x \sin 2\pi y]$$

4. Find the deflection $u(x, y, t)$ of a rectangular membrane $(0 \le x \le a,\ 0 \le y \le b)$ and $\left(\frac{\partial u}{\partial t}\right)_{t=0} = 0$. The deflection at $t = 0$ is $xy(a^2 - x^2)(b^2 - y^2)$.

$$\left[\begin{array}{l} \textbf{Ans.}\, u(x,y,t) = \sum_{m=1}^{\infty} \sum_{n=1}^{\infty} \frac{144a^2b^2}{m^3n^3\pi^6} \sin\frac{m\pi x}{a} \sin\frac{n\pi y}{b} \cos cp_{mn}t \\ \text{where } \quad p_{mn}^2 = \pi^2\left(\frac{m^2}{a^2} + \frac{n^2}{b^2}\right). \end{array}\right]$$

5. Find the deflection $u(x, y, t)$ of a membrane $(0 \le x \le a,\ 0 \le y \le b)$ when the initial velocity is zero and the initial deflection is given by $\sin\dfrac{2\pi x}{a}\sin\dfrac{\pi y}{b}$.

$$\left[\textbf{Ans. } u(x,y,t) = \sin\frac{2\pi x}{a}\sin\frac{\pi y}{b}\cdot\cos\pi ct\sqrt{\frac{4}{a^2}+\frac{1}{b^2}}\right]$$

5.8 EQUATIONS OF TRANSMISSION LINES

Suppose the resistance, inductance and capacitance vary linearly with x in transmission line, where x is the distance from the source of electricity, measured along the line. If $V(x, t)$ denote the potential at a point on the cable, $i(x, t)$ is current there R is the resistance of the cable per unit length and L is inductance per unit length.

Then the equations

$$\frac{\partial V}{\partial x} = -L\frac{\partial i}{\partial t}, \frac{\partial i}{\partial x} = -C\frac{\partial V}{\partial t}$$

$$\left.\begin{aligned}\frac{\partial^2 V}{\partial x^2} &= RC\frac{\partial V}{\partial t}\\ \frac{\partial^2 i}{\partial x^2} &= RC\frac{\partial i}{\partial t}\end{aligned}\right\} L = 0 \qquad \ldots(A)$$

and

$$\left.\begin{aligned}\frac{\partial^2 V}{\partial x^2} &= LC\frac{\partial^2 V}{\partial t^2}\\ \frac{\partial^2 i}{\partial x^2} &= LC\frac{\partial^2 i}{\partial t^2}\end{aligned}\right\} R = 0 \qquad \ldots(B)$$

The set (A) is known as 'Telegraph equations', and (B) is known as 'Radio equations'.

Example 34: Neglecting R and G obtain potential $V(x, t)$ in a line l km. long, after t seconds the ends are suddenly grounded, if initially $i(x, 0) = I_0$ and $V(x, 0) = V_0 \sin \dfrac{\pi x}{l}$. Also current,

(U.P.T.U. 2008)

Solution: We have

$$\frac{\partial^2 V}{\partial x^2} = LC\frac{\partial^2 V}{\partial t^2} \qquad \ldots(1)$$

Let $\qquad V(x, t) = X(x)T(t) \qquad \ldots(2)$

$$\therefore \quad \frac{\partial^2 V}{\partial x^2} = T\frac{d^2 X}{dx^2} \text{ and } \frac{\partial^2 V}{\partial t^2} = X\frac{d^2 T}{dt^2}$$

Putting these values in equation (1), we obtain

$$T\frac{d^2X}{dx^2} = LCX\frac{d^2T}{dt^2} \Rightarrow \frac{1}{X}\frac{d^2X}{dx^2} = \frac{LC}{T}\frac{d^2T}{dt^2} = -p^2$$

$$\therefore \quad \frac{1}{X}\frac{d^2X}{dx^2} = -p^2 \Rightarrow \frac{d^2X}{dx^2} + p^2X = 0$$

$$\Rightarrow \quad X = c_1 \cos px + c_2 \sin px$$

and

$$\frac{LC}{T}\frac{d^2T}{dt^2} = -p^2 \Rightarrow \frac{d^2T}{dt^2} + \frac{p^2}{LC}T = 0$$

$$\Rightarrow \quad T = c_3 \cos\frac{pt}{\sqrt{LC}} + c_4 \sin\frac{pt}{\sqrt{LC}}$$

From equation (2), we get

$$V(x, t) = (c_1 \cos px + c_2 \sin px)\left(c_3 \cos\frac{pt}{\sqrt{LC}} + c_4 \sin\frac{pt}{\sqrt{LC}}\right) \quad ...(3)$$

The boundary conditions are

(*i*) $V(0, t) = 0$ (*ii*) $V(l, t) = 0$ | As the ends are suddenly grounded.

The initial conditions are

(*iii*) $V(x, 0) = V_0 \sin\frac{\pi x}{l} \Rightarrow \frac{\partial V(x, 0)}{\partial t} = 0$

(*iv*) $i(x, 0) = I_0 \Rightarrow \frac{\partial i(x, 0)}{\partial t} = 0$

At $x = 0$, $V = 0$, equation (3) gives

$$0 = c_1\left(c_3 \cos\frac{pt}{\sqrt{LC}} + c_4 \sin\frac{pt}{\sqrt{LC}}\right)$$

$\Rightarrow$ $c_1 = 0$ [otherwise $V(x, t) = 0$, which is impossible]

$\therefore$ From (3), we have

$$V(x, t) = c_2 \sin px\left(c_3 \cos\frac{pt}{\sqrt{LC}} + c_4 \sin\frac{pt}{\sqrt{LC}}\right) \quad ...(4)$$

At $x = l$, $V = 0$, equation (4) gives $\boxed{p = \frac{n\pi}{l}}$.

$\therefore$ From equation (4), we have

$$V(x, t) = \sin\frac{n\pi x}{l}\left(c_2c_3 \cos\frac{n\pi t}{l\sqrt{LC}} + c_2c_4 \sin\frac{n\pi t}{l\sqrt{LC}}\right)$$

The general form of above solution is

$$V(x, t) = \sum_{n=1}^{\infty} \sin\frac{n\pi x}{l}\left(a_n \cos\frac{n\pi t}{l\sqrt{LC}} + b_n \sin\frac{n\pi t}{l\sqrt{LC}}\right) \quad ...(5)$$

$$\frac{\partial V}{\partial t} = \sum_{n=1}^{\infty} \sin\frac{n\pi x}{l}\left(-a_n \cdot \frac{l\sqrt{LC}}{n\pi}\sin\frac{n\pi t}{l\sqrt{LC}} + b_n \cdot \frac{l\sqrt{LC}}{n\pi}\cos\frac{n\pi t}{l\sqrt{LC}}\right)$$

At $t = 0, \quad \frac{\partial V}{\partial t} = 0$

$\therefore$ Above equation gives

$$0 = \sum_{n=1}^{\infty} \sin\frac{n\pi x}{l} \cdot b_n \frac{l\sqrt{LC}}{n\pi} \quad \Rightarrow \quad \boxed{b_n = 0.}$$

From (5), we obtain

$$V(x,\ t) = \sum_{n=1}^{\infty} a_n \sin\frac{n\pi x}{l}\cos\frac{n\pi t}{l\sqrt{LC}} \qquad \text{...(6)}$$

and at $t = 0,\ V = V_0 \sin\frac{\pi x}{l}$

$$\Rightarrow \quad V_0 \sin\frac{\pi x}{l} = \sum_{n=1}^{\infty} a_n \sin\frac{n\pi x}{l}$$

or $$V_0 \sin\frac{\pi x}{l} = a_1 \sin\frac{\pi x}{l} + a_2 \sin\frac{2\pi x}{l} + \ldots..$$

$$\Rightarrow \quad \boxed{a_1 = V_0} \quad \text{and } a_2 = a_3 = a_4 = \ldots\ldots = a_n = 0.$$

Putting the value of a_1 in equation (6), we get

$$V(x,\ t) = V_0 \sin\frac{\pi x}{l}\cos\frac{\pi t}{l\sqrt{LC}}. \qquad \textbf{Ans.}$$

Second part: We know that

$$\frac{\partial V}{\partial x} = -L\frac{\partial i}{\partial t}$$

So $$\frac{\partial}{\partial x}\left(V_0 \sin\frac{\pi x}{l}\cos\frac{\pi t}{l\sqrt{LC}}\right) = -L\frac{\partial i}{\partial t}$$

$$\Rightarrow \quad \frac{\pi}{l}V_0 \cos\frac{\pi x}{l}\cos\frac{\pi t}{l\sqrt{LC}} = -L\frac{\partial i}{\partial t} \qquad \text{...(7)}$$

Integrating (7), we obtain

$$i = -V_0\sqrt{\frac{C}{L}}\cos\frac{\pi x}{l}\sin\frac{\pi t}{l\sqrt{LC}} + \phi(x) \qquad \text{...(8)}$$

But At $t = 0$, $i = I_0$, equation (8), gives

$$I_0 = \phi(x)$$

Putting the value of $\phi(x)$ in equation (8), we get

$$i = I_0 - V_0\sqrt{\frac{C}{L}}\cos\frac{\pi x}{l}\sin\frac{\pi t}{l\sqrt{LC}}. \quad \textbf{Ans.}$$

EXERCISE 5.6

1. Find the potential $V(x, t)$ in a line of length l, after t seconds the ends were suddenly grounded, given that $i(x, 0) = i_0$ and $V(x, 0) = V_1 \sin\frac{\pi x}{l} + V_5 \sin\frac{5\pi x}{l}$.

   $$\left[\textbf{Ans. } V(x,t) = V_1 \sin\frac{\pi x}{l}\cos\frac{\pi t}{l\sqrt{LC}} + V_5 \sin\frac{5\pi x}{l}\cos\frac{5\pi t}{l\sqrt{LC}}\right]$$

2. Neglecting R and G find the e.m.f. $e(x, t)$ in a line one km long, t seconds after the ends were suddenly grounded, if initially $i(x, 0) = I$ so that $\frac{\partial e}{\partial t} = 0$ at $t = 0$ and $e(x, 0) = E_1 \sin\frac{\pi x}{a} + E_2 \sin\frac{2\pi x}{a}$.

   $$\left[\textbf{Ans. } e(x, t) = E_1 \sin\frac{\pi x}{a}\cos\frac{\pi t}{a\sqrt{LC}} + E_2 \sin\frac{2\pi x}{a}\cos\frac{2\pi t}{a\sqrt{LC}}\right]$$

3. Obtain the solution of the radio equation $\frac{\partial^2 V}{\partial x^2} = LC\frac{\partial^2 V}{\partial t^2}$ when a periodic e.m.f. $V_0 \cos pt$ is applied at the end $x = 0$ of the line. $[\textbf{Ans. } V(x, t) = V_0 \cos\{pt - px\sqrt{LC}\}]$

4. A transmission line 1000 km. long is initially under steady state conditions with potential 1300 volts at the sending end ($x = 0$) and 1200 volts at the receiving end ($x = 1000$). The terminal end of the line is suddenly grounded, but the potential at the source is kept at 1300 volts. Assuming the inductance and leakance to be negligible find the potential $V(x, t)$.

   $$\left[\textbf{Hint: Apply } \frac{\partial^2 V}{\partial x^2} = RE\frac{\partial E}{\partial t}; \text{ in steady state } \frac{\partial E}{\partial t} = 0\right]$$

   $$\left[\textbf{Ans. } V(x, t) = 1300 - 1\cdot 3x + \frac{2400}{\pi}\sum_{n=1}^{\infty}\frac{(-1)^{n+1}}{n}\sin\frac{n\pi x}{l}e^{-\frac{n^2\pi^2 t}{l^2 RC}}\right]$$

5. A steady voltage distribution of 20 volts at the sending end and 12 volts at the receiving end is maintained in a telephone wire of length l. At time $t = 0$, the receiving end is grounded. Find the voltage and current t second later. Neglect leakage and inductance.

   $$\left[\begin{aligned}\textbf{Ans. } V(x, t) &= \frac{20(l-x)}{l} + \frac{24}{\pi}\sum\frac{(-1)^{n+1}}{n}\sin\frac{n\pi x}{l}e^{-\frac{n^2\pi^2 t}{Rcl^2}} \\ i(x, t) &= \frac{20}{lR} + \frac{24}{lR}\sum(-1)^n\cos\frac{n\pi x}{l}e^{-\frac{n^2\pi^2 t}{Rcl^2}}\end{aligned}\right]$$

OBJECTIVE TYPE QUESTIONS

A. Pick the correct answers of the given choices below:

1. Solving P.D.E. $4u_x + u_y = 3u, u(0, y) = e^{-5y}$ the function $u(x, y)$ has its value.

   (i) e^{2x} (ii) e^{2y} (iii) e^{2x-5y} (iv) e^{2x+5y}

   [U.P.T.U. (Special) 2009]

2. Using the method of separation of variables in the equation $\frac{\partial u}{\partial x} = 2\frac{\partial u}{\partial t} + u$, if $u = X(x) \cdot T(t)$, then $X(x)$ is equal to

   (i) $X = c_1 e^{kx}$ (ii) $X = c_1 e^{-kx}$ (iii) $X = c_1 e^{x^2}$ (iv) $X = c_1 e^{kx^2}$

3. The solution of $\frac{\partial u}{\partial t} = c^2 \frac{\partial^2 u}{\partial x^2}$, when $k = 0$ is

   (i) $u = (c_1 - c_2 x)$ (ii) $(c_1 + c_2 x)c_3$ (iii) $(c_1 + c_2 x^2)$ (iv) None of these

4. In steady state condition the solution of heat equation $\frac{\partial u}{\partial t} = c^2 \frac{\partial^2 u}{\partial x^2}$ is equal to

   (i) $u = c_1 x + c_2$ (ii) $u = c_1 x - c_2$ (iii) $u = (c_1 + c_2 x)e^x$ (iv) $u = 0$

5. The solution of $\frac{\partial u}{\partial x} = 4\frac{\partial u}{\partial y}$ by method of separation of variables is equal to

   (i) $u = c_1 c_2 e^{\left(x + \frac{y}{4}\right)k}$ (ii) $u = e^{\left(x - \frac{y}{4}\right)k}$ (iii) $u = c_1 c_2 e^{\left(x - \frac{y}{4}\right)k}$ (iv) $u = c_1 c_2 e^{\frac{xy}{4}k}$

B. Fill in the blanks:

1. Applying the method of separation of variables techniques the solution of the P.D.E. $3u_x + 2u_y = 0$ is ________

   where $u_x = \frac{\partial u}{\partial x}, u_y = \frac{\partial u}{\partial y}$. (U.P.T.U. 2009)

2. For the wave equation $\frac{\partial^2 y}{\partial t^2} = c^2 \frac{\partial^2 y}{\partial x^2}$ if $y = X(x) \cdot T(t)$ then $X =$ ________.

3. For the steady state condition the solution of heat equation $\frac{\partial u}{\partial t} = c^2 \frac{\partial^2 u}{\partial x^2}$ is equal to ________.

4. The equation $\frac{\partial^2 u}{\partial x^2} + \frac{\partial^2 u}{\partial y^2} = 0$ is known as ________.

C. Indicate True or False for the following statements:

1. With usual symbols P.D.E. $u_{xx} + u^2 u_{yy} = f(x, y)$ is non-linear in 'u' and is of second order. *(U.P.T.U. 2009)*
2. The small transverse vibrations of a string are governed by one dimensional heat equation.
3. Two dimensional steady state heat flow is given by Laplace's equation $u_t = a^2(u_{xx} + u_{tt})$.
4. The P.D.E. of all sphere whose centre lie on z-axis and given by equations $x^2 + y^2 + (z - a)^2 = b^2$ a and b being constants are governed by $xz_y - yz_x = 0$.
5. The Laplace's equation in three dimensions is $\dfrac{\partial^2 u}{\partial x^2} + \dfrac{\partial^2 u}{\partial y^2} - \dfrac{\partial^2 u}{\partial z^2} = 0$.
6. The telegraph equations are $\dfrac{\partial^2 v}{\partial x^2} = RC\dfrac{\partial v}{\partial t}$ and $\dfrac{\partial^2 i}{\partial x^2} = RC\dfrac{\partial i}{\partial t}$.
7. The radio equations are $\dfrac{\partial v}{\partial x} = LC\dfrac{\partial^2 v}{\partial t^2}$ and $\dfrac{\partial i}{\partial x} = LC\dfrac{\partial^2 i}{\partial t^2}$.
8. In steady state condition always $\dfrac{\partial u}{\partial t} \neq 0$.

ANSWERS TO OBJECTIVE TYPE QUESTIONS

A. Pick the correct answers:

1. *(iii)* **2.** *(i)* **3.** *(ii)* **4.** *(i)* **5.** *(i)*

B. Fill in the blanks:

1. $u(x, y) = ce^{\frac{k}{6}(2x-3y)}$ **2.** $X = c_1 e^{\sqrt{kx}} + c_2 e^{-\sqrt{kx}}$

3. $u = c_1 x + c_2$ **4.** Laplace equation.

C. True or False

1. T **2.** F **3.** F **4.** T **5.** F

6. T **7.** F **8.** F.

Examination Paper 2009

Uttarakhand University

B.Tech. (I Year) II Semestar

MATHEMATICS–II

Time: 3 Hours] [*Total Marks: 100*

1. Attempt any **four** parts: **5 × 4 = 20**

(a) Solve the differential equation

$$(1+y^2)dx+(x-e^{\tan^{-1}}y)dy=0\,.$$

(b) Solve the differential equation

$$(D^2+2D+1)y=x\cos x.$$

(c) Solve the homogeneous differential equation

$$x^2\frac{d^2y}{dx^2}-x\frac{dy}{dx}-3y=x^2\log x.$$

(d) Solve the equation

$$x\frac{d^2y}{dx^2}-\frac{dy}{dx}-4x^3y=8x^3\sin x^2.$$

(e) Solve the following differential equation

$$\frac{dx}{dt}+4x+3y=t,\frac{dy}{dt}+2x+5y=e^t.$$

(f) An alternating e.m.f. ***E sin wt*** is applied to an inductance ***L*** and a capacitance ***C*** in series. Show that the current in the circuit is

$$\frac{E_w}{(n^2-w^2)L}(\cos wt-\cos nt),\text{ where } n^2=\frac{I}{LC}.$$

2. Attempt any **two** parts: **10 × 2 = 20**

(a) Solve in series the differential equation

$$2x^2 \frac{d^2y}{dx^2} - x\frac{dy}{dx} + (1-x^2)y = x^2.$$

(b) Prove that

$$P_n(x) = \frac{1}{n!\,2^n}\frac{d^n}{dx^n}(x^2-1)^n$$

(c) Prove that when n is a positive integer $J_n(x)$ is the coefficient of z^n in the expansion of $e^{x\left(z-\frac{1}{z}\right)/z}$ in ascending and descending powers of z. Also prove that $J_n(x)$ is the coefficient of z^{-n} multiplied by $(-1)^n$ in the expansion of above expression.

3. Attempt any **four** parts: **5 × 4 = 20**

(a) If $L\{F(t)\} = f(s)$ and $G(t) = \begin{cases} F(t-a), & t > a, \\ 0, & t < a, \end{cases}$

then find the value of $L\{G(t)\}$.

(b) Find the Laplace transform of $\sin\sqrt{t}$.

(c) If $erf(t) = \frac{2}{\sqrt{\pi}}\int_0^t e^{-x^2}\,dx$, then find the value of $L\{erf(\sqrt{t})\}$

(d) Evaluate: $L^{-1}\left\{\frac{5s-2}{s^2(s+2)(s-1)}\right\}$

(e) State convolution theorem and hence evaluate

$$L^{-1}\left\{\frac{s}{(s^2+4)^3}\right\}$$

(f) Using Laplace transform solve the following differential equation

$$\frac{d^2y}{dt^2} + 9y = \cos 2t, \text{ if } y(0) = 1,\ y(\pi/2) = -1.$$

4. Attempt any **two** parts: **10 × 2 = 20**

(a) Find series of sines and cosines of multiples of x which will represent $x + x^2$ in the interval $-\pi < x < \pi$.

Hence show that $\frac{\pi^2}{6} = 1 + \frac{1}{2^2} + \frac{1}{3^2} + \ldots\ldots$

Also find the sum of the series when $x = \pm\,\pi$.

(b) Find the Fourier series to represent

$$f(x) = \begin{cases} \pi x, & \textit{for} \quad 0 \le x < 1 \\ 0, & \textit{for} \quad x = 1 \\ \pi(x-2), & \textit{for} \quad 1 < x \le 2 \end{cases}$$

(c) Solve the following partial differential equation:

$$(D^2 - D'^2 - 3D + 3D')z = xy + e^{x+2y}.$$

5. Attempt any **two** parts: **10 × 2 = 20**

(a) A string is stretched between the fixed points (0, 0) and (l, 0) and released at rest from the initial deflection given by

$$f(x) = \begin{cases} \frac{2k}{l}x, & \textit{when}\; 0 < x < \frac{l}{2} \\ \frac{2k}{l}(l-x), & \textit{when}\; \frac{l}{2} < x < l \end{cases}$$

Find the deflection of the string at any time t.

(b) Solve the equation

$$\frac{\partial^2 v}{\partial x^2} = \frac{\partial v}{\partial t}, \text{ given that } v = 0$$

When $t = \infty$ and when $x = 0$ and $x = l$

(c) Solve $\frac{\partial^2 u}{\partial x^2} + \frac{\partial^2 u}{\partial y^2} = 0$, for $0 < x < \pi$, $0 < y < \pi$,

$$u(x, 0) = x^2, u(x, \pi) = 0$$

$$\left(\frac{\partial u}{\partial x}\right)(0, y) = \left(\frac{\partial u}{\partial x}\right)(\pi, y) = 0.$$

INDEX